PATENTS FOR CHEMICALS, PHARMACEUTICALS, AND BIOTECHNOLOGY

FUNDAMENTALS OF GLOBAL LAW, PRACTICE, AND STRATEGY

Fifth Edition

by

PHILIP W. GRUBB

and

PETER R. THOMSEN

OXFORD
UNIVERSITY PRESS

Great Clarendon Street, Oxford OX2 6DP

Oxford University Press is a department of the University of Oxford.
It furthers the University's objective of excellence in research, scholarship,
and education by publishing worldwide in

Oxford New York

Auckland Cape Town Dar es Salaam Hong Kong Karachi
Kuala Lumpur Madrid Melbourne Mexico City Nairobi
New Delhi Shanghai Taipei Toronto

With offices in

Argentina Austria Brazil Chile Czech Republic France Greece
Guatemala Hungary Italy Japan Poland Portugal Singapore
South Korea Switzerland Thailand Turkey Ukraine Vietnam

Oxford is a registered trade mark of Oxford University Press
in the UK and in certain other countries

Published in the United States
by Oxford University Press Inc., New York

Fifth edition first published 2010

British Library Cataloguing in Publication Data
Data available

Library of Congress Cataloging in Publication Data
Data available

Typeset by Glyph International, Bangalore, India
Printed by
CPI Antony Rowe

ISBN 978-0-19-957523-7

1 3 5 7 9 10 8 6 4 2

PREFACE TO THE FIFTH EDITION

Another five years have gone by since the previous edition, and this version covers developments up to August 2009. In the course of these five years I have retired from Novartis, and therefore am no longer closely involved with patent matters on a daily basis. For this reason, my former colleague Dr Peter Thomsen has joined me as a co-author, and I welcome his experience and knowledge particularly in the fields of biotechnology and of EU law.

We had hoped to be able to write about the long-overdue conversion of the USA to a first-to-file system, but this, like the European Community Patent, remains a mirage on the horizon. Even the Supreme Court opinion in *in re Bilski* comes too late for our deadline. Nevertheless, we are able to report on many new and important decisions of the courts in the USA, the UK, and elsewhere.

In order to accommodate the additional material while keeping the book approximately the same size, we decided to eliminate the chapter on Patents and Information (Chapter 20 of the 4th Edition). Much of the contents were outdated and neither of the authors is an expert in this field.

The wording of the text is perhaps not as gender-neutral as some might wish, but one simply cannot write 'or she' after each occurrence of 'he'. Readers will have to accept that, except when an individual male person is referred to, the pronoun 'he' is to be taken as standing for a person of either sex.

My thanks are due to my former colleague Mel Kassenoff for his help on questions of US law and practice, and of course to my wife Kay for her continued patience and support.

Again, I gratefully acknowledge the permission of United Media to use the quote at the head of Chapter 16.

Philip Grubb
November 2009

Around the time when my interest in the field of patents had been first awakened it happened that I visited a friend who was working on his PhD thesis in Cambridge, England. As often on such travels, I went to a local bookshop to browse through their assortment. I was excited when I found there a green book that really attracted my attention. It was Philip Grubb's *Patents for Chemicals, Pharmaceuticals and Biotechnology*. I really wanted to have this book and therefore I bought it. After having read it I was even more attracted to the field of intellectual property and patents because the book explained the basics of how patents work in the pharmaceutical and chemical industry in a way that I was able to understand without yet having had any specialized training in patents.

At that time I could of course not have dreamed that a couple of years later after having studied IP I would become a colleague of Philip Grubb in the IP department of Novartis and again a couple of years later I would take over from him some training activities for patent attorney candidates. I have always recommended the book to scientists and patent attorney candidates who wanted to have an understandable but still comprehensive introduction. Beyond the basic concepts the book shed light on how patents work together with other areas of law such as competition law or regulatory law. That makes the book a valuable source even for already trained patent attorneys who come from a different technical field and start to work in the pharmaceutical area.

Thus, I felt honoured when Philip asked me whether I could help him in producing the 5th edition which after some hard work of updating and supplementing you can hold now in your hands.

I would like to thank Philip for the opportunity to work with him on this book and for the great collaboration. Thanks also to my family who had to accept my limited availability during certain time periods. Finally, I would also express my thanks to Oxford University Press who made the new edition possible.

Peter Thomsen
November 2009

CONTENTS

LIST OF FIGURES AND TABLES

Tables

Australia

Belgium

Dominican Republic

France

Germany

Switzerland

United States

European Commission Decisions

European Court of Justice and Court of First Instance

European Patent Office

TABLE OF LEGISLATION

NATIONAL LEGISLATION

EC Directives

EC Regulations

INTERNATIONAL LEGISLATION

LIST OF ABBREVIATIONS

used in the text

AIPA	American Inventors Protection Act of 1999 (US)
AIPLA	American Intellectual Property Law Association
AMA	American Medical Association
ANDA	abbreviated new drug application (US)
ARIPO	African Regional Industrial Property Organization
ASEAN	Association of Southeast Asian Nations
ATCC	American Type Culture Collection
BA	Board of Appeal (EPO)
BGH	*Bundesgerichtshof* (German Federal Supreme Court)
BPAI	Board of Patent Appeals and Interferences (US)
BPatG	*Bundespatentgericht* (German Federal Patent Court)
BPD	Biotechnology Patenting Directive (EU)
CAFC	Court of Appeals for the Federal Circuit (US)
CBD	Convention on Biological Diversity
CCP	*Certificat complementaire du protection* (complementary protection certificate, earlier French version of the SPC)
CCPA	Court of Customs and Patents Appeals (US)
CDA	confidential disclosure agreement
CEIPI	*Centre d'Études Internationales de la Propriété Industrielle* (Centre for the International Study of Industrial Property)
CFI	Court of First Instance (EU)
CHO	Chinese hamster ovary
cip	continuation in part (US)
CIPA	Chartered Institute of Patent Attorneys (UK)
CIS	Commonwealth of Independent States
CMC	case management conference (UK)
CP	Community patent (EU)
CPA	Chartered Patent Agent/Attorney (UK)
CPA	continued prosecution application (US)
CPC	Community Patent Convention
CPR	Civil Procedure Rules 1998 (UK)
CSIR	Council of Scientific and Industrial Research
DAS	*Deutsche Auslegeschrift* (German published specification)
DG	Directorate General (EC Commission)
DNA	deoxyribonucleic acid
DOS	*Deutsche Offenlegungschrift* (German publication document)
DPS	*Deutsche Patentschrift* (German published patent)

DSB	Disputes Settlement Body (WTO)
EBA	Enlarged Board of Appeal (EPO)
ECB	European Central Bank (EU)
ECJ	European Court of Justice
ED	Examining Division
EEA	European Economic Area
EMEA	European Medicines Evaluation Agency
EMR	exclusive marketing rights
EPC	European Patent Convention
epi	Institute of Professional Representatives before the EPO
EPLA	European Patent Litigation Agreement
epo	erythropoietin
EPO	European Patent Office/Organization
EQE	European qualifying examination
ESD	examination support document (US)
EST	expressed sequence tags
ETH	*Eidgenössische Technische Hochschute* (Federal Institute of Technology, Switzerland)
EU	European Union
FACS	fluorescence-activated cell sorting
FCIPA	Fellow of the Chartered Institute of Patent Attorneys (UK)
FDA	Food and Drug Administration (US)
FIPCO	fully integrated pharmaceutical company
FTC	Federal Trade Commission (US)
GATT	General Agreement on Tariffs and Trade
GCC	Gulf Cooperation Council
G-CSF	granulocyte stimulating factor
GDR	German Democratic Republic
GSP	Generalized System of Preferences (US)
HBV	hepatitis B virus
hESC	human embryonic stem cell
HSC	hematopoietic stem cell
HUGO	Human Genome Organization
IDA	international depository authority
IDS	information disclosure statement (US)
IIB	*Institut Internationale des Brevets* (now part of the EPO)
IND	investigative new drug (US)
INN	international non-proprietory name
IP	intellectual property
IPB	Intellectual Property Bureau (China)
IPC	international patent classification
IPE	international preliminary examination (PCT)
IPEA	International Preliminary Examining Authority (PCT)
IPER	international preliminary examination report (PCT)

IPO	initial public offering
IPRP	international preliminary report on patentability (PCT)
ISA	International Searching Authority (PCT)
ISR	international search report
ITC	International Trade Commission (US)
ITMA	Institute of Trademark Agents (UK)
IV	invention value
JMOL	judgment as a matter of law (US)
JPO	Japanese Patent Office
LBA	Legal Board of Appeal (EPO)
LCM	life cycle management
MAb	monoclonal antibody
MPEP	Manual of Patent Examining Procedure (US)
MRC	Medical Research Council (UK)
MSC	mesenchynal stem cell
MSF	*Médecins sans Frontières* (Doctors without Borders)
MTA	materials transfer agreement
NAFTA	North American Free Trade Agreement
NCE	new chemical entity
NDA	new drug application (US)
NGO	non-governmental organization
NIH	National Institutes of Health (US)
NRDC	National Research and Development Corporation (UK)
NSAI	non-steroidal anti-inflammatory
OAPI	*Organisation Africaine de la Propriété Intellectuelle* (African Intellectual Property Association)
OAU	Organization for African Unity
OBRA	Omnibus Budget Reconciliation Act (US)
OD	Opposition Division
OHIM	Office for Harmonization in the Internal Market (EU)
OMPI	*L'Organisation Mondiale de la Propriété Intellectuelle* (WIPO)
OPI	open to public inspection
PAT	Patent Appeals Tribunal (UK)
PCC	Patents County Court (UK)
PCR	polymerase chain reaction
PCT	Patent Cooperation Treaty
PF	participation factor
PHOSITA	person having ordinary skill in the art (US)
PIP	paediatric investigation plan
PLT	Patent Law Treaty
POC	proof of concept
PSH	Public Health Service Act of 1946 (US)
PTA	patent term adjustment (US)
QC	Queen's Counsel (UK)

PART I

INTRODUCTION AND BACKGROUND TO THE MODERN PATENT SYSTEM

1

THE NATURE AND ORIGINS OF PATENT RIGHTS

> One man should not be afraid of improving his possessions, lest they be taken away from him, or another deterred by high taxes from starting a new business. Rather, the prince should be ready to reward men who want to do these things and those who endeavour in any way to increase the prosperity of their city or their state.
>
> Niccolò Machiavelli, *The Prince* (1514)

What is a Patent?

A patent may be defined as a grant by the state of exclusive rights for a limited time in respect of a new and useful invention. These rights are generally limited to the territory of the state granting the patent, so that an inventor desiring protection in a number of countries must obtain separate patents in all of them. The name 'patent' is a contraction of 'letters patent' (from the Latin, *litterae patentes*, meaning 'open letters'), which means a document issued by or in the name of the sovereign, addressed to all subjects and with the Great Seal pendant at the bottom of the document so that it can be read without breaking the seal.

Letters patent are still used in the UK, for example to confer peerages and to appoint judges, but are no longer used to grant patents for inventions. Prior to 1878, letters patent for inventions were engrossed on parchment and bore the Great Seal in wax. Subsequently, paper was used and a wafer seal of the Patent Office replaced the Great Seal, but the wording of the letters patent document (not to be confused with the printed patent specification) was still very impressive. As can be seen from Fig. 1, British patents were still being granted in 1981 in the form of a command from the Queen to her subjects to refrain from infringing the patent under pain of the royal displeasure. Unfortunately for the patentee with a taste for this sort of thing, under the Patents Act 1977, all that he gets is a very unimpressive certificate of grant from the Patent Office, now renamed the Intellectual Property Office.

Exclusionary Right

It is important to realize that the rights given by a patent do not include the right to practise the invention, but only to exclude others from doing so. The patentee's freedom to use his own invention may be limited by legislation or regulations having nothing to do with patents, or by the existence of other patents. For example, owning a patent for a new drug clearly does not give the right to market the drug without permission from the responsible health authorities, nor does it give the right to infringe an earlier existing patent. In the very common situation in which A has a patent for a basic invention and B later obtains a patent for an improvement to this invention, B is not free to use the invention without the permission of A, and A cannot use the improved version without coming to terms with B. A patent is neither a seal of government approval, nor a permit to carry out the invention. We often hear it said that 'This patent allows Company X' to do something or other. It does not: it only allows them to stop someone else from doing it. The right to prevent others from carrying out the invention claimed in a patent may be enforced in the courts; if the patent is valid and infringed, the courts can order the infringer to stop its activities, as well as provide other remedies, such as damages.

It is also important to distinguish between ownership of an invention or a patent and ownership of goods that incorporate the invention or fall under the patent. The question of who owns the goods is completely different from that of who owns the patents. Unlike the situation with regard to copyright, infringing goods do not become the property of the patentee, and even if the patentee has manufactured the goods, once he has sold them, he usually can retain no control over their subsequent use or resale. The fundamental distinction between the ownership of patents and the ownership of things that are patented is often misunderstood or deliberately misrepresented, so that, for example, patents granted for transgenic animals are

Patent No. 1575423

Foreign Application
23 September 1976

Date of Patent 20 September 1977
Date of Sealing 21 [illegible] 1981

Elizabeth the Second by the Grace of God of the United Kingdom of Great Britain and Northern Ireland and of Her other Realms and Territories, Queen, Head of the Commonwealth, Defender of the Faith: To all to whom these presents shall come greeting:

WHEREAS a request for the grant of a patent has been made by

SANDOZ LTD., of Lichtstrasse 35, 4002 Basle, Switzerland, a Swiss Body Corporate,

for the sole use and advantage of an invention for

Cationic monoazo dyestuffs based on pyridone coupling components:

AND WHEREAS We, being willing to encourage all inventions which may be for the public good, are graciously pleased to condescend to the request:

KNOW YE, THEREFORE, that We, of our especial grace, certain knowledge, and mere motion do by these presents, for Us, our heirs and successors, give and grant unto the person(s) above named and any successor(s), executor(s), administrator(s) and assign(s) (each and any of whom are hereinafter referred to as the patentee) our especial licence, full power, sole privilege, and authority, that the patentee or any agent or licensee of the patentees and no others, may subject to the conditions and provisions prescribed by any statute or order for the time being in force at all times hereafter during the term of years herein mentioned, make, use, exercise and vend the said invention within our United Kingdom of Great Britain and Northern Ireland, and the Isle of Man, and that the patentee shall have and enjoy the whole profit and advantage from time to time accruing by reason of the said invention during the term of sixteen years from the date hereunder written of these presents: AND to the end that the patentee may have and enjoy the sole use and exercise and the full benefit of the said invention, We do by these presents for Us, our heirs and successors, strictly command all our subjects whatsoever within our United Kingdom of Great Britain and Northern Ireland, and the Isle of Man, that they do not at any time during the continuance of the said term either directly or indirectly make use of or put in practice the said invention, nor in anywise imitate the same, without the written consent, licence or agreement of the patentee, on pain of incurring such penalties as may be justly inflicted on such offenders for their contempt of this our Royal Command, and of being answerable to this patentee according to law for damages thereby occasioned:

PROVIDED ALWAYS that these letters patent shall be revocable on any of the grounds from time to time by law prescribed as grounds for revoking letters patent granted by Us, and the same may be revoked and made void accordingly:

PROVIDED ALSO that nothing herein contained shall prevent the granting of licences in such manner and for such considerations as they may by law be granted: AND lastly, We do by these presents for Us, our heirs and successors, grant unto the patentee that these our letters patent shall be construed in the most beneficial sense for the advantage of the patentee.

IN WITNESS whereof We have caused these our letters to be made patent as of the twentieth day of September one thousand nine hundred and seventy-seven and to be sealed.

Comptroller-General of Patents Designs, and Trade Marks.

FIG. 1. Letters Patent granted under the Patents Act 1949

described as giving ownership of 'life' and patents for isolated human genes are talked of as though they give property rights over human beings.

Property Right

A patent is nevertheless a piece of property, and may be a very valuable one. Although intangible property, it may be dealt with in the same sort of ways as tangible property, such as real estate. Just as the owner of a house may live in it, sell or rent it to another, mortgage it, or even have it demolished, so a patentee may keep the patent rights, assign the patent to someone else, grant someone else a licence to do something covered by the patent, mortgage the patent (that is, use the patent as security for a loan), or, of course, abandon the patent to the public. Abandonment of patent rights is very common because, in the great majority of countries, renewal fees must be paid each year to keep a patent in force; these renewal fees often rise steeply as the age of the patent increases, and only those patents that are of real commercial importance are kept alive for their full term. For US patents applied for before 12 December 1980, no renewal fees were payable and positive action had to be taken in order to 'dedicate to the public' such a US patent. Some of these patents may still be in force, and no doubt for many of them the patentee has no further commercial interest; they may nevertheless still impede the commercial activities of others. For US patents applied for after that date, however, renewal fees are payable, as is the case in practically every country in the world.

Limited Duration

In any event, no patent can go on indefinitely. It is a point that is fundamental to the whole concept of patents that the exclusive rights are granted only for a limited period of time and that the general public is free to use the invention once this term has expired. British patents used to be granted generally for a term of 14 years, this being the time required for two generations of apprentices to be trained in the invention; in 1919 the term became 16 years. Extensions were possible in exceptional circumstances, for example if the patentee was unable to exploit the invention because of wartime conditions, or if the patentee had a particularly deserving invention on which, through no fault of his own, he had not made sufficient profit. A combination of these two grounds enabled the British patent for a pioneering invention relating to colour television to be extended to a total term of 32½ years: unfortunately for the patentee, his infringement action against producers of colour television sets was unsuccessful, the patent being finally held invalid.[1]

In the USA and Canada the term of a patent was until fairly recently 17 years from the date of grant, which meant that the longer the Patent Office took to grant the patent, the later was the expiry date. In most other countries, the term ran from the date of application and so the expiry date was fixed irrespective of how long the

[1] *Valensi v. British Radio Corp. Ltd* [1973] RPC 337 (CA).

process of grant might take. A term of 20 years from the filing date was set for European patents by the European Patent Convention, which came into force in 1978, and the Patents Act 1977, which came into force on the same day, set the same term for British national patents. This has become the international standard set by the Trade-Related Aspects of Intellectual Property Rights (TRIPs) Agreement under the General Agreement for Tariffs and Trade (GATT) (see Chapter 2) and is now adhered to by nearly all countries. When the US law was changed to implement TRIPs, all granted patents and pending applications were given a term of 17 years from grant or 20 years from filing, whichever was longer, whereas applications filed after 20 October 1995 received only the standard 20-year term.

The term of available patent protection is more important in some industries than others. In the pharmaceutical industry, for example, in which it takes many years for a product to reach the market and in which the same product, once introduced, can usually be sold for 20 years or more, it is vital to the patentee to obtain as long a patent term as possible. On the other hand, in an industry in which products can be brought to the market quickly but are rapidly replaced by newer products, the inventor or his assignee is more interested in obtaining rapid grant of an enforceable patent than in prolonging patent term. For products that require a long approval process before marketing, such as pharmaceuticals and agrochemicals, it is now possible to obtain extensions of the standard patent term in the USA, Europe, and Japan.

Patents and Monopolies

So far, in discussing the basic nature of patents, we have avoided using the word 'monopoly', a word that has acquired many negative connotations. This is intentional, because there is a clear distinction to be made between a monopoly of an existing commodity and the exclusive rights given by a patent for a new invention. The old definition of monopoly given in *Blackstone's Commentaries* is still a good one:

> a licence or privilege allowed by the King, for the sole buying and selling, making, working or using of anything whatsoever; whereby the subject in general is restrained from that liberty of manufacturing or trading which he had before.

According to this definition, exclusive patent rights are not a monopoly because, being for a new invention, they cannot possibly take from the public at large any right that the public previously had. Even when the term 'monopoly' is given the broader meaning of any exclusive right to make, use, or sell, the distinction between a monopoly in an existing commodity and a patent monopoly in respect of a new invention should be kept in mind. The two were clearly distinguished in English law as long ago as 1624, although today it is still being alleged, for example, that patents on new products obtained from the neem tree will stop Indian peasants from using traditional neem remedies.

Early History in England

Elizabethan Monopoly Grants

In the reign of Queen Elizabeth I, monopolies in commodities such as salt, coal, playing cards, and many others were frequently granted by letters patent either in return for a cash payment as a means of raising revenue, or as a convenient method of rewarding royal favourites at the public expense. Such monopolies were a continuing cause of unrest, since not only were prices of everyday articles artificially raised, but the patent holders were also given wide powers of enforcement of their rights, including powers to search premises for infringing articles and to levy fines on the spot. The popular outcry against these depredations reached such a pitch that in 1601 Elizabeth, who always knew how to give way gracefully when no other course was open to her, issued a proclamation revoking the majority of grants of monopoly. Perhaps more importantly, whereas previously the grant of monopolies had been a matter of royal prerogative that could not be challenged by the subject, the proclamation of 1601 allowed matters concerning such grants to be contested in the common law courts.

The next year, Edward Darcy, who had been granted by letters patent a monopoly in the importation, making, and selling of playing cards, attempted to enforce his right in the courts against an infringer named Allein. The court held that the monopoly was illegal and the patent was declared invalid. In the course of this case, it was clearly stated that patents for new inventions should form an exception to the general rule against monopolies:

> When any man by his own charge and industry, or by his own wit and invention doth bring any new trade into the realm or any engine tending to the furtherance of a trade that was never used before; and that for the good of the realm; in such cases the King may grant him a monopoly patent for some reasonable time, until the subjects may learn the same, in consideration of the good that he doth bring by his invention to the Commonwealth.[2]

This classic statement of the law introduces the legal concept of the 'consideration' for the patent grant.

In common law, a contract between two parties will normally be valid only if there is consideration on both sides. For example, in a contract of sale, A transfers property to B in consideration of a sum of money paid by B to A; the consideration for B's payment is the property transferred. It may be considered that a patent for an invention is in the nature of a contract between the inventor and the state in which the state ensures that the inventor will have exclusive rights for a limited time in consideration for the benefit to the state that is expected to arise from the invention.

[2] *Darcy v. Allein* [1602] 1 WPC 1 (Court of Queen's Bench).

The Statute of Monopolies

In spite of the judgment in the case of Darcy's patent, illegal monopolies continued to be granted by King James I, and to curb this continuing abuse, Parliament enacted on 25 May 1624 the Statute of Monopolies, which formed the basis of the law on patents in England for over 200 years. It consisted of a general prohibition on the grant of monopolies, qualified by certain specific exceptions. Section 6 exempted patents for new inventions from the general prohibition, in the following words:

> Provided also, and be it declared and enacted, that any declaration before mentioned shall not extend to any letters patent and grants of privileges for the term of fourteen years or under, hereafter to be made, of the sole working or making of any manner of new manufactures within this realm, to the true and first inventor and inventors of such manufactures, which others at the time of making such letters patents and grant shall not use, so as also they be not contrary to the law nor mischievous to the state, by raising prices of commodities at home, or hurt of trade, or generally inconvenient.

It will be seen that section 6 contains exceptions to the exception, providing that patents that raised prices, hurt trade, or were generally inconvenient could be declared invalid. Nowadays, we would call such provisions 'anti-trust' or 'abuse of monopoly' provisions.

This piece of legislation was not innovative, but simply declarative of the common law as it had already been established. It did not alter the fact that an inventor had no automatic right to a patent for his invention; the grant of a patent was still an act of royal prerogative that had to be sought by petition and which could be refused at will.

Where the Statute of Monopolies speaks of an 'inventor', the term means not only an inventor in the modern sense of the originator or creator of a new idea, but also extends to a person who brings something new into the country for the first time. Many of the early patents granted before the Statute of Monopolies had been to inventors of this type: for example the first and second English patents for invention (in 1449 and 1552) both related to glass-making techniques that were known in continental Europe, but not established in England. Here, the consideration for the grant of the patent was the establishment of the new industry in England, or, in modern terms, the transfer of technology to a developing country. The basic objective was to improve the balance of trade by reducing imports and encouraging new industries that could generate exports.

Disclosure of the Invention

The first English patent granted to an inventor in the modern sense of the word appears to have been that to Giacopo Acontio in 1565 for a new type of furnace. Early English patents for inventions contained no more description of the invention than the title, and as long as the pace of technological progress remained snail-like and the number of patents granted was small, this was no doubt sufficient. As the number of patents increased, however, and as patents began to be granted for specific improvements rather than for the setting up of whole new industries, it became common to add a short description of the invention to the letters patent. By the early 18th century, it had become the rule that patents were granted on condition that the patentee filed a detailed description of the invention within a fixed period after grant.[3] Gradually, the concept arose that the disclosure of the invention in the patent specification was the consideration for the grant, a concept that is still much in vogue.

According to this view of the patent system, an inventor has the choice of keeping his invention secret or of applying for a patent. In the first case, he may succeed in keeping the invention secret for a very long time, but if it becomes known to others, or if others invent it independently, he has no redress. In the second alternative, the state guarantees the inventor a monopoly for a limited time; afterwards, anyone is free to carry out the instructions published in the specification and practise the invention, to the general benefit of the economy.

This theory presupposes that the technological development of the society in question is sufficiently advanced that there are enough people able to put the invention into effect on the basis of a written description—a condition that is by no means always met in many countries that grant patents. Furthermore, while it can logically be advanced in respect of an invention such as a process, which can be kept secret within the walls of a factory, it is clearly not valid where the invention is a new article or a new chemical compound that is published to the world as soon as it is sold. The question of the consideration for the grant is more complex than would appear from this simple 'disclosure theory'.

Early History in Continental Europe

Venice

At the time when the nature of monopolies for new inventions, as distinct from existing commodities, was still being worked out in England, there was already in the Republic of Venice a decree on the protection of inventions that still sounds very modern today despite dating from 1474.[4] In the following extract, the notes indicate the modern concepts that we would apply to the provisions:

There are in this city, and also there come temporarily, . . . men from different places[1] and most clever minds, capable of devising and inventing all manner of ingenious contrivances. And should it be provided, that the works and contrivances invented by them, others having seen them could not make them and take their honour, men of such kind would exert their minds, invent and make things which would be of no small utility and benefit to our State.[2] Therefore, . . . each person who will make in this city any new and ingenious contrivance, not made heretofore in our dominion,[3] as soon as it is reduced to perfection,[4] so that it can be used

[3] *Liardet v. Johnson* (1778) 1 WPC 53.
[4] 'Venetian patents 1450–1550' [1948] JPOS 511.

and exercised,[5] shall give notice of the same[6] . . . . It being forbidden to any other in any territory and place of ours to make any other contrivance in the form and resemblance[7] thereof, without the consent of the author up to ten years.[8] And, however, should anybody make it, the aforesaid author and inventor will have the liberty to cite him before any office of this city,[9] by which office the aforesaid who shall infringe be forced to pay him the sum of one hundred ducats[10] and the contrivance be immediately destroyed.[11] Our government shall be at liberty to take and use in his need any of said contrivances, provided that no others than the authors shall exercise them.[12]

1 Rights not limited to local nationals.
2 General economic benefit as consideration for the grant.
3 Local novelty requirements.
4 Reduction to practice required.
5 Sufficiency requirements.
6 Disclosure a condition of patenting.
7 Infringement not limited to exact copies.
8 Fixed term of protection.
9 Infringement action before administrative bodies.
10 Damages for infringement.
11 Delivery up and destruction of infringing goods.
12 Limited government use provisions (only the inventor can supply).

In 1594, Galileo was granted a Venetian patent for an irrigation machine. By this time, the length of the patent term had increased to 20 years and it was required that the machine actually be constructed within one year, in effect replacing a requirement for actual reduction to practice before grant by compulsory working provisions. In spite of this high degree of sophistication, however, the Venetian patent system fell into disuse as the power and importance of Venice declined, whereas the English system has remained continuously in effect to the present day.

Germany

Historical research has shown that during the 15th and 16th centuries there was a well-established system of inventors' privileges analogous to patents granted either by the Emperor or by local princes within the German states of the Holy Roman Empire, even though no codified laws seem to have existed. There are even records of what were essentially infringement actions taken by a jeweller named Claudio vom Creutz in Nuremberg between 1593 and 1604 under an imperial patent relating to polishing of gemstones. In one case, at least, the infringer was imprisoned, had to pay costs, and was banished from the city.[5] But the chaos caused by the Thirty Years' War in the early 17th century destroyed this early patent system along with much else, and patent laws in various German states developed only slowly during the first part of the 19th century.

5 P. Kurz, 'Historische Patentprozesse (I)' [1996] *Mitteilungen der deutschen Patentanwälte* 65.

France and the Netherlands
In pre-revolutionary France, there were grants of monopoly privileges for economic and tax reasons, similar to those in Elizabethan England, and no true patents for new inventions. During the Revolution in 1789, all privileges and monopolies were abolished, but a new modern patent law was enacted only two years later. In the Netherlands, on the other hand, patents were granted by the States General as well as by the individual provinces for new and useful inventions, the emphasis being more on utility than novelty. During the 200 years leading up to the Napoleonic wars, over 600 such patents were granted by the States General. In the next century, however, the Dutch were more hostile to patents, as we shall see in the next chapter.

Early History in North America

The development of patents in North America was, understandably, based largely on concepts developed in England. Although, before independence, the colonies lacked the sovereign power to grant letters patent, they nevertheless had legislation, such as that of Massachusetts in 1641, giving exclusive rights for limited periods to persons introducing new industries to the colony. After independence, South Carolina, for example, introduced a statute (1784) dealing with inventions on the same basis as artistic copyright and providing a 14-year term. On the other hand, Article 39 of the Maryland State Constitution of 1776 stated that 'monopolies are odious, contrary to the spirit of free government and the principles of commerce; and ought not to be suffered'. Perhaps the Elizabethan type of monopoly was what Maryland had in mind.

The US Constitution
The Articles of Confederation, which preceded the Constitution and which were, in any case, not ratified until 1781, only delegated certain specific powers to the US Congress and the power to grant patents was not among them. Finally, in 1788, the US Constitution was ratified, containing Article I, section 8:

> The Congress shall have Power . . . To promote the Progress of Science and useful Arts, by securing for limited Times to Authors and Inventors the exclusive Right to their respective Writings and Discoveries.

Congress could have chosen to promote the progress of science and useful arts, for example by granting bounties and cash awards, but these had the disadvantage that they would cost the government money, whereas the grant of exclusive rights would cost nothing.[6]

[6] E.C. Walterscheid [1998] JPTOS 11.

Grant to Inventors

There are two points worth noting in this brief statement. First, although it merely gave Congress power to enact a patent law without seeming to place any restrictions on what form such a law might take, nevertheless the wording 'to inventors' is probably the reason why today the USA is the only country in which a patent must be applied for by the inventor and not by an assignee, such as the inventor's employer. It is for this reason that, as we shall see, correct designation of inventorship plays such an important role in US patenting, whereas in most other countries it has little or no effect on patent validity although it may be important for other reasons, such as compensation for employee inventors.

First to File

Furthermore, these words of the Constitution appear to be the basis for the practice that whereas in most countries questions of precedence between two patent applications claiming the same invention are resolved on the simple basis that the first to file an application has priority, in the USA the patent is granted, subject to certain conditions, to the person who first made the invention. Because this is by no means an easy matter to sort out, a lengthy and cumbersome procedure known as 'interference' has had to be developed to resolve priority in conflicting applications.

It has been suggested that, because of the wording of the Constitution, any change in the US law in these respects would require a constitutional amendment. Other authorities discount this view, pointing out that copyright in the USA may be applied for by an assignee even though artistic copyright is covered by the same section of the Constitution as is patent protection, and that the Constitution does not specify how the term 'inventor' is to be defined. It has been argued[7] that the positions of the USA and the UK are not as far apart as is generally believed. In the UK, the Statute of Monopolies talked of patents being granted to the 'true and first' inventor, which implies a first-to-invent system, but later cases defined the first inventor as the first to bring the invention to the public by filing a patent for it. In interference proceedings in the USA, there is a rebuttable presumption that the first applicant is the first inventor, and all that would be necessary would be to make the presumption irrebuttable. Whether this change will ever be made is still an open question.

The second point of interest in the constitutional provisions on patents is the statement that the purpose of granting exclusive rights to inventors is to promote scientific and technical progress. This brings us again to the question of the consideration for the grant, which is here expressed not as the narrow exchange of protection in consideration of disclosure, but rather as the broad concept that a patent system encourages progress. Although the mechanism of how it is supposed to do so is not stated, the association in the USA, over a long period of time, of a strong

[7] T. Roberts, 'The case for first to file' [2004] CIPA 15.

patent system with an enormous degree of scientific and technological development appears to confirm the view of the framers of the Constitution.

Consideration for the Grant of a Patent

Consideration for the grant of a patent should not be regarded for individual patents in isolation. It is not the establishment of new industry, although in a few very rare cases a single invention will base an entire new industry. It is not the disclosure of the invention, since in most cases the invention will be made public if and when it is commercialized. It is not the working of the invention, since it is only commercially feasible to work 10 per cent or less of the inventions that result in patents.

The consideration for the granting of patents, in general, is the benefit that results to the state by technological progress as represented by the commercialization of inventions. The connection between the granting of patents and the commercialization of inventions is simply that the existence of patent rights removes part of the risk involved in investment in a new development. Who, after all, would be willing to invest large sums of money in a new project knowing that an imitator could copy the product as soon as it was marketed, without incurring any research costs? The justification for the patent system is that it provides an incentive for investment in new ideas, without which technological development would be much slower and more difficult.

2

HISTORICAL DEVELOPMENTS

I began with the Queen upon the Throne. I ended with the Deputy Chaff-wax.
Charles Dickens, 'A poor man's tale of a patent' (1850)

The UK 1800–2009

It may be more than a coincidence that the Industrial Revolution began in England, a country in which there had been for over two hundred years an uninterrupted tradition of patents for new inventions. But although by the early 19th century the pace of industrialization had accelerated enormously, particularly in textile manufacture and transportation, and the economic importance of patent protection was becoming clear, the British patent system had not kept pace with technical progress. By the first half of the 19th century, the system for granting patents in England had indeed advanced to the point at which a specification describing the invention had to be filed within a certain time after grant. The actual process of getting the patent, however, was an incredible rigmarole of petty bureaucracy immortalized by Dickens in his 1850 short story, 'A poor man's tale of a patent'. The sequence involved (at one stage or another) obtaining the signatures of the Home Secretary, one of the two Law Officers, the Sovereign, and the Lord Chancellor, and sealing various documents with the Signet, the Privy Seal, and, finally, the Great Seal on the letters patent document itself. All of this took time and money: approximately six weeks and £100 (which is equivalent to approximately £7,500 today). What is more, the patent extended only to England and Wales, and separate patents had to be obtained for Scotland and Ireland so that the total cost of UK patenting was over £300, a sum that only a wealthy man could then afford.

The UK Patent Office

Two years after Dickens' satire, in 1852, a major reform of the patent system was enacted, in which the Patent Office was set up and empowered to grant a single patent covering the whole of the UK. For the first time, a description of the invention had to be filed on applying for a patent. This could be a complete specification, giving a full description; or a provisional specification, giving merely an outline to be completed later, within a fixed period after grant. The costs of obtaining a patent were greatly reduced, but renewal fees had to be paid in order to keep the patent in force for its maximum term. An important step forward was that the patent was dated from its application date, so that a disclosure of the invention during the application procedure would no longer invalidate the patent.

Now that specifications describing the invention were required as part of the application procedure and not as an afterthought, more care began to be given to the drafting of the descriptions, and, in particular, to pointing out what were considered the new and important parts of the invention. As more and more patents were granted, it became necessary to clarify what the patentee thought was the crux of the invention, over which he claimed a monopoly, and the practice grew up of doing this by means of a separate part of the specification referred to as the 'claims'. At that time, infringement actions were still heard before a jury, who had to determine the scope of the monopoly on the basis of the specification, the claims serving only to point the jury in the right direction.

Patent Claims

Claims grew in importance with the reorganization of the courts in 1875, which transferred jurisdiction in patent cases to the Chancery Division of the High Court, and with the Patents Act of 1883, which required specifications to contain at least one claim. Even so, the claim could be of the type 'the . . . substantially as herein described', which in effect abdicated any responsibility for defining the invention and left the question for the court to determine as before. Gradually, however, it became settled law that the patentee set the boundaries of the monopoly by the wording of the claim and that 'what was not claimed was disclaimed'.[1]

We shall discuss in more detail later (see Chapter 19) how claims are drafted and interpreted, and what protection is given by various types of claim. At this point, all that need be mentioned is that a patent claim may claim a product (for example a machine, a manufactured article, a chemical compound, or a composition comprising a mixture of substances) or a process (which may be a process for manufacturing an article or synthesizing a compound, or may be a method of using a product). Until recently, the law differed widely from country to country as regards what types of claim were permissible and what legal effect they had; thus in the UK a process claim was infringed not only by carrying out the process in the UK, but also by selling in the UK the product of the process, whereas in the USA the claim did not extend to the product of the process. The TRIPs Agreement has now mandated that, as in the UK, a process claim must cover the direct product of the process, and most countries have now adhered to this standard (see Chapter 3).

Provisional and Complete Specifications

Further developments made by the 1883 Act included the provision that a complete specification had to be filed before the patent was granted, the establishment of a Register of Patents, and the possibility for third parties to oppose the grant of a patent (for example on the ground that the invention was not new). At the same time application costs were further reduced and it became possible to file an application with a provisional specification for the sum of £1. This nominal application fee amazingly enough remained unchanged until 1978, being probably the only

1 *EMI v. Lissen* [1938] 56 RPC 23 (HL), *per* Lord Russell.

example in history of any charge, official or otherwise, remaining stable for 95 years. When the Patents Act 1977 came into force an application could be filed for a fee of £5, which was still cheaper in real terms than it was in 1883. However, this fee did not have the permanence of its predecessor: by 1997 it had reached £25 and was expected to increase further, but to everyone's surprise the fee was abolished completely in October 1998, only to reappear in December 2007 at a level of £30. As in the European Patent Office, a filing date will still be given even if the application fee is not paid.

The increasing importance and complexity of drafting patent specifications and claims encouraged certain consulting engineers to specialize in this new art, and to set up in business as agents for inventors in the drafting and procuring of patents. In 1882, an Institute of Patent Agents was founded, and in 1888 a Register of Patent Agents was established by the Board of Trade under the control of the Institute, which obtained its Royal Charter in 1891. The profession of patent agency in Britain is still controlled by the Chartered Institute, which sets examinations for admittance to the Register of Patent Agents.

Examination

Throughout the 19th century, patents were granted in the UK without anything more than a purely formal examination. Only in 1902 was a novelty examination provided for, and not until five years later could the Patent Office actually refuse an application on the ground that the invention was not new. Even then, the search was limited to British patent specifications not more than 50 years old, and only after 1932 were foreign publications considered. Such publications, however, could be considered only if they were shown to be published within the UK—an approach consistent with the old concept that patents could be granted for bringing into the country an invention already known elsewhere, as well as for making a completely new invention.

These various changes to the law on patents were made in a piecemeal manner, and although the Patents and Designs Act of 1907 was a more-or-less comprehensive codification of the law, in line with the common law approach to legislation, it did not attempt to replace the existing case law on patents that had been built up by judicial decisions over the centuries.

Abolition of Chemical Product Protection

In the 1870s, Britain had come close to abandoning the patent system altogether, because it was considered to be protectionist in nature and was opposed by free trade advocates, including *The Economist*.[2] Having passed that crisis, the patent system in the UK increased gradually in strength, until it took its first backward step

[2] M. Coulter, *Property in Ideas: The Patent Question in Mid-Victorian England*, Thomas Jefferson University Press, 1991.

in 1919. Up to that time, patents had been granted as a matter of course for new chemical substances. But the British chemical industry felt itself technologically inferior to that of Germany, which, for some years before the First World War, had dominated the dyestuffs market. British industry pressed for the abolition of patent protection for chemicals as such, and limitation of patent protection to that for specific processes for the preparation of chemicals. In this way, British firms hoped to be free to imitate a German dyestuff appearing on the British market so long as they could find an alternative process for its preparation—a task that was a good deal easier than inventing a new and better dyestuff themselves. At the same time, the German dyestuff manufacturers themselves were happy with a system that had only process protection for chemicals in Germany.

This change in the law was duly made in 1919, together with the further weakening of patent protection for pharmaceuticals by allowing compulsory licences to be granted virtually on demand for patents relating to medicine. The first of these retrograde steps was abolished after the Second World War in 1949; the second, only in 1977. One small positive development in 1919 was that the term of a patent was increased from 14 to 16 years.

Patents Act 1949

The Patents Act 1949 was also primarily a codification of gradual changes in the law that had been brought about by the courts and, apart from the restoration of protection for chemical substances per se, contained little that was really new. For the first time, all of the grounds on which a patent could be declared invalid by the Patent Office or by the court (a wider set of grounds) were set out in full and priority dates of claims were defined.

The priority date of a claim was essentially the date on which the subject matter of the claim was first disclosed to the British Patent Office or to a foreign patent office in an application from which priority was properly claimed. No publication of this subject matter could affect the validity of the claim so long as it took place later than the priority date of that claim.[3]

Patents Act 1977

The Patents Act 1977 introduced major changes to British patent law, primarily to bring the national law into harmony with the European Patent Convention (EPC; see below). Like the EPC itself, the Patents Act 1977 came into force on 1 June 1978. Among other changes, this Act lengthened the term of patents from 16 to 20 years, adopted the simplified grounds for invalidity as set out in the EPC, defined precisely what constitutes infringement, strengthened the examination procedure of the British Patent Office by allowing examiners to raise objections of lack of inventive step, and introduced provisions for compensation to employee inventors

[3] Section 5(1), PA 1949.

in some circumstances. Earlier British Patents Acts had generally codified existing common law practice and made minor changes as required, but the Patents Act 1977 was enacted to recognize and implement a new European system of law for the grant of patents, which itself had been drafted *ab initio* on the civil law principle, and it made far-reaching changes that destroyed or reduced the effect as precedent of a great many decided cases in British patent law.

Later Developments

Subsequent legislation in the UK has had relatively minor effect. The Copyright, Designs and Patents Act 1988 was primarily concerned with the first two mentioned types of intellectual property. It deregulated to some extent the profession of patent agents (see Chapter 17), set up the Patents County Court (see p. 195) as an alternative tribunal to the Patents Court (itself set up by the Patents Act 1977), and made some minor amendments to the Patents Act 1977 . The Competition Act 1998 revoked sections 44 and 45 of the Patents Act 1977 (see p. 482). The Patents Act was amended in 2004 by the Regulatory Reform (Patents) Order 2004,[4] which relaxed formal requirements as required by the Patent Law Treaty (see p. 50). This came into effect on 1 January 2005, whereas the Patents Act 2004, which amended the 1977 Act so as to enable the UK to ratify the new version of the European Patent Convention (EPC 2000; see p. 31), did not enter into force until December 2007.

The Patent Office moved from London to Newport, South Wales, in 1991, and in 2007 adopted the 'trendy but pointless'[5] operating name of the 'UK Intellectual Property Office', but for the purposes of legislation the name 'Patent Office' remains.

The USA 1790–2009

Patent Laws 1790–1870

The constitutional provisions on the protection of inventions first took form in the Patent Act of 1790. This established a very strict examination system under which all patent applications had to be scrutinized by a committee of three cabinet ministers, consisting of the Secretary of State, the Secretary of War, and the Attorney General, at least two of whom had to be present. Incidentally, the first US patent, bearing the signature of George Washington, was for a chemical invention relating to the manufacture of pearl ash (potassium carbonate).

At that time, Thomas Jefferson was Secretary of State. Himself an inventor, he initially took a keen personal interest in the examination and granting of patents, and, with the possible exception of Albert Einstein (who was once an examiner in

4 SI 2004/2357.

5 According to Jacob LJ in *Du Pont v. UK Intellectual Property Office* [2009] EWCA Civ 966.

the Swiss Patent Office), must surely be the most distinguished patent examiner in history. It soon, however, became a burden that he no longer relished. As he wrote to a friend in 1792:

Above all things he prays to be relieved from it, as being, of everything that was ever imposed on him, that which cuts his time up into the most useless of fragments and gives him from time to time the most poignant mortification. The subjects are such as would require a great deal of time to understand and do justice by them, and not having that time to bestow upon them, he has been oppressed beyond measure by the circumstances under which he has been obliged to give undue & uninformed opinions upon rights often valuable and always deemed so by the authors.[6]

When only 55 patents had been granted in three years under this impractical system, the USA moved to the other extreme. The Patent Act of 1793 allowed the grant of a patent upon request without any examination, even after the Patent Office was set up in 1802. The Patent Office was first located in four rooms of a building known as Blodgett's Hotel, where it survived the minor unpleasantness of the British troops burning Washington in 1814, the Superintendent of Patents successfully pleading with the British commander not to 'burn what would be useful to all mankind'. However, the Patent Office burned down of its own accord in 1836, with the loss of all of its records.

The confusion caused by the grant of a great many overlapping patents under the law of 1793 led to a Congressional report, on the basis of which a systematic examination system was introduced in 1836 under the direction of a Commissioner of Patents. This system, set up at a time when the Dickensian ritual described earlier was the only way of getting a patent in England, is still the basis of US patent law today.

The 1836 Act retained the original patent term of 14 years, but enabled extensions of seven years to be obtained. A Register of Patents was established and an appeals procedure was set up. The applicant was required to supply a model of the device and these models were on display at the Patent Office. This general requirement to supply models was not abolished until 1880 and although many thousands of models were destroyed by fire or otherwise lost, many are now in museums or private collections, and give a fascinating insight into patenting activities in the mid-19th century.

Various amending Acts were passed in the years after 1836, which were consolidated in the Patent Act of 1870. Among these was the extension, in 1861, of the patent term to 17 years. Under the 1870 Act, the Commissioner was given power to issue regulations for the administration of the patent law.

Supreme Court Decisions

The development of US patent law has been strongly influenced by decisions of the courts—particularly those of the US Supreme Court. As an example of this, we may

6 Thomas Jefferson, Letter to Hugh Williamson, 1 April 1792.

consider the difficult question of how much invention is needed to support a patent. On the one hand, it is clearly wrong that patents should be granted for improvements so minor that any competent mechanic or chemist could make them as a matter of course. On the other, it is also wrong that patents should be granted only for outstanding inventions that revolutionize society. Many inventions are ingenious and at least potentially useful without being world-shattering, and one great merit of the patent system is that the value of the patent grant is left to be determined by market forces. It is not as if the state, in granting a patent, guarantees to the patentee that he will profit by it:[7] by and large, a patent for a poor invention will not be very valuable.

The extreme position that only outstanding inventions should be patentable was nevertheless adopted by the Supreme Court in 1941,[8] when Justice Douglas condemned the grant of patents for 'gadgets' as being contrary to the constitutional requirement that the grant of patents should 'promote science' and thus 'push back the frontiers of knowledge', clearly forgetting that the Constitution also spoke of promoting 'the useful arts'. After this and similar cases,[9] a very high standard of inventiveness was set by the courts for US patents. A patent could not be granted unless there was 'invention' and there was no invention if there was not a 'flash of genius'.

Patent Laws 1952–2002

The new US Patent Law enacted in 1952 attempted to reverse this trend by shifting discussion from the word 'invention' to the word 'obvious'. Under the new law, an invention not only must be new, but also must not be obvious over the 'prior art' (a term used for all earlier publications or knowledge that can be cited against a later patent application). To some judges the concept of an obvious invention appeared to be a contradiction in terms, but the change in the law did eventually lead to the replacement of a purely subjective criterion of invention (did the inventor have a flash of inspiration, or was it routine work?) by a more objective and less stringent criterion (would the inventive step have been obvious to an average skilled workman or chemist in possession of all the prior art?).

More recently there have been a range of far-reaching changes in US patent law, which are discussed in detail in later chapters. At the end of 1980 an Act was passed that, for the first time, introduced renewal fees as a condition for the maintenance in force of US patents. It also made it possible for the patentee or another party to request re-examination of a granted US patent on the basis of prior art that had not been considered during prosecution. Other provisions dealt with the ownership of

[7] Inventors have not always realized this. Early last century, one inventor in the UK, who wrote his own patent specification, ended it with the words: 'What I claim is: one thousand pounds.'

[8] *Cuno Engineering Corp. v. Automatic Devices Corp.* 314 US 84 (Sup Ct, 1941).

[9] For example, *Great Atlantic and Pacific Tea Co. v. Supermarket Equipment Corp.* 87 USPQ 303 (Sup Ct 1950).

rights in inventions made with the help of US government funding.[10] A minor change made in 1975 was that the US Patent Office changed its name to the 'US Patent and Trademark Office' (USPTO).

In 1982 the US federal court system was modified by the creation of a new Court of Appeal for the Federal Circuit (CAFC), which took over the functions of the Court of Customs and Patent Appeals and some of the functions of the Court of Claims. More importantly, it assumed the jurisdiction previously held by the 12 regional courts of appeal to hear appeals on patent matters from the federal district courts.

In October 1984 a number of changes were made, including a relaxation of the strict rules on joint inventorship, the formation of a new Board of Appeals and Interferences, and provision for 'statutory invention registrations', a type of defensive patent that would constitute prior art against later patent applications, but give no monopoly rights. These have not proved popular, and very few have actually been issued, but the provision remains in effect.[11]

More important than any of these was the Drug Price Competition and Patent Restoration Act of 1984, commonly known as the Hatch-Waxman Act, which provided for extensions of patent term for human drugs, food additives, and medical devices, the commercialization of which had been delayed by regulatory procedures (in the Food and Drug Administration). At the same time, the Act made registration easier for competitors when patent protection expired and provided that activities necessary for obtaining regulatory approval did not amount to patent infringement (see p. 170).

There have been a number of legislative changes to the US patent law since then, most of them fairly minor, but the most important was certainly the Uruguay Round Agreements Act of 1994 (URAA), which made the changes that were necessary in order to bring US law into line with the TRIPs Agreement (see below). Since then, the American Inventors Protection Act of 1999 (AIPA) has introduced the possibility of extension of patent term to compensate for prosecution delays in the USPTO, publication of pending applications at 18 months from filing, *inter partes* re-examination, and certain changes to the law on novelty. All of these provisions were so complex that it was said that a better name for the law might have been the 'American Patent Attorneys Full Employment Act'. Some further amendments to the AIPA were made in November 2002, relating to re-examination and the prior art effect of Patent Cooperation Treaty (PCT) filings.

The USPTO is organizationally part of the US Department of Commerce, and until 2001 was headed by a Commissioner of Patents and Trademarks. The Patent and Trademark Efficiency Act of 1999, which came into force in January 2001, replaced the office of Commissioner with that of Undersecretary for Intellectual

[10] The University and Small Business Patent Procedures Act of 1980, known as the Bayh-Dole Act; see p. 414

[11] 35 USC 157.

Property and Director of the USPTO. Whether or not this has improved the efficiency of the USPTO is, however, controversial.[12]

New Proposals
Patent Law Reform Bills were introduced in Congress in 2007 and again in 2009; the 2007 Bill came to nothing and, at the time of writing, it seems that the same fate awaits the new Bill. This is a disappointment for those who would like to see the USA join the rest of the world in adopting the first-to-file system, which was a feature of both Bills. Strangely enough, abolition of the first-to-invent system was not particularly controversial; the main problem appears to be a fundamental disagreement between the two main users of the patent system, namely the pharmaceutical/biotech industry and the electronics industry, on the issue of the penalties for infringement. For the pharmaceutical industry, it is vital to be able to stop an infringer by means of an injunction, which should be granted more or less automatically if a valid patent is held to be infringed. For the manufacturer of a complex piece of electronic equipment, for which a hundred or more third-party patents may be relevant, it is unacceptable that a product might have to be taken off the market if it is found to infringe one claim of one of these many patents. Other contentious issues include the possibility of what would essentially be opposition proceedings at any time after grant, to be heard by an already overburdened Board of Patent Appeals and Interferences.

Other Industrialized Countries

Germany
Prior to the unification of Germany in 1871 there were 29 different patent laws in the various independent German states; even in Prussia, the most industrially developed state, patents were extremely difficult to obtain and to enforce. The first unitary patent law was passed in 1877, largely as a result of representations made by the industrialist and inventor Siemens, and despite the lack of interest shown by Chancellor Bismarck. The chemical industry at that time campaigned successfully against patent protection for chemical substances as such, and in the new law, only processes for their production were patentable. At first, such process patents were not considered to be infringed by the sale of the product of the process, but in 1888 BASF successfully sued the Swiss firm Geigy for infringement by selling in Germany a dye produced in Switzerland by a BASF-patented process,[13] and in 1891 this form of indirect product protection was codified in the German patent law. Product protection itself did not come until 1968.

[12] See, e.g., the many patent blogs by Greg Aharonian.

[13] P. Kurz 'Historische Patentprozesse (II)' [1996] *Mitteilungen der deutschen Patentanwälte* 368.

There were for over 40 years two German laws: those of the Federal Republic of Germany and the German Democratic Republic (GDR). Since 1990, only the patent law of the Federal Republic survives, but nevertheless patents granted for the GDR prior to reunification were automatically extended at that time to cover the whole territory of Germany. An important development in German patent law was the introduction, in 1957, of a standard system for the remuneration of employee-inventors.

France

The French Patent Law of 1791 established the concept of intellectual property, which belonged by right to the inventor. However, the law granted patents without any examination of their novelty and this remained the basis of the French patent system for many years, during which only formal examination was carried out. Since 1969, the French Patent Office has carried out novelty searches on patent applications and, for some years, it was up to the applicant to decide whether he wished to amend or withdraw the application as a result of the search. Only recently has the Patent Office had the power to refuse the application if the search shows the invention to be old.

The Netherlands

After the re-establishment of Dutch independence in 1815, a very inadequate patent law introduced in 1817 was repealed in 1869. No patents were granted from then until 1912, when the patent system was reintroduced with a strict examination system; in 1995, it switched to the opposite extreme of a system without any substantive examination at all.

Switzerland

Switzerland, like the Netherlands, had no patent law at all for the greater part of the 19th century and was at that time a 'patent piracy' country in which the products of the German chemical industry were safely imitated. A move to introduce patents in 1882 was strongly opposed by J. Geigy-Merian (founder of Geigy AG, later incorporated into Ciba-Geigy and Novartis), who denounced patents as a 'paradise for parasites' (by which he seems to have meant patent lawyers rather than inventors).[14] Until the case of *BASF v. Geigy* in 1888,[15] the products could also be imported and sold in Germany, thanks to the short-sighted refusal of the German industry to accept product protection. A Swiss patent law was enacted in 1888, but only those inventions that could be demonstrated by a model were patentable, thus excluding chemical substances and processes from protection. A modern law was finally introduced in 1907 and product protection for chemicals came into effect only in 1978. A new Swiss patent law is currently being introduced.

[14] G. Dutfield, *Intellectual Property Rights and the Life Science Industries*, Ashgate Press, 2003, p. 53.

[15] *Entscheidungen des Reichsgerichts* (1888) vol. 22 pp. 8–19.

Japan

The history of Japan is not one of continuous development, but one of a long sleep followed by a sudden awakening. For centuries under the shogunate, Japanese society had been rigid and unchanging, and new technology such as firearms had been suppressed, for example by the 'ban on novelty' decree of 1721. The Meiji Restoration then ushered in a period of rapid change, during which Japan borrowed much from Western countries, including, in 1885, its first patent law, closely modelled on the German patent law of 1877. Japanese patent law still retains many features of German law, including, for example, deferred examination, and the separation of validity and infringement issues. Japanese patent law has in turn served as a model for laws of other countries, such as South Korea.

Developing Countries

The Colonial Legacy

The majority of developing countries began their independence with patent laws that had been written by their former colonial rulers, for example the Indian Patents and Designs Act of 1911, or were closely based on such laws. Later, however, many developing countries considered that such laws, which generally gave strong patent protection, were not appropriate to their stage of development. Indeed, patents were often perceived as a tool of the ex-colonial powers to continue exploitation by other means, and many developing countries had the mistaken idea that the grant of patents in their countries was an impediment, rather than an aid, to industrial development and the transfer of technology. This ignored the fact that the actual number of patents that were filed in developing countries was so small that it was hard to believe that the existence of patents could be responsible in any way for lack of development. On the contrary the low level of development, and thus the small market size, was the reason why so few patents were filed there.

Erosion of Patent Protection in the 1970s

During the 1970s, there was a general weakening of patent protection in developing countries, particularly in the pharmaceutical field, for which protection was totally or effectively abolished in a number of countries. A classic example of this is the Indian Patents Act 1970.

The Indian Act reduced the patent term generally from 16 to 14 years, and provided for government use without compensation and even government expropriation of patents. Furthermore, all patents for chemical inventions were automatically endorsed 'licences of right' three years after grant. Patents granted for processes for the manufacture of any food or medicine had a maximum term of only five years from grant, and the royalty payable on a licence of right granted on such a patent was limited to 4 per cent of the bulk sales price of the product. For this class of inventions, not only was the patent life reduced to a point at which the patent would normally have expired well before the product came on the market, but also, if there was any patent term left, a competitor could operate under the patent on payment of a nominal royalty.

India accordingly had the worst patent law in the world, with the possible exception of the Dominican Republic, where not only are patents ineffective against importation, but they are also actually invalidated if the patentee imports the patented products.[16]

Erosion of patent rights was also very marked in Latin America, for example in Brazil, where patent term was reduced and all patent protection for pharmaceutical inventions was abolished. In Argentina, court decisions rather than a change in the law were responsible, and in Mexico, the patent law of 1976 allowed only weak certificates of invention to be granted for pharmaceuticals. In the Andean Pact countries, Decision 85 of the Cartagena Agreement proposed a model patent law that prohibited pharmaceutical patent protection.

Proposals were also made to amend the Paris Convention in ways that would abandon the principle of equal treatment on which the whole convention was based, replacing it with the principle of a bias in favour of developing countries. For example, one proposal was to permit the grant of exclusive compulsory licences, which would have been equivalent to the expropriation of patent rights. The industrialized countries found themselves having to fight a rearguard action to prevent such extreme proposals being adopted.

Transfer of Technology

One of the major concerns of international politics has been the economic relationship between the industrialized countries and the developing countries, often referred to as the North–South dialogue. Transfer of technology to developing countries is badly needed, and is in the long-term interests of the industrialized countries themselves, because higher per capita income in the developing countries will give rise to more international trade. The problem is that the technology is not in the possession of governments who can simply transfer it as if it were cash. Technology, whether patented or in the form of know-how, is primarily in the possession of industrial companies existing within the framework of the free market economy, and such companies will normally transfer technology only in the context of a licence agreement, for which enforceable patents form a good legal basis.

Restrictions on Licence Agreements

In the 1970s, a parallel development to the weakening of patent protection in developing countries was the imposition of severe limitations on the freedom of contracting parties to reach a mutually acceptable licence agreement, either by national

[16] At least, this is what the law says. It is claimed that the provisions of TRIPs automatically overrule this.

laws, such as that of Brazil in 1975, or by international agreements through United Nations (UN) organizations, such as the UN Conference on Trade and Development (UNCTAD). This body tried to establish a Code of Conduct for the Transfer of Technology (the TOT-Code), which would have the aim of preventing abuse of monopoly by licensors, and generally shifting the balance of rights and obligations in favour of the developing countries. The fallacy in this approach is immediately apparent: anyone can write a contract that is grossly biased in favour of one of the parties, but it is not so easy to find another party willing to sign it. The result was simply that companies refused to enter into any agreements containing such terms, and technology transfer was effectually halted.

New Developments in the 1980s and Early 1990s

The tide turned in the 1980s, when the Western industrialized countries, led by the USA and encouraged by the pharmaceutical industry, finally took action to defend their own interests and refused to give way to unreasonable demands. The US government decided to switch its attention from UN agencies and concentrate on trade issues, initially by threatening to remove from preferred status under the US Generalized System of Preferences (GSP) those countries that refused to give adequate patent protection for inventions. The resulting bilateral negotiations were able to obtain stronger protection in countries such as South Korea and Taiwan. Mexico improved its patent law as part of the negotiations leading up to the North American Free Trade Agreement (NAFTA), and finally, the Uruguay Round of the GATT negotiations incorporated the TRIPs Agreement (see below), which entered into force in 1995. The debate today is couched in terms of 'globalization', which has been encouraged by the World Trade Organization (WTO) and TRIPs agreements. Some see this as the great hope for further worldwide economic development;[17] others see it as a conspiracy of the rich countries against the poor.[18]

International Developments

The Paris Convention

The International Convention for the Protection of Industrial Property (known as the Paris Convention) was signed in Paris in 1883, originally by 11 countries. The UK acceded in 1884. The Convention is now adhered to by the majority of the countries of the world that have any form of patent protection and, because the Convention also deals with trademarks and designs, even by some countries that have no patents. Interestingly, two of the original members, Switzerland and the Netherlands, had no patent system themselves at that time. The basis of the Paris

17 See, e.g., T.L. Friedman, *The World is Flat*, Penguin, 2006.

18 See, e.g., the statement of India's Trade Minister, Kamal Nath, after the breakdown of WTO talks in Geneva, 31 July 2008.

Convention is one of reciprocal rights, so that an applicant or patentee from one Convention country shall have the same rights in a second Convention country as those of a national of that second country.

The most important practical result of the Convention is the possibility of claiming Convention priority for applications made outside one's home country. The system[19] is such that if an application for a patent is properly made in one Convention country, corresponding applications may be filed in other Convention countries within one year from the first filing date; if certain conditions are met, these later applications will be entitled to the priority date of the first application. This means that they will be treated as if they were filed on the same day as the first application, so that a publication of the invention after the first filing date but before the filing date of the later application, will not invalidate the later filing.

For example, suppose that XYZ Ltd files a patent application for a new product A in the UK on 1 June 2009. At any time before 1 June 2010 it may file corresponding applications in France, Germany, Japan, USA, and as many other Convention countries as it chooses, claiming priority from the UK application. If in the meantime a description of product A is published by another company (or, as is more likely, by XYZ itself), this publication will not affect the validity of its patent rights. On the other hand, if before 1 June 2010 XYZ Ltd decides that it has no real interest in selling the product, it can save itself the trouble and expense of filing in other countries. If it were not for the Convention, a decision whether or not to file in, say, ten countries would have to be taken at the time of first filing, and a great deal of money and effort would be wasted on protecting inventions that, within a few months, turned out to be unpatentable or commercially uninteresting.

It used to be the case that the 12-month term for Convention priority had to be rigidly kept. If the period was exceeded even by one day, priority would be lost and any intervening publication of the invention would invalidate the foreign application (unless the Convention year ended on a Saturday, Sunday or public holiday, in which case the application could be filed on the next working day).[20] Now, however, countries that are bound by the Patent Law Treaty (see below) must allow priority to be claimed even if the later application was filed up to 14 months after the first application, provided that the applicant can show that failure to file within 12 months was unintentional (or, alternatively, in spite of due care having been taken).[21] The UK has now implemented this provision, applying the less strict criterion of unintentionality.[22]

The later (Convention) application need not be an exact equivalent of the original (priority) application; thus XYZ Ltd could expand the scope of its foreign

19 Article 4, Paris Convention.

20 Article 4C(3), Paris Convention.

21 Article 13(2), PLT.

22 Section 5(2A)–(2C), PA 1977, added by the Regulatory Reform (Patents) Order 2004; r. 7, PR 2007.

applications to cover the related product B, but as a general rule claims to B would not be entitled to the priority date of the UK application, which disclosed only A. It is noteworthy that both the UK and US patent systems have for some time allowed a form of 'internal priority' by which a later patent application could take the date of an earlier application in the same country. In the UK, under the old patent law, this was done by filing an application with a provisional specification, followed by a complete specification up to 12 months later, extensible to 15 months on paying an extra fee. Claims 'fairly based' on the provisional specification were entitled to its filing date; new matter took the date of filing the complete specification. A similar system still applies under the 1977 Act, although there are now two separate applications instead of a single application with two specifications.

In the USA, a 'continuation-in-part' application (known as a 'cip') may be filed containing the same description as an earlier application, together with new matter. A cip may be filed at any time before the earlier application is granted, but there are complex rules that could lead to claims based on the new matter being invalid if filing has been left too late. When the USA adapted its patent term in view of the TRIPs Agreement, a new system of provisional applications was introduced (see Chapter 5).

All of these systems allow improvements to be incorporated in a patent application within a certain time and are much to be preferred to a system in which one is held to the original form of one's first filing. In countries where there is no such internal priority, a resident applicant may end up with worse protection in his home country than in all others, in which the Paris Convention has allowed expansion of the original disclosure. Switzerland was in this situation until the Swiss patent law was changed in 1996, and for this reason some Swiss companies made their first filings in Germany or the UK instead of in Switzerland.

The European Patent Convention

In 1963 a number of European countries signed the Strasbourg Convention, which recommended certain common standards for novelty, inventiveness, and the type of invention that may be patented. This later formed the basis for the European Patent Convention (EPC) of 1973, which has led to the establishment of the European Patent Organization, consisting of the European Patent Office (EPO), which grants European Patents, and the Administrative Council, which supervises the EPO.[23] The Administrative Council is made up of representatives of all contracting states, and these are usually the heads of the national patent offices. The EPO must therefore be the only commercial organization in the world that is run by a committee of its main competitors.

The EPC was originally negotiated by 19 countries, not all of which ratified the Convention initially. It started with only seven contracting states (Belgium, France,

23 Article 4, EPC.

Germany, Luxembourg, the Netherlands, Switzerland, and the UK), but by 2000 had been adhered to by all of the countries of the European Union (EU), together with Turkey, Cyprus, Monaco, Switzerland and Liechtenstein—that is, 20 states in all. All states that wish to become members of the EU must join the EPC and the majority of the countries seeking to become EU members had already joined the EPC by early 2004. Cyprus joined in 1998; Bulgaria, the Czech Republic, Estonia, Slovenia, and Slovakia in 2002; Hungary and Romania in 2003; Poland, Iceland, and Lithuania in 2004; Latvia in 2005; Malta in 2007; Norway and Croatia in 2008; and Macedonia and San Marino in 2009, giving a total of 36 member states. Furthermore, European patents may be extended to certain states that are not members of the EPC. Currently these are Albania, Bosnia, and Serbia and Montenegro.

The EPC set up a new and self-contained system of law providing for the grant of patents in any or all of the contracting states by means of a single patent application examined by the EPO in Munich. We shall, in a later chapter, look at how the EPC works, but the main point to bear in mind is that the European patent is not a single unitary patent, but is more like a bundle of national patents in each of the countries that the patentee has chosen. Because these national patents are subject to the national laws as regards validity and infringement, it is obviously desirable for the contracting states of the EPC to make their national patent laws conform to a common standard with respect to these topics.

By the time the EPO opened its doors to applications on 1 June 1978, most of the contracting states had already changed their laws to the extent necessary to provide standard grounds for invalidity (according to the Strasbourg Convention as adopted by the EPC), a 20-year patent term from filing, and product protection per se for all chemical substances, including pharmaceuticals. In the UK, as we have seen, this was achieved by means of the Patents Act 1977.

The EPC was amended by a diplomatic conference in November 2000 and the new version (EPC 2000) entered into force after a long ratification process on 13 December 2007. It transfers many procedural matters, including most time limits, from the Convention itself to the Implementing Regulations, thus allowing them to be changed in future by a simple decision of the Administrative Council, rather than by a new Diplomatic Conference. It has become easier to restore the situation if a time limit is missed, and there is now a new centralized procedure allowing the patentee to request revocation or limitation of the patent at any time during the life of the patent. The provisions of EPC 2000 will be described in more detail in later chapters.

Now, 30 years after its opening, the EPO is generally seen as a success. The standard of examination is felt to have struck a reasonable balance between liberality and restrictiveness, the numbers of oppositions has remained within reasonable limits, and although the decisions of the opposition divisions sometimes seem arbitrary, the body of jurisprudence built up by the Boards of Appeal is, on the whole, consistent and sensible. There have indeed been disagreements about the breadth of scope granted in some cases, the difficulty of challenging excessive breadth of scope, irregularities of procedure in oral proceedings, and tendencies of Boards of Appeal to avoid taking a position on controversial issues. A more serious problem is the excessive delay in deciding oppositions and appeals in the EPO, which adds to legal uncertainty and costs for the parties involved (see p. 129). Nevertheless, these problems are of relatively minor importance when compared with the convenience and cost savings of a central and reasonably reliable granting procedure for nearly all European states. This is not to say that the cost could not be further reduced: national patent offices take too large a share of the renewal fees, and the requirement to translate the European patent into the national language in nearly all designated member states imposes large and unnecessary costs—a requirement that has been relaxed to some extent as a result of the London Agreement (see p. 128).

One consequence of the success of the EPO is that some national patent offices are finding it difficult to survive. The Dutch Patent Office, for example, imposed very strict standards of examination and was very expensive, so it was not surprising that the number of national applications declined to the point at which it was no longer possible to employ enough examiners to be able to examine in all technical fields. As a consequence, the Netherlands has changed from what was probably the strictest examination system in the world to a system with no substantive examination whatsoever. *Sic transit gloriae mundi.*

The Community Patent

Efforts to provide a single unitary patent for the entire EU, covering all member states just as a US patent covers all of the 50 states of the Union, began even before the EPC was adopted. The Community Patent Convention (CPC) was signed in 1975, but was never ratified by a sufficient number of states to allow it to come into force. One major problem with the CPC was the requirement to provide a translation of the full text into an official language of each member state, which meant that the Community patent (CP) would have been as expensive as a normal European patent covering all EU states. Moreover, it could have been revoked for the whole EU by a non-specialist court in any EU country. Not surprisingly, potential users were very reluctant to endorse such a system.

The project languished for a long time, with various proposals of the European Commission failing to obtain the necessary consensus among member states. Some political leaders had constitutional objections to the idea that a document such as a patent could have a legal effect in their country without being translated into the local language; others seemed to think that cultural identity or national pride were important issues. Most observers had given up hope that a reasonable compromise could be found, until a draft Regulation to establish a Community patent was proposed in 2004. It would allow the applicant to choose from: a national patent; a European patent, which may not cover all of the EU states; or a CP. This approach would require amendment of the EPC to allow the accession of the EU as a whole. A designation of 'EU' in a European patent application would then be a request for a CP to be granted by the EPO.

The draft provided that, after the grant of a CP, the patentee must file a translation of the claims only into an official language of each member state, except for states that have renounced their right to receive such translations. Litigation, including both infringement and validity issues, would be heard at first instance by a special Community Patent Court attached to the European Court of First Instance (CFI) in Luxembourg, with appeals to the CFI; until the new Court were set up, national courts would have jurisdiction. Unlike the previous proposals, this appeared to be the basis for a system that industry could live with, but unfortunately it has again been brought to nothing by the intransigence of certain states about translation requirements, and the chances of a CP ever becoming reality seem to be minimal. The latest proposals (2008) suggest as a solution that machine translations could be used purely for information purposes, with more accurate translations being required only in the event of litigation. It is not yet clear whether this suggestion has any chance of approval.

Independently of these discussions, negotiations on a possible unified patent litigation system for Europe are currently in progress (see Chapter 10).

Other Regional Patent Organizations

The European Patent Organization is not the only, nor even the first, organization set up to grant patents covering more than one country. The first was the African Intellectual Property Organization, known by its French acronym OAPI, set up by the Libreville Agreement of 1962, as modified by the Bangui Agreement of 1977. This grants a single regional patent covering 14 former French colonies in West and Central Africa. The English-speaking counterpart to this is the African Regional Industrial Property Organization (ARIPO), also covering 14 states, and dating from 1976. In both cases, the economic levels of the member countries are such that there is little interest in obtaining patents through these regional offices, although this may change for ARIPO if South Africa should eventually become a member.

A more recent creation is the Eurasian Patent Convention, under which a single application filed in Moscow can give protection for the Russian Federation and for eight of the former Soviet republics. Its appeal remains limited because a number of the more economically important of these states, such as Georgia, Ukraine, and Uzbekistan, remain outside the Convention.

Even more recent is the setting up in 1998 of the Patent Office of the Cooperation Council for the Arab States of the Gulf (the Gulf Cooperation Council, or GCC) in Riyadh, Saudi Arabia. This can grant a unified patent for Kuwait, Qatar, Oman, Saudi Arabia, Bahrain, and the United Arab Emirates (UAE), but national filings are also possible.

The Andean Pact is an association adhered to by Bolivia, Colombia, Ecuador, and Peru.[24] It does not have a central patent office, but decisions of the Andean Pact Council are binding upon member states in certain areas, including patent law.

24 Venezuela withdrew its membership status from the Andean Pact in 2006.

The Patent Cooperation Treaty

An important development in international patenting is the Patent Cooperation Treaty (PCT), which entered into force in January 1978 and which, at the time of writing, had been ratified by 141 countries, including all EPC states, the USA, Japan, China, and Russia. The PCT, like the Paris Convention, is administered by the World Intellectual Property Organization (WIPO), a UN organization with its headquarters in Geneva. It is not a supranational patent office, although patent applications can be filed there. The tasks of searching and examination are delegated to other offices; WIPO does not grant a 'world patent', but at most provides non-binding opinions on patentability. Its main aim is to simplify the process of filing patent applications simultaneously in a number of countries. The old procedure, which can of course still be used, requires a completely separate application in each country, which has to be translated into the local language before filing, or at best shortly afterwards. Under the PCT, a single international application may be filed in one of the official receiving offices, or at the WIPO itself, and potentially gives rights for all PCT contracting states. An initial 'international phase', during which a search and possibly also a preliminary examination is carried out, is followed after 30 months by a 'national phase', during which selected national or regional patent offices conclude the examination process and grant (or refuse) the patent. The international application may claim priority from an earlier national, regional, or PCT application if filed within the normal 12-month period of the Paris Convention.

During the international phase, the application is passed to an International Searching Authority (ISA), which carries out a search for relevant prior art, and the application is published, together with the search report, 18 months after the priority date. This corresponds to the original PCT Chapter I procedure and nothing further need be done during the international phase. The applicant will also receive an 'international preliminary report on patentability' (IPRP) based on the search report, but if he wishes to have an opportunity to contest this, he must file a Chapter II demand and pay an additional fee. The final IPRP is then produced by an International Preliminary Examination Authority (IPEA), which may or may not be identical with the ISA. (For further details of the PCT procedure, see Chapter 6.)

In order to enter the national phase, a copy of the application, together with the IPRP, is sent to the chosen national or regional patent offices together with any required translation. This must normally be done within 30 months from the priority date, although some offices, including the EPO, allow entry up to 31 months from priority.

From the point of view of the applicant, the chief advantage of the PCT is that the major expense of national fees and translations is postponed for 30 months from the international filing date, after the search report and IPRP have been received, and after the applicant has had more time to think about the commercial value of the invention. At this point he may abandon the application altogether, or file in a more limited list of countries than originally planned, and no large amount of money will have been wasted. This is particularly advantageous when filing in a large list of countries is being considered. Although the PCT procedure is somewhat more expensive than the national filing route, for many applicants, the extra cost is more than covered by the savings incurred when a case is dropped before entering the national phase.

The PCT procedure is also useful when a decision to file in foreign countries cannot be taken until a short time before the end of the priority year, so that there would be insufficient time under the normal procedure to prepare the necessary translations. Until recently, its formalities were complex and inflexible, and it could not be used for filing in all countries of interest to most applicants, but procedural matters have now been greatly simplified, it is no longer necessary to designate specific countries on filing, and with 141 contracting states, very few countries of commercial interest are not included (Taiwan being the most notable exception).

The annual number of PCT filings has risen exponentially: from 2,625 in 1978, the first full year of its existence, to 7,952 in 1986, 54,422 in 1997, and no fewer than 158,400 in 2007, 34 per cent of which originated in the USA, 18 per cent in Japan, 11 per cent in Germany, 4.5 per cent in Korea, 4.1 per cent in France, and only 3.5 per cent in the UK. These top six countries together account for 75 per cent of all PCT filings. Applications in English still predominate (60 per cent), followed by Japanese (16 per cent), German (11 per cent), and French (3 per cent). It is noticeable that the number of filings from Asian countries has increased markedly during the last five years.

The TRIPs Agreement

The General Agreement for Tariffs and Trade (GATT) was set up in 1948 to deal with multilateral trade issues. The latest complete round of GATT negotiations, the Uruguay Round, was concluded in April 1994 and led to the establishment of the World Trade Organization (WTO), which became operational on 1 January 1995. The agreement on Trade-Related Aspects of Intellectual Property Rights (somewhat inaccurately rendered by the acronym TRIPs) was adopted as an integral part of the Final Act of the Uruguay Round, so that all countries that become members of the WTO must accept the provisions of TRIPs as part of the deal. The TRIPs Agreement covers a whole range of intellectual property issues, including patents, trademarks, geographical indications, industrial designs, integrated circuits, copyright, and trade secret protection, as well as general provisions about basic principles, enforcement, and dispute resolution, and generally requires strong patent protection, including a term of at least 20 years from filing, and protection for all inventions, including product protection for pharmaceuticals. Currently 153 states and territories are members of WTO, including Taiwan, which is unable to adhere to the Paris Convention or the PCT because its sovereignty is not recognized by most countries. The provisions of TRIPs are dealt with in the next chapter. In effect, the TRIPs Agreement has pre-empted the endless discussions in WIPO that always tended in the direction of weakening patent protection.

The Convention on Biological Diversity

An important element in the ongoing dialogue between developed and developing countries is the Convention on Biological Diversity (CBD), which was negotiated at the UN Conference on Environment and Development (UNCED) held in Rio in 1992 and which entered into force on 29 December 1993. Currently, 191 states are parties to the Convention, that is, nearly all countries in the world except the USA, which signed the Convention but has never ratified it. The CBD is an international agreement that has as its objectives the conservation of biological diversity, its sustainable use, and benefit sharing by access to resources and transfer of technology.[25] This may seem to have nothing to do with patents, but it is highly relevant to the question of whether patents can be obtained for inventions related to biological material obtained from another country.

The Patent Law Treaty

The Patent Law Treaty (PLT), concluded in June 2000, harmonizes the formal requirements for patent filings in the contracting states. The treaty has been signed by 43 states and ratified by 18, and entered into force in April 2005 (see p. 50).

The London Agreement

The London Agreement (see Chapter 7) is a voluntary agreement entered into in 2000 among certain contracting states of the EPC whereby these states will not require translations of European patents into their national language under certain conditions. Entry into force required ratification by eight states, including the UK, France, and Germany; the long-delayed ratification by France allowed the Agreement to come into effect on 1 May 2008 and, at the time of writing, 14 contracting states have acceded.

The Substantive Patent Law Treaty

In contrast to the agreement reached on formal matters in the PLT, little progress has been made in negotiations on harmonizing substantive patent law—that is, on a Substantive Patent Law Treaty (SPLT). The main stumbling block has been the reluctance of the USA to give up its unique first-to-invent system.

Outlook

This chapter has set out international developments in patent law as they have occurred, without going into the question of how these changes came about. It is of course clear that such developments, which have generally acted to strengthen patent protection, did not simply happen of their own accord, and that political, diplomatic, and industry lobbying activities have played a larger role than any objective

25 Article 1, CBD.

analysis of the economic and social benefits of the patent system. In particular, industry's 'forum switching' from WIPO and the UN Conference on Trade and Development (UNCTAD) to GATT resulted in the success of the TRIPs Agreement. At the same time, however, it brought patent matters to the attention of a much wider circle of interested parties, including many non-governmental organizations (NGOs), which were opposed to a strong patent system for a variety of reasons. The result has been a vocal backlash against TRIPs and against patents in general, which is now increasingly being accepted as the new reality by public and politicians alike. The next chapter goes into these aspects in more detail.

3

HARMONIZATION OF PATENT LAW

> I have seen with real alarm, several recent attempts in quarters carrying some authority to impugn the principle of patents altogether—attempts which, if practically successful, would enthrone free stealing under the prostituted name of free trade.
>
> John Stewart Mill, Speech in the House of Commons (c. 1860)

Introduction

The last 30 years has seen a steady trend towards harmonization of patent laws, both as regards formal matters and as regards substantive patent law. Thirty years ago, every country had its own individual combination of formal requirements, novelty standards, examination procedure, patent term, and scope of protection. Now, although some differences remain, there has been enormous progress towards a standard system for filing patent applications and obtaining patents. On the formal side, the Patent Cooperation Treaty (PCT) has provided a single system for filing applications in up to 139 countries and further progress has been made by the Patent Law Treaty (PLT, see below). Although the Substantive Patent Law Treaty (SPLT) has not yet been concluded, a great measure of harmonization of levels of patent protection has been achieved by the TRIPs Agreement. Further harmonization has been achieved by various bilateral and multilateral free trade agreements, as well as by the Trilateral Cooperation (TLC) between the three major patent offices: the European Patent Office (EPO); the Japanese Patent Office (JPO); and the US Patent and Trademark Office (USPTO).

The TRIPs Agreement

Basic Principles

Articles 1–8 of the TRIPs Agreement include the basic principles of 'national' treatment and 'most favoured nation' treatment: each member must give to the nationals of other members treatment no less favourable than that given to its own nationals, and must give to the nationals of all members the same privileges as are given to the nationals of any member. Thus, subject to certain exemptions, bilateral intellectual property (IP) agreements between members that restrict the benefits of the agreement to the two parties should no longer be permitted. These Articles are of particular importance because their implementation, unlike other provisions of TRIPs, could not be delayed beyond 1 January 1996 by any member country, whatever its state of development.

Provisions Relating Specifically to Patents

Articles 27–34 of the TRIPs Agreement require World Trade Organization (WTO) member states to introduce strong patent protection, the most important elements of which are that:

(a) patents are to be available under essentially the same criteria of patentability as in the European Patent Convention (EPC) for all fields of technology, including product patents for pharmaceuticals (Article 27);
(b) patent rights are to be without discrimination as to whether products are locally made or imported (Article 27);

(c) there are provisions defining what constitutes infringement, including importation of a patented product (Article 28.1(a)) and using, selling or importing the direct product of a patented process (Article 28.1(b));
(d) compulsory licences are to be allowed only under strict conditions (Article 31);
(e) an opportunity for judicial review of any decision to revoke a patent is required (Article 32);
(f) the patent term is to be at least 20 years from the filing date (Article 33), which, according to the transitional provisions (Article 70.2), should also apply to patents that are already granted; and
(g) there is to be reversal of the onus of proof for process patents (Article 34).

Provisions on Enforcement of Intellectual Property Rights

The whole of Part III of the TRIPs Agreement (Articles 41-61) is devoted to the enforcement of IP rights. Members are obliged to provide enforcement procedures that are effective, fair, and equitable, and not unnecessarily costly. There must be due process, including the right to legal representation, the right to comment on all evidence on which the decision is based, and the right to judicial review of administrative decisions.

In IP cases, judicial authorities must have the power to order the production of evidence, and to give orders for injunctions, costs and damages, and the destruction or confiscation of infringing goods. Judges must be able to make orders for preliminary or interlocutory injunctions, in appropriate cases without the defendant being notified (as in the case of search orders in the UK). Members must adopt procedures to enable customs authorities, upon request, to prevent the importation of counterfeit trademark or pirated copyright goods (but not necessarily, however, of patent infringements). Criminal penalties must be provided for wilful trademark counterfeiting or copyright piracy on a commercial scale; criminal penalties for other types of IP infringement are optional.

Summary

According to TRIPs, practically all countries of the world are obliged to have patent systems in which compounds, including pharmaceuticals, can be patented per se for a term of at least 20 years, with no local working requirements and no routine granting of compulsory licences, with importation of a product and sale of the product of a process being clearly defined as infringement, and with clear standards for the enforcement of patent rights. When the WTO came into existence on 1 January 1995, it was probably true to say that not a single member country had a patent law that was completely in accordance with TRIPs. For example, many countries, both developing and industrialized, had patent terms that were, or at least could be, less than the 20 years from filing mandated by TRIPs. In the case of some countries such as India, Argentina, Brazil, and Turkey, patent protection for pharmaceuticals was totally or effectively lacking, and major revisions were needed. And even industrialized countries such as the USA, the European Union (EU) countries, and Japan had laws that did not conform to TRIPs in more subtle ways, such as the conditions for compulsory licensing. All of these laws had to be changed, but not necessarily right away.

Deadlines for Implementation

No country was required to change its laws to make them conform to TRIPs until one year after the establishment of the WTO, that is, until 1 January 1996, but two major classes of country were allowed to postpone the changes for longer periods. According to the transitional arrangements, developing countries and those countries in transition from a centrally planned to a free market economy and facing special problems in the preparation and implementation of IP laws could delay implementation of TRIPs for a further period of four years, that is, to 1 January 2000.

Countries in the first category (but not the second) that on 1 January 2000 still did not provide product protection for certain areas of technology (for example pharmaceuticals) could delay the introduction of product protection in these areas (but not the other TRIPs obligations) for another five years, that is, to 1 January 2005. And finally, the least-developed countries that are members of WTO were entitled to delay implementation of all TRIPs provisions other than Articles 3–5 until 1 January 2006—and even this has now been extended to 2016.

Black Box Applications

An important concept of the transitional provisions of TRIPs, relating to the protection of existing subject matter, was the provision of so-called 'black box' filings.[1] Where a member did not make available, as of the date of entry into force of the Agreement establishing the WTO, patent protection for pharmaceutical and agricultural chemical products, that member had to provide as from that date (that is, from 1 January 1995 or the date on which that country became a member, if later) a means by which applications for such inventions could be filed.

It was to apply to these filings, as of the date of application of this Agreement (that is, the date, as set out above, on which the obligations of Part II, section 5, of TRIPs applied to that country), the criteria for patentability set out in TRIPs as if these criteria were being applied on the date of filing in that country.

It was then to provide patent protection as from grant for the remainder of the patent term (that is, 20 years from the filing date), if the invention was patentable by TRIPs criteria. Furthermore, a product for which a 'black box' filing had been made had to be given exclusive marketing rights for five years from market approval in

[1] Article 70.8, TRIPs.

the country in question, or until a product patent was granted or rejected, whichever was shorter, provided that, for that product, in another member country:

(a) a patent application had been filed subsequent to 1 January 1995;
(b) a patent had been granted; and
(c) marketing approval had been obtained.[2]

These provisions are complex and have been differently interpreted in different countries.

TRIPs Implementation: Changes to Patent Laws

Dispute Procedure

A very positive feature of the Uruguay Round of GATT is that for the first time there is a dispute procedure that can lead to economic sanctions against the member that is in violation of the GATT. Previously, the dispute procedure was not binding and an adverse finding could safely be ignored. The same general dispute procedure applies to TRIPs.

If a member has a complaint, it must first consult with the other member; if this is unsuccessful, it may request that a panel be set up. The WTO Disputes Settlement Body (DSB) will then set up a panel, which will make a report within six months. Appeals lie to an Appellate Body, the appeal being limited to issues of law. Once the panel report or the Appellate Body report is adopted by the DSB, the party concerned must notify its intentions with respect to implementation of the recommendations. If the recommendations are not implemented within a reasonable time, the DSB may grant authorization to the other party to suspend concessions or other GATT obligations to the member that is in breach, and in extreme cases, this may include the imposition of punitive import tariffs on products from the country found to be in breach. Such tariffs often cause economic harm to industries that were not parties to the original dispute, for example in connection with the unjustified refusal of the EU to allow imports of beef from hormone-treated cattle, the USA imposed tariffs on imports of Roquefort cheese.

TRIPs does not, however, give directly enforceable rights to natural or legal persons. Thus if a company or an individual is aggrieved because of non-observance of the TRIPs Agreement in another country, the only remedy available is for the aggrieved party to ask its own government to take up the matter. If the matter is considered important enough, the dispute resolution provisions of TRIPs can then be used. Within the EU, the Commission itself can take action, and this may be initiated by a party making a formal complaint under the provisions of the Trade Barrier Regulation.[3]

[2] Article 70.9, TRIPs.
[3] Council Regulation (EC) No. 3286/94 of 22 December 1994.

Patent Term

The adjustment of patent term to meet the TRIPs provisions was a relatively easy step to take, and the majority of WTO member states had already made this change within the first two years, including Argentina, Brazil, Indonesia, and the Philippines. But not all of the countries that extended their patent term made the extension applicable to existing patents, as is required by TRIPs.[4] Among those countries that did not extend the term of existing patents are Malaysia, Brazil, Pakistan, and Argentina, although it is now possible to request an extension of old patents in Argentina on an individual basis. Even where the term extension is made to apply to existing patents, any delay in the implementation of the law meant that patents expired that should have been extended. There is no obligation under TRIPs to restore dead patents to life[5] and indeed to do so would be unfair to the general public. This situation arose, for example, in Korea (see Chapter 8).

Black Box Filings

The complicated 'black box' provisions described above were very important because they gave two important rights even at a time when a country had not yet implemented product protection for pharmaceuticals. The first is that applications could be filed that might eventually give rise to product protection; the second is that they should have allowed the applicant to obtain temporary exclusivity by way of exclusive marketing rights (EMR). Both of these proved difficult in practice.

All WTO members had to provide for such filings as of 1 January 1995, or their accession to the WTO, if later. Such applications for pharmaceuticals were not to be examined (other than perhaps a formal examination to establish that the application would be entitled to a filing date) until the country in question had changed its law to be in conformity with the TRIPs provisions on the patentability of pharmaceutical products. This is after all why such applications have been nicknamed 'black box' filings. Some countries failed to introduce the possibility of black box applications until well after the deadline, while others allowed applications to be filed, but did not give them special status.

Exclusive marketing rights

EMR was generally expected to mean that if a black box filing had been made and the other conditions met, the exclusivity of the patent holder would be guaranteed by the refusal of the regulatory authorities to grant further marketing approvals to generic companies. In most countries this was never put to the test, because the period during which black box applications could be filed was usually not more than two or three years, and within this short time it was hardly likely that a product could progress from patent filing to marketing approval. In India, the TRIPs

[4] Article 70.2, TRIPs.
[5] Article 70.3, TRIPs.

provisions were interpreted differently, and EMR was seen as a special kind of IP right that had to be applied for at the Patent Office and enforced in the courts. Furthermore, the long delay in implementing product protection in India, together with the importance of its generics industry, meant that the problem became a practical one.

Status in Individual Countries

India India is a country in which there is a serious political conflict between those politicians and companies who wish to open up the Indian economy to the outside world and see membership of WTO as an important part of this goal; and those elements that, with considerable popular support, wish to retain the old protectionism and delay or nullify any changes to the patent law. Accordingly, India ratified the GATT on 31 December 1994 by governmental decree, without seeking the approval of the Indian Parliament, and at the same time amended the patent law by a Patent Amendment Ordinance so as to allow black box filings.

The Ordinance was later allowed to lapse, and the bill that was to replace it failed to get parliamentary approval, with the result that, although black box filings were still accepted, they were no longer officially recognized, and their legal status was unclear. Dispute proceedings under the GATT were called for by the USA, alleging that India was in breach of its obligations under TRIPs by not officially providing for black box filings. The DSB and the Appellate Body both held against India, but the USA agreed to give India until April 1999 to implement measures that should have been in place in January 1995.

The gradual implementation of TRIPs in India has proceeded by piecemeal amendment of the Patents Act 1970, rather than by writing a new law, and the result is rather like a car built by adding BMW parts to a Ford Model-T chassis. The Patents (Amendment) Act 1999 finally provided for black box applications and for EMR, but made no other changes. After three years of intensive debate, the Patents (Second Amendment) Act 2002 implemented a number of TRIPs obligations, including a 20-year patent term, although the result still did not comply with TRIPs in a number of respects. In particular, whereas it would have been easy for the Act to provide that product protection for pharmaceuticals would come into effect on 1 January 2005, it did not do so, and a third amendment Act was required. The draft Bill published in December 2003 made the necessary provision for product protection and introduced certain other changes, such as the early publication of pending applications and post-grant rather than pre-grant opposition. Unfortunately the final version of the Patents (Amendment) Act 2005 reintroduced pre-grant (in addition to post-grant) opposition, and made some further retrograde changes, leaving the Indian law still non-compliant with TRIPs in some respects.

One of the first EMRs issued by the Indian Patent Office was granted to the Swiss company Novartis for the anti-cancer drug Glivec®. Initial attempts to enforce the EMR were successful, and injunctions were granted against some imitators. But as soon as the new law came into force in January 2005, the black box application for

Glivec was examined on a priority basis and was refused because new crystal forms were defined as not being patentable inventions.[6] An attempt to challenge the constitutionality of this provision of the law predictably failed, and as of writing, an appeal against the refusal itself is still pending. Even if the patent were to have been granted, however, it would have been of limited value because, under the provisions of the new law, patents granted on black box applications could not be enforced against infringers who were already on the market at the time that the patent was issued.

Pakistan Pakistan refused to allow black box filings until 1998 because no implementing legislation had been passed. Dispute proceedings were threatened by the USA and resolved when Pakistan passed an implementing Bill. Not only did this retroactively validate unofficial black box filings made since 1 January 1995, but it also permitted the filing up to 4 February 1998 of new black box applications corresponding to filings made in other WTO countries between 1 January 1995 and 4 February 1998; the Pakistan filing was entitled to the same date as the original application in the other country. Product protection was made available to pharmaceuticals from 1 January 2005.

Turkey Turkey accepted filings for pharmaceutical compounds, but although the law had not yet been amended, the applications were not put into a black box, but were subjected to the same exorbitantly expensive search as applications in other fields of technology. This at least, however, had the positive effect that a change of law would not result in a large backlog of unexamined applications. Initially Turkey proposed to postpone the grant of even process patents for pharmaceuticals for five years and of product patents for ten years, but the discussion was pre-empted by the somewhat unexpected accession of Turkey to the EPC in 2000.

Brazil Brazil represents a success story in the TRIPs saga. Black box filings were accepted from the outset without difficulty, and a new patent law, giving product protection for pharmaceuticals, entered into force on 15 May 1997. Since then, examination of the black box filings has proceeded under the new law. This law was in full conformity with TRIPs except in that it still contained local working requirements. Indeed in two respects it went beyond what was mandated by TRIPs in that it provided for pipeline protection and also made it possible to prevent parallel imports.

Argentina In Argentina the executive branch of the government appeared to be generally sympathetic to improving patent protection, but the legislature was heavily influenced by the lobbying of the local 'pharmaceutical industry', that is,

[6] Section 3(d), Patents Act, as amended 2005.

a small group of companies owned by wealthy individuals who make their money simply by selling generic drugs or copies of patented drugs that are imported from other countries such as India. There is essentially no local manufacturing capability, and very little value is added within the country by such companies.

The result was that, after an endless succession of new laws, presidential vetoes, Decrees, and 'corrective Bills', among other things, a new law was enacted and came into force in January 1997. Not only did this law postpone the introduction of product protection until 1 January 2001 (in other words, Argentina was to be regarded as a developing country), but it was contrary to TRIPs in a number of other respects. Black box filings could be made, but there was no provision for marketing exclusivity, which is particularly important in a situation in which there would be a long delay before the implementation of product protection.

The Middle East and North Africa The Middle Eastern and North African countries, for example Egypt, Jordan, Tunisia, and Algeria, have generally had weak patent laws, but in June 2002 Egypt enacted a patent law providing a 20-year term and product protection from 2005, and at the same time joined the PCT. Iran and Saudi Arabia have patent systems in which patents granted in other countries may be registered (see p. 98), but there is little experience of how such registration patents may be enforced. It is of interest that the concept of IP is not regarded as being in any way in conflict with Islamic law (sharia), although commercial operations such as lending money for interest may be so.[7]

Grant and Enforcement of Patents

It is of little use to have a strong patent law on paper unless patent applications are actually processed to grant and the resulting patents can be enforced. Many developing countries have recently improved their patent laws, mainly as a result of TRIPs, and these laws often require substantive examination of the novelty and inventive step of patent applications. This puts these countries in a difficult position: because of their improved laws, their patent offices receive many more applications than before, but they lack the necessary technical staff and infrastructure with which to examine the applications in the way that the laws seem to require. The problem should not be underestimated: to be able to carry out a proper substantive examination in all technical fields, a patent office must have a large technically qualified staff and enormous resources of scientific and patent literature, and access to electronic databases, among other things. There are probably not more than ten patent offices in the world capable of this task, and as we have seen, even the Netherlands Patent Office has decided not to be one of them. Not only is it far beyond the means of most developing countries, but it would also be an absurd

[7] Azmi, 'Basis for the recognition of intellectual property in the light of the shari'ah' [1996] 27 IIC 649.

waste of their resources to duplicate work that has in most cases already been performed by the USPTO, EPO, or JPO.

With the help of organizations such as the EPO and the World Intellectual Property Organization (WIPO), patent offices in developing countries are being helped to deal with the flood of new applications, wherever possible by means that will avoid the need to carry out independent technical examination and which will grant patents based on the allowance of corresponding patents in one of the major patent offices, or, where the country in question has acceded to the PCT, on the basis of the PCT International Preliminary Report on Patentability. Singapore is a good example (see p. 137).

Effective enforcement requires that there be competent courts with judges who understand patent law; perhaps even more necessary is a legal environment in which patent infringement is seen as something to be seriously discouraged with heavy financial penalties, not as a local growth industry, the activities of which should be ignored or even encouraged. One problem with the TRIPs provisions on the enforcement of IP rights is that it is sufficient to comply with TRIPs if the laws contain corresponding provisions, even if in practice these are not implemented.

Mexico, for example, has had an excellent patent law since 1992, but for some time, no patents were being granted because no implementing regulations were in existence. This has since been rectified and patents are being granted, but the enforcement of such patents against local imitators has proved extremely difficult. In the pharmaceutical field, enforcement should be made much easier by a Decree of June 2003 according to which an applicant for marketing approval for a patented drug must show that it is the patentee or has a licence.

China also presents a serious problem. Although it has a perfectly adequate patent law, it lacks the tradition of resolution of disputes by litigation. Patent infringement cases may be brought before the People's Courts, or be dealt with administratively by the Intellectual Property Bureaus. In the past, the infringing company often turned out to be owned by the local administrative authority or another state body, which made it unlikely that effective action to stop infringement would be taken. However, matters are steadily improving, and although trademark counterfeiting remains a problem, there seems to be a real effort on the part of the Chinese government to provide effective enforcement of patent rights. In a recent WTO dispute procedure, the USA was partially successful in its complaint of inadequate trademark and copyright enforcement in China, but no complaint was made regarding patent enforcement.

Compulsory Licensing

TRIPs provides that any compulsory licences must be:

(a) considered on their individual merits;
(b) granted only if a licence on normal terms has been requested and refused;
(c) of limited scope and duration;

(d) non-assignable other than with the business;
(e) primarily for domestic supply rather than for export;
(f) capable of being terminated if circumstances change;
(g) subject to adequate remuneration; and
(h) subject to judicial review.

Compulsory licences to enable the working of a dependent patent may be granted only if the invention claimed in the later dependent patent involves an important technical advance of considerable economic significance, and the compulsory licence, if granted, shall be non-assignable except with assignment of the dependent patent. When reviewed by the European Commission in late 1995, the laws of no single EU member state complied fully with all of these provisions, so it is hardly surprising that the laws of developing countries were often less than perfect in this respect. What is clear-cut is that the discriminatory compulsory licensing of pharmaceutical patents must be stopped. It is also a concern that certain of the TRIPs provisions relating to compulsory licences are not applicable to compulsory licences granted to remedy anti-competitive practices. This is a loophole that may enable evasion of these provisions simply by means of an administrative body declaring the conduct of the patentee to be 'anti-competitive'. In the worst-case scenario, an imitator offers the patented product at a lower price, the patentee who tries to stop the infringement is regarded as anti-competitive, and the infringer gets a compulsory licence.

Access to Medicines and Compulsory Licences for Export

The entry into force of the TRIPs Agreement in 1995 came at a time when the extent of the AIDS epidemic in Africa was becoming clear. Patented antiretroviral drugs were being used in the USA and Europe to control the progression of the disease, but these drugs were very expensive and unaffordable for the vast majority of patients in developing countries. The perception arose, fostered by NGOs such as Médecins sans Frontières (MSF) and Oxfam, that patents were the reason why these drugs were not being made available to AIDS sufferers at an affordable price.

In fact, patents are not the problem. Not only are there no patents for most of these AIDS drugs in most African countries, but there are also no patents in any countries for most of the drugs on the World Health Organization (WHO) Essential Drugs List. Why then are these essential drugs not readily available to patients in poor countries? The answer is simply lack of money to buy even cheap medicines, and lack of social and medical infrastructure to deliver them. The terrible truth is that if AIDS could be cured by a glass of clean water, there would still be millions who would have no access to the cure. Unfortunately, patents and the 'greedy' pharmaceutical companies make a much easier target than the miserly rich country governments and the corrupt poor country governments who together make up the real problem.

The pharmaceutical industry also did a lot of damage to its own image, however: notably by its legal action in the South African courts against the South African government over a draft law for the compulsory licensing of some pharmaceuticals. The proposed law was contrary to TRIPs and arguably unconstitutional, but the spectacle of the pharmaceutical industry suing Nelson Mandela (who was then still president and was the first name on the list of defendants) in order to maintain control of drug supplies was a public relations disaster of the first magnitude.

As a result, the industry found itself in a very weak position when in November 2001 the WTO General Council, meeting in Doha, issued a Declaration on the TRIPs Agreement and Public Health, Chapter 6 of which required the TRIPs Council to find a solution under the international patent system to enable the least-developed countries to have access to cheaper medicines and to report back to the General Council by the end of 2002. Proposals were accordingly made to change the compulsory license provisions of TRIPs in order to facilitate this. TRIPs already allowed compulsory licenses to be granted more easily in cases of national emergency,[8] such as the AIDs epidemic, but it was pointed out that compulsory licences were not a useful remedy for a poor country that had no manufacturing industry to take advantage of a licence and which probably had no relevant patent to be licensed in any case. The proposed remedy was to allow compulsory licences to be granted in a more developed country, such as India, to enable drugs to be exported to the poorest countries, a derogation from the TRIPs principle that compulsory licences are to be granted for domestic use only and not for export.

The pharmaceutical industry disliked this proposal for two reasons. Firstly, the main beneficiaries would be the Indian generics companies, which with no research overheads could still make a lot of money by selling drugs cheaply. Secondly, and more importantly, there was the considerable danger that medicines intended for sale to poor countries would be diverted to rich country markets, and undercut the price of patented products there. For a time, agreement in the WTO on this issue was blocked by the USA, but at the end of August 2003, shortly before the WTO meeting at Cancun, agreement was reached at the WTO General Council on the basis of an earlier decision of the TRIPs Council referred to as the 'Motta Text', accompanied by a Statement written by the chairman of the TRIPs Council (Mr Menon, of Singapore). The Decision and Statement, referred to as the 'Motta/Menon Text', is in the form of a waiver from the requirements of Article 31(f) of the TRIPs Agreement. It placed certain limitations on the original proposal and the Menon Statement insisted on the need for measures to prevent diversion. Subsequently, a protocol to amend the TRIPs Agreement to reflect this decision by adding a new Article 31*bis* together with an interpretative Annex was agreed in Geneva on 6 December 2005. Its entry into force requires ratification by two-thirds of the member states, and as of July 2008, only 15 had done so. The first 'Doha

[8] Article 31(b), TRIPs.

licence' under these new rules was granted by Canada in August 2007 to allow Apotex, a generic pharmaceutical company, to export an anti-AIDS drug to Rwanda.

The agreement on this issue had the positive effect that the subsequent breakdown of the Cancun summit could not be blamed on the pharmaceutical industry. Many countries have now introduced legislation to implement the agreement by allowing export compulsory licences to countries that do not have their own pharmaceutical manufacturing capability. But it will be seen in a few years that 'Doha licences' will not provide the solution to access to medicines at affordable prices. What then? Will the real problems of lack of money and infrastructure in the poorest countries be tackled, or will it all be the fault of patents and the pharmaceutical industry yet again?

Further Initiatives on Harmonization

The Patent Law Treaty

The PLT, in force since April 2005, simplifies the requirements for obtaining a filing date, and allows the specification to be filed in any language, with translation provided later if necessary. It makes it easier for the applicant to restore his rights if certain time limits are missed (including the previously inviolable 12-month limit for claiming Convention priority), and should generally reduce costs and increase legal certainty.

The Substantive Patent Law Treaty

Discussions have been taking place within the WIPO for some time with the aim of negotiating a treaty to harmonize matters of substantive patent law, rather than the formal matters concluded in the PLT. Stumbling blocks have been the inability of the USA to change to a first-to-file system, following the failure of a Patent Reform Bill, and the insistence of developing countries upon a number of points, such as the automatic invalidation of a patent for incorrect disclosure of the geographical origin of genetic material, which were unacceptable to developed countries. An open forum meeting on the SPLT held by the WIPO in March 2006 degenerated into what was described as 'a sort of one-sided political rant on all sorts of matters to do vaguely with patents'.[9]

This fiasco had two consequences: firstly, all further discussion of an SPLT within the WIPO was put on hold in 2007, although in 2008 the Standing Committee on the Law of Patents was said (rather circularly) to be 'working towards a work programme'; secondly, the group B (developed) countries of the WIPO began their own discussions within a so-called 'B+' group, consisting of the group B (developed) countries plus the EU, the EPO, and certain additional EPC contracting

[9] Australian comment, reported by Intellectual Property Watch, 6 November, 2007.

states. The intention was to evolve a common set of proposals that could then be brought back into the WIPO, but the very considerable disagreements even within this group make it unlikely that much progress will be made any time soon.

Free Trade Agreements

It has already been mentioned that, prior to the TRIPs Agreement, bilateral or regional free trade agreements had played a part in strengthening patent protection in many developing countries. Even after TRIPs, agreements of this kind continued to be negotiated by the USA, and there has been considerable criticism that, by negotiating from a position of strength, the USA forces its partners to strengthen patent protection well beyond what is required by TRIPs. Provisions that are frequently objected to include requirements for regulatory data protection, and requirements that marketing approval be not granted if the approved product would infringe a third-party patent (as was adopted by Mexico, see above). In fact, regulatory data protection is required by TRIPs,[10] although many countries ignore or circumvent this. The second requirement is analogous to some of the provisions of the US Hatch-Waxman Act (see p. 189), but in the absence of a sophisticated legal system, is certainly liable to abuse. Recently, the Korean Free Trade Agreement extended the grace period in Korea from six to 12 months, not a major issue, but an irritation for those who, like the authors, believe that grace periods have no place in a first-to-file system.

The Trilateral Cooperation

The TLC is a relatively informal cooperation between the world's three most important patent offices: the USPTO; the EPO; and the JPO. This started as a simple project to exchange data and reduce duplication (or rather triplication) of effort, but has developed into a mechanism for the harmonization of patent examination, and may prove to be a more fruitful route to the harmonization of substantive patent law than the SPLT.

The TLC began in 1983, largely as an initiative of Gerald Mossinghoff, then US Commissioner of Patents and Trademarks. At that time, the exchange of information between patent offices still largely involved the mailing of paper copies; what was needed was a unified system for the storage, retrieval, and exchange of data in electronic form. A WIPO committee was working towards this goal, but was hampered by the need to obtain the unanimous agreement of all participating countries. Mossinghoff preferred the less diplomatic, but more effective, method of obtaining agreement among the three main patent offices to set standards that other patent offices would then be obliged to follow.

This approach worked well, and by the mid-1990s, hundreds of millions of pages of patent documents had been scanned using a mutually accessible system; by 1999

[10] Article 39.3, TRIPs.

a common system for the electronic filing of patent applications had been developed. The TLC was also responsible for the standard system of sequence identifiers (SIs) required for patent applications containing amino acid or nucleotide sequences, as well as for the PATENTIN software used to generate SIs (see p. 369).

Of particular interest to patent attorneys is the series of TLC projects comparing the prosecution of applications in new technological fields in the three offices, which indicate whether particular hypothetical claims would be considered allowable, or what types of objection would be raised against them. Frequently one finds that certain types of claim would be refused by all three offices, albeit on different grounds. This gives a level of legal certainty to applicants and leads to a degree of harmonization in examination, which may eventually help to bring about harmonization in substantive patent law.[11]

Later projects of the TLC include the 'Patent Prosecution Highway', which is intended to allow the rapid grant of applications in one office if a corresponding application has been granted in another. This may eventually lead to a situation in which each participating office would give full faith and credit to a determination by another participating office that an application was allowable, although there must be some further harmonization of substantive patent law before this could happen. A pilot scheme between the USPTO and the EPO was started in September 2008.

The Convention on Biological Diversity

Although the conservation of biodiversity is stated to be a common concern of humankind,[12] the Convention on Biological Diversity (CBD) repudiated the view that previously had been generally held, namely, that the natural world was the 'common heritage of mankind', and replaced this with the principle that nation states have sovereign rights over their own biological resources.[13] This does not create any new form of property right, and it is not stated that countries own the genetic resources available within their territories. What it does say is that each country has the right to control access to its genetic resources, and to determine the conditions under which this will be allowed. At the same time, it must be remembered that plants and animals do not respect national frontiers, and the concept of sovereignty over genetic resources is one that is easier to apply to a rare plant that is unique to one particular country than it is to plants or animals found in a number of different countries, and makes little sense when one is considering a ubiquitous microorganism.

Article 16 of the CBD is of particular importance, because it raises the issue of transfer of technology to developing countries, and specifically mentions patents in

[11] P.W. Grubb, 'The Trilateral Cooperation' [2007] 2 JIPLP 397.

[12] Preamble, para. 3, CBD.

[13] Preamble, para. 4 and Art. 3, CBD.

this connection, which is why the CBD is highly relevant to a discussion of patent rights in developing countries. Article 19 also relates to access to the results and benefits of biotechnology by the countries that provide the genetic resources. Some industrialized countries had concerns that these provisions would be used to bring back the discredited ideas of compulsory transfer of technology without adequate compensation and for this reason the USA initially refused to sign the CBD. President Clinton subsequently signed, adding (as did the EU) an interpretative note on this issue, but the CBD, like many other international treaties, has not been approved by the US Senate, and accordingly has not been ratified by the USA.

That these concerns were not without justification may be seen from the general perception that biodiversity exists mainly in the developing countries of the South, that this biodiversity has historically been exploited by the industrialized nations of the North, and that the CBD provides a means of redressing the balance, permitting patent rights to be ignored in the process. It is often alleged that 90 per cent, or some such figure, of all drugs are obtained or derived from sources in the South, and although the provisions of the CBD are clearly not retroactive,[14] it is suggested that compensation, in the form of royalties, should now be paid for this. In 1998, for example, a committee of the Organization for African Unity (OAU) proposed a model patent law for African countries stating that ownership (whatever that means) of new compounds obtained from African natural products should rest with local indigenous communities for 'all times and in perpetuity'.[15] In view of the fact that the worst enemy of many African indigenous communities is their own national government, a certain degree of scepticism seems appropriate. It has even been seriously proposed that Northern countries should now be paying compensation to the descendants of the Incas for the potatoes stolen from them by the conquistadores. Twenty years, for these activists, is far too long for a patent, but 400 years is not long enough for their novel form of rights.

In real life, things are not quite so simple as this, as is illustrated by the fact that one of the world's most successful drugs, the immunosuppressant cyclosporin (sold as Sandimmune® or Neoral®), was first isolated from a fungal microorganism found in a soil sample from Norway, which is about as far 'North' as one is likely to get. Subsequently, the same microorganism has been found in practically every country in the world, but if it were first to have been found in Africa or Latin America, cyclosporin would now be held up as another example of 'biopiracy'. Furthermore, many developing countries obtain economically important products from plant species that were originally taken from other developing countries. It seems unlikely, for example, that Malaysia now intends to compensate Brazil for its use of the rubber tree.

[14] Article 15.3, CBD.

[15] [1998] 392 Nature 423.

Opposition to TRIPs

Possible Amendments to TRIPs

Article 27.3(b) of TRIPs allows members to exclude plants and animals from patentability, but provides that plant varieties must be protectable either by patents or by special forms of protection, such as plant breeders' rights. This subparagraph was to be reviewed by the TRIPs Council four years after the entry into force of the Agreement Establishing the WTO in 1999 (see Chapter 14), but as of 2009 this has still not happened. Although it seems clear that the intention was that this review should be strictly on the narrow issue of whether the exclusion of plants and animals from patentability should continue to be allowed, some developing countries wish to use this review not only to reverse Article 27.3(b) by compelling, rather than permitting, the exclusion of plants and animals from patentability, but also as a 'foot in the door' to renegotiate whole areas of the TRIPs Agreement.

In particular, reference is made to the previous paragraph,[16] which states that members may exclude from patentability 'inventions, the prevention within their territory of the commercial exploitation of which is necessary to protect *ordre public* or morality, including to protect human, animal or plant life or health or to avoid serious prejudice to the environment'. In this way, the anti-patent forces hope to be able to amend TRIPs to deny patentability, for example, to human genes ('immoral'), to anything to do with the screening of plants or microorganisms for useful compounds ('biopiracy'), and to anything that could conceivably have adverse effects on the environment. Oddly enough, the same people are now arguing that patents should not be granted for 'green technology', without noticing any logical inconsistency.

Neither Article 27.2 nor any other part of TRIPs other than Article 27.3(b) is subject to compulsory review, and indeed the review process cannot itself result in an amendment of the text of TRIPs, but may at most produce suggestions as to how the TRIPs Agreement might be modified at a later date, for example in a new WTO negotiating round.

It was also suggested that there should be a review of TRIPs in 2000. However, the relevant provision of TRIPs[17] refers to a review of implementation and it made no sense to start to renegotiate the TRIPs obligations before there had been full implementation of those initially agreed upon. Conversely, there would have been no point in a developing country implementing its TRIPs obligations if it had reason to believe that these obligations would soon be significantly relaxed by an amendment of the Agreement. Independently of this, however, the US government announced its support for a new 'Doha Round' of WTO negotiations, at which TRIPs could be strengthened (or alternatively, weakened). Despite many diplomatic

16 Article 27.2, TRIPs.
17 Article 71.1 (first sentence), TRIPs.

conferences, however, no progress has been made in these negotiations, and as of 2009, it seems that the whole system of the WTO is in danger of becoming irrelevant, especially when the economic recession makes protectionism politically attractive.

As stated in the last chapter, from the point of view of industry, bringing patent matters out of UN agencies such as the WIPO, and into the context of the GATT and the WTO, was a successful strategy in so far as it led to the adoption of the TRIPs Agreement, but had the downside that patents came out of the shadows and into the spotlight of the anti-globalization protesters. As a result, NGOs and developing countries, sometimes aided by the WHO, have for the last 15 years been trying their best to block the implementation of TRIPs and to weaken it by amendment. Three main types of argument are being presented, based respectively on: emotion and disinformation; the premise that TRIPs is incompatible with the CBD; and (rarely) serious economic grounds.[18]

Disinformation

A considerable amount of the argumentation that is presented is what can best be described as 'disinformation', which includes incorrect and misleading statements that are made in good faith by those who are misinformed, as well as deliberate lies told by those who know better. In the former category belong statements such as 'an American company has patented the name "basmati rice"', arising out of confusion between patents, trademarks, and indications of origin. The statement that 'the poor Indian peasant cannot use his neem tree because W.R. Grace has patented it'[19] most likely belongs in the latter.

Compatibility of TRIPs with the CBD

It is frequently argued that the TRIPs Agreement is incompatible with the CBD, and because the CBD is the earlier agreement, its provisions take priority over those of TRIPs. In fact, according to the Vienna Convention on the Law of Treaties, this is incorrect; the earlier treaty is to be interpreted in the light of the later, since this was negotiated in full knowledge of the earlier treaty.[20] In this instance, however, there is no such conflict. The CBD deals with a body of law separate from that which is the subject of TRIPs and, in the areas in which they come in contact, primarily Articles 16 and 19 of the CBD, account is taken of the distinct nature of IP rights.

In particular, the transfer of technology provisions of the Convention state that access and transfer of technology shall be provided on terms that recognize and are

[18] Often mixed with political arguments: see, e.g., E. Henderson: 'TRIPs and the Third World: The example of pharmaceutical patents in India' [1997] 11 EIPR 651. For some reason, Australian academics are particularly opposed to TRIPs.

[19] Vandana Shiva, speaking at a conference with the unlikely title 'Patents, Genes and Butterflies', Bern, October 1994.

[20] Article 3(a), Vienna Convention.

consistent with the adequate and effective protection of IP rights.[21] Parties are supposed to ensure that, subject to national legislation and international law (for example TRIPs), IP rights are supportive of, and do not run counter to, the objectives of the Convention.[22] The CBD and TRIPs are perfectly compatible with each other, and neither is to be applied in such a way as to undermine the objectives of the other. Strong patent protection is an incentive to the transfer of technology, including that which is to be encouraged by the CBD.

Economic Arguments

One simple argument that is often used against patents is that they raise the price of products to the consumer. The primary concern of a developing country is to get the drugs that its people need at the lowest price, and many developing country health officials are firmly of the opinion that prices of all pharmaceutical products would go up steeply if patent protection were to be introduced. Some academics argue that there is a human right to access to medicine, that the TRIPs Agreement leads to higher drug prices, and therefore that the WTO is in conflict with international human rights law.[23]

This simple picture is incorrect for a number of reasons. Firstly of course new patent laws have no effect upon products that are already on the market. In particular, the great majority of the products on the WHO Essential Drugs List are no longer patented in any country. If there is any problem, it can apply only to the new innovative drugs that form a relatively small part of the total market in developing countries. It has been estimated, for example, that in 1993 the drugs that would have been patented in India if India were to have had a 'Western' patent law represented only about 10 per cent of the total drug market in India[24] and this is unlikely to change dramatically now that India finally does have product protection for pharmaceuticals.

Even for new innovative drugs, the situation is by no means clear-cut. In the great majority of cases, alternative therapies are available, and if a patented drug were priced too highly, an alternative would be used. In any case, drug prices are generally controlled by governmental regulation or the market power of a monopsony purchaser, and even when an innovative drug is open to competition from imitators because no patent protection is available, the imitator's product is seldom sold at a significantly lower price than that of the originator. What patents achieve for the innovator is not the power to charge an artificially high price, but the ability to prevent the imitator from diverting sales and profits to its own pocket.

[21] Article 16.2, CBD.

[22] Article 16.5, CBD.

[23] H.P. Hestermeyer, *Human Rights and the WTO: The Case of Patents and Access to Medicines*, Oxford University Press, 2009.

[24] H. Redwood, *New Horizons in India*, Oldwicks Press, 1994.

Another argument is that weakening or abolishing patent protection is advantageous to local industry, because it gives local companies freedom to copy. As we have seen, many European countries have taken this view in relatively recent times, and it is asked why what was considered appropriate, say, for Switzerland when it was a developing country in the late 19th century should not be appropriate for an Asian developing country today.

This argument is not without merit, but there are two strong counter-arguments. The first is that the world is a very different place now than it was over a hundred years ago, and that even if the levels of development may be comparable, the global environment is not. The extent of international trade today is such that no country can act as if it were in isolation. In particular, the countries that are members of the WTO have voluntarily accepted a whole package of trade terms under the GATT of which TRIPs is only one small part. If a developing country feels that it is disadvantaged by TRIPs, it must nevertheless consider that it has obtained benefits and concessions in other areas, for example in the export of textile goods to developed countries. If TRIPs were to be renegotiated, then these other provisions of the GATT should also be open for discussion.

The second counter-argument is that Switzerland was just as mistaken in its views on patents a hundred years ago as many developing countries are today. Any short-term advantage to local industry gained by the weakening of patent protection is outweighed by long-term disadvantages. When the local industry has reached the point at which it can manufacture high-tech products under licence, it will be unprotected against cheap imports and will not be able to develop as it should, and when it becomes capable of innovation, there will be no incentive to invest when other companies can copy with impunity.

This is why many of the larger Indian generics companies have now come around to support a stronger patent law in India. They are now able to undertake innovative research, and do not want their results to be copied by less competent Indian companies.

Summary

It is not a self-evident proposition that a strong patent system such as that which now exists in industrialized countries is in the public interest in these countries; still less is it apparent that such a system is good for all countries in whatever stage of industrial development. The economic tools to carry out such a cost-benefit analysis simply do not exist. Nevertheless, such evidence as there is tends to show that there is a positive correlation between strength of patent protection and level of industrial development, and, more convincingly, that in countries where patent protection in certain areas was abolished or severely weakened, the results were the opposite of what might have been expected. On the other hand, no country can be shown to have been harmed by the introduction of patent protection and developing countries, such as Hong Kong and Singapore, which have traditionally had strong patent protection, have had the best economic results.

It is a fact, for example, that the rapid industrial progress of the USA in the latter half of the 19th century followed the setting up of a strong patent system, although this far from establishes any cause-and-effect relationship. Japan's rapid emergence since the Second World War as one of the world's leading industrial nations coincides with the establishment of a reasonably strong patent system, which has since been strengthened further. Of course, many other factors are involved and we cannot know how these countries would have developed in the absence of patents. Nevertheless, as illustrated by Japan and elsewhere, when a country makes the transition from being predominantly an importer of technology, or an imitator of others' efforts, to being primarily an innovator rather than a copier, its patent laws are strengthened.

This suggests that it is in the best interests of the developing countries to accept and implement TRIPs rather than to try to avoid its obligations or to emasculate it by amendment.

PART II

PATENT LAW AND PROCEDURE

4
WHAT CAN BE PATENTED

Is there any thing whereof it may be said, See, this is new? it hath been already of old time, which was before us.

Eccles. i. 10

God hath made man upright; but they have sought out many inventions.

Eccles. vii. 29

Introduction

In this chapter, we shall consider what kinds of invention are patentable. The question of how an invention that basically is patentable can be protected by a valid patent, which meets conditions such as sufficiency of description, is dealt with in Chapters 11 and 18.

The Requirements of the European Patent Convention and European National Laws

There are three simple requirements for a patentable invention, as set out in TRIPs,[1] the European Patent Convention (EPC), and in the laws of EPC member states, for example in the British Patents Act 1977. These are that:

(a) the invention must be new;
(b) it must involve an inventive step; and
(c) it must be capable of industrial application.

The same three requirements are met with in one form or another in the USA, Japan, and indeed in practically every country that has a patent system at all. There are in addition certain matters that are specifically excluded from patent protection both in the EPC and in the Patents Act 1977, and which are allowed by the provisions of TRIPs to remain excluded. These exclusions are not necessarily to be found in the laws of other countries, for example the USA, and more extensive lists of excluded subject matter may be found in the laws of countries that have not yet fully adapted their laws to TRIPs.

Novelty

The first and clearest requirement is that nothing can be patentable that is not new. If a patent were to be granted for something already known, then, on the classical theory of patent protection in exchange for disclosure, the patentee would be receiving something for nothing: there would be no consideration for the grant. Quite independently of this theory of the patent system, if the general public, or even any part of it, were to know of the information and be free to use it, then the grant of a patent in respect of this information would deprive the public of rights that it previously had. Such a patent monopoly would be unjust for the same reason as were Queen Elizabeth I's monopolies on salt and playing cards.

Considering that the concept of novelty is so basic to patentability, it may seem odd that there are several different concepts of novelty that have been applied to inventions. The most straightforward is that of 'absolute novelty' applied by the EPC and by the national patent laws of all EPC member states, that is, that an invention is new if it is not part of the 'state of the art', which is defined as everything that was available to the public by written or oral publication, use or any other way, in any country in the world, before the priority date of the invention.

Although some countries, for example France, have applied the absolute novelty criterion for many years, in 1977 the concept was new to British patent law. Under previous Patents Acts, 'local novelty' was the rule, which meant that a prior publication or use had to occur within the UK in order to damage the novelty of a patent

1 Article 27.1, TRIPs.

application. This concept goes back to the early days of patents in England, when, as we have seen, patents were frequently granted for inventions that, although known abroad, were brought into the kingdom for the first time by the patentee. Under the 1949 Act, it was still possible to apply for a patent as an inventor by importation, so that a person in the UK to whom an invention was communicated from abroad had the right to apply for a patent.

Under the old British rule of local novelty, it was fairly well established that it was enough to destroy the novelty of a later patent application if only one person in the UK was in possession of information amounting to a description of the invention and if that person was free to do what he liked with the information. This, but without the limitation to the UK, remains true under the 1977 Act and is also the law as applied in the European Patent Office (EPO) by the Boards of Appeal.[2] Now, it seems clear that a prior publication in, for example, a local newspaper in Lhasa would destroy the novelty of a British patent application even if the publication were never read outside of Tibet—a situation just as anomalous as that under the old law, according to which a printed US patent could be an effective publication only from the day on which copies of it or its abstract arrived in the UK.

Some countries also have, or have had, a system that is somewhere between absolute and local novelty. According to this 'mixed novelty' system, which is still the law in the USA, a later patent application is rendered invalid by written publication anywhere in the world, but by oral publication or use of the invention only in the home country: that is, prior use or a public lecture in a foreign country would not invalidate if there was no written description. This has sometimes given rise to problems when patent applications have been filed in the USA relating to aspects of traditional knowledge in countries such as India. A classic example is the US patent[3] that was granted (to two Indian nationals resident in the USA) for the use of turmeric in wound healing, something that had been known in India for thousands of years. The use of the invention in India was not itself prior art in the USA, and although there must have been some written descriptions of the invention, the search carried out by the USPTO examiner did not find them. The patent was revoked in re-examination proceedings instituted by the Indian Council of Scientific and Industrial Research (CSIR), but there was no good reason for the Indian government to spend its taxpayers' money on attacking a patent that had no effect in India and no commercial value anywhere else.

Japan has had all three types of novelty requirement within a relatively short time. Up to 1987, local novelty was the rule, then a mixed novelty system applied, and finally since January 2000 the law requires absolute novelty. China has also gone over from a mixed-novelty system to an absolute-novelty system since 1 October 2009.

[2] See, e.g., T 482/89 (OJ 1992, 646) *TELEMECHANIQUE/Power supply unit.*

[3] USP 5,401,504.

Under the absolute-novelty system that is now the law in EPC countries, prior use of an invention anywhere in the world would invalidate a national or European patent application, if that use were to make the invention available to the public. The situation is clear if the invention is a machine, a gadget, or a chemical compound or composition that can be analysed and reproduced by the skilled person 'without undue burden'. In this case, sale makes the invention available to the public, and it is immaterial whether in fact anyone did investigate the workings of the machine or analyse the compound, or even whether or not anyone would have any motivation for doing so.[4] But the use, or even the widespread sale to the public, of a complex mixture that cannot be precisely analysed may not be held to make the invention that it represents available to the public. For example, it was held by a Technical Board of Appeal that a control program stored on a microchip had not been made available to the public, although the chip had been sold, because it would take many years of effort to analyse it.[5] In a later case, however, it was held that the concept of 'undue burden' did not properly belong to the determination of novelty, and that even if the commercial product could not be precisely analysed, it would destroy novelty if the analysis were to enable the production of anything falling within the claim.[6]

A special situation is that of so-called 'selection inventions' in which an earlier publication discloses a broad class and the invention is or is characterized by a narrower subclass. This situation may occur in mechanical inventions in which the class is a group of structural elements, one of which is selected as being particularly useful. More usually, however, selection inventions are found in the field of chemistry, in which a narrow group of compounds is selected from a known broad group. As long as no members of the narrow subgroup are *specifically* disclosed in the publication, it is generally considered, at least in the UK, USA, and by the EPO, that the compounds are novel, even though they may have been described in general terms. In Germany, it was until recently very difficult to obtain patents for selection inventions, because it was considered that, as long as the broad class was not too large, disclosure of the class was implicitly a disclosure of every member of the class. But German practice now seems to be more in line with that of the UK and the EPO.[7] Selection inventions in the chemical field are discussed further in Chapter 12.

An important question in considering novelty and inventive step is the position of earlier patent applications that were not published at the priority date of a later application. Unpublished patent applications are not available to the public; on this basis, one would expect that they should not be considered as part of the state of the art. It has, however, been a principle of patent law from the earliest times that not more than one patent should be granted for the same invention, because if this were

[4] G 1/92 (OJ 1993, 277) *Availability to the public.*

[5] T 461/88 (OJ 1993, 295) *HEIDELBERGER DRUCKMASCHINEN/Microchip.*

[6] T 952/92 (OJ 1995, 755) *FISONS/Prior use.*

[7] *Olanzapine*, X ZR 89/07, Mitt. 3/2009, 119 (BGH 2008).

not the rule, licensees could be forced to pay twice over to obtain the same rights and the term of patent protection for one invention could be extended beyond the statutory period.

In the old British law, this problem of 'double patenting' was dealt with by making it a separate ground of invalidity of a patent if the invention as claimed had been claimed in a granted British patent of earlier priority date. This approach was sound in theory, but in practice gave rise to a great deal of uncertainty. The situation often arose in which the disclosure in the specification of an earlier application would clearly have been an anticipation if it were to have been published before the priority date of a later application, but where the invention was claimed in somewhat different terms in the two applications. There were a number of cases in which the courts held that there had to be substantial identity between claims in order for this objection of prior claiming to be established, which greatly reduced the effectiveness of this approach.

Under the EPC, the 'whole contents' approach is adopted. Under this system, the whole contents (not only the claims) of an earlier unpublished application are considered. This is brought about simply by defining the state of the art to include unpublished patent applications of earlier date. This greatly simplifies the situation, although the new system is in some ways inequitable. Thus whereas under the old British 'prior claiming' law, the application containing the prior claim had to become a valid granted patent in order to be effective; under the 1977 Act, all that is necessary is for the earlier application to have been published even if it is later withdrawn or refused. In this latter situation, refusal of the later application is no longer necessary in order to avoid double patenting, but it is refused anyway.

In the EPO, a European application filed, but not yet published, before the priority date of a later application is prior art against that application, provided that the earlier application is not withdrawn before publication. Under the original version of the EPC, the prior art effect applied only to states that had been validly designated in the earlier application so that the earlier application could be novelty-destroying in respect of some states designated in the later application, but not against others. This was changed by EPC 2000 and an earlier application is now prior art in respect of all countries designated in a later application, whether or not these were designated in the earlier. Unpublished European patent applications designating the UK can be prior art against a later British national application and an earlier unpublished British application is prior art against a granted European patent (UK) under British law. In the EPO, however, earlier unpublished national applications are not considered as prior art against European applications.

Although under the whole-contents approach the earlier unpublished application is considered to be part of the state of the art, this applies only to considerations of pure novelty, and not to the question of whether or not there is an inventive step. The existence of an earlier unpublished application can destroy the novelty of an invention, but cannot be used to argue that the invention is obvious. This possibility means that, under the Patents Act 1977, lack of novelty and obviousness must be clearly distinguished from each other. In early reported cases in England, this was often not the case and many patents were found invalid for lack of novelty or 'anticipation', whereas the real reason was lack of an inventive step. Nowadays, the term 'anticipation' is normally used to mean lack of novelty and is considered to occur when a piece of prior art (that is, a publication, use, or public oral disclosure that was part of the state of the art before the priority date of the patent application in question) either is or describes something that would be an infringement of one or more of the claims in the application. The test for anticipation is therefore essentially the same as that for infringement, and is met in the case of a written publication if the publication clearly describes something having every feature of the claim, or gives instructions to do something that if carried out would give something falling within the scope of the claim. Thus a claim to a chemical compound may be anticipated by a description of a process if carrying out the process will inevitably give that compound, even if the compound itself was not described. But anticipation requires 'more than a signpost upon the road to the patentee's invention . . . the prior inventor must be clearly shown to have planted his flag at the precise destination'.[8]

In order to constitute an anticipation, there must not only be a disclosure of the invention, but there must also be enablement: that is, the person skilled in the art must be able to perform the invention on the basis of the information given, together with his own common general knowledge. The test for enablement of a prior art disclosure is the same as that for the sufficiency of a patent specification.[9]

Japan also has a system in which earlier unpublished Japanese applications are part of the state of the art, but with the difference that earlier unpublished applications of the same applicant are excluded. In the USA, a pending patent application of earlier date is *a priori* prior art against a later application, unless the later applicant can show that he had an invention date earlier than the date of filing of the earlier application. If it is prior art, it can be applied to attack both novelty and inventive step (see below).

A disclosure formed by combining two documents together is not novelty-destroying, although it may be relevant to the question of inventive step. Indeed it is not even permissible in the EPO to attack novelty by combining two different embodiments described in the same document, unless the document itself indicates that they should be combined.[10] The same principle has also been applied in the USA, where it was held that, in order to anticipate, the prior art document must not only disclose all of the elements of the claim, but must also disclose the claimed arrangement of these elements.[11] Nevertheless, the prior art document must be interpreted in the light of the common general knowledge of the skilled worker in

[8] *General Tire v. Firestone* [1972] RPC 457 (CA).
[9] *Synthon v. SmithKline Beecham* [2005] UKHL 59.
[10] T 305/87 (OJ 1991, 429) *GREHAL/Shear.*
[11] *Net Moneyin v. Verisign*, 545 F.3d 1359 (Fed Cir 2008).

the relevant field as of the date of publication of the document. Needless to say, there is a grey area between what is clearly common general knowledge (for example something in a standard reference book used by everyone in the field) and what is simply another publication, and there have been many decisions on this point both in the UK courts and in the EPO.

Inventive Step

The concept of novelty should be basically a simple matter, which should be capable of being tested rather easily once the claim in question has been 'construed': that is, once it has been logically analysed to determine its scope. The question of whether or not something for which a patent is applied for involves an inventive step is one that is intrinsically much more difficult, because to some extent judgement of what is or is not obvious must be a subjective matter.

Because the question is such a contentious one, there have been a great many patent cases in which obviousness has been at issue, and a great many judges have tried at various times to define what is meant by 'obviousness', or to pose questions such as 'Is the solution one which would have occurred to everyone of ordinary intelligence and acquaintance with the subject matter who gave his mind to the problem?',[12] or, more bluntly, was it 'so easy that any fool could do it'?[13]

The leading English case on obviousness is probably *Pozzoli vs. BDMO*,[14] in which a test for determining obviousness from the much earlier *Windsurfing* case[15] was restructured and redefined. In the *Pozzoli* case, the Court of Appeal identified four steps that must be taken in the analysis of obviousness, as follows.

(1) (a) Identify the notional 'person skilled in the art'.
    (b) Identify the relevant common general knowledge of that person.
(2) Identify the inventive concept of the claim in question or, if that cannot readily be done, construe it.
(3) Identify what, if any, differences exist between the matter cited as forming part of the 'state of the art' and the inventive concept of the claim or the claim as construed.
(4) Viewed without any knowledge of the alleged invention as claimed, do those differences constitute steps that would have been obvious to the person skilled in the art, or do they require any degree of invention?

With all respect, this circumlocution merely brings us back to the phrase 'would have been obvious', and does not greatly help us to decide when something is obvious and when it is not.

[12] *Parks-Cramer v. Thornton* [1966] RPC 407 (CA).

[13] *Edison Bell v. Smith* (1894) 11 RPC 457, 497 (CA).

[14] *Pozzoli SPA v. BDMO SA* [2007] EWCA Civ 588.

[15] *Windsurfing International Inc. v. Tabur Marine (GB) Ltd* [1985] RPC 59 (CA).

Finally, the matter should come down to the simple dictionary definition of obvious, that is, 'very plain', bearing in mind that the reason for requiring the presence of an inventive step before granting a patent is that the ordinary worker in that field should remain free to apply his normal skills to making minor variations of old products.

Thus the person to whom the invention must be non-obvious if it is to be patentable is 'the person skilled in the art', that is, a competent worker, but without imagination or inventive capability. In the USA, this hypothetical person is often referred to as 'PHOSITA' for 'person having ordinary skill in the art'. In the days when the great majority of patents were for relatively simple mechanical devices, it was common to describe the person skilled in the art as an 'ordinary workman'. This is no longer appropriate in view of the increasing technical sophistication of industry. For chemical patents, the person skilled in the art may normally be considered as the average qualified industrial chemist, and for complex inventions, such as in the field of biotechnology, the notional addressee of the patent specification, that is, the 'person skilled in the art', may be considered to be a team of highly qualified scientists.

It does become something of a legal fiction to suppose that such a team could be competent, but non-inventive, considering that its members would if employed in industry be expected by their company to make inventions as part of their normal duties, and if academic scientists would be expected by their university to produce original scientific work, which amounts to much the same thing. The point is that 'obviousness' should be judged by someone with average qualifications and imagination for those in the field. It is tempting for a party attacking a patent on the ground of obviousness to use an expert witness with the highest possible qualifications, but it is not very helpful to have a Nobel laureate testify that something is obvious. It may be obvious to a genius, but is it obvious to the normal worker in the field?

One of the principles that have been established in the course of the many cases on obviousness is that there is no quantitative restriction on the size of the inventive step; thus the invention is patentable if it involves any inventive step, no matter how small. How the invention was made, whether as a result of planned research, a flash of inspiration, or even pure chance, is not relevant to the question of obviousness. An invention may be simple without being obvious; indeed producing a simple solution to what appears to be a complex problem is often highly inventive. It is often very easy to reconstruct an invention with the benefit of hindsight, as a series of logical steps from the prior art, but it does not necessarily follow that the invention was obvious, especially if there is evidence that the invention was commercially successful, or supplied a need. The question 'If the invention was obvious, why did no one do it before?' is usually a relevant one to ask, although there may often turn out to be a good reason why no one would bother to try. As in the USA (see below), it is not enough to establish obviousness that an invention is obvious to try; there must also be a 'fair expectation of success'.[16]

[16] *Conor v. Angiotech* [2008] UKHL 49.

The practice in the EPO is to apply the 'problem and solution approach' to inventive step.[17] This derives from Rule 42(1)(c), which states that the invention is to be disclosed in such a way that the technical problem (even if not expressly stated as such) and its solution can be understood. Having established what is the closest prior art, the examiner is supposed to determine what was the technical problem solved by the invention, and then to judge whether or not the solution would have been obvious to the person skilled in the art. This procedure is supposed to make the evaluation of inventive step more objective and to rule out *ex post facto* analysis,[18] but the difficulty is that the 'problem' is determined with hindsight in full knowledge of the invention, as well as of the prior art, and may have had nothing to do with the problem that the inventor was trying to solve. From the British point of view, the determination of this hypothetical 'problem' seems an unnecessary additional step in the simpler analysis of what is the difference between the prior art and the invention, and whether or not this difference (whether or not it is to be considered as a problem) would be obvious to the person skilled in the art. Nevertheless, the problem-and-solution approach is the main test for inventive step applied in the EPO, although some Board of Appeal decisions have held that this is only one possible way of judging inventive step and that its use is not a *sine qua non*.[19]

In considering obviousness, anything in the state of the art, other than unpublished earlier patent applications, may be taken into account. Under the Patents Act 1977, documents can be combined together in considering obviousness if a man skilled in the art would naturally consider them in association; thus it may be enough if they simply relate to the same technical field. The jurisprudence of the EPO is similar: it is permissible to combine documents in assessing inventive step only if it would have been obvious for the skilled person to do so at the time of filing.[20]

Industrial Application

The third basic EPC requirement is that the invention should be capable of industrial application; this requirement is also stated in the UK Patents Act 1977.[21] Industrial application is broadly defined, and includes making or using the invention in any kind of industry, including agriculture. This concept of 'industrial applicability' of an invention replaces the old and rather vague concept of 'manner of manufacture' that was applied in British patent law before the 1977 Act. In none of the earlier Patents Acts was it stated what constituted an invention; the criterion was one developed by the courts, which asked 'Is the invention a manner of manufacture for which patents could be granted under the Statute of Monopolies?' (see p. 9).

[17] T 1/80 (OJ 1981, 206) *BAYER/Carbonless copying paper*; T 24/81 (unpublished) *BASF/Metal refining*.

[18] T 645/92 (unpublished) *SHELL/Lactones*.

[19] T 465/92 (OJ 1996, 32) *ALCAN/Aluminium alloys*.

[20] T 552/89 (unpublished) *TETRA PAK/Cold cathode*.

[21] Section 4(1), PA 1977.

In the early years of the 20th century, this approach led judges to adopt a restrictive attitude as to what type of invention could be patentable. The word 'manufacture' had acquired the connotation of the production of something tangible and many cases arose of patents being denied for inventions that, although industrially applicable, did not produce or restore a 'vendible product'. Later, the courts recognized that the increasing diversity of technology required a less rigid approach and the new definition of industrial applicability in the 1977 Act coupled with the specific exceptions to patentability (discussed below) represented very little change from the practice immediately before the Act came into force. Nevertheless, the requirement that the invention be industrially applicable is not a trivial one, as Human Genome Sciences found when its patent to a novel cytokine was held invalid on this ground in the Patents Court.[22] Although the compound was correctly identified as a new member of the tumour necrosis factor (TNF) superfamily, there was no indication of what the cytokine did or of how it might be useful in medicine.

Specific Exceptions

The EPC and the Patents Act 1977 make certain specific exceptions to patentability that apply whether or not the invention is capable of industrial application. Artistic works and aesthetic creations are not patentable, and generally are not industrially applicable either; scientific theories and mathematical methods, the presentation of information, business methods, and computer programs are also unpatentable, although they may very well be applied in industry and in practice the prohibition on computer programs is not absolute (see Chapter 16). The practice of the EPO, based on decisions of the Boards of Appeal, requires that, for an invention to be patentable, it must be technical in nature. This is a very sensible approach, which prevents the patenting in Europe of absurd 'inventions', such as that of a method of playing a tennis stroke, for which a patent was granted in the USA.[23] But it must be said that there seems to be no actual basis in the EPC for this requirement. Indeed, in the Convention itself, the word 'technical' appears only in the context of the qualifications (legal or technical) of members of the Boards of Appeal. Rules 42 and 43 of the Implementing Regulations state, however, that the invention must relate to a technical field and solve a technical problem, and that the claims shall define the invention in terms of its technical features. The problem is that the term 'technical' is nowhere defined and has to be interpreted in specific cases by the Boards of Appeal.

Animal and plant varieties are not patentable in countries adhering to the EPC, although, in the USA, plants may be protected either by normal utility patents or by special plant patents. In the UK and certain other European countries, new plant varieties, although not patentable, can be protected by plant breeders' rights granted

[22] *Eli Lilly & Co. v. Human Genome Sciences Inc.* [2008] EWHC 1903 (Pat).
[23] USP 5,993,336.

under the Convention of the International Union for the Protection of New Varieties of Plants (known by its French abbreviation, UPOV). The USA has introduced an equivalent right for sexually reproducing plant varieties under the Plant Variety Protection Act of 1970.[24] A major problem is presented by the fact that neither the EPO nor the Patents Act 1977 give a definition of what constitutes a 'variety'. Transgenic plants and animals are in principle patentable only if they do not constitute a variety; as we shall see in Chapter 15, this gives rise to a difficult and uncertain situation.

A further exception to patentability under the 1977 Act is constituted by inventions 'the commercial exploitation of which would be contrary to public policy or morality'.[25] This replaces the previous wording: '[. . .] the publication or exploitation of which would be generally expected to encourage offensive, immoral or antisocial behavior' The new version corresponds to EPC 2000, which, however, uses the French term '*ordre public*'. It is not clear that 'public policy' and '*ordre public*' mean the same thing, but at least the wording makes it absolutely clear that the provision relates only to exploitation of the invention, and has nothing to do with any allegedly immoral behaviour of the applicant. The Guidelines for Examination in the EPO state that an invention should be excluded from patentability on this ground only if it was something that the public at large would consider abhorrent: for example an improved anti-personnel mine.[26]

The EPC specifically states that an invention does not necessarily fall under this exclusion 'merely because it is prohibited by law or regulation in some or all of the Contracting States'. If the use of an invention is illegal, it is not necessarily unpatentable; *a fortiori* if its use is completely legal, there should be no basis on which this exclusion from patentability could apply. Such a basis is, however, given by the EU Biotechnology Patenting Directive 1998,[27] which sets out certain categories of invention, such as 'the use of human embryos for industrial or commercial purposes', which are defined as being unpatentable under Article 53(a) of the EPC; these have since been codified in the EPC Implementing Regulations.[28] Recently, this provision was interpreted by the Enlarged Board of Appeal to refuse a patent for human stem cells because at the priority date the only process for obtaining such cells involved the destruction of human embryos.[29] The embryos were left over from *in vitro* fertilizations and would have been discarded in any event, and the experiments were perfectly legal in most EU countries. Article 53(a) is being used even more broadly than this by persons opposed to any patenting of animals, plants, genes, cells, and other biotechnological inventions, on the vague ground that this is

[24] See 7 USC §§2321–2582.
[25] Section 1(3), PA 1977, as amended.
[26] Guidelines C IV 4.1
[27] Directive 98/44/EC.
[28] Rule 28, IR.
[29] G 2/06 (OJ 2009, 306) *WISCONSIN ALUMNI RESEARCH FOUNDATION/Stem cells.*

Grace Periods

A 'grace period' may be defined as a period of time before the filing date of an application during which certain types of prior art do not invalidate the application. Depending upon the applicable law, it may refer to any publications of the invention deriving directly or indirectly from the applicant, or be restricted to exceptional situations such as publication in breach of confidence.

The first type of grace period, that is, allowing a valid application to be filed even after deliberate publication by the inventor, is found, for example, in Australia, Brazil, Canada, Malaysia, Mexico, and Philippines (12 months), and in Japan, Korea, and Taiwan (six months). In some of these, for example Brazil, Japan, and Korea, the benefit of the grace period must be claimed on filing, or shortly after; in the others, this is not necessary. In Mexico, the grace period applies before the priority application is filed, so that it is possible to publish an invention in the UK, file a British patent application 11 months later, and file a valid Mexican application claiming the UK priority date within the following year. In the other countries, this is not possible: the national filing must itself be made within 12 or six months of the publication.

In the EPC and in the national laws of most EPC member states, including the UK, a six-month grace period is available only for publication resulting from display at certain certified international exhibitions (of which there are very few) and publications resulting from a breach of confidence. Where such grace periods exist, it is often unclear whether they apply to a period before the priority date, or only to a period before the national or EPO filing date. In two cases heard in 1995, the German Federal Supreme Court held that the national filing date is relevant,[50] while the Dutch Hoge Raad considered that the grace period runs before the priority date.[51] In 2000, the Enlarged Board of Appeal of the EPO held[52] that the six-month period of Article 55(1) of the EPC applied to the period before the European filing date and not to a period before the priority date; this ruling should now be followed by national courts in all EPC contracting states.

It has from time to time been proposed that a grace period for the inventors' own publications should become a feature of the European patent system. It is felt that the present system is unduly harsh to individual inventors and academic scientists who may publish their results before realizing that they may be commercially interesting. But the possibility that patent applications could still be validly filed even after an invention had been published would greatly increase the difficulty of estimating whether a manufacturer's proposed action would infringe any other party's patent rights. Applicants publishing their inventions before filing in reliance on a grace period would be faced with complex requirements, different deadlines in

50 *Coriolis Force*, (OJ 1998, 263) (BGH 10th Civil Senate).
51 *Follicle Stimulating Hormone II*, (OJ 1998, 278) (Hoge Raad).
52 G 3/98 (OJ 2001, 62) *Grace period*.

different countries, and a loss of rights in countries in which grace periods did not exist. The present system may be harsh, but it is at least clear and simple: if you publish, then you perish.

It is frequently said that the US patent system works well with a grace period, but what the USA has is something different. If the critical date for prior art purposes is the invention date and not the filing date, then a publication by the inventor is not prior art at all. A grace period excludes something from the prior art that a priori is within it; what the USA has is a statutory bar that after 12 months makes something prior art that originally was not. This is logical in the context of a first-to-invent system, but makes no sense in a first-to-file system. In the course of negotiations on the Substantive Patent Law Treaty (SPLT, see p. 50), it is being suggested that introduction of a grace period in all countries should be offered in return for the USA giving up the first-to-invent system. Indeed, the disadvantages of a grace period would be well worth accepting if this goal could be achieved. But if the USA is not about to change its system, there is no reason to introduce a grace period in Europe. It has rightly been said that the general adoption of a grace period would effectively replace both 'first to file' and 'first to invent' by 'first to publish' as a basis for the grant of patent rights.

Special Categories of Invention

Chemical Compounds, Manufacturing Processes, and Uses

Although the great majority of countries now grant patents for chemical compounds per se, and this is one of the main requirements of TRIPs, this has been a relatively recent development in many cases. Even countries such as Germany, Japan, the Netherlands, and Switzerland made this change only between 1968 and 1978, and the Scandinavian countries, Austria, Spain, Portugal, and Greece did so even later. A patent for a new compound covers the compound however it is made and for whatever purpose it is used, and is thus a very strong form of protection.

It used to be that in many countries the only form of protection available for the invention of a new chemical compound was a patent claiming a process for its manufacture, and such patents were not very useful if alternative processes could be found.

The value of process patents depends chiefly upon two factors: whether or not there is derived product protection; and whether or not there is reversal of the onus of proof. Where there is derived product protection, sale of the product will infringe the patent if it is has been produced by the claimed process. The difficulty here is that the patentee is usually not in a position to prove that the product has been manufactured by its process—unless, perhaps, careful analysis can detect traces of a characteristic starting material or by-product. This is where reversal of the onus of proof is extremely valuable. Normally in a patent infringement action, the onus is on the patentee to prove to the court that the patent has been infringed; reversal of the onus of proof means that where the compound is new, the court will assume that it has been produced by the patented process unless the person accused of infringement can prove otherwise. This is now one of the requirements of TRIPs.[53] Now, of course, compound protection will normally be available for new chemical compounds, but there may be situations in which a compound is novel but unpatentable because it lacks inventive step, and here, reversal of onus of proof for a process patent may still be important.

In any case, no matter how strong the process patent may be, it will normally cover only one method of making the compounds in question and cannot be used to stop anyone from making the compounds by a completely different method. If the inventor can think of ten different processes, then to cover all of these would, in most countries, require ten separate patents—and even then someone else will come along with an eleventh. For this reason, complete protection was seldom possible in a country that did not grant product per se protection and this is why TRIPs was so important to the chemical industry.

A new use of a known chemical compound will, if an inventive step is present, be a patentable invention. Currently some developing countries refuse to grant patents for new uses[54] and claim that this is not contrary to TRIPs, because TRIPs, Article 27.1, talks only of products and processes, not uses. This is mere semantic hair-splitting: a 'use' can equally well be described as a 'process for using'; the real reason is a desire to restrict as much as possible the right of inventors to patent their inventions.

For more on chemical compounds, manufacturing processes, and uses, see Chapter 12.

Pharmaceuticals and New Pharmaceutical Uses

Before TRIPs, special rules applied in many countries for inventions of a new chemical compound useful as a pharmaceutical. An understandable concern for public health often led to the conclusion that patents for medicines were contrary to public policy, and that medicines would be cheaper and more readily available if they could not be patented. As we shall see, such evidence as there is tends to the opposite conclusion.

Some countries that normally had product per se protection allowed only process and derived product protection for pharmaceuticals. Such was the situation, for example, in Canada, Spain, Norway, and Finland. At the other extreme, some countries had no protection whatsoever for pharmaceutical inventions: for example Italy (until 1978), Brazil (until 1997), and Turkey (until 2000).

Countries that allowed only process protection for pharmaceuticals often had further special provisions, such as shorter terms for pharmaceutical patents or

[53] Article 34, TRIPs.

[54] For example, s. 3(d), Indian Patent Act 1970.

compulsory licences. Until the 1977 Act came into force in Britain, compulsory licences could be granted at any time on pharmaceutical patents. But the licences that were granted had the royalty set by the courts at high levels, which took into account the patentee's investment in research. More to the liking of the imitator were the provisions in Canada, where the royalty rate for compulsory licences was fixed at a maximum 4 per cent of net sales price of the finished product, and in India, where the figure was 4 per cent of the ex-factory bulk price of the active ingredient, which is practically nothing.

The patenting of new uses of pharmaceutical compounds causes problems in countries in which methods of medical treatment are not patentable, since it can be argued that such a use is equivalent to a method of medical treatment. The problem is sometimes solved, or evaded, by the use of special claim wording, as described in Chapter 15.

For more on pharmaceuticals and new pharmaceutical uses, see Chapter 13.

Microbiological Inventions

Although the TRIPs Agreement allows the exclusion of patents for plants and animals, this exclusion does not extend to patents for microorganisms and microbiological processes.[55] Nevertheless, microbiological inventions that involve the use of a new strain of microorganism, whether this is found in nature, selected from organisms produced by artificially induced random mutation, or transformed by recombinant DNA technology, present special problems. This is because, for reasons that we shall discuss in a later chapter, the requirements of a full and sufficient disclosure of the invention are interpreted by most countries to mean that the new strain must be deposited in a recognized culture collection and made available to the public. From the point of view of the inventor, this loss of control of the strain may outweigh the advantages of obtaining patent protection in certain cases. For inventions involving recombinant DNA technology, a written description of the relevant DNA sequences can be given and this usually makes deposition of a transformed organism or vector unnecessary. Information on DNA and amino acid sequences must, however, generally be provided in electronic form in a standard format, to allow computerized searching.

For more on microbiological inventions, see Chapter 14.

Computer-Related and Business Method Inventions

Although computer hardware can of course be patented in the same way as any other electro-mechanical invention, computer programs as such are excluded from patentability both by the EPC[56] and by the Patents Act 1977.[57] Nevertheless, software-related inventions have been held patentable when they are considered to

[55] Article 27.3(b), TRIPs.
[56] Article 52(2)(c), EPC.
[57] Section 1(2)(c), PA 1977.

provide a technical effect that goes beyond the usual interaction of the software with the computer hardware: for example to control an external technical process, or even to increase the working memory of the computer running the program. Whereas a computer programmed in a certain way would be patentable if there were such an effect, it was not generally possible to patent the actual program itself, whether in abstract form or as written onto a disc. The situation in the USA was more favourable to the patenting of software inventions. Following court decisions holding such inventions patentable, the USPTO brought out new guidelines in 1996 that, among other things, enable discs and CD-ROMs carrying a computer program to be patented as articles of manufacture. Subsequently, software itself, as well as business methods, have become generally patentable. Some decisions by the US Federal Appeal Court in 2008 and 2009 have, however, reintroduced limitations on the patentability of business methods and software-implemented processes, although the US Supreme Court may have a final word on the issue.[58] The EPO is now moving in the same direction and, although business methods generally remain unpatentable, claims to computer programs on recording media have been granted. Recent UK court decisions have led to a divergence of patentability criteria between the UK and the EPO in this area, which is a highly undesirable development.

Just as in the early days of biotechnological inventions, the lack of experience of many examiners with this new field has led to many software patents being granted that on their face appear to be excessively broad, and in particular to cover many ways in which the Internet is currently being used. Attempts by some companies to enforce such patents will no doubt give rise to court decisions establishing their valid scope, but there will be some years of confusion before the position becomes clear.

For more on computer-related and business method inventions, see Chapter 16.

[58] *In re Bilski* 545 F3d 943 (Fed Cir 2008, *en banc* certiorari granted); 1365 (Fed Cir 2007, revised *en banc* 2009); *In re Ferguson* Fed Cir 2007-1232 (2009).

5
FILING A PATENT APPLICATION

> The Inventions Office is stuffed with plans for labour-saving processes. . . . And why don't we put them into execution? For the sake of the labourers; it would be sheer cruelty to afflict them with excessive leisure.
>
> Aldous Huxley, *Brave New World* (1932)

Should an Application be Filed?

Once an invention has been made, a decision must be taken whether or not a patent application should be filed, and if so, when. A necessary first step, of course, is that the inventor or his employer realizes that an invention has been made at all. Very often, work that is done in university laboratories, or in the production or customer service departments of industrial companies, may give rise to patentable inventions that are not recognized as such. In the former case, the academic inventors may not be aware of possible commercial applications of their work (although the business acuity of academics has markedly increased in recent years); in the latter, the inventors may think that only what comes out of the company research department can be an invention. In such cases, it frequently happens that, by the time attention has been drawn to the possibility of patent protection for the invention, publication has already taken place either in the form of a paper in a scientific journal, or by sale of a product or disclosure of a process to a customer.

Assuming, however, that a piece of work is recognized as being potentially patentable and that no publication has occurred, should a patent application be filed, and when? Because a first patent application in one's own country is inexpensive, and because in most countries priority in case of conflict goes to the first applicant (but not in the USA, see below), it would seem that the answers are quite clear: 'yes', and 'at once'.

Some qualification is nevertheless required. It may be clear from the outset that patent protection is not needed. Thus it could be that all that is required is freedom to use the new development oneself, rather than any possibility of excluding others. In such a situation, a rapid publication of the results will meet the case, because, once a publication has occurred, in most countries no one else will be able to obtain a patent. Alternatively, the invention may be of a kind such that infringement would be very difficult to detect and a patent would therefore be almost impossible to enforce, but would merely give away useful information. In such instances, it may be preferable to keep the invention as secret know-how. Furthermore, the invention may be so close to the prior art or of such doubtful commercial value that a preliminary evaluation will lead to a decision that even the small effort and outlay of filing a priority application would not be justified.

As a general rule, however, if an invention appears patentable and may be of some commercial interest, a priority application should be filed. The applicant then has one year under the Paris Convention before significant amounts of time and money must be invested in filing in other countries. In this time, the invention can be further evaluated and considered, and if it is decided not to proceed further, the application can simply be dropped. No publication of the invention occurs and nothing has been lost except the relatively small cost of filing the first application. If, however, the decision is made to file in other countries, or even to proceed to obtain a patent in the home country, the first filing gives a priority date that can be relied upon later. In general, then: when in doubt, file an application.

When to File

The second question is when to file. We have already offered the answer 'at once', but this needs qualification. Whereas it may be possible in the case of a simple invention for the inventor to explain the idea to a patent attorney one day, the patent attorney to draft an application the next, and the application to be on file at the patent office on the third day, this sort of urgency is seldom practicable and usually undesirable. Indeed when one of the authors was a young and enthusiastic trainee, he once did just that, only to be told by the inventor on the fourth day that the invention had now been tested and did not work. It is normal not to file an application until the invention has actually been tried—that is, a prototype built or a compound synthesized—and it appears that it may be useful. This is, however, more of a practical matter than a requirement of patent law. It is perfectly possible to file a patent application on the basis of a prediction, and if the prediction is correct and the description is sufficient, the fact that the invention was not tried out before the application was filed is no reason why the patent should not be valid.

The other main problem is that it is seldom desirable to describe and claim an invention in the way in which the inventor first describes it. What the inventor has invented can usually be described as one or more embodiments of the invention: for example a particular device, in the case of a mechanical invention; in the chemical context, a small group of new chemical compounds that may have pharmaceutical properties, a new detergent formulation, or a way of getting improved yield from a chemical process by control of temperature and pressure. It is the first task of the person drafting the patent application to define what is the invention, as distinct from the embodiment of it that has been produced so far. Judgement must be used to decide how broadly the invention can be claimed and this is important even at this early stage. It would be easy merely to write a description of the particular compounds, formulations, and reaction conditions that the inventor has made or tried, but such a description could probably not serve as basis for claims broad enough to give adequate protection.

To take the example of the detergent formulation: the inventor may have tried combinations of four ingredients A, B, C, and D, and have varied the proportions of the main component A between 50 per cent and 70 per cent with success, while leaving the relative proportions of B, C, and D in the remaining 50–30 per cent unaltered. The patent attorney would in addition wish to know whether all four components are essential, whether additional ingredients could be added, within what ranges each component could be varied, and whether each component could be defined as a class of compounds or only as a single compound.

The inventor may be able to answer some of these questions from experience, or at least be able to give an educated guess without further experiment, but in many cases some further experimental work will be necessary before a clear picture of the invention emerges.

It is also very desirable to establish at the time of first filing not only the outer limits of the invention, but also what are the preferred narrower limits to which the application can fall back if the broader scope turns out to be indefensible. It is important for the drafter to have an idea of the prior art in the field of the invention and to draft the text accordingly, but because he must always bear in mind that closer prior art may be found after filing, he will usually draft the specification broadly and give himself basis to restrict later, if need be.

Consideration of all of these matters takes time and the carrying out of additional experimental work takes even more. A balance must be struck in each case between the conflicting requirements of having enough information to draft a good text and filing as early as possible to obtain the earliest priority date. A lot depends upon the extent to which competitors are known to be active in the same area: there are many cases in which a single day's advantage in priority date has made all the difference in cases of concurrent invention. While it is impossible to give a firm rule for all cases, as a general guideline, one or two months between invention and filing may be realistic. In a field in which more urgency is desirable, a subsequent application can be filed when more facts are known and the two combined later (see p. 89).

Inventions Made in the USA

It has already been mentioned that the USA does not as yet give priority to the first applicant, but rather to the first to invent, and that interference proceedings are set up by the US Patent and Trademark Office (USPTO) to determine priority of inventorship in cases of conflicting applications. Furthermore, publication of the invention before the filing date is not fatal to a US patent application, as long as the publication was not made more than one year before the filing date, or (by another) before the applicant's invention date. For these reasons, there is not so much pressure upon a US inventor who is interested primarily in protection in the USA to get a patent application on file as quickly as possible. The inventor knows that, as long as he has fully documented the invention date (and has not suppressed or abandoned the invention in the meantime), he can rely upon this date in any conflict with another inventor and to avoid prior art published within a year before the actual filing date. The tendency in the USA is therefore to take longer to define and to refine an invention before committing it to paper and filing it at the USPTO than would be the case for an invention made in Europe.

This course of action is understandable in view of US patent law, but is not without danger. Even in the USA itself, if an interference should be declared between two conflicting applications, the senior party (who has the earlier filing date) has considerable procedural advantages, because the burden of proof lies on the junior party. Even more important, however, is the possible effect of delay in filing upon corresponding applications in countries outside the USA. The US applicant is sometimes apt to forget that the invention date, which is all-important within the USA, is of no importance whatsoever anywhere else. In every other country in the world, all that matters is the US filing date, from which the inventor claims priority. What is more, a publication before the US filing date may not affect the validity of the US application, but will be fatal to applications in most other countries. It is by no means uncommon for US applicants to publish their own inventions before filing, in reliance on the one-year 'grace period', only to discover later that they have thereby destroyed any chance of getting patent protection abroad.[1]

Thus, even as long as the USA retains the 'first to invent' principle, any party owning rights to an invention made within the USA who is at all interested in the possibility of obtaining protection in other countries should act as if the USA were a 'first to file' country, and file as early as reasonably possible.

[1] The best-known example is perhaps the Cohen/Boyer gene-splicing patent—see Chapter 14.

One point that must be remembered in connection with inventions originating in the USA is that a licence from the USPTO is required in order to file any application on such an invention outside the USA, unless a corresponding US application was filed more than six months previously. This is similar to the UK provisions described below, but is more strictly applied: whereas, in the UK, it is sufficient if the earlier British application was in respect of the same invention, in the USA, any significant new matter added at the foreign filing stage will require a foreign filing licence. Particular care must be taken where there is collaborative work between US and European inventors: even if the invention is seen as originating primarily in Europe, a foreign filing licence may be needed in respect of any contribution from the USA. It is also likely that a licence may be required if information about an invention made in the USA is sent abroad to enable the drafting of a patent application to be outsourced to India or elsewhere. The advice of a US patent attorney should always be taken.

Keeping Records of Inventions

Just as a US applicant with international interests should file early, so a non-US applicant who has any interest in protection in the USA should keep all necessary records to enable the date of invention to be established in the event of a US interference. For the purposes of US interference proceedings, invention is separated into the distinct elements of 'conception' and 'reduction to practice'. 'Conception' is usually defined as a complete mental realization of the invention; 'reduction to practice' consists of physically completing the invention and proving that it is useful for a particular purpose. Thus the preparation of a new substance is not reduction to practice without some test results to prove its utility. Because the uncorroborated statements and records of the inventor are not admissible as evidence, everything must be corroborated by an independent witness.

Prior to 1 January 1996, a date of conception or actual reduction to practice could be established only if these acts took place in the USA,[2] and a non-US applicant could only rely upon the 'constructive reduction to practice' constituted by filing a US patent application, or a foreign patent application from which priority is validly claimed. Because this was clearly contrary to the 'national treatment' provisions of the TRIPs Agreement, it was changed by the Uruguay Round Agreements Act of 1994 at the same time as changes were made to the term of a US patent. Now, dates of conception and reduction to practice later than 1 January 1996 in any World Trade Organization (WTO) member country may be used both in interference proceedings and in order to antedate a prior art reference less than 12 months before the first filing date (see p. 74).

[2] Under the provisions implementing the North American Free Trade Agreement (NAFTA), inventors in Canada or Mexico could establish an invention date as early as 8 December 1993.

But this non-discrimination is not without its disadvantages. We have already seen that it is not enough for the inventor merely to produce a laboratory notebook to show the date of the invention (for example when he synthesized a new compound). It is necessary for scientists to date and sign each page of their laboratory notebooks, and to have each page read and witnessed by a colleague who will not be a co-inventor, in order to provide corroboration of their experimental work. And even if the synthesis of a new compound, for example, can be adequately corroborated, this still does not constitute reduction to practice until it has been tested and found to be at least potentially useful for something. The tests and test results must of course be corroborated, just as was the initial synthesis. Furthermore, when diligence must be shown, then evidence of some degree of continuous activity over the relevant time must be given. Thus although non-US companies were discriminated against, they were at least spared the large number of non-productive man-hours that had to be spent on record-keeping and witnessing of notebooks in a US research organization.

Unfortunately this is no longer the case and non-US companies must now go to the same degree of trouble, or risk losing to a competitor, who may be from any WTO country, in a US interference. The Americans call this 'having a level playing field', but most companies, including many in the USA, would prefer to be playing a different game—that is, first to file.

Where to File

Many countries require that the first patent application for an invention made in the country be filed in that country, at least unless special permission is obtained. A UK resident may not, without written permission of the Comptroller, file outside the UK a patent application relating to military technology, or potentially damaging to national security or public safety, unless a corresponding UK application has been filed at least six weeks previously.[3] This is an improvement over the previous situation, in which the prohibition applied to all applications made by UK residents, but the draconian penalty of up to two years' imprisonment can still be applied if the applicant or his agent contravenes the law knowingly or recklessly.

Because of such laws and because of practical convenience, most applicants file their priority application in their home country. Applicants not subject to such restrictions, however, may choose their country of filing, taking into consideration matters such as the cost, the possible language of filing, the availability of a search report within the priority year, and the possibility of 'internal priority', that is, of filing a later application in the country of first filing claiming priority from the first application and thereby not losing up to a year of patent protection in that country. The last of these points, for example, led some Swiss companies to file priority

[3] Section 23, PA 1977, as amended by s. 7, PA 2004.

applications in Germany or in the UK rather than in Switzerland, where up until 1995 it was not possible to claim internal priority.

Cost saving has become less of an issue, since the countries that are members of the Patent Law Treaty (PLT) are obliged to grant an official filing date without payment of any fees.[4] Whereas some countries still require the applicant to pay an application fee in order to have an official filing date from which they can claim priority, a filing date is secured for a European patent application in line with the PLT if the application form and specification are filed without payment.[5] If the application and search fees are not paid within the allowed time, the application is considered as withdrawn, but it still has a regular filing date, and according to the Paris Convention,[6] the subsequent fate of a regularly filed application is irrelevant as far as the priority date is concerned. Although the European Patent Organization (EPO) is not a party to the Paris Convention, a European patent application is equivalent to a national application[7] and can therefore base a claim to Convention priority. There are similar provisions in the UK regarding establishing an official date of filing.[8]

Of course, if it is desired to have a search during the priority year so that the foreign filing decision and the drafting of the foreign filing text may be done with knowledge of the prior art, it will be necessary to pay the application and search fees in the EPO or in the UK. Because the EPO can deliver a reasonably good search well within 12 months, it is becoming increasingly popular to make the priority filing a European application and 14 per cent of all European applications in 2008 were first filings with no claim of priority. Any applicable national security requirements may be met by filing the European application through a national patent office.

Even though the worst discrimination against non-US applicants has now been removed, it can still be advantageous in some situations to have as early a US filing date as possible, and for this reason, it is sometimes recommended (unsurprisingly, usually by US patent attorneys) that all applicants should file their priority applications in the USA. The relatively high cost of a normal US filing makes this rather unattractive, but the alternative possibility of filing a provisional application is cheaper and the application may be filed without claims. A provisional application may be abandoned and replaced by a regular filing within 12 months of its filing date, the regular filing taking the priority date of the provisional filing. Alternatively, the status of the application itself may be converted from 'provisional' to 'regular'.[9] This procedure is not advisable, however, since not only is it more expensive, but it also leads to a loss of patent term of up to one year.

[4] See Art. 5, PLT.
[5] Article 80, r. 40, EPC.
[6] Article 4A(3), Paris Convention.
[7] Article 66, EPC.
[8] Section 15(1)(d), PA 1977.
[9] 37 CFR 1.53(c)(3).

The non-US applicant must be careful if he wants to be able to rely in the USA upon the priority date of a first filing in another country. The US patent law has rather strict requirements about what constitutes a sufficient disclosure of an invention in a patent application. In particular, the application must disclose how to make and to use the invention, and must give the 'best mode', or the best way known to the inventor, of carrying out the invention. US courts have ruled that, for a US application to have the benefit of priority from a foreign application, that foreign application must meet the same requirements of 35 USC 112(1), as for a US application.[10] The effect is that a priority application, whether in the UK, Germany, Switzerland, or Japan, must be drafted with the requirements of US patent law very much in mind—at least if there is any chance at all that a corresponding application will later be filed in the USA.

The same strict requirements apply to a US provisional application. Although it is not required to have claims, it must meet the normal requirements for written description, sufficiency of disclosure, and best mode, otherwise a later US filing will not be able to claim the benefit of its priority date.[11]

Also within the EPO, the correct claiming of priority is becoming of increasing importance. Under the old British system, which was initially followed by the EPO,[12] a publication during the priority year of subject matter contained in the priority application could not be used to attack the validity of any claim in a European application, even if that claim related to matter added only at the foreign filing stage. The idea was that filing a priority application gave the applicant freedom to disclose his invention to others, without adverse effect on the later filing (known as the 'umbrella theory'). But this approach was overruled by the Enlarged Board of Appeal,[13] with the result that any publication of the original invention may invalidate any claim in the final application that is not entitled to the priority date. Not only that, but the requirements for claiming priority have also been made stricter by a later Board decision[14] to the effect that priority can be validly claimed only if the skilled person can derive the subject matter of the claim directly and unambiguously from the previous application. This may be a particular problem for US-originating biotech applications in relation to which a first filing, made shortly before publication in a scientific journal, is often insufficient to base claims of reasonable scope, but the publication will render invalid any claims that are not entitled to the priority date. Ideally there should be no publication of the invention during the priority year, but in an academic environment, it is almost impossible to insist on this. If the priority filing is a European application, it should always be withdrawn

[10] For example, *Kawai v. Metlesics* 178 USPQ 138 (Fed Cir 1973).

[11] *New Railhead Manufacturing LLC v. Vermeer Manufacturing Co. et al.* 63 USPQ 2d 1843 (Fed Cir 2002).

[12] T 301/87 (OJ 1990, 335) *BIOGEN/Recombinant DNA.*

[13] G 3/93 (OJ 1995, 18) *Publication in priority year.*

[14] G 2/98 (OJ 2001, 62) *Priority claiming.*

before publication. If it is allowed to publish and if priority is lost, it becomes prior art against the later application.[15]

The Priority Year

Once the first patent application has been filed, the 12-month period provided for by the Paris Convention begins to run. During this priority year, work on the invention will normally continue: for a chemical invention, for example, further compounds will be made and tested, new formulations compounded, or new process conditions tried. All of this material can be used in preparing the patent applications to be filed abroad and, where possible, a subsequent application in respect of the home country. It is also possible to file new patent applications for further developments that are made in the priority year and then at the foreign filing stage to combine these together into a single application. The advantage of this is that the new developments will then have an earlier priority date (the date of filing of the new applications) than they otherwise would have (the date of filing of the foreign text). There are, however, some disadvantages:

(a) costs are somewhat increased, mainly because copies of all priority applications have to be provided in most countries when priority is claimed, and in some countries translations will be required;
(b) administration and records become more complex; and
(c) extra care must be taken over inventorship.

When one or more additional applications are filed during the priority year, it is preferable that these do not claim priority from earlier applications in the series: that is, if the first filing is P1 and there is a second filing P2 six months later, and then a final application F will be filed 11 months from P1, claiming priority from both P1 and P2, it is better if P2 is filed without claiming priority from P1. Such a claim to priority brings no advantage, and at least in the EPO, it could give rise to problems. One Technical Board of Appeal has held that a claim to priority can be exercised only once, so that if P2 claims priority from P1, F can no longer do so.[16] Although this decision is not followed by the EPO and later case law has taken the opposite position, it is best to avoid any risk that it might be applied in future.

The extent to which new filings should be made during the priority year obviously depends upon the importance of the invention and the degree of competition in the field; for certain pharmaceutical inventions, it may be justifiable to file a new application for every new compound or group of compounds made and tested, even if this means combining five or six applications upon foreign filing. In most areas, however, it would be unusual to file more than one extra application during the priority year and the majority of patent applications are based upon a single priority

[15] T 1443/05 (unpublished) *THOR/Synergistic biocide composition.*
[16] T 998/99 (OJ 2005, 229) *L'OREAL/Skin equivalent.*

document. The record for the highest number may be held by the European application of a Japanese company that claims priority from no fewer than 41 Japanese applications.

It may also be desirable to file more than one application at the same time, particularly when the invention is a complicated one. Suppose that the inventor has carried out a limited amount of experimental work, which is sufficient to define an invention of fairly narrow scope, but at the same time, sees the possibility of extending the idea to a wider area that he has not yet had time to test. In a case like this, a single first application restricted to the narrow scope would be inadequate if by the end of the priority year the broader scope of the invention were to have been tested and confirmed. But if the broad scope is described in the first filing and then only the narrow part can be substantiated, then foreign applications can indeed be filed limited to the narrow scope; at some stage, however, the priority application will be published and there will then be no chance of filing a later patent application directed to the broader concept.

The answer is to file two first applications of different scope on the same day, and then to claim priority from one and abandon the other. The abandoned application never sees the light of day and so one can choose the scope of the foreign applications, claim priority validly, and not publish any more than is necessary.

The Foreign Filing Decision

Although the period for claiming Convention priority is 12 months, a decision on whether or not to 'foreign file' cannot be left too long. If it is likely that applications will be made in countries in which it is necessary to translate the text before filing, it is clear that it must normally be sent off at least a month, and preferably five or six weeks, before the anniversary of the filing date. When one adds to this the time required for preparing the text of the application, including reviewing further developments with the inventor and considering any new prior art that has been found during the priority year, then a decision to foreign file taken any later than three months before the end of the Convention year may lead to time pressure. A better date to aim at is eight months from the first priority date.

In many cases, the need for immediate translation may be eliminated by filing at the EPO or by pursuing the PCT route. Even when filing a national application in Japan, it is now possible to file initially in English and to provide a translation later. All of these possibilities reduce the time pressure; nevertheless, rush decisions should be avoided whenever possible.

At the foreign filing stage, discussions should be held in order to reach a decision on what should be done with the application. There are four main possibilities that are open to the applicant:

(a) to abandon the application;
(b) to abandon the application and refile;

(c) to obtain a patent in the home country only; or
(d) to file corresponding applications in one or more foreign countries.

Let us look at each of these in turn.

Abandonment

The simplest course of action, in cases in which there is no commercial interest in the invention at all or a search has shown that the invention is old or unpatentable, is to do nothing. It is usually unnecessary to write to the patent office deliberately to abandon the application. Sooner or later, a fee must be paid or some action taken to keep the application in being, and when this is not done, the application will lapse. A letter of express abandonment has the disadvantage that it is normally irrevocable; inventors and their employers frequently change their minds.

An application that is allowed to lapse at this stage will not have been published and so the invention will remain secret. If interest in it revives at a later date, it can be the subject of a later application, provided that no one else has published or patented it in the meantime. If the applicant wants to ensure that no one else can patent the invention, he should have it published, either by continuing an application in his home country or elsewhere long enough for it to issue as a published application (see below), or by sending it to a journal, such as *Research Disclosure*, in which any disclosure may be rapidly published for a reasonable fee. Publications on the Internet are less reliable because it may be difficult later on to provide clear evidence that the disclosure was publicly available via the Internet before a certain date.[17]

In the USA, the application may be converted to a statutory invention registration (SIR) and published without examination on payment of a printing fee.[18] This has the potential advantage that it can be placed in interference with a third-party application for a later invention in order to prevent the issuance of a patent to that party.

Refiling

It frequently appears that eight or nine months after the first filing date is altogether too short a time in which to reach a decision on whether or not to invest time and money in the foreign patenting of a particular invention. It may be that further testing has to be done; it may be that commercial interest is low at the time, but could increase; it may be simply that the inventors have been busy with other things and have done no more work on the invention since the first application was filed. In such cases, the best solution is to start from the beginning again. The existing

[17] See, e.g., decision of German Federal Patent Court 17 W (pat) 1/02 *Computernetzwerk-Information*, GRUR 2003, 323.

[18] 35 USC 157.

application is specifically abandoned; a new application, which may be identical with the old one, is filed and the 12-month countdown restarts.

This is a very convenient solution, but there are three points that must be kept in mind. Firstly, the correct procedure must be followed. According to Article 4C(4) of the Paris Convention, priority may be validly claimed from a refiled application only if the original application is abandoned, without any rights outstanding and without priority having been claimed from it, before the second application is filed (in the same country). In theory, the right to claim priority from the abandoned application remains, since the Paris Convention says that once a filing date has been established the subsequent fate of the priority application is irrelevant. But this cannot be regarded as a 'right outstanding' in the sense of Article 4C(4), the last sentence of which clearly states that the withdrawn application shall not thereafter serve as a basis for claiming a right of priority.

The second point is that refiling always entails a loss of priority, usually of between eight and ten months, and the risk is always present that someone else may have filed a patent application for the invention during this time. If this happens, the refiled application cannot lead to a valid patent. Consequently, in a field in which competitors are known to be active, the loss of priority may involve an unacceptable risk.

Finally, if there has been any known publication of the invention since the priority date, abandonment and refiling is ruled out. Such publication most frequently arises from the inventor himself or from his employers. The inventor may have published the results in a scientific journal, or in a technical information sheet; the employer may have given samples of a product embodying the invention to potential customers to test, or may even have openly sold the product. Most inventors, if they know anything at all about patents, know that they should not publish inventions before a patent application is filed. It is not, however, so generally realized that publication within the priority year can also be damaging: not only can such publication destroy the validity of any parts of the final patent that are not entitled to the first priority date, but it also limits one's options by excluding the possibility of abandonment and refiling.

Home-Country Patenting

In some cases, an applicant may be an individual or a small company having no commercial interests or prospects of licensing outside the home country. For such applicants, the expense of foreign filing would be wasted and the most that they will wish to do is to obtain a patent in their own country. Even if the applicant is a larger company that would normally file any commercially interesting case in several countries, individual applications may be of such low interest that protection in the home country is all that is needed. The only question then is, if the country is one that allows internal priority, whether a new application should be filed incorporating any improvements made in the priority year, or whether the original text is sufficient. For those few countries not allowing internal priority, of course, there is no choice in the matter, if the original priority is to be kept.

Foreign Filing
Finally, if an invention appears likely to be commercially important, the decision may be to file corresponding applications in a number of other countries. This decision is best approached on purely commercial terms. Filing and prosecuting a patent application in a reasonable number of countries costs a considerable amount of money, and this should be regarded as an investment that, like any other, should be expected to show a profit. It is of course extremely difficult to estimate the cash value of a patent even if it is granted and covers a commercial product, and trying to do so for a product not yet launched and a patent not yet applied for is an exercise in guesswork. Nevertheless, something of the kind must be attempted if the foreign filing decision is to be taken on a rational basis, and the type of question to be asked relates to the expected turnover and profitability of the product in question, and how these would differ if competitors could or could not be excluded.

For the pharmaceutical industry, one can assume that patent protection would certainly be worth the outlay for any new product that actually reaches the market and the question is rather what is the chance that the product in question will progress that far. Usually it is impossible to make any meaningful prediction at the time that the foreign filing decision must be taken and the applicant must rely upon some rule of thumb such that, if the product is being developed further, foreign filing should be carried out as a matter of course. High patenting costs are a necessary part of the high research overheads of the pharmaceutical industry. In industries in which the development time is short and products rapidly become obsolescent, it is possible that, by the time a patent issues, the product will have already been superseded. In such a situation, it will be only the exceptional invention that will be worth foreign filing.

The Choice of Countries When the decision to foreign file has been taken, the next question is in which countries to file. Here, the points to be considered are which countries are major markets or manufacturers of the type of products in question, how important the product is likely to be, and what is the relative value-to-cost ratio for the countries to be considered. Of course, for particular applicants, there may be special reasons for filing in a particular country, such as the existence of a subsidiary company there, or licence agreements with a company in that country. Leaving aside such individual reasons, however, if an application is worth foreign filing at all, it is probably sensible to seek patent protection at least in the major industrial countries, such as the USA, Japan, Germany, France, the UK, and Italy. With increasing importance of the invention, filing could be carried out in further countries, which will very much depend on the nature of the technical field. For most inventions, one might, for example, consider (in the following order): the remaining European Patent Convention (EPC) countries that have a significant market size; Canada, Australia, New Zealand, and South Africa (which have strong patent protection that is not expensive); China, Taiwan, and South Korea (major markets in which patents are difficult to obtain and enforce, and expensive); major Latin American countries (which have relatively weak, but inexpensive, protection); India and Association of Southeast Asian Nations (ASEAN) countries; minor EPC countries; and, finally, Russia and Commonwealth of Independent States (CIS) countries.

For a company large enough to have several foreign filings per year, a great deal of repetitive discussion can be avoided by having fixed filing lists of countries for most normal situations; only cases that are out of the ordinary need then be discussed individually.

Non-Convention Filings

It should not be forgotten that the end of the Convention year is not necessarily the last chance for filing corresponding applications. If a foreign filing has been decided on and carried out in, say, six countries, and shortly thereafter it is felt that ten would have been more appropriate, then applications can still be filed in the remaining four countries, although Convention priority can no longer be claimed. This means that if the invention has been published, such new 'non-Convention' filings are pointless and, because most countries and the PCT now publish pending patent applications shortly after 18 months from the priority date, the last possible date for further non-Convention filings is generally 18 months from the priority date, or about six months after foreign filing.

European Patent Applications

If it is intended to foreign file in any of the EPC contracting states, it must next be considered whether to file a single European application, which now by a single fee designates all contracting states, or to file separate applications at the national patent offices of those countries in which protection is desired. There are two main criteria for this decision: the relative costs of the two different routes and the relative strength of the resulting patents.

Considering cost first: it has been the aim of the EPO from the beginning that an application for a European patent designating more than three or four countries (when countries could be designated separately) should be cheaper than a series of applications in the corresponding national patent offices, taking into account attorneys' charges, as well as official fees.

It does depend very much upon which three or four countries are of interest. At one extreme, an applicant who requires protection only in Luxembourg, Belgium, and Switzerland can do so by the 'national route' quite cheaply, particularly because a single French-language text can be used in all three countries. Here, a European application would clearly be more expensive. But the average English-speaking applicant seeking to file in the Netherlands, Germany, and Sweden would find the European route cheaper.

It is also necessary to specify who is the applicant. An individual or small company must rely completely upon the services of independent patent practitioners in each country. A larger company, which employs its own patent specialists, will normally prepare its own patent applications at least for its home country. It is possible for an international company with a large patent department to prepare applications in different languages and to file these directly in some European countries. For such an applicant, which has already greatly reduced the cost of national filings, there may be little or no further savings to be made by filing a European application. But with the EPC now covering 36 states (see p. 31), the European route is very much cheaper if all possible states are of interest.

The other main factor to be considered is the strength of the patent protection that will be obtained. At present, there is no effective examination in a number of EPC countries, including Belgium, Luxembourg, Slovenia, Switzerland, and now even the Netherlands. The fact that a patent has been granted in one of these countries therefore does not help in judging whether or not it may be valid. On the other hand, a granted European patent for one of these countries will have been subject to a strong examination and is more likely to be valid. The negative side, of course, is that if the EPO decides that the application is not patentable, or if it is successfully opposed, protection is lost for all designated states.

A national application that is not subject to substantive examination will normally be granted more quickly than a European application, and if it is valid, will give rise to enforceable rights at an earlier stage. Thus in some fields it may be advantageous to file not only a European patent application, but also a national application in one or more EPC states. Dual protection, however, is not allowed in many countries, and after expiry of the opposition period of a granted European patent at the latest, the applicant must choose which form of protection will be retained. In the UK, if the applicant wishes to maintain the British national patent instead of the European patent, the designation of the UK must be withdrawn before the European patent is granted. If this is not done, any corresponding national UK patent will be revoked[19] when the opposition period expires, unless as a result of the opposition the European patent is revoked (at least in respect of the UK) or limited in scope so that the patents are no longer for the same invention.[20]

Now that a single designation fee is payable, which covers all contracting states,[21] the choice of which countries to pursue can be left until the European patent is granted. The applicant must then decide in which countries to enter the national phase and supply translations where necessary.

[19] Section 73(2), PA 1977.

[20] Section 73(3), PA 1977.

[21] Article 79(1), EPC 2000; r. 39, IR, as amended April 2009.

A European application can be filed in any of the three official languages of the EPO: English; French; or German.[22] The language in which the application is filed then becomes the language of the proceedings for the examination of the application.

In 1979, the first full calendar year of operation of the EPO, approximately 12,000 European applications were filed, of which 45 per cent were in English, 40 per cent were in German, and 15 per cent were in French. The proportion of applications filed in English was less than had been expected, reflecting the fact that while most major German companies enthusiastically adopted the European system from the beginning, UK and US companies were generally more cautious. These companies tended at first to use a selective approach in which only their better applications were filed at the EPO, whereas those with which difficulties might be expected were filed by the national route. A further large factor is that Japanese companies, which might be expected to file predominantly in English, were initially very reluctant to file at the EPO.

Of the EPO applications filed in 1979, 45 per cent related to chemical inventions, compared with 33 per cent mechanical and 22 per cent electrical inventions. The high figure for chemical applications reflects the fact that it was the pharmaceutical companies, for whom strong patent protection in many countries is particularly important, which showed the most interest in the European patent system in its early stages.

Twenty-nine years later, in 2008, the number of applications for a European patent had risen to over 145,000, counting both direct European applications and PCT applications entering the regional phase. Of these, 49 per cent originated in EPC member states and 51 per cent in the rest of the world The USA had become the largest single source of applications, with 26 per cent, compared with 18 per cent from Germany, 16 per cent from Japan, 6.2 per cent from France, 5 per cent from the Netherlands and only 3.5 per cent from the UK, a similar number to those from South Korea.[23]

International Applications

As is clear from the above statistics, a further possibility open to an applicant is to file an international (PCT) application. The PCT used to require individual designations of those countries for which the applicant wanted protection, but, since March 2004, a single application fee includes the designation of all 139 contracting states. The PCT application includes an EPO designation, allowing the applicant to obtain a European patent for any EPC countries, all of which have now ratified the PCT.

22 Article 14(1), EPC.
23 See EPO, *Annual Report*, 2008.

In the great majority of cases, applicants will use the PCT route, coupled with regional phase entry to the EPO for EPC countries. In future, it may also be possible to obtain a Community patent by the same route. There are now few countries of any commercial importance that are not PCT member states (Taiwan and some Latin American countries being the most significant) and if, at the foreign filing stage, a single PCT application is filed, then, effectively, the decision as to the list of countries for foreign filing is postponed from eight or nine months from the priority date to 27 or 28 months from that date. By this time, more information about the importance and patentability of the invention is available, and in some cases, it may be decided not to proceed with the national phase of the PCT application at all. If this happens only once in ten applications, it will probably offset the additional costs of the PCT route. Conversely, if the initial decision would have been to file in only a few countries and the importance of the invention has increased, any number of additional PCT states can be added to the list. If it is already clear at the foreign filing stage that protection will definitely only be sought in two or three countries, filing national or regional applications such as a US, European, and Japanese application will be cheaper and quicker than selecting the PCT route.

Registrations and Patents of Importation

The fact of having a granted patent in one country can occasionally give rise to rights in other countries. There are isolated cases in which, by special agreement, a patent granted in one country may extend automatically to another: for example, although Liechtenstein is a sovereign state, a Swiss patent extends also to Liechtenstein, and a patent in San Marino will automatically cover Italy (and vice versa). It may also be possible validly to file patent applications in certain countries even after the invention has been published. One method is the registration of granted British patents in certain British dependent territories and ex-colonies. Some of these countries additionally have their own independent patent laws, but in others, registration of a British patent is the only form of patent protection that is available. Registration must normally be done within three years of grant and the registration lasts for as long as the British patent remains in force. Generally no renewal fees are payable.

In recent years the number of such registration countries has decreased and those remaining are of little commercial importance. For example, it is now no longer possible to register a British patent in Singapore, which has its own independent patents. Since the return of Hong Kong to China, both British and Chinese patents may be registered there, but instead of being able to wait until five years after the grant of the British patent, the first step must now be taken on publication of the British application (or of a European patent application designating the UK), and the Hong Kong registration must be completed shortly after the grant of the British or European patent. Cyprus, which was formerly a country in which British patents could be registered, is now a contracting state of the EPC in its own right. The old form of registration patents still exists, for example in Bermuda or in the Channel Islands, which, although possessions of the British Crown, are part of neither the UK nor the EU, and to which a British patent does not automatically extend.[24]

A different group of countries, mainly in Latin America and the Middle East, had for a certain period a special type of patent known as a 'patent of importation' or 'patent of revalidation', which like a registration is based on a patent already granted in another country. Novelty of such revalidation patents is not destroyed by written publication of the invention of the basic patent, although it normally is destroyed by prior local sales of the patented product. Revalidation patents are normally granted only to the original patentee, or his assignees, and have a term equal to the remaining term of the basic patent. The conditions under which such patents are granted vary considerably from country to country, and in countries such as Argentina and Uruguay, which have adapted their laws to TRIPs, the status of registration patents is unclear.

Patents of importation have some similarities with the so-called 'pipeline protection' that was made available in some countries as a transitional measure at the time when product protection for pharmaceuticals was first introduced. Because of the long development time for pharmaceutical products, the benefit of such a change in the law would normally not be felt for a period of ten years or so; only after this time would patented products actually be on the market. To compensate for this delay, it was decided, in some countries, for example Korea, Mexico, Ecuador, and Brazil, that compounds still in the development pipeline could be protected in these countries, even though patents for these compounds had been published in other countries. This may be regarded as a temporary derogation from the normal rules of novelty, pipeline protection only being invalidated by actual use, usually only in the country granting the protection (for example for Mexico), but sometimes in any country (for example for Brazil). The term of such protection is normally linked to the term of the base patent in another country, which is similar to the provisions for patents of importation. In some countries, for example Brazil and the Dominican Republic, the constitutionality of pipeline patents has been challenged retroactively because the patent allegedly took something away from the public domain. This led, for example, to a complete invalidation of all pipeline patents in the Dominican Republic in 2006.[25]

Petty Patents

A large number of countries provide intellectual property (IP) rights that differ from patents in that the terms are shorter (typically, between six and ten years instead of 20) and the criteria for validity, particularly regarding inventive step, are

24 For more details, see EPO OJ 2004, 179.

25 *Merck v. Libertador Marketing*, Decision of Dominican Republic Supreme Court of 12 July 2006.

less strict. Typically, they are granted very quickly after only a formal examination, but although they are prima facie very weak, they can often be troublesome to the competition because, unless they clearly lack novelty, they are difficult to challenge.

In Germany, the corresponding IP right is the *Gebrauchsmuster*, or 'Utility Model', which has a ten-year term, and can be granted for all inventions other than processes and methods, including for chemical substances. Similar rights are obtainable in Austria and many other European countries. The Dutch short-term patent, however, has recently been abolished (see p. 138).

In Australia, the innovation patent, which replaced the earlier petty patent in 2002, is a useful tool for 'innovators' who want to have exclusive rights without actually making an invention. The innovation patent is granted for a term of eight years after a purely formal examination lasting only a month or two, and although it must be searched and 'certified' before being enforced against an infringer, the fact that the validity criterion is the presence of an 'innovative step' rather than an inventive step makes it almost impossible to invalidate as long as bare novelty is present. In one case, the Full Federal Court held valid and infringed an innovation patent for a road marker post, the innovation consisting of an obvious combination of old features.[26] At a time when other jurisdictions are tightening inventive step requirements to stop patenting of trivial inventions, this seems a step in the wrong direction.

[26] *Dura-Post (Aust) Pty Ltd v. Delnorth Pty Ltd* [2009] FCAFC 81.

6

OBTAINING A GRANTED PATENT: PATENT COOPERATION TREATY PROCEDURE

If the PCT were a computer program, it would crash.
Software engineer studying for European Qualifying Examination (2000)

Patent Cooperation Treaty procedure

General

The Patent Cooperation Treaty (PCT) is a special agreement under Article 19 of the Paris Convention and is open only to states that are members of the Convention. As mentioned in Chapter 2, it does not grant an 'international patent', because patent grant remains the responsibility of the national or regional patent offices. These offices are, however, assisted in their task by the provision of an international search report (ISR) and a written opinion on patentability on each PCT application, which are provided by a major patent office acting as international searching authority (ISA). There may in addition be an international preliminary report on patentability (IPRP) provided by the same or a different patent office acting as international preliminary examining authority (IPEA). Despite these multiple possibilities for search and preliminary examination, the main purpose of the PCT remains that of making the patent application process simpler and cheaper. This is achieved by having a single set of formal requirements, and a single search and publication procedure, and by postponing translation requirements until 30 months from the priority date.

Originally, the PCT tended to be used only as an emergency procedure when a foreign filing decision was taken at the last minute. It was not user-friendly, formalities were complex, and it was easy to lose rights through minor mistakes. Further, many countries of interest were not PCT contracting states. The system has now improved in a number of ways, as follows.

(a) Entry into the national phase was at 20 (or 21) months from the priority date, unless a demand was made for international preliminary examination (IPE) under PCT, Chapter II. Since 2004, national phase entry is at 30 (or 31) months, whether or not Chapter II is used.
(b) Previously, designations of individual countries were required and it was not possible to enter the national phase in any country that was not designated on filing. Now, there is automatic designation of all contracting states.
(c) Filing in any language is permissible if the filing is made at or transmitted to the International Bureau of the World Intellectual Property Organization (WIPO).
(d) Essentially all countries of commercial interest are now PCT contracting states, Taiwan being the most notable exception.

Chapter I: Filing an International Application

An international application (that is, a PCT application) is a patent application filed under the provisions of and with reference to the PCT. The application filed contains, by default, the designation of all states for every kind of protection available. It has the effect of a regular national filing (including the establishment of a priority date) in each designated state and the international filing date is the filing date in each designated state.

The applicant may be any natural or legal person (for example the inventor or an assignee) and it is possible to have different applicants for different contracting states. Thus the inventor(s), and not any assignee, must be the applicant in respect of the designation of the USA. At least one applicant must be a national or resident of a PCT contracting state.[1]

The PCT application may claim priority under the Paris Convention from an earlier national, regional, or international application, and where priority is claimed, the priority date is the usual starting point for calculating time periods under the PCT. If no priority is claimed, the terms will run from the international filing date.

The Receiving Office

Each PCT contracting state specifies one or more Receiving Offices (ROs) at which its nationals or residents may file a PCT application. Normally the patent office of the state will act as RO and where the state is party to a regional patent organization,

[1] Rule 18.3, PCT.

the regional patent office may also be used. In any event, an international application may always be filed at the WIPO International Bureau in Geneva. Thus, for example, a sole applicant of British nationality and resident in the UK may file a PCT application at the UK Patent Office, the European Patent Office (EPO), or at the International Bureau, although if the application does not claim priority from an earlier application filed at the UK Patent Office, the security provisions of the Patents Act 1977 may require that the PCT filing be at the UK Patent Office (see p. 86).

If there are multiple applicants, or if a single applicant has different nationality and residence, there may be additional possible Receiving Offices from which a choice can be made. If this is the case, care must be taken with respect to the language of filing and the desired ISA, because each RO stipulates one or more permissible languages in which an international application must be filed and also specifies which ISA(s) may be used. Thus, for example, an international application filed at the Danish Patent Office as RO may be filed in Danish, Norwegian, Swedish, English, French, or German, but for applications filed in a Scandinavian language, the competent ISA is the Swedish Patent Office; for the other languages, the ISA must be the EPO. In many countries, it is not possible to file an international application in the official language of that country: for example, in Indonesia the filing must be in English, and in Vietnam, in English or Russian. In Greece and Italy, a PCT filing must be filed in one of the official languages of the EPO, because the EPO must be the ISA, but for national security reasons, a translation into Greek or Italian must be provided unless priority from an earlier application is claimed. The International Bureau will, however, accept PCT filings in any language and, if an application is filed at an RO in a non-permitted language, that office will forward the application to the WIPO, where it will be given the original filing date.[2] A translation into a language permitted by the chosen ISA must be supplied within a certain time. In practice, the great majority of PCT applications (60 per cent in 2007) are filed in English, followed by Japanese (16 per cent), and German (11 per cent).

Table I shows the permissible filing languages and the competent ISA and IPEA for the 15 ROs at which the greatest number of applications were filed in 2007.

Fees

Within one month of the international filing date, three types of fee are payable to the receiving office: a transmittal fee, which is set by and retained by the Receiving Office; a search fee, set by and transferred to the ISA; and an international fee, which is transferred to the International Bureau. The latter consists of a basic fee, which is a fixed sum—in 2009, 1,320 Swiss francs (CHF)—plus a charge for each page in excess of 30 pages. Fees charged by the International Bureau are reduced by 90 per cent for applicants from a defined list of low-income countries.

[2] Rule 12, PCT.

Table I *PCT—filing languages*

Receiving Office	*Permissible Filing Languages*	*Competent ISA*	*Competent IPEA*
International Bureau (WIPO)	All	*	*
EPO	E, F, G	EPO	EPO
Patent office of:			
Australia	E[2]	Australia, Korea	Australia, Korea
Canada	E, F	Canada	Canada
China	C, E	China	China
Finland	E, Fi, Sw	Sweden or Finland (E, Fi, Sw) or EPO (E)	Sweden, Finland or EPO
France	F	EPO	EPO
Germany	G	EPO	EPO
Israel	E	USPTO or EPO	USPTO (if ISA) or EPO (if ISA)
Japan	J, E	JPO(J) or EPO(E)	JPO(J) or EPO(E)
Korea	E, J, K	JPO(J), Korea (K), Australia or Austria (E)	JPO(J), Korea (K), Australia,[1] Austria (E)
Netherlands	Du, E, F, G	EPO (will search in Du)	EPO
Sweden	Da, E, Fi, N, Sw	Sweden (all) or EPO (E)	Sweden or EPO
UK	E	EPO	EPO
USA	E	USPTO or EPO	USPTO or EPO (if ISA)

Notes:
* Depends on nationality or residence of applicant(s)
1. Only for residents or nationals of Korea
2. Or in any language together with an English translation
Abbreviations: C—Chinese; Da—Danish; Du—Dutch; E—English; F—French; Fi—Finnish; G—German; J—Japanese; K—Korean; N—Norwegian; Sw—Swedish

Designations

In recent years, the cost of designating all countries in a PCT application has been considerably reduced. Between 1996 and 1998, it was necessary to pay 11 designation fees, each of CHF185, but by 2002, this had gone down to five fees each of CHF140, a reduction from CHF2,035 to CHF700. A PCT application now automatically designates all contracting states without the payment of any separate designation fees being required. This includes the designation of both patent protection and utility model protection, where this is available, and also the designation of both national and regional protection, for example, for European Patent Convention (EPC) member states. For certain of these countries, however, it is not possible to obtain national patents by the PCT route. If Belgium, Cyprus, France, Greece, Ireland, Italy, Latvia, Malta, Monaco, the Netherlands, or Slovenia is designated in a PCT application, this is treated as a designation of the EPO and a European patent for these countries will be obtained.[3] All of these are countries that do not carry out

3 Article 153(1), second sentence, EPC.

substantive examination in their national patent offices; the reason why they have chosen to close the national route for PCT applications is to ensure that patents granted for these countries on the basis of PCT applications will have received substantive examination at the EPO and thus have a higher presumption of validity. A further exception to the comprehensive designation of all contracting states is that the applicant may opt out of the designation of Germany, Japan, Korea, or the Russian Federation for applications claiming priority from a national filing in the same country because of requirements of the national laws of these countries relating to self-collision.

Processing by the Receiving Office

The RO first checks whether it is competent to act as such and, if it is not, forwards the application to the International Bureau. If it is competent, it carries out a formal examination to ensure that the basic requirements are met, and if they are, accords the application an international filing date. In order to receive a filing date, the applicant must not clearly lack the right to apply (that is, by not being a national or resident of a PCT contracting state); the application must contain the name of the applicant, an indication that it is intended to be an international application, designation of at least one state (now automatic), a part that prima facie appears to be a description, and a part that prima facie appears to be a claim (or claims).[4] This wording is more vague than, for example, the EPC requirement that the application must contain at least one claim, but the UK Patent Office, acting as RO, has held that a statement of invention within the specification did not meet this requirement and a filing date was denied.[5]

In addition, the RO checks whether any drawings referred to are included, whether a translation of the application is required, and whether there are any formal defects that do not affect the filing date. It collects the fees stipulated by the RO, the ISA, and the International Bureau, and ensures that these are paid on time. It makes a formal check on any priority claims and obtains national security clearance, if this is required by law. Finally it forwards the record copy of the application to the International Bureau and the search copy, along with any required translation, to the ISA.

If minor formal defects exist, the applicant will be required to correct them within a stated time limit (usually two months). If this time limit is missed, the RO must take a decision that the application is deemed to be withdrawn, although the application will not lapse if the correction is filed after the time limit has expired, but before the RO issues the decision. If the RO issues a decision that the application is deemed withdrawn, there is no remedy or appeal possible in the international phase. The applicant must immediately enter the national phase and seek whatever

[4] Article 11(1), PCT.

[5] *Penife's International Application*, BI 0/382/03 (UK Patent Office, 5 November 2003, unpublished).

remedies are available under national or regional law (for example further processing before the EPO).

International Search

As of 2009, thirteen patent offices worldwide can act as ISAs and also as IPEAs: namely the patent offices of Austria, Australia, Canada, China, Finland, Japan, Korea, Russia, Spain, Sweden, and the USA, as well as the EPO and the Nordic Patent Institute in Copenhagen. The Indian Patent Office has been accepted as an ISA and IPEA, but has not yet commenced operations. Table II lists the ISA and IPEA offices, along with the languages used and the percentage of the total PCT searches and preliminary examinations carried out by each office in 2008.

Table II *International searching authorities*

ISA	*Languages used*	*% of searches*	*% of IPEs*
Austrian Patent Office	English, French, German	0.7	0.6
Australian Patent Office	English	2	5
Canadian Patent Office	English, French	1.5	3
Chinese Intellectual Property Office	Chinese, English	4	2
EPO	English, French, German, Dutch	48	55
Finnish Patent Office	English, Finnish, Swedish	0.4	0.8
Japanese Patent Office	Japanese, English	17	12
Korean I.P.O.	Korean, English	12	2
Nordic Patent Institute	Danish, English, Icelandic, Norwegian, Swedish	0.1	<0.1
Russian Patent Office	Russian, English, French, German, Spanish	0.5	0.5
Swedish Patent Office	Danish, English, Finnish, French, Norwegian, Swedish	1.4	3
Spanish Patent Office	Spanish	0.7	0.6
USPTO	English	13	16

The ISA is obliged to establish an ISR and written opinion on patentability[6] within three months from the date of receipt of the search copy by the ISA, which is usually about 16 months from the priority date if priority is claimed. Firstly, however, the ISA checks whether or not there is unity of invention. The PCT requires that an international application must relate to only one invention, or to two or more inventions so linked as to form a single inventive concept.[7] If the ISA considers that these criteria are not met, it will communicate its reasoned conclusion to the applicant and request the payment of one month's additional search fees corresponding to the number of additional distinct inventions considered to be present. The first-claimed invention ('main invention') is always searched; further inventions are

[6] Rule 43*bis*.1(a), PCT.

[7] Rule 13.1, PCT.

searched only if additional search fees are paid. Failure to pay additional fees does not lead to the application being refused or deemed to be abandoned, but the additional inventions will not be searched, the written opinion will not contain a preliminary opinion on the unsearched claims, and in Chapter II, the claims relating to unsearched inventions need not be examined by the IPEA.

In certain circumstances, the ISA may refuse to carry out any search at all: for example, if all claims are directed to subject matter that the ISA is not required to search (such as methods of medical treatment, if the EPO is ISA), or to clearly unpatentable subject matter (such as a scientific theory or an abstract concept), or if the application is drafted so badly that it is impossible to determine what the invention is supposed to be. As in the case of a partial search, lack of any search does not directly affect the status of the application, but no IPE can be made if there is no ISR.

The applicant may choose to pay one or more additional search fees under protest, in which case, the ISA, at the same time as carrying out the search on the additional invention or inventions, will review the invitation to pay additional fees. In certain ISAs (Austria, China, Korea, and the EPO), this review is subject to payment of a protest fee. If, upon review, the ISA concludes that the protest was justified, the additional search fees paid will be refunded either totally or partially; the protest fee will be refunded only if the ISA finds that the protest was entirely justified. If, upon review, the ISA concludes that the invitation was justified, the protest is denied. The applicant will be given detailed reasons for the denial. The applicant may request that the text of the protest and of the decision be notified to the designated offices, which may require that the applicant furnish a translation. (For procedure in cases in which the EPO is the ISA, see p. 388.)

There are no provisions for filing divisional applications during the international phase. This may only be done during the national phase in accordance with national or regional law.

The ISR is in standard format. It starts by indicating the international patent classification (IPC) of the subject matter of the application and the fields in which the search was made, states whether unity of invention was considered to be present, and if not, indicates which of the claims have been searched. It will also give an indication of any finding that a meaningful search could not be carried out in respect of certain (but not all) claims. Either the title and abstract supplied by the applicant are indicated to be approved, or new ones are established by the ISA.

The most important part of the report lists the prior art documents considered to be relevant. Each document is categorized by a code letter, of which the most important are: 'X' (damaging to novelty or inventive step when taken alone); 'Y' (damaging to inventive step when combined with one or more other documents); and 'A' (defining general state of the art, not of particular relevance). The letter 'P' indicates a document published between the claimed priority date and the international filing date. In each case, the specific part of the document that is considered most relevant and the specific claims to which it is relevant are indicated. Less usual are references designated 'O' (document referring to an oral disclosure or use) and 'T' (document filed after the priority date that helps to understand the principle or theory underlying the invention).

The initial written opinion is an initial, preliminary non-binding opinion on the novelty, inventive step, and industrial applicability of the invention. It will be sent to the applicant and to the International Bureau at the same time as the ISR, but it will not be published together with the application. It will normally be little more than a rewording of the ISR, that is, if there are one or more X citations, the written opinion will state that the application is considered to lack novelty. There is no formal procedure available for applicants to reply to the written opinion, but applicants may submit informal comments on it to the International Bureau, which will communicate such comments to designated offices, together with the Chapter I IPRP if and when it is sent. The IPRP may be issued under either Chapter I or Chapter II of the PCT. If the applicant does not file a timely demand for IPE, the International Bureau will establish the IPRP (Chapter I) as having the same content as the written opinion of the ISA.[8] The IPRP and any informal comments made by the applicant are sent to designated offices and are made publicly available after 30 months from the priority date, but are not 'published' like the international application and the ISR.

International Publication

The International Bureau publishes the application, together with the ISR, as soon as possible after 18 months from the priority or filing date, which normally means on the first Thursday after the 18 months are complete, unless the International Bureau is closed on that day, in which case, the following Thursday will be the publication day. Earlier publication is possible upon express request by the applicant. If the ISR is already available, no fee is charged for this; if not, a fee is payable. International publication may be prevented by withdrawal of the application and may be postponed by withdrawal of the earliest priority claim, in each case, before the technical preparations for publication are complete. These preparations are normally completed 15 calendar days before the planned publication date. The consequences of postponing publication by dropping a priority claim (apart from the possibility of invalidation by intervening prior art) are that all PCT time limits calculated on the basis of the earliest priority date and which have not yet expired are recalculated on the basis of any remaining priority date or of the international filing date.

International publication is now made only in electronic form, although the applicant will be provided with a paper copy upon request. Paper publication of the *PCT Gazette* ceased on 1 April 2006 and an electronic version is now available online at http://www.wipo.int/pctdb. This is a fully searchable database of all

[8] Rule 44*bis*.1(a), PCT.

published international applications, with the latest bibliographical data, weekly indexes, and official notices. All PCT published applications contain a front page with: bibliographical data and abstract; description, claims, and drawings, if any; and the ISR. Where applicable, they may also contain amended claims and any accompanying statement under Article 19, and any relevant data on deposited biological material. Figure 2 illustrates an example of the title page of a published PCT application.

Article 19 of the PCT gives the applicant the right to amend the claims (but not the description or drawings) after receipt of the ISR. Any such amendment, which must not go beyond the content of the application as filed, must be filed directly with the International Bureau within two months of the date of transmittal of the ISA and written opinion to the applicant by the ISA, and may be accompanied by a statement. The amended claims are then published, together with the original claims. A PCT application is always published with an English language title, abstract, and search report, but the description may be in English, Arabic, French, German, Russian, Chinese, Spanish, or Japanese. In 2008, 66 per cent of all PCT applications were published in English, compared with 16 per cent in Japanese, 11 per cent in German, 3 per cent in each of Chinese and French, and less than 1 per cent in any of the other languages. If the original text is not in one of the permitted publication languages, for example an application filed in Danish at the Danish Patent Office, an English translation may be supplied by the applicant, or may alternatively be prepared by the ISA (for which the applicant is charged a fee) and the translated text is published. At the time of publication, copies of the application are sent by the WIPO to all designated countries that have not renounced their right to receive copies.

Chapter II: The International Preliminary Report on Patentability

International Preliminary Examination

IPE is an optional procedure that is requested by filing directly with the IPEA a 'demand' that contains the automatic 'election' of all contracting states that are bound by the provisions of Chapter II of the PCT, which now includes all PCT contracting states. It results in a preliminary, non-binding opinion by the IPEA on the novelty, inventive step, and industrial applicability of the claimed invention, which is not an opinion on the patentability of the invention according to the national laws of the various elected states. This opinion is referred to interchangeably as the IPRP (Chapter II) or as the international preliminary examination report (IPER).[9] Further, entering Chapter II gives the applicant the opportunity to amend the entire international application before entering the national phase in the elected offices.

[9] See title, r. 70, PCT.

PCT WORLD INTELLECTUAL PROPERTY ORGANIZATION
International Bureau

INTERNATIONAL APPLICATION PUBLISHED UNDER THE PATENT COOPERATION TREATY (PCT)

(51) International Patent Classification 6 : A61N 5/00, H01J 37/147	A1	(11) International Publication Number: WO 95/32021 (43) International Publication Date: 30 November 1995 (30.11.95)

(21) International Application Number: PCT/US95/06140

(22) International Filing Date: 19 May 1995 (19.05.95)

(30) Priority Data:
08/246,860 20 May 1994 (20.05.94) US

(71)(72) Applicant and Inventor: MILLS, Randell, L. [US/US]; Suite 208, 1860 Charter Lane, Lancaster, PA 17601 (US).

(74) Agents: LESTER, Michelle, N. et al.; Cushman, Darby & Cushman, L.L.P., 1100 New York Avenue, N.W., Washington, DC 20005 (US).

(81) Designated States: CA, JP, European patent (AT, BE, CH, DE, DK, ES, FR, GB, GR, IE, IT, LU, MC, NL, PT, SE).

Published
With international search report.

(54) Title: APPARATUS AND METHOD FOR PROVIDING AN ANTIGRAVITATIONAL FORCE

(57) Abstract

A method for producing an antigravitational force comprises an electron source (100) including electrons (113), an electron guide (109) for forming the electrons (113) to be negative curvature; the gravitation body (113) is comprised of matter of positive curvature where opposite curvatures provide a mutually repulsive antigravitational force. The electrons (113) are given negative curvature of an electron beam (113) from atoms such that negatively curved electrons (113) emerge. The emerging beam of negatively curved electrons (113) experience an antigravitational force. The antigravitational force of the electron beam (113) is transferred to a negative charged plate (121).

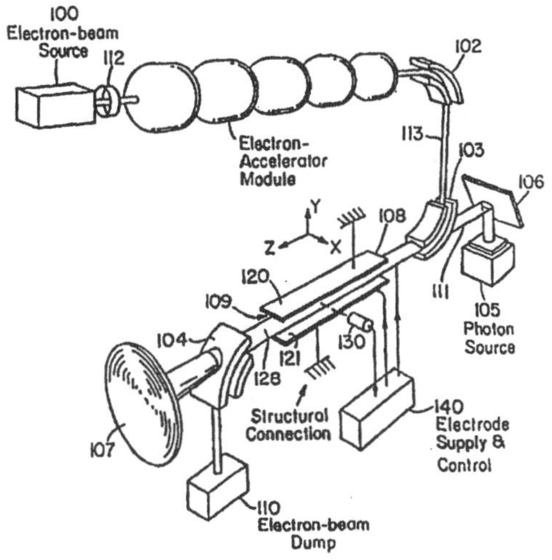

FIG. 2. Front page, PCT published application

The Chapter II demand must be filed within 22 months of the first priority date or three months from the date of transmittal of the ISR and written opinion, whichever is the later. IPE will not start before expiration of this time limit, unless the applicant specifically requests an earlier start.[10] If no demand is filed, the IPRP will be issued by the ISA.

The applicant may file 'informal comments' to the International Bureau, but must in any event enter the national phase within 30 months (31 months for the EPO) of the first priority date if he wishes to proceed. If the applicant does enter the national phase, copies of any 'informal comments' are sent with the application to the designated offices.

The demand must be filed at the competent IPEA, as specified by the RO, or one of them, if there is a choice. Normally the same office that acted as ISA will act as IPEA, but this is not always the case. The demand must be accompanied by a preliminary examination fee and a handling fee, both of which are payable to the IPEA, which forwards the handling fee to the International Bureau. If the demand is filed at an IPEA that is not competent to carry out the examination of the application, the demand will be sent to the International Bureau, which will forward it to the correct IPEA, without loss of date.

If the applicant files a demand, but takes no further action, the IPEA will issue an IPRP that will normally be identical with the written opinion already given by the ISA, so that the applicant will have paid significant fees for nothing. It used to be the case that the applicant would, by filing a demand and entering Chapter II, at least obtain a ten-month postponement of the entry into the national phase, but national phase entry at 30 or 31 months from the priority date is now the rule whether or not Chapter II is used.

Under PCT Article 34, all parts of the international application, and not only the claims, may be amended at any time during Chapter II, between the filing of the demand and the establishment of the IPRP. The IPEA is not, however, obliged to consider any amendment submitted after it has begun to draw up a written opinion or the final report.[11] Preferably, any amendments should be filed at the same time as the demand, to ensure that the examination will be based on the application as amended and to give the maximum time for meaningful IPE.

An applicant filing a demand for IPE should at the same time file a response to the written opinion of the ISA, giving arguments for patentability. The IPEA may then issue a second written opinion, to which the applicant may make a further response, and the IPRP will then be issued. Alternatively, the IPEA may already issue the IPRP in place of a second written opinion, and then no further response is possible. The IPRP must be established not later than 28 months from the priority date, so the procedure is short compared with the duration of the examination

[10] Rule 69.1(a), PCT.
[11] Rule 66.4*bis*, PCT.

process in most patent offices that carry out substantive examination, which means that the examination is necessarily superficial in nature.

A summary of the PCT procedure in the international phase is given in Fig. 3.

Entry into the National Phase

The applicant must then decide with which contracting states he wishes to proceed and, in the case of EPC countries for which there is a choice, whether to use the national or the European route. Entry into the national or regional phase must be done no later than 30 (or 31) months from the priority date. It is worth noting that this can be delayed by up to 12 months if the claim to priority is abandoned, but if the deadline has already been passed, abandonment of priority will not help. Entry into the national phase (or regional phase for Euro-PCT applications) means that the formal requirements of each state must be met, all national fees paid, and translations made into local languages, as necessary. Further amendments may be made to the description, claims, or drawings on entry to the national phase.

For entry into the regional phase at the EPO, the time limit is 31 months from the priority date, whether or not Chapter II is used. The application must be translated into English, French, or German unless already published in one of these languages, in which case, the language of the proceedings at the EPO will be the language of publication. At the time of writing, however, there is a case pending before the Enlarged Board of Appeal in which the applicant wished to change the language of the proceedings from French (the language of PCT publication) to English, and this was refused by the EPO.[12]

For entry into the US national phase, two different routes are available. The applicant may enter the national phase in the normal way no later than 30 months from the priority date. Alternatively, since the PCT filing is equivalent to a regular national application, a continuation or continuation-in-part application based upon it may be filed within 30 months from the priority date, and the PCT application may then be abandoned. The national, search, and examination fees must be paid, and an inventor's oath or declaration must be filed. In the UK although a filing date may be obtained without payment of any application fee, a fee of £30 is payable to ensure entry of a PCT application into the national phase. If this were not the case, international applications published in English and designating the UK would enter the national phase automatically with no further action by the applicant being necessary. This is not always what the applicant wants, especially because PCT (GB) publications are prior art in the UK as of their priority date.

How the PCT application is further examined also depends upon each separate national phase patent office. In some countries, for example Singapore, a patent will automatically be granted on a PCT Chapter II application without further

[12] Pending as G 4/08 (for referral, see OJ 2009, 167) *MERIAN/Languages*.

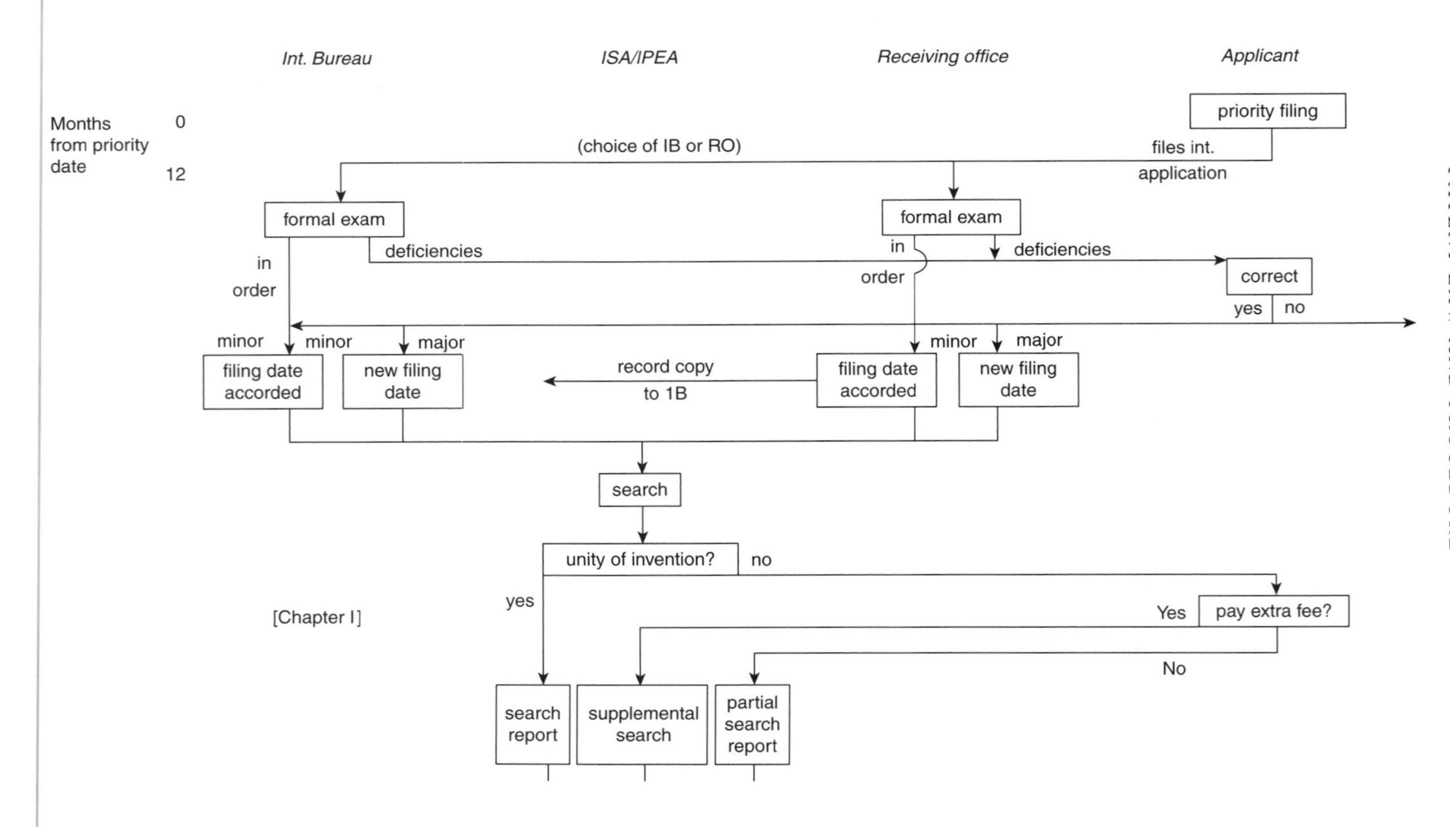
Int. Bureau
ISA/IPEA
Receiving office
Applicant
Months from priority date
0
12
priority filing
(choice of IB or RO)
files int. application
formal exam
formal exam
in order
deficiencies
in
order
deficiencies
correct
yes
no
minor
minor
major
filing date accorded
new filing date
record copy to 1B
minor
major
filing date accorded
new filing date
search
unity of invention?
no
yes
[Chapter I]
Yes
pay extra fee?
No
search report
supplemental search
partial search report

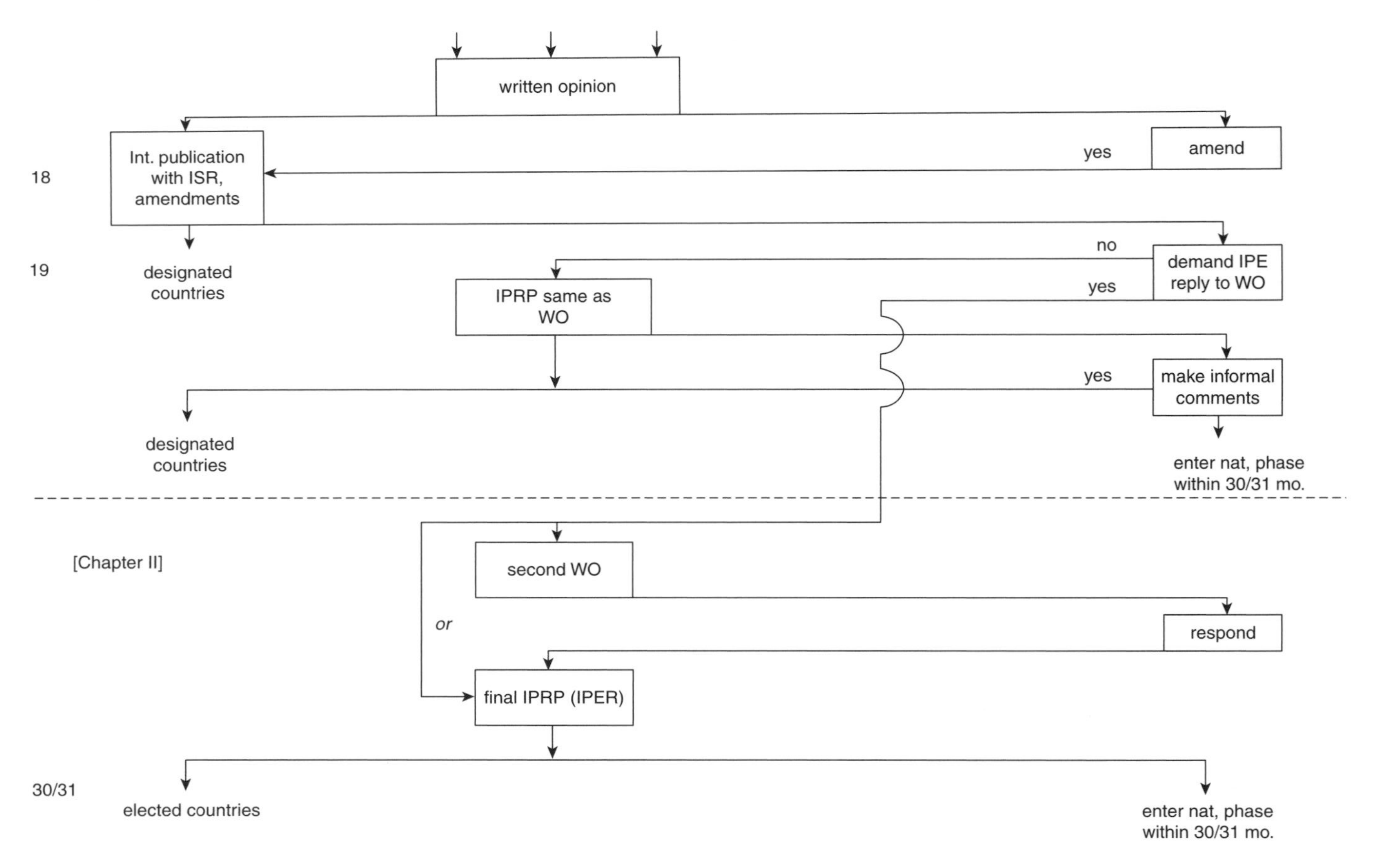

FIG. 3. PCT International phase procedure

examination, even if the IPRP is totally negative (see p. 137). In the USA, the impression is that the IPRP will be ignored, at least if it is issued by an authority other then the US Patent and Trademarks Office (USPTO). In other countries, the IPRP will be used as the starting point for examination and it may be difficult to overcome a negative report.

7

OBTAINING A GRANTED PATENT: EUROPEAN PATENT OFFICE PROCEDURE

In America, there's a failure to appreciate Europe's leading role in the world.
Barack Obama, Strasbourg (2009)

Procedure in the European Patent Office

General Principles

There are certain basic principles governing procedure in the European Patent Office (EPO). No decisions may be taken unless the parties have had the opportunity to comment[1] and oral proceedings must be held if at least one party so requests.[2] Article 125 of the European Patent Convention (EPC) states that, in the absence of specific provisions, the EPO shall take into account the principles of procedural law generally recognized in the contracting states. This presents some difficulties in view of the differing legal systems in these states, but, for example, the German rule of *Treu und Glauben* (good faith) is applied so that an applicant should not be

[1] Article 113(1), EPC,
[2] Article 116(1), EPC.

disadvantaged by relying upon what he has been told by the EPO, even if this turns out to be misleading.[3]

When the applicant is faced with a possible loss of rights, for example the refusal or withdrawal of the application, he (almost) always has the possibility to recover his position.

Loss of rights may occur either as a result of an adverse decision by the EPO, such as refusal of the application, or of the applicant failing to take some necessary action before the expiry of a time limit.

An adverse decision may be appealed, and if the applicant has failed to act in time, possibilities to recover include: extensions and grace periods, where provided; further processing under EPC Article 121; restoration of rights under Article 122; and requesting an appealable decision under Rule 69(2).

Filing

Generally speaking, a European patent application may be filed directly at the EPO, either at Munich or The Hague, at the EPO branch office in Berlin, or at the national patent office of a contracting state. In the latter case, the date of receipt by the national office counts as the EPO filing date. In the Netherlands, however, a European patent application must be filed at the EPO and not at the Dutch Patent Office. In many EPC contracting states, such as the UK, France, or Italy, the secrecy provisions of the national patent laws prohibit the filing of a patent application abroad unless a corresponding national application has been filed, or special permission has been obtained.[4] This means, for example, that a European application by a British resident made without claim of Convention priority must be filed at the UK Patent Office, although in the more usual case, in which priority is claimed from an earlier British application, the applicant has a choice.

A European patent application can be filed in any language in order to get a filing date accorded, but an applicant who has not filed an application in one of the three official languages of the EPO, that is, English, French, or German, has to provide a translation into one of the three official languages within two months from filing. If an applicant is a national or resident of a contracting state that has an official language that is different from the three official EPO languages and the application is filed in such an official language of the contracting state, then the EPC provides for a 20 per cent reduction of the application fee. For example, a Swedish applicant who files an application in Swedish will receive such a discount, but must later provide a translation into one of the three official languages.

The language of the application or the translation becomes the language of the proceedings. Any amendments to the specification must, for obvious reasons, be in

[3] For example, J 3/87 (OJ 1989, 3) *MEMTEC/Membranes*.

[4] A summary can be found in ch. 2 of the EPO brochure, *National Law Relating to the European Patent Convention*, 13th edn, 2006.

the language of the proceedings. The applicant may, however, communicate with the EPO in any of its three official languages.[5]

In order to obtain a filing date, the application, which should be on the standard printed form supplied by the EPO, must contain information allowing the EPO to contact or identify the applicant, indicate that a European patent is sought, and include a description of the invention, which may be replaced by a simple reference to an earlier filed patent application with a valid filing date accorded by a patent office. The application should also contain a title and abstract, a claim or claims, and a designation of the inventor. Fees are charged for the application, for the search, and for each claim in excess of 15, the fees being due within one month of filing. If the fees are not paid in time, the EPO will send a notification to the applicant and the fees may still be validly paid within one month of the notification. If the application fee and the search fee are not both paid within the prescribed time, the applicant will receive a notification from the EPO that the application will be deemed to be withdrawn if the applicant does not pay the fees (with a 50 per cent surcharge) within two months from the date of such notification. Irrespective of whether any fees are paid, the application can nevertheless (if it is the first application for the invention) base a later claim to priority.

The EPO now encourages online electronic filing of applications by reducing the filing fee from €180 to €100 for online filings, including Patent Cooperation Treaty (PCT) regional phase entries, made after 1 April 2008.

The European patent application may claim priority at the time of filing from one or more earlier European, PCT, or national applications filed within the preceding 12 months. In the case of national applications, these may have been made in any member state of the Paris Convention or any member state of the World Trade Organization (WTO).[6] Although the EPO is not a member of the WTO, WTO member states (such as Taiwan) that are not members of the Paris Convention may recognize priority claims based on European applications.[7]

Designations

Prior to April 2009, a European patent was not automatically granted for all of the EPC states; the applicant had to designate those countries for which a European patent was required and pay a designation fee for each country, whereby the payment of seven designation fees (each of €80) covered the designation of all contracting states. At the date of filing of an application, all member states of the EPC are automatically designated.[8] The applicant formerly had to indicate for which states he intended to pay designation fees; now, a single designation fee of

[5] Rule 3(1), IR.

[6] Priorities from WTO countries that are not members of the European Patent Convention are recognized for EPO filings on/after 13 December 2007.

[7] Information from the EPO, 21 January 2008.

[8] Article 79(1), EPC 2000.

€500 (in 2009) pays for the designation of all contracting states. Payment of the designation fee is due within six months after the publication of the search report on the application. If the fee is not paid in time, the applicant will be notified that the application is deemed to be withdrawn,[9] starting a two-month period in which the fee can still be validly paid with a 50 per cent surcharge. Generally, almost any missed deadline or missed payment of a fee can be cured within two months of a notification from the EPO by requesting 'further processing',[10] performing the omitted act and paying a special fee, or (in the case of a missed payment) by payment of the missed fee plus a 50 per cent surcharge. The applicant is free expressly to withdraw any designation at any time up to the grant of the European patent application. This may be advisable if an applicant would prefer to keep a parallel national patent rather than to have the European patent with the same filing or priority date to become valid in an EPC member state that prohibits the double protection of an invention by a national and a European patent.[11]

It is possible to have different applicants for different countries, although there must be a single representative responsible for communications with the EPO.

Formal Examination, Search Report, and Publication

The application is sent to the Receiving Section, which first checks to see that the requirements for receiving a filing date have been met and that the fees have been paid, then carries out a formalities examination. The applicant is given a time of between one and two months in which to correct minor objections made at this stage.

The application then passes to the Search Division, which produces the European search report on the invention. As soon as possible after 18 months from the filing date or earliest priority date, the application is published (previously by photolithography, now by electronic means), together with the search report and abstract. An example of the title page of a published European patent application is shown as Fig. 4.

There was previously a strict geographical separation of functions at the EPO, spelled out in Articles 16–18 of the EPC. The Receiving Section and also the Search Division, for historical reasons—that is, their association with the former International Patent Institute (known by its French abbreviation, IIB)—were located at The Hague, while the Examining Division was in Munich. After some time, the EPO became convinced that it was more efficient to have a single person acting both as search examiner and as the primary examiner in the Examining Division. This was gradually implemented over a number of years under the cheerful acronym BEST ('Bringing Examination and Search Together'), but because of

[9] Rule 39(2), IR.

[10] Article 121, EPC.

[11] For example, the UK, where the Comptroller may revoke a national patent to the extent that it covers the same invention as a European patent with the same priority/filing date under s. 73, PA 1977.

(19) **Europäisches Patentamt**
European Patent Office
Office européen des brevets

(11) Publication number: 0 283 442 A1

(12) EUROPEAN PATENT APPLICATION

(21) Application number: 88830018.3

(22) Date of filing: 18.01.88

(51) Int. Cl.[4]: **A 61 K 9/54**
A 61 K 37/54

(30) Priority: **21.01.87 IT 4754687**

(43) Date of publication of application:
21.09.88 Bulletin 88/38

(84) Designated Contracting States:
AT BE CH DE ES FR GB GR LI LU NL SE

(71) Applicant: **SANDOZ S.A.**
Lichtstrasse 35
CH-4002 Basel (CH)

(72) Inventor: **D'Amico, Paolo**
152 Viale Cortina d'Ampezzo
I-00135 Roma RM (IT)

(74) Representative: **Bazzichelli, Alfredo et al**
c/o Società Italiana Brevetti S.p.A. Piazza Poli 42
I-00187 Roma (IT)

(54) **Process for the manufacture of gastro-resistant and enterosoluble small spheres of digestive enzyme and pharmaceutical preparation so obtained.**

(57) A process for the preparation of pancrelipase micro-spheres is described, starting from seeds consisting of pancrelipase granules of specific particle size, obtained by means of precompression, followed by dry granulation and sphering.

FIG. 4. Front page, European patent application

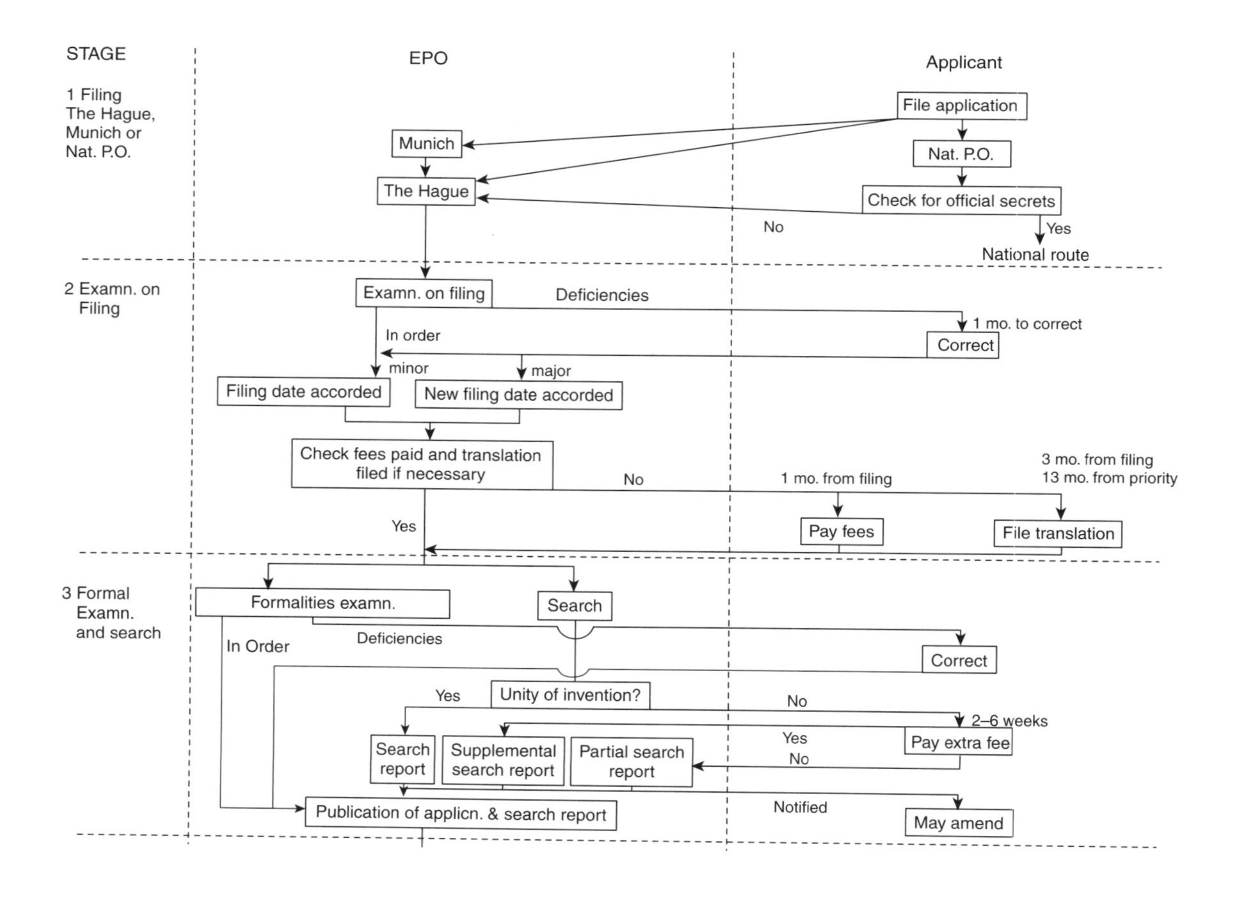
STAGE
EPO
Applicant
1 Filing The Hague, Munich or Nat. P.O.
File application
Munich
Nat. P.O.
The Hague
Check for official secrets
No
Yes
National route
2 Examn. on Filing
Examn. on filing
Deficiencies
1 mo. to correct
In order
Correct
minor
major
Filing date accorded
New filing date accorded
Check fees paid and translation filed if necessary
No
1 mo. from filing
3 mo. from filing
13 mo. from priority
Yes
Pay fees
File translation
3 Formal Examn. and search
Formalities examn.
Search
Deficiencies
In Order
Correct
Yes
Unity of invention?
No
2–6 weeks
Yes
No
Pay extra fee
Search report
Supplemental search report
Partial search report
Notified
Publication of applicn. & search report
May amend

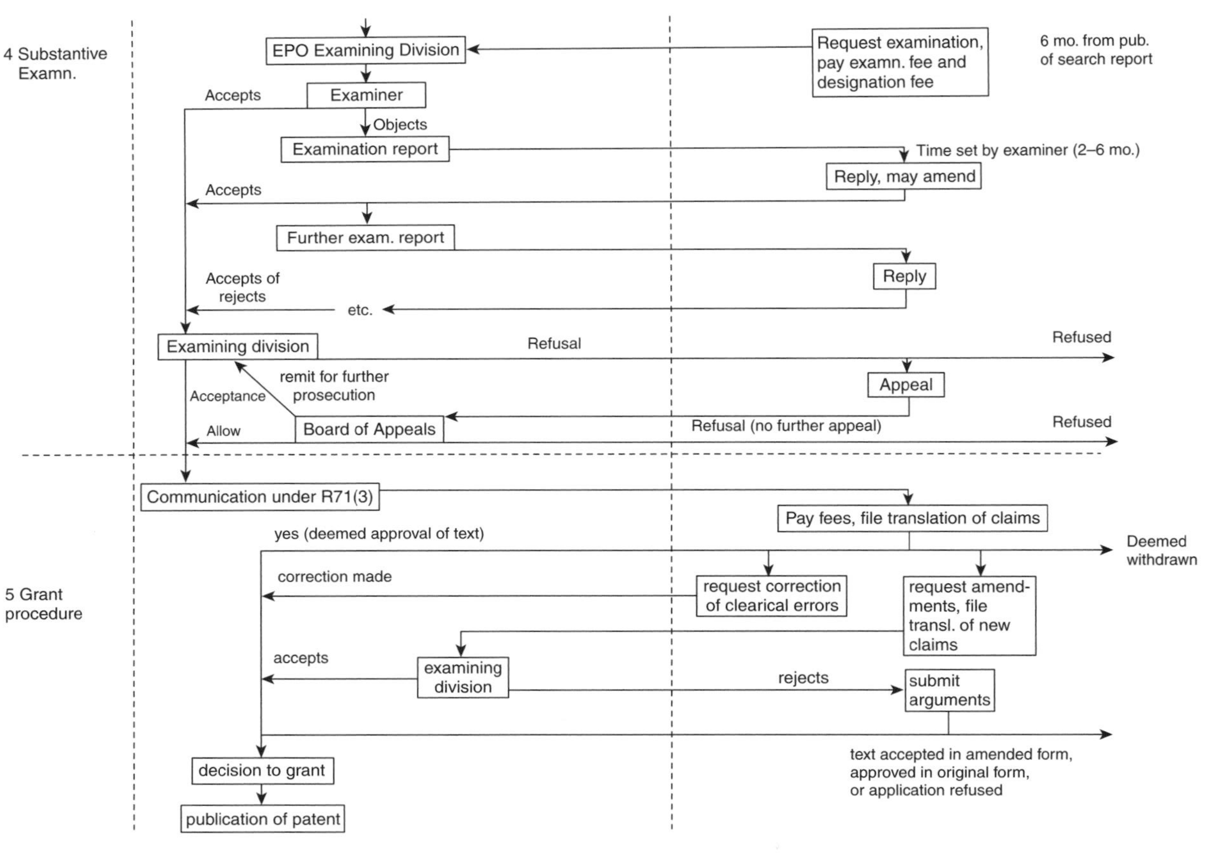

FIG. 7. European patent office procedure

A contracting state may require that a European patent does not become effective in the country unless a translation of the full specification and claims into an official language of that country is provided.[25] Because all contracting states except Luxembourg and Monaco insisted on this requirement, this meant that the expense of translation associated with foreign filing was not avoided by the European procedure, but only postponed. The question of the translation of the specification and the associated costs is a hotly disputed one. On the one hand, there is certainly an argument that individuals or small companies in each country have a right to be able to read in their own language a document that may limit their freedom of action; on the other, the major users of the European patent system are tired of paying large sums of money to produce translations that are never read by anyone. Not only are the translation costs themselves burdensome, but most national patent offices also charge a fee for publication of the translation, which, for a 50-page specification, ranged in 2004 from €10 in Italy to an outrageous €1,241 in Austria.

The situation has now been improved by the entry into force of the London Agreement (see p. 36). Those EPC states that have ratified the London Agreement and which have an official language that is one of the three languages of the EPO (that is, France, Germany, Liechtenstein, Luxembourg, Monaco, Switzerland, and the UK) do not require any translation to be made when the European patent enters the national phase. Latvia, Lithuania, and Slovenia require only that the claims be translated into the national language; Croatia, Denmark, Iceland, the Netherlands, and Sweden additionally require a translation of the specification into English, unless the European patent is published in English. This is certainly a major improvement, but there are still many EPC states requiring translation of the entire specification into the local language, including Italy, Spain, Greece, and Portugal.

It has been the experience of most patent practitioners that the EPO is very flexible in its formal requirements and does its best to help the applicant to overcome any procedural difficulties. This is sometimes carried to extremes, as in the case in which the Legal Board of Appeal (LBA) held that, despite the clear wording of the then current version of Rule 25 that a divisional application may only be filed up to the time of approval of the text, it was permissible to file a divisional at any time up to grant of the parent case.[26] The EBA[27] had to point out that Rule 25 did in fact mean what it said, although the Rule has subsequently been changed.

Opposition

When a European patent is granted by the Examining Division, the patentee has still to face a further hurdle: the European opposition procedure. Within nine months of the publication of notice of grant, third parties may oppose the patent on the grounds that the subject matter was not patentable (that is, that it was not novel,

[25] Article 65(1), EPC.

[26] J 11/91 and J 16/91 (consolidated) (OJ 1994, 28) *DOW/Deadline for filing divisional application.*

[27] G 10/92 (OJ 1994, 663) *Divisional application.*

lacked an inventive step, or was not capable of industrial application), that the disclosure was insufficient, or that new matter was added during prosecution. Oppositions are heard by a three-member Opposition Division, one of whom may have been a member of the Examining Division, and appeals lie to the Technical Board of Appeals (TBAs), with the possibility of a point of law being referred to the EBA, but no further. The opposition may result in the patent being upheld unchanged, amended (in which case, the amended specification is republished), or revoked. Revocation or amendment normally applies to all designated countries, unless the application was filed before 13 December 2007 and the successful ground of opposition was lack of novelty over an earlier unpublished application that did not designate all countries designated in the opposed patent. More details of EPO opposition procedure are given in Chapter 10.

It was originally supposed that, in the EPO, where a single opposition could knock out a patent in a number of countries simultaneously, the opposition rate would be even higher than in Germany; a level of 25 per cent was predicted. In fact, the percentage of European patents that are opposed is, at present, only just over 5 per cent overall, although in the pharmaceutical field it is significantly higher than this. Approximately 33 per cent of oppositions lead to revocation of the patent; in the remaining two-thirds of the cases, either the opposition is rejected or the patent is maintained in amended form. It might be thought that because as few as 2 per cent of granted patents are revoked in opposition proceedings, opposition is not an important issue, but this is not the case. It must be remembered that it is actually, or potentially, commercially important patents that tend to be opposed, and that whereas an unsuccessful opponent may later attack validity in the national courts, a patentee whose patent is finally revoked in opposition proceedings has lost all rights permanently in all EPC countries.

Even if the patentee is ultimately successful, opposition proceedings are almost always very slow: it will normally take five years to get a second-instance decision and it is not unknown for this period to be twice as long. During this time, the patentee's rights are uncertain. He may sue an infringer in the national courts, but in Germany, for example, the *Landgericht,* used to having validity determined in a different tribunal, will usually stay proceedings until the opposition is resolved. The UK Courts will not normally stay proceedings in view of the long delays involved and may revoke a European patent (UK) even while the opposition is pending. In one case, the argument that the courts had no power to do so was rejected by the Court of Appeal, and the patent was duly revoked.[28] The existence of a pending opposition will certainly, however, make it more difficult to obtain an interlocutory injunction, especially if the patent has been revoked at first instance and this decision is on appeal.

[28] *Beloit v. Valmet* [1996] FSR 718 (CA).

Added to the problems of uncertainty due to delay are the concerns expressed by some patentees that the standards of evidence in opposition proceedings are not strict enough. For example, it is said that dubious evidence of prior use by the opponent is sometimes accepted at face value without having to be given in sworn statements or being tested in cross-examination.[29]

Appeals in the European Patent Office

Legal Basis

It was accepted from the beginning of discussions on the EPC that there must be a second instance to hear appeals from the EPO. The question was whether this should be an international court or a body within the EPO. This gave problems to Germany, since the German Constitution requires the possibility of review of all administrative decisions by a court; it was for this reason that the *Bundespatentgericht* (BpatG, or Federal Patent Court) had to be set up in the 1960s. The problem is solved by the provisions that the Boards of Appeal always have at least one legally qualified member, and that they are independent and not subject to instructions from the EPO. The EPC states[30] that Board members are appointed for five years and can be removed only by the Administrative Council on a proposal from the EBA. If they are not reappointed, the service regulations state that they must be employed at the same salary elsewhere in the EPO. Members of the Board may not be members of the first-instance Divisions. They are bound only by the EPC, which means that they can overrule the guidelines. The Board can also in effect declare Implementing Regulations to be void as in conflict with the EPC—thus a Board held that Rule 54, in its original form, could not be reconciled with the EPC and the Rule was subsequently changed.[31]

The EPO has both Legal and Technical Boards of Appeal (LBAs and TBAs), although these terms are not used in the EPC. Any Board the members of which are all legally qualified constitutes an LBA; all others are TBAs. A TBA normally consists of one legally qualified and two technically qualified members. All legally qualified Board of Appeal members are members of the LBA and also of all TBAs. There is also an EBA, which decides points of law referred by Board of Appeal and gives opinions on points of law referred to it by the EPO President in the event of conflicting decisions of different Boards; it does not, however, act as a court of third instance.

Unlike the procedure in the UK or US patent offices, there is no further appeal from the EPO to the courts, for example to the European Court of Justice (ECJ), so

[29] P.B. Archer, letters in [1996] *CIPA* 678 and [1997] *CIPA* 73.

[30] Article 23, EPC.

[31] J 22/86 (OJ 1987, 280) *MEDICAL BIOLOGICAL/Disapproval.*

any decision of a Board of Appeal is final.[32] Under EPC 2000, there is the possibility of a petition to the EBA for review of a Board of Appeal decision in exceptional circumstances: for example if a member of the Board should have been excluded; if there has been a violation of the right to be heard, or other fundamental procedural defect; or if a criminal act may have had an impact on the decision.[33] By mid-2009, a number of such petitions had been heard, but in no case did the EBA grant review of the decision.

The question has been raised whether the EPO is in conformity with Article 32 of the TRIPs Agreement, which requires judicial review of any decision to revoke a patent, and doubts have been expressed as to whether the Boards of Appeal are really judicial in nature (the Opposition Divisions are clearly administrative). In one case in the UK, a patentee whose European patent had been revoked in opposition and who had lost his appeal tried to contest the effect of this decision in the English courts.[34] The judge held that the EPO Boards of Appeal were judicial in nature, their members being judges in all but name, and that there was no violation of TRIPs. Furthermore, the matter was so clear that no reference to the ECJ would be made.

Proposals are currently being discussed to clarify the independence of the Boards of Appeal and Enlarged Board of Appeal by separating them from the European Patent Office and constituting them as a third body within the European Patent Organization, which would then consist of the Administrative Council, the European Patent Office, and the Court of Appeals of the European Patent Organization, corresponding to the legislative, executive, and judicial branches of certain governments. This would of course require further amendments to the EPC and it is not clear when, if ever, this will happen.

Article 32 of TRIPs seems to provide that there must always be a second instance in proceedings in which validity is in issue, even if the first instance is itself judicial. It has been pointed out by some commentators that the strict wording goes further than this: since a patent may be upheld by the lower court and revoked on appeal, there would then have to be a third instance. This would lead to the absurd situation in which the House of Lords could not revoke a patent if the Patents Court and the Court of Appeal had found it valid; as the UK Court of Appeal has held, this cannot be what was intended.[35] All that seems to be necessary is that the EPO Boards of Appeal should generally remit to the Opposition Division cases in which new grounds or evidence of invalidity are admitted at the appeal stage, so as to ensure that the case is fully heard at two instances.

[32] G 1/97 (OJ 2000, 322) *ETA*.

[33] Article 112a, EPC 2000.

[34] *Lenzing's Application* [1997] RPC 245 (Pat Ct).

[35] *Pozzoli v. BDMO* [2007] EWCA (Civ) 588.

Substantive Law on Appeals

Appeals may only be made from decisions and not from any finding or opinion that is not a decision. Appeal suspends the operation of the first-instance decision, so that a patent revoked by the Opposition Division can still be enforced as long as the decision is on appeal—although, in such circumstances, the national courts might stay the case. An interim decision cannot be appealed separately from the final decision, except when the decision specifically allows it. The most common example is when in an opposition the Opposition Division decides to grant the patent in amended form. The proceedings are not terminated until the amended patent is granted, by which time, the patentee must have provided translations of the new claims and paid a printing fee. To save this having to be repeated after appeal, separate appeal is allowed. A decision on the admissibility of an opposition might also be appealed separately. But if a separate appeal is allowed but not made, the point cannot be raised in the appeal on the final decision. Clearly, if the point is decided in a separate appeal, it cannot be raised a second time.

Any party to the proceedings who is 'adversely affected' by the decision may appeal. A party is adversely affected unless that party's requests have been met in full. All other parties to the proceedings are parties to the appeal as of right, whether or not they are adversely affected. Thus when there are two opponents and one appeals, the other is automatically a party and does not need to pay the appeal fee. But if the party who did pay the appeal fee withdraws, the other opponent has no right to continue on his own.[36] If the opponent as sole appellant withdraws the appeal, the proceedings are automatically terminated, even if the Board of Appeal considers the patent invalid and even if the patentee wishes to continue.[37]

Two decisions of the EBA in 1994 established the principle that, in appeals from the Opposition Division, a party cannot put himself in a worse position by appealing (known as the principle of *reformatio in peius*).[38] Thus if, in an opposition with a single opponent, the Opposition Division decides to maintain the patent in a narrower scope, both parties are adversely affected and may appeal. If either party appeals, the other is a party to the proceedings. But:

(a) if only the patentee appeals, the patent may be maintained in its original scope, or as limited by the Opposition Division, but cannot be revoked entirely; and
(b) if only the opponent appeals, the patent may be revoked or maintained as limited by the Opposition Division, but cannot be restored to its original scope.

Of course, if both parties appeal, all outcomes are possible.

36 G 2/91 (OJ 1992, 206) *KROHNE/Appeal fees.*

37 G 8/93 (OJ 1994, 887) *SERWANE/Withdrawal of opposition.*

38 G 9/92, G 4/93 (both at OJ 1994, 875) *BMW(MOTOROLA)/Non-appealing party.*

Appeals Procedure

Notice of appeal must be filed, together with the appeal fee, within two months of the notification of the appealed decision, and must contain a statement identifying the decision and the extent to which its amendment or cancellation is requested. No details need to be given at this stage. The next stage is that a complete statement of grounds must be filed within four months of the notification of the decision. A separate statement can be dispensed with if the grounds are stated in or can be inferred from the notice of appeal. When the appeal relates to technical matters, however, a real statement of grounds, with facts and arguments, is essential. There have been many cases in which appeals were held inadmissible because of inadequate statements of grounds.[39]

If the appeal is filed in time and the fee is paid, the appeal goes to the first-instance department the decision of which is contested. It first considers whether the appeal is admissible and, if it considers it both admissible and well founded, it must rectify its decision. This does not apply if another party is involved (for example in an opposition). If decision is not rectified within three months, the appeal is passed on to the appropriate Board of Appeal without comment on its merits. It was, at one time, the practice that revision would be granted only if the decision appealed was wrong. Later practice was that revision should be granted if the appellant offered amendments that would clearly remove all objections to grant, for example by limiting the claims in a way previously requested by the Examining Division, but so far resisted by the applicant. The present position is that revision should be granted if the reply deals with the objections that based the decision, even if other objections remain.[40]

In the next step, the Board of Appeal checks again if the appeal is considered admissible and, if so, it proceeds to consider whether it is allowable. The Board then decides the appeal. It may do anything that the first-instance department could have done, or it may remit the case to that department, which shall be bound by the Board's decision. Thus if the Board has issued a decision that certain claims are unpatentable, this issue cannot be reopened either by the Examining Division or by the Board of Appeal in any subsequent appeal proceedings. The Board must normally remit the case to the first instance if fundamental deficiencies are apparent in the first-instance proceedings. If the Board disagrees with an earlier Board decision, it must give its reasons for so doing (unless, by so doing, it is following a ruling of the EBA).[41] Although in the event of disagreement within the Board, a vote is taken to determine the decision, decisions are usually given, both orally and in writing, as if they were unanimous. Unlike the procedure in the Board of Patent

[39] For example, T 220/83 (OJ 1986, 249) *HULS/Grounds for appeal*; T 145/88 (OJ 1991, 251) *NICOLON/Statement of grounds*.

[40] T 139/87 (OJ 1990, 68) *BENDIX/Governor valve*.

[41] Article 15, Rules of Procedure for the Boards of Appeal.

Appeals of the US Patent and Trademark Office (USPTO), dissenting opinions are not normally made known. There have, however, been two decisions of the EBA in which a separate minority opinion was set out.[42]

The appeal fee may be reimbursed if two conditions are met: firstly, the appellant must win, whether by interlocutory revision or before the Board of Appeal; and secondly, there must have been a substantial procedural violation on the part of the first-instance Division. This may happen, for example, if the first-instance decision is not adequately reasoned.[43] The fee is not refunded if grounds are filed too late, or if an appeal is withdrawn before the statement of grounds is filed. This is to discourage the filing of 'precautionary appeals'.[44]

[42] G 3/92 (OJ 1994, 607) *LATCHWAYS/Unlawful applicant*; G 9/92 (see fn. 38 above).
[43] T 278/00 (OJ 2003, 546) *ELI LILLY/Naphthyl compounds.*
[44] T 89/84 (OJ 1984, 562) *TORRINGTON/Reimbursement of appeal fee.*

8

OBTAINING A GRANTED PATENT: NATIONAL PROCEDURES

For how do I hold thee but by thy granting?
And for that riches where is my deserving?
The cause of this fair gift is me is wanting,
And so my patent back again is swerving.

William Shakespeare, 'Sonnet 87'

Types of Examination

Formal

Once a patent application has been filed, it must at some stage be examined. Every country that grants patents has some type of examination, however rudimentary.

At the bottom of the scale are countries in which the only examination is a purely formal one, to check that the papers are in order and the fee has been paid. Belgium, for example, had such a registration system until very recently; now, a search report is required, but even so, a patent will be automatically granted for anything for which an application is made and all questions of validity are left to the courts.

Non-Unity

Some countries, while having essentially a registration system, do at least check that the application describes an invention of a type that is legally patentable, and rather more go one step further and examine for unity of invention. It is a common feature of nearly all patent systems that a single patent should not be granted for more than one invention. This is not so serious a situation as two patents being granted for the same invention, but is objected to perhaps because the patent office will receive only one set of fees instead of two or more. The interpretation of what constitutes a single invention varies considerably from country to country. Typically, however, objections may be raised if in the chemical field the applicant attempts to claim end-products and intermediates in the same application, or claims a new use of a group of compounds at the same time as claiming per se a subgroup of compounds that are themselves new.

The European Patent Convention (EPC) and the UK Patents Act 1977 allow more than one invention to be present in an application provided that they are linked together to form part of the same inventive concept.[1] Once a patent has been granted, its validity cannot be attacked on the grounds that it contains more than one invention. In the Patents Act 1997, this is stated explicitly;[2] in the EPC, it follows from its absence from the grounds of opposition[3] and national revocation.[4]

When two or more distinct inventions are present and a patent office raises a well-founded objection of non-unity of invention, the remedy is for the applicant to file a divisional application. This is a new application, which may have the same text as the original ('parent') application or which may have a shortened text, but which must not include any new matter—that is, anything that was not present in the original specification. Only the claims will be changed, so as to claim the matter that is being 'divided out'. The requirements as to when the divisional application must be filed vary: in some countries, a fixed term is set from the date of the official objection of non-unity; in others, a divisional application can be filed at any time before the parent application is allowed or granted. Whenever it is filed, however, and provided that no new matter is introduced, it will be treated as if it had the same filing date as the parent application, so that publication of the invention after the filing date of the parent and before that of the divisional will not affect validity.

[1] Article 82, EPC; s. 17(6), PA 1977.
[2] Section 26, PA 1977.
[3] Article 100, EPC.
[4] Article 138, EPC.

Novelty

Approximately half of the countries that grant patents carry out some form of substantive novelty examination, in which the prior art, or part of it, is searched for anything that could be an anticipation of the invention. Some patent offices, for example the European Patent Office (EPO) and the German Patent Office, carry out a search entirely by themselves; others may require the applicant to notify them of prior art that has been cited during examination in other specified countries, and even to provide copies of official actions and cited references. From 1 July 2004 the UK Patent Office may issue a 'disclosure request' when issuing search reports, inviting applicants to provide copies or details of search reports that other patent offices have issued on any corresponding applications. This information is supposed to be provided no later than when responding to the first examination report or within two months of the examiner reporting that the application complies with the Act. It does not appear, however, that non-compliance with the request would have any serious consequences. A further group of countries—France, Turkey, and Ireland, for example—do not carry out searches themselves, but, for national applications, require a search to be carried out by an international searching office such as the EPO or accept the results of a novelty search carried out in another country. As mentioned in Chapter 3, there are also countries in which although the law provides for a novelty examination, the lack of technically trained examiners means that in practice such searches cannot be carried out. Countries in this category are being encouraged by the EPO and other bodies to grant patents based on corresponding patents granted, for example in the EPO, the US Patent and Trademark Office (USPTO), or the Japanese Patent Office (JPO). Alternatively, accession to the Patent Cooperation Treaty (PCT) will enable them to grant national patents based on a search and international preliminary report on patentability, made by a competent authority.

An interesting approach is that of Singapore, which joined the PCT in 1995 and enacted its own patent law (previously, British patents were registered). Although Singapore is technically advanced in many fields, it was rightly felt that it would be a waste of technically trained personnel to duplicate searches of patent applications. For Singapore national applications, searches are carried out either by the Austrian or Australian patent office, while for PCT applications entering the national phase in Singapore, a patent is automatically granted if an international preliminary report on patentability (IPRP) has been made, even if this is negative. The applicant may amend the patent application before grant to correct deficiencies noted in the IPRP, but is not obliged to do so. It is left up to the patentee whether he wants to have a patent that the IPRP says lacks novelty and whether he wants to try to enforce it before the Singapore courts.

This is not unlike the system in the Netherlands, where if the applicant wants a national patent, a novelty search has to be carried out, even though no substantive examination is based upon it. If the fee is paid, the patent is then granted with the search report attached, even if the search report shows that the invention is old.

Previously, if no search report was paid for, a short-term patent would be granted with a term of only six years. Dutch short-term patents were, however, phased out in 2008.

The USA is in a special position as regards prior art searches. Although a number of countries may require the applicant to state the prior art cited by certain other patent offices, there is usually no obligation to give anything more than is asked for; this is not so in the USA. Even before 1977, it was clear that the applicant had to bring to the notice of the USPTO all prior art relevant to novelty or inventive step of which he was aware; since 1977, this requirement has been codified in the rules. Deliberate concealment of relevant prior art is regarded as 'inequitable conduct' and will result in the patent granted as a result of such conduct being incurably unenforceable, as well as the possibility of the attorney responsible being disbarred and the patentee being sued for anti-trust violations (see p. 499). An information disclosure statement (IDS) should be filed, setting out in standard form all of the prior art known to the inventor and the attorney, preferably within three months of the US filing date or before the first substantive official action issues, unless the filing is a provisional application or a continuation application for which the prior art citations are carried over from the parent case. Later filing of the IDS may be done on the payment of fees. Although there is no actual compulsion to file an IDS, the consequences of failing to do so could be very serious. The obligation to notify does not end there, but continues up to the date of grant of the patent.

Although the EPO used to keep searching and examination separate from each other, it has now combined the two (see p. 118). In contrast, the JPO now outsources searching to commercial search companies and the USPTO is considering doing the same. It is feared that this will result in increased costs to the applicant and possibly lower the quality of the searches.

Inventive Step

Only a relatively small number of countries go beyond a novelty examination and look into the question of whether or not the invention is obvious over the prior art. Among those that do are Germany, Japan, the Scandinavian countries, the USA, and the UK. The EPO also examines for the presence of an inventive step. The degree of difficulty in overcoming obviousness objections varies. It is usually relatively easy to persuade the UK Patent Office that an invention is non-obvious; the EPO is somewhat more troublesome; and countries such as Germany and the USA present more serious difficulties (see Chapter 19).

Deferred Examination

The idea that the applicant could voluntarily postpone examination of the application for a period of years originated in the Netherlands as a method of relieving some of the pressure upon the overworked Dutch Patent Office. The idea was that postponing examination would not only relieve the immediate overload, but would also have a long-term effect in that many applications would never need to be examined because the applicants would have lost interest in them by the date at which they would finally have to decide whether to request examination or abandon. The system was introduced in 1964, and was followed by Germany in 1968 and Japan in 1971. Some other countries have systems with similar features.

The Dutch system is no longer in effect since the Dutch Patent Office has given up substantive examination altogether, but the German and Japanese systems are still in place. In Japan, for example, the application is received, subjected to a formal examination only, and published as filed 18 months from the filing date, or from the priority date if priority is claimed (known as the *kokai* publication). No further action is taken, and no fees are payable, until the applicant or a third party requests examination and pays the examination fee, which must be done within three years from filing or the application will lapse. For applications filed before 1 October 2001, examination could be deferred for up to seven years. The German system is similar to that of Japan, except that examination is in two stages: a prior art search, followed by a substantive examination in which an examiner applies the cited prior art to determine whether or not the invention is patentable. These two stages may be carried out quite separately, so that an applicant may request a search upon filing the application and then wait seven years before requesting substantive examination of it. Unlike Japan, annual maintenance fees are required in Germany to keep the application in being during this seven-year period. In Canada, a request for examination may also be deferred for up to five years from the filing date.

When a patent is granted after deferred examination, there may well be only ten years or less of the 20-year term of the patent remaining, because this is counted from the filing date and not from the date of grant. Nevertheless, many inventions are not commercialized, or imitated, until after several years from the time that they were made and it is only then that a granted patent is needed. In the pharmaceutical industry, in particular, in which registration of a new drug often takes seven to ten years, it used to be standard practice for applicants to defer examination for as long as possible. If in this time development of the drug was abandoned, then not only was the patent office saved the trouble of examining the case, but the applicant was also saved the cost of the examination fee and the effort of prosecuting the application in that country. This has changed somewhat in view of the Japanese law on extension of term, which may penalize an applicant whose patent is granted only after clinical testing of the product has begun in Japan (see p. 172).

Under the EPC,[5] examination must be requested no later than six months after the publication of the search report, that is, about 22 months after the first filing date.

[5] Article 94(2), EPC.

application as a DOS; then, publication of the accepted application as a *Deutsche Auslegeschrift* (DAS); and finally, after the opposition stage, as a *Deutsche Patentschrift* (DPS). From the beginning of 1981, the procedure was simplified so that the DAS stage is omitted and belated opposition is possible after publication of the DPS. Similarly, in Japan, there were three publications: the *kokai* and the *kokoku* (corresponding to DOS and DAS) being published before the patent was granted, and opposition being possible after publication of the *kokoku*. Japan changed its law in 1996 so as to provide for post-grant opposition and in 2004 oppositions were incorporated into the existing system of 'invalidation trials' before the JPO. From the point of view of the patent owner, the great advantage of post-grant opposition is that, while the opposition or invalidation proceedings are going on, he has a granted patent that, in theory at least, can be enforced in the courts against an infringer. Post-grant opposition or invalidation trial at any time after grant is now becoming the general rule, except in some countries where the law is still based on the old UK law, such as Australia, New Zealand, and Israel. In some Latin American countries, what is called 'opposition' is really an intervention by a third party in the examination process after publication, but before substantive examination begins. India has the misfortune (from the point of view of the patent applicant) of having both pre-grant and post-grant opposition proceedings.

In the UK, the 1977 Act allows revocation by the Patent Office at any time after grant, the grounds now being the same as those on which the courts can order revocation.[8] The number of oppositions in the UK in the last few years before the Patents Act 1977 came into force had sunk considerably and involved less than 1 per cent of the applications that were published. The reason for this was probably that the Patent Office and the Patents Appeal Tribunal (PAT)—that is, a High Court judge to whom appeals from the Patent Office could be made—were strongly inclined to give the applicant the benefit of the doubt in any opposition proceedings, the theory being that the patent, if granted, could always be attacked further in the High Court. In practice, this meant that it was practically impossible to knock out a patent completely in opposition (or belated opposition) proceedings; usually all that happened was that the applicant made some necessary amendments and ended up with a stronger patent than he would otherwise have had.

Under the 1977 Act, the number of belated oppositions (that is, revocation actions in the Patent Office) has shrunk even more rapidly. In recent years, there have typically been only between five and ten such proceedings per year, resulting in between one and four revocations.

In contrast, in Germany there have always been a relatively high percentage of oppositions. Approximately 20 per cent of all granted applications were opposed in the 1970s and about 10 per cent in 1995. Since January 2002, oppositions have been

[8] Section 72(1), PA 1977.

heard by the *Bundespatentgericht* (BPatG, or Federal Patent Court) rather than by the German Patent Office. This was supposed to be for a three-year period only, in order to relieve the backlog of oppositions, but was continued in 2005 and remains the situation in 2009. Perhaps as a result of this change, the percentage of patents opposed dropped to only 4 per cent in 2003.

The 1980 amendment to the US patent law allowed a third party to ask for re-examination of a granted US patent, which is also a form of belated opposition proceedings. More recent revisions to the law have created *inter partes* re-examination, in which the third party can play a more active role (see p. 217).

Procedure in the UK

Filing

In order to establish a date of filing for a patent application in the UK, the applicant must file at the UK Patent Office a description of the invention, and must indicate the identity of the applicant and that a patent is requested.[9] Filing may be by direct delivery of the documents to the office, or by post or fax. Electronic filing is also possible, using the EPOonline software and a 'smart card'. If the description is not accompanied by the standard application form (Form 1), this must be supplied subsequently. The description may be a very simple one and need not contain any claims; it may therefore be like a provisional specification under the old law. The applicant receives a filing receipt, which states the filing date and the application number. The application number is a five-digit number following the last two digits of the year; thus the first application filed in 2009 has the number '0900001'. The application fee must be paid within two months from the filing date, or 12 months from the priority date, whichever is later: that is, as in the EPO, a filing date may be obtained before any fees are paid.

Preliminary Examination and Search

Within the same time limit of two months from the filing date, or 12 months from the priority date, the applicant must supply claims and an abstract, and must request, and pay the fee for, the preliminary examination and search, if he wishes to continue with the application. This is done by filing Patent Office Form 9 with a fee of £130 (as of 2009). The applicant may also file a new application claiming priority from the first one and proceed with that instead of the earlier one, thereby taking the opportunity to add new matter and broaden the scope of the original disclosure. There is no need to wait 12 months before requesting preliminary examination and search; this may be done on filing so that the applicant can consider whether or not to foreign file on the basis of the search report. If he then files a second UK application

[9] Section 15(1), PA 1977.

claiming priority from the first, the new application must also have a preliminary examination and search report; the closely related search that has already been made will not normally be accepted as sufficient. For an application claiming priority from an earlier filing in the UK or elsewhere, the time limit for paying the search and examination fee is 12 months from the priority date or two months from the filing date, whichever is later.

The preliminary examination deals with various formal matters and also with unity of invention; if the examiner considers that two or more inventions are present, he may search only one of them, or may require additional search fees to be paid for a complete search. The search report provides a list of prior art considered relevant, divided into the categories of prior art published before the filing date and UK applications of earlier filing date that were not prior publications, but which will be part of the state of the art under the whole contents approach. At this stage, no comments on the cited documents are normally made by the examiner. From 1 April 2004, however, selected applications receive an 'examination opinion', together with the search report. This is intended to highlight major examination issues at an early stage and to encourage applicants to amend the application before the first official action is issued. The Patent Office is particularly targeting very broad speculative claims and complex independent claims with overlapping scope relating to a number of separate inventions. About one month after the search report issues, the applicant is informed of the date set for early publication of the application, which will be as soon as feasible after 18 months from the filing date. At the same time, the application is allocated a serial number that will in due course be the patent number; this is a seven-digit number, the first digit of which is '2'.

A further requirement before publication is that, unless all inventors are also applicants, a statement of inventorship must be submitted no later than 16 months from the filing date. A copy of this is then sent by the Patent Office to all of the named inventors in British-originating applications who are not themselves applicants.[10]

If the application originally claimed priority from one or more earlier UK or foreign applications, then the time limits for supplying abstract, claims, and statement of inventorship, and for requesting preliminary examination and search, run from the earliest priority date, except that the abstract and claims may be supplied up to two months from the filing date. In practice, applications filed with a claim to priority will normally contain an abstract and claims on filing, and the statement of inventorship and request for preliminary examination will be filed at the same time as the application. Publication occurs about 18 months after the filing date of the first application from which priority is claimed.

[10] Section 13(2), PA 1977; r. 14, PR 1995.

Publication

After receiving the search report and up to the receipt of the first examination report, the applicant is free to amend the specification as many times as he wants[11] (for example to reduce the scope so as to distinguish the invention from the cited prior art) or, if the invention turns out to be totally anticipated, he may decide to abandon it altogether. If the early publication of the specification should include the amendment, however, or if publication should be avoided altogether, the applicant must act quickly if, as is generally the case with applications claiming priority, the search was requested towards the end of the possible period. Once the letter announcing the date of publication has been sent—and it is generally sent within a month or so of the issue of the search report—it is difficult, if not impossible, to have amendments made to the published specification or to prevent publication by abandonment.

The specification is then printed as a published UK patent application, bearing the seven-digit serial number, followed by an 'A'. Copies of the specification may then be purchased from the Patent Office and the file is open to public inspection. The published application also lists the prior art cited in the search report. An example of the front page of such an application is shown in Fig. 8. Publication is not purely electronic and paper copies may be requested.

Once the application has been published, it is of course prior art against any application filed later than its publication date and also against applications filed after its filing date, under the whole contents approach. Furthermore the applicant has certain rights against infringers as from the publication date.

At any time after the publication of the application until the applicant receives a letter of notification of grant, any person may send to the Patent Office observations on the patentability of the invention.[12] The observations, which may be made anonymously, will be considered by the examiner during the substantive examination and a copy will be sent to the applicant, who can comment upon them. The person making the observations does not become a party to the proceedings, so that he cannot make any further comments and is not informed officially what use, if any, the examiner makes of his contribution. Such a person can, however, check this for himself, since, after the publication date, the file of the application is open to public inspection at the Patent Office.

Substantive Examination

The next stage in the proceedings is that the applicant must, within six months of the publication date, file a request for substantive examination on Form 10 and pay a further fee (£70 in 2009). If the request is not made and the fee not paid in that time (or within one additional month on payment of a substantial additional fee),[13] the application is treated as withdrawn and can be reinstated only with difficulty.

[11] Rule 36(3), PR 1995.

[12] Section 21, PA 1977.

[13] Rule 110(3), PR 1995.

(12) UK Patent Application (19) GB (11) 2 021 437 A

(21) Application No **7911926**
(22) Date of filing **5 Apr 1979**
(23) Claims filed **5 Apr 1979**
(30) Priority data
(31) **14321/78**
(32) **12 Apr 1978**
(33) **United Kingdom (GB)**
(43) Application published
5 Dec 1979
(51) **INT CL²**
B01J 1/00
(52) Domestic classification
B1X 20
(56) Documents cited
None
(58) Field of search
B1X
F4S
F4X
(71) Applicants
Sandoz Ltd., 35 Lichtstrasse, CH-4002 Basle, Switzerland
(72) Inventors
Ludwig Hub, Tomas Kupr
(74) Agents
B. A. Yorke & Co

(54) **Reactor simulator**

(57) A reactor simulator comprises a reactor vessel 1, a heat exchange liquid jacket 2a around the vessel through which heat exchange liquid may be circulated and means 3 for adjusting the level X—X of heat exchange liquid in the heat exchange jacket, the heat exchange area and hence the heating/cooling capacity of the reactor varying accordingly to enable the ratio of heat/cooling capacity of a large scale reactor to the corresponding reactor volume to be accurately simulated in a small scale reactor.

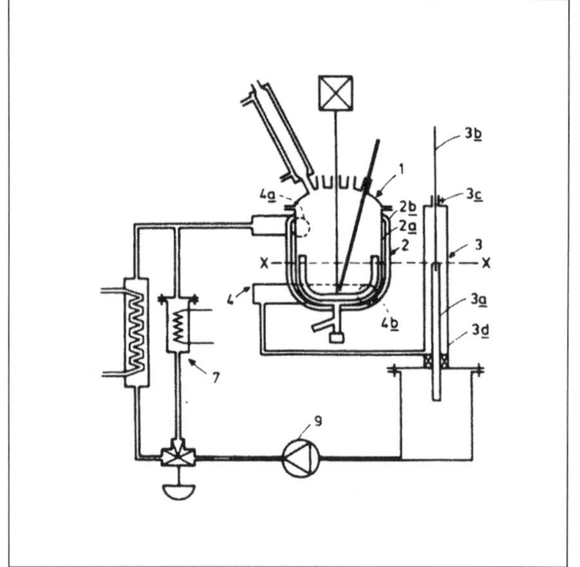

GB 2 021 437 A

FIG. 8. Front page of a published British patent application under the Patents Act 1977

To avoid the serious consequences of missing this deadline, it may be advisable to request substantive examination at the time that the request for preliminary examination and search is made, that is, on filing, in the case of an application claiming priority. If these two requests are made simultaneously, the practice now is that the office will carry out a single combined search and examination. This has the advantage, if an early grant is desirable, that the procedure is fast; the results of the combined search and examination should be available within four months of filing, and the applicant's response will be dealt with without delay. If rapid grant is not particularly required, however, a specific request may be made on Form 10 to keep the search and substantive examination separate, even if both Forms 9 and 10 are filed at the same time. In this case, as when the forms are filed separately, the applicant may decide after receiving the search report not to proceed with substantive examination. The fee for substantive examination will then be refunded, provided that the examiner has not already begun work.

The examiner carries out the substantive examination based upon the prior art cited in the search report, but not necessarily limited to this. He then issues an official letter in which is listed whatever objections there may be to the grant of a patent on the application. These may range from the serious (for example an allegation that the invention is not new or is obvious in view of the prior art), to the trivial (for example a requirement that a registered trademark used in the description be identified as such). In the fairly numerous cases in which the search report cited no relevant prior art, the official letter will normally concern itself only with formal matters, and with the relationship of the claims to each other and to the description. If an examination opinion has been issued with the search report and the applicant has not filed any amendment, the first official letter will be in the form of an 'abbreviated examination report' that essentially repeats the contents of the examination opinion and sets a two-month period for response, which if not met can lead to the refusal of the application. Abbreviated examination reports may also be issued even if there was no examination opinion, but in such cases, the term for reply will be longer.

The first official letter may occasionally raise no objections, but indicate that the application is in order for grant; in this case, however, the official letter will not be sent until at least three months after publication, to allow time for any third-party comments. Even after a positive examination report, the applicant still has one opportunity for voluntary amendment of the specification, to be made within two months of the report.

In the more usual situation in which objections have been made, the UK Patent Office fixes a term, generally of three to six months, within which the applicant must reply. There is also an overall time limit of four-and-a-half years from the first priority claimed, or from filing, if no priority was claimed, within which the application must be in order for acceptance.[14] This term is extensible by one month on

[14] Section 20(1), PA 1977; r. 34(1)(a)(i), PR 1995.

payment of a fee; further extensions are at the discretion of the Comptroller, and are granted only for good reasons. In any event, the time within which the application must be in order for acceptance cannot expire earlier than 12 months from the date of the first official action.[15] This may be important in the case of a PCT application going into the national phase in the UK already two-and-a-half years after the priority date. If the first official action were to be delayed for more than a year, the overall time limit would be very short.

In answering the official letter, the applicant may argue that the objections are not valid ones, or may amend the specification in any way necessary to meet the objections, except that care must be taken not to introduce any new matter not previously disclosed. The introduction of new matter, even if not objected to by the examiner, could lead to the granted patent being invalid. The applicant has one opportunity at this time to make further voluntary amendments not in consequence of the objections raised; afterwards, such amendments can be made only with the permission of the examiner. Minor amendments may be made by the examiner upon written request, but for any changes involving more than a few words, retyped pages must be submitted. If the examiner feels that the objections have not been fully answered by the reply to the official letter, a second official letter may be issued; if disagreement persists, the applicant may ask for a hearing in the Patent Office and has the right of appeal to the Patents Court if the Office decides against him. A further appeal to the Court of Appeals is possible if leave is granted by the Patents Court or by the Court of Appeals. The British procedure is summarized in diagrammatic form in Fig. 9.

The applicant may request accelerated search and examination, and this will be granted if there are sufficient reasons: for example threatened infringement, or the need to have a granted patent to secure investment. As from May 2009, prosecution may also be accelerated on request for applications relating to 'green technology', a somewhat vague term for inventions to protect the environment, reduce carbon emissions, or the like.

The UK Patent Office has also been involved in the 'Patent Prosecution Highway' programme (see p. 52) with the JPO since July 2007 and with the USPTO since September of that year. This is, however, of more symbolic than practical importance, because it is highly unlikely that a patent would be ready for grant in Japan or the USA before substantive examination of the corresponding UK application had even begun.

Grant

Assuming, however, that all objections are met within the four-and-a-half-year period, the applicant will be sent a letter notifying grant (the date of the letter being the administrative date of grant) and the effective date of the patent rights will be

[15] Rule 34(1)(a)(ii), PR 1995.

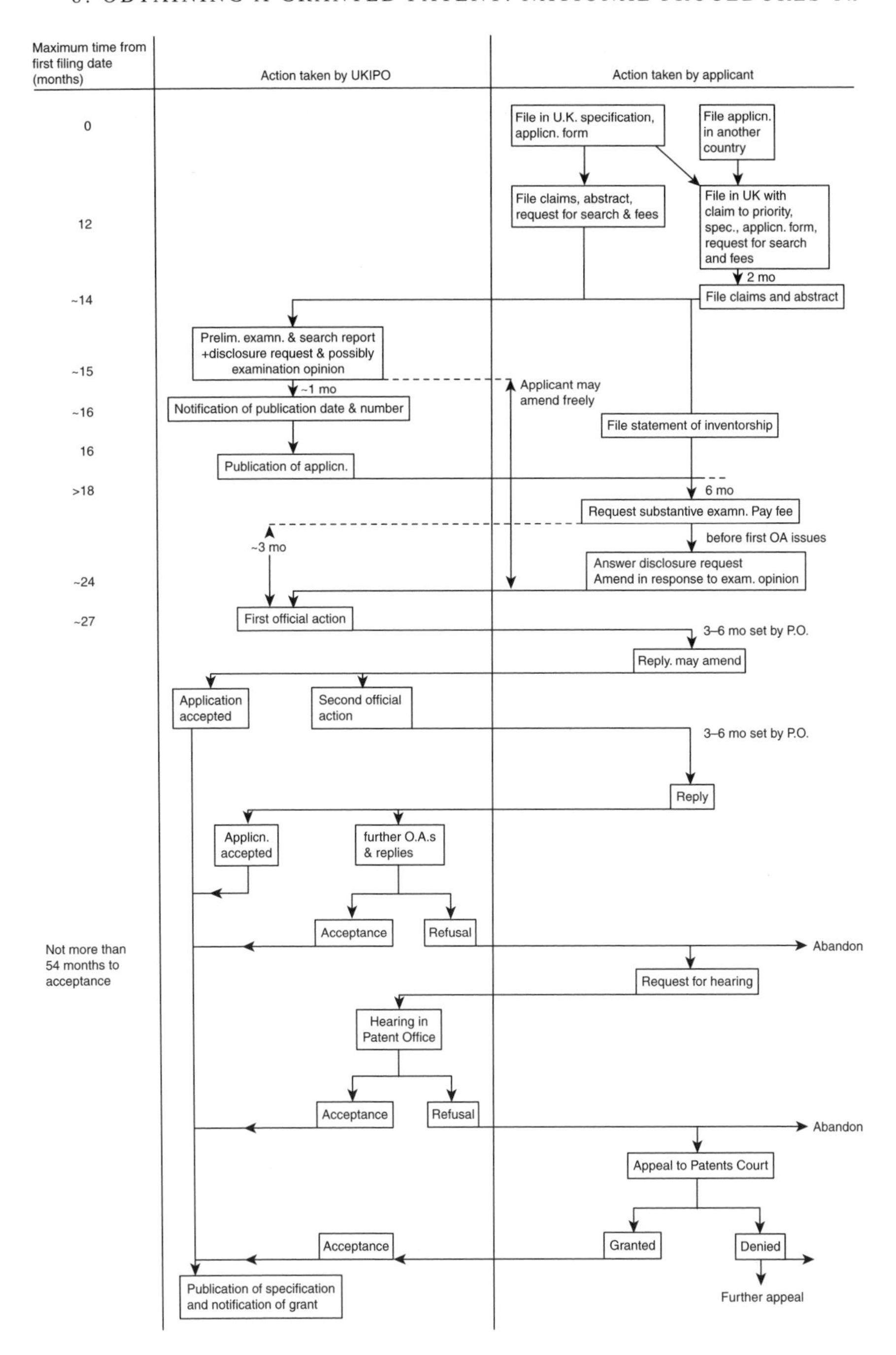

FIG. 9. British Patent Office procedure

the date of publication of the notice of grant in the *Official Journal* (*Patents*), which occurs several weeks later.

The patent specification is then published for a second time, incorporating any amendments that have been made during prosecution. This time, the text is not set in print, but is reproduced directly from the applicant's typescript by photolithography, a procedure that produces a bulkier and less attractive document, but which avoids the possibility of errors in printing or electronic scanning. There is a printed title page, headed by the seven-digit publication number followed by the letter 'B', an example of which is given as Fig. 10. The patentee receives a 'certificate of grant' stating that a patent has been granted, but there is no longer any letters patent document such as that illustrated as Fig. 1 in Chapter 1. Indeed, it would be true to say that the patent itself, as distinct from the patent specification, has only a notional existence.

Procedure in the USA

Filing

A regular patent application in the USA (as distinct from a provisional application, see p. 87) must be made by the inventor or inventors, even if the rights have been assigned to an employer. Each inventor must sign a declaration, which is normally attached to the specification and which sets out, among other things, that he believes himself to be the first inventor, and acknowledges his duty to disclose all information known to be material to the patentability of the invention. The declaration usually includes an authorization of an agent to act for the inventor before the USPTO. If the inventor does not understand English, he must sign a declaration written in a language that he does understand. It is important that the inventor reads the specification carefully and does not regard signing the declaration as a mere formality. It is now possible to file a US application without a declaration by the inventor and to file the declaration subsequently, on payment of an extra fee.

Once the specification and basic filing fee (reduced in December 2004, from \$850 to \$330, and not raised since) have been lodged at the USPTO, the application is given a filing date and an application number, and is then allocated on the basis of its subject matter to a 'technology centre', formerly known as an 'examining group'. If the declaration was not attached to the application, the applicant will be required to supply it within one month of the official notification, or two months from the filing date, whichever is later. These periods may be extended on payment of a fee.

Publication

Publication of a regular US application will take place 18 months from the first claimed priority date, unless at the time of filing (and not even one day later) a non-publication request is submitted, because no corresponding application is being

(12) UK Patent (19) GB (11) 2 038 375 B

(54) Title of invention

Improvements in or relating to dyed and shrinkproofed wool

(51) INT CL^3: **D06P 3/16**

(21) Application No
7943111

(22) Date of filing
14 Dec 1979

(30) Priority data

(31) **12834/78**

(32) **18 Dec 1978**

(31) **9366/79**

(32) **18 Oct 1979**

(33) **Switzerland (CH)**

(43) Application published
23 Jul 1980

(45) Patent published
12 May 1982

(52) Domestic classification
D1B 2F

(56) Documents cited
GB 1553811
GB 1264683
GB 837950
GB 835267
GB 1006787

(58) Field of search
D1B
B2F

(73) Proprietor
Sandoz Ltd
Lichtstrasse 35
4002 Basle
Switzerland

(72) Inventors
Oskar Ammen
Hermann Egli
Karl Zesiger

(74) Agents
B. A. Yorke & Co.,
98, The Centre, Feltham,
Middlesex, TW13 4EP

LONDON THE PATENT OFFICE

FIG. 10. Front page of a granted British Patent under the Patents Act 1977

filed in any early publication country or region. If subsequently it is decided to file in such a country or region, then the non-publication request must be withdrawn within 45 days of the foreign filing, and at the same time, a notice of the foreign filing must be made to the PTO. The publication itself is in electronic form only. An example of the first page of such an early published US application is shown as Fig. 11.

Examination

The first reaction from the examiner in charge of the case will often be a 'restriction requirement', which is an objection that the application claims more than one invention. Such restriction requirements are now usually made by telephone. If a written restriction requirement is made, the applicant is normally given 30 days in which to reply; although he may argue against the examiner's position, the applicant must at least provisionally elect one invention, as defined by one group of claims, for further prosecution. He may later decide to drop the other group or groups, or may decide to file a divisional application to cover them; however, this may be left until prosecution of the parent case is complete. In this situation, a first office action 'on the merits' of the invention will issue within three to four months of the restriction requirement.

If there has been no restriction requirement, a first office action on the merits will issue in due course. There is now a significant backlog of unexamined applications in most technology groups. The USPTO publishes each week, in the *Official Gazette*, a list of the average filing date of applications receiving a first office action in the previous three months. A survey in November 2008 showed that the average time to a first office action was about 43 months for pharmaceutical formulation inventions and 34 months for applications relating to organic chemistry. Telephonic restriction requirements are not counted, because these are not considered to be office actions. Such delays will result in significant patent term adjustments (see p. 169).

At this stage the examiner will have carried out a search and will normally cite anything from one to ten pieces of prior art; usually, these are US patents, but sometimes they may be foreign patents or papers from scientific journals. It is standard practice for all of the claims to be rejected on the first official action and the examiner will carefully list all of the grounds upon which each claim is rejected, with reference to the appropriate section of the patent law.

The applicant must answer the official action within an inextensible term of six months. He is, however, well advised to answer within three months if possible, because additional fees must be paid if the reply is made later than this. As of 2009, a reply made in the fourth month costs $110, in the fifth month, $490, and in the sixth month, $1,110, which is more than the cost of refiling the application. The 'free' term of three months is quite short if the applicant is in Europe or Japan and correspondence must pass over two agents.

The reply to the official action may take the form of an amendment to the claims, typically to reduce the scope in order to distinguish more clearly from the cited

(19) **United States**

(12) **Patent Application Publication** (10) **Pub. No.: US 2003/0083544 A1**

Richards et al. (43) **Pub. Date: May 1, 2003**

(54) **METHOD AND APPARATUS FOR FINDING LOVE**

(76) Inventors: **Catherine Richards, (US); W. Martin Snelgrove, (US)**

Correspondence Address:
W. MARTIN SNELGROUE
Suite 700
312 ADELAIDE ST. W.
Toronto, ON M5V 1R2 (CA)

(21) Appl. No.: **10/279,731**

(22) Filed: **Oct. 25, 2002**

Related U.S. Application Data

(60) Provisional application No. 60/330,566, filed on Oct. 25, 2001.

Publication Classification

(51) **Int. Cl.**[7] **A61F 5/00**

(52) **U.S. Cl.** **600/38**

(57) **ABSTRACT**

An apparatus is presented which, carried by or embedded in a lonely or socially inept individual, communicates with like devices in such a way as to divine the likelihood of attraction due to relative sexual, social, intellectual or spiritual interests of the bearers. It may either be programmed explicitly by a trusted body, or suspect compatibility by observing and mining patterns of behaviour, environment and physiological response in the users of the said devices. The users are signalled or led to initial interaction in such a way as to maximize the likelihood of prolonged and deepened contact.

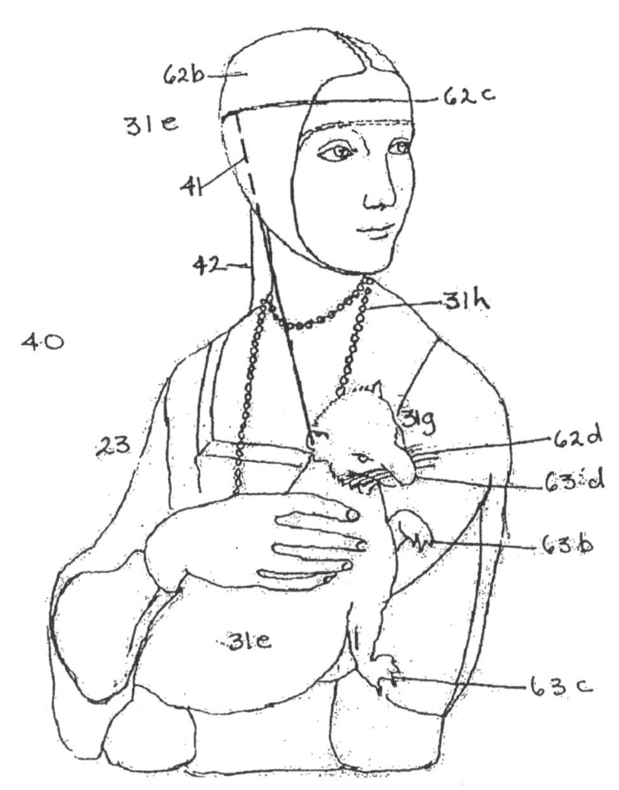

FIG. 11. Front page of a published US Patent application

prior art, coupled with arguments as to why the examiner's rejections should be withdrawn. The applicant may also put in evidence in the form of an affidavit or declaration, either to establish an invention date earlier than the effective date of a piece of prior art published less than one year before the filing date, or to overcome an allegation of obviousness by showing surprising advantages for the invention over what is described in the prior art (see Chapter 20). He should also at this stage file an IDS mentioning any prior art of which he has become aware since filing the original IDS and which is more relevant than anything already of record relating to any aspect of the claimed invention.

The examiner may decide to allow all of the claims at this stage, in which case the applicant will, in due course, be sent a notice of allowance. Alternatively, as long as at least some of the claims are still rejected, the examiner will issue a further official action. If the issues have not yet been clarified and, in particular, if any new prior art is cited or new grounds of rejection are introduced, this second action will be similar to the first. If no new issues are raised, however, the second official action will normally be 'made final' and the claims will be 'finally rejected'. Despite its name, a final rejection does not mean that the proceedings are at an end; it does, however, severely restrict the applicant's opportunities for further prosecution. The applicant may again argue, or offer evidence, but although he may propose amendments, the examiner can refuse to permit them if he feels that they do not put the case in condition for allowance or improve its position on appeal. The examiner may then withdraw the final rejection and allow the application, but if he maintains his rejection of any of the claims, the examiner will, shortly after the applicant's response, issue an advisory action.

The advisory action states that the response has been considered, but does not put the case in condition for allowance. It may state that some claims are allowable or would be allowable if amended further, or may conclude that all claims are still rejected. It will also state whether or not any amendments proposed by the applicant in the last response have been entered.

It is desirable to answer a final rejection within two months if possible, because this may reduce the cost of further extensions of time: an advisory action does not give rise to a new term of reply and further action must be taken by the applicant within the statutory six-month period from the date of the final rejection. If the original response to the final rejection was made three months or more after it was issued, the further response after the advisory action becomes quite expensive.

In response to the advisory action, the applicant must do one of three things in order to maintain the application:

(a) if some claims are allowable, accept what has been allowed, and respond by cancelling the non-allowable claims and making whatever further amendments are needed;
(b) lodge an appeal;
(c) file a request for continued examination (RCE, see below), or a new application as a continuation in part (cip) of the old;

(d) do nothing, in which case, the application automatically becomes abandoned when the time limit for response runs out.

Appeal

The procedure for appeal is that the applicant files a notice of appeal within the time for response to the final rejection. Within two months of filing the notice of appeal, an appeal brief must be filed, which sets out in full the claims on appeal, the rejections applied against them, and the arguments why the rejections should be reversed. The examiner considers this and will normally write a response brief setting out his reasons for disagreeing with the applicant's arguments, although the examiner may, at this stage, give way and decide to allow some or all of the claims. If the examiner's answer raises any new issues, the applicant may respond to these in a further reply brief.

Since 2005, there have been 'filtering mechanisms' in place designed to prevent rejections that are clearly wrong from having to be dealt with by the Board. A patent applicant may now request an internal review of the examiner's final rejection at the technical centre level before an appeal brief is filed. Even if no such request is made, the examiner's case for a rejection is usually reviewed at the technical centre level before the examiner is allowed to file a responsive brief. As a result, the proportion of rejections reversed by the Board has dropped dramatically: from about 40 per cent as recently as 2005, to only 20 per cent in the first two months of 2009. Nevertheless, the annual number of appeals reaching the Board has doubled over the same time period, while the number of pending appeals has increased fivefold. Some commentators say that this is the consequence of an attitude on the part of examiners in recent years that improving patent quality means rejecting all applications on principle, no matter how meritorious they may be.

If the appeal goes forward, it is then scheduled to be heard by a three-person panel of the Board of Patent Appeals and Interferences, which is a board of senior USPTO officials having the position of administrative law judges. The Board may consider the appeal purely on the written arguments, but the applicant has the right to request an oral hearing at which he or his agent or attorney can argue briefly in person before the Board and answer any questions that the Board may put. Because of the large number of appeals to be heard, it now may be about two to four years from the writing of the appeal brief until the Board gives its decision. It may reverse the examiner's decision in whole or in part, in which case, the application will be returned to the examiner for further prosecution or allowance; alternatively, it may affirm the examiner, in which case, the final rejection stands.

If the Board of Patent Appeals and Interferences has decided against the applicant, he may, within two months, request rehearing, although it is understandably rare for the Board to reverse itself. The applicant may alternatively, or after a request for reconsideration has been unsuccessful, either appeal to the Court of Appeals for the Federal Circuit (CAFC) or bring a civil action in the District Court for the District of Columbia. The CAFC is a specialist court, the judges of which are, or have become, experienced in patent matters, and it often produces decisions that appear more sensible than those of the Board of Appeals. The District Court is not a specialist court and it is less often used as a route for appeals from the Patent Office, but has the advantage that it can consider fresh evidence, whereas the CAFC must decide on the basis of the record. From the CAFC, a final appeal could be taken by writ of certiorari to the Supreme Court, but only if the Court itself decides to hear the case because, for example, an important point of law is involved. Appeal from the District Court lies initially to the CAFC, then to the Supreme Court, as above.

Refiling

It has been mentioned that an alternative procedure is that of refiling the application. This can be done at any time during prosecution and can be a very useful strategy to employ. In a continuation application, the text is refiled substantively unchanged and the claims may be altered by amendment either concurrently with the filing or at a later time. The continuation application has the effective filing date of the original (parent) application, so that nothing is lost by the refiling. Because no new matter is added to the description, the inventor does not have to sign a new declaration, a copy of the declaration made in the parent case being filed with the specification for the new application, which, in all cases prior to rule changes made in 1997, was given a new application number. By filing a continuation application, amendments to the claims can be made that the examiner would not permit in the parent case after a final rejection, or time may be gained in which to carry out comparative tests that may be necessary to establish patentability.

From 1 December 1997 until 14 July 2003, it was possible to refile under Rule 53(d) as a continued prosecution application (CPA). Unlike other types of continuation and cip applications, neither a new specification nor even a copy of a declaration was required, and the new application had the same application number, as well as the same filing date as the old. This procedure has now been replaced by an RCE, which is analogous in some respects to a request for further processing in the EPO, but costs the same as a new filing (but with no claims fees). If an RCE is filed, the finality of the office action is withdrawn, but not the office action itself, which still requires a response. It is still possible to file a normal continuation application.

A cip application contains subject matter disclosed in the parent case, but also contains new matter. This may be substantial, for example the scope of the claims may be broadened, and new descriptive matter and examples added to support the new claims, or it may involve correction or clarification of mistakes or ambiguities in the parent text. The matter originally disclosed in the parent application retains the original effective filing date; the new matter, and any claim based upon it, has the effective filing date of the cip application. For a cip application, a new declaration must be executed by the inventor.

When a continuation or cip application has been filed, the parent case may or may not be abandoned and the new application will, in any event, be examined. But if the parent case had been finally rejected and essentially the same claims are again presented in the new application, or by way of an RCE, the examiner may make the first official action a final rejection. By a combination of appeals and refilings, the prosecution of a difficult case may last up to ten years or even more. This used to be to the advantage of the applicant (unless he was interested in obtaining early grant in order to enforce the patent rights as soon as possible), since the 17-year term of patent protection ran from the date of grant and any delay in prosecution gave additional patent term. Since the US Uruguay Round Agreements Act (URAA) of 1994 set the term of a US patent at 20 years from filing, however, refiling does not extend the patent term, since the 20-year term runs from the filing date of the parent application, not from that of the continuation or cip. Nevertheless, patent office delays may give additional duration of protection by way of patent term adjustment under the AIPA.

When, at any stage of the proceedings, the examiner decides that all claims on file are allowable, he issues a form letter telling the applicant that prosecution on the merits is closed, although formal matters may remain. Once these have been dealt with, which may be done by telephone, if appropriate, a notice of allowance is issued. The applicant then has three months in which to pay the issue fee; when this has been done, the patent is printed a few months later and granted as of its date of publication. The US prosecution process is illustrated in simplified form in Fig. 12, and the front page of a US patent is shown as Fig. 13.

At any time while the application is still pending, a divisional application may be filed to claim matter originally claimed in the first (parent) application, but then withdrawn following a restriction requirement, or to claim matter originally disclosed, but not claimed in the parent. The form that the divisional application takes is the same as for a continuation application, unless new matter is added at the same time. Further continuations, cips, or divisionals may be filed as long as one application in the series is still pending. Under the present US law, the patents granted on all such applications will expire on the same date as the patent application, that is, 20 years from its filing date. Previously, they could be used effectively to extend the patent term, because each would expire 17 years from its own date of grant, unless it claimed matter that the examiner felt was essentially the same invention as claimed by the parent application. In this case, the later patent would be granted subject to a 'terminal disclaimer', which meant that it would cease to have effect when the first patent expired.

This can be of great importance for the patentee. For example, Eli Lilly owned a patent derived from a divisional application and having an expiry date two years later than the patent on the parent case, which was due to expire in 2001. In August 2000, the CAFC held in effect that there should have been a terminal disclaimer and that the patent would also expire in 2001, not 2003.[16] This may not sound like a

[16] *Eli Lilly and Co. v. Barr Laboratories Inc. et al.* 55 USPQ 2d 1609 (Fed Cir 2000); 55 USPQ 2d 1869 (Fed Cir 2001).

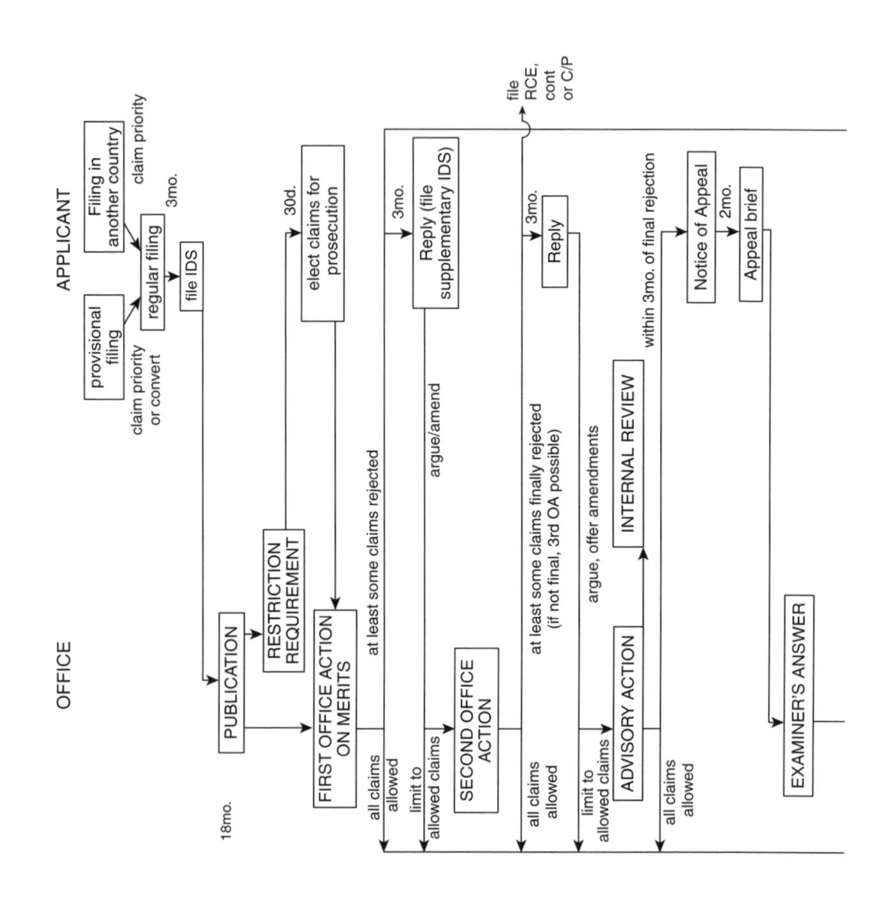
OFFICE
APPLICANT
provisional filing
Filing in another country
claim priority or convert
claim priority
regular filing
3mo.
file IDS
18mo.
PUBLICATION
RESTRICTION REQUIREMENT
30d.
elect claims for prosecution
FIRST OFFICE ACTION ON MERITS
at least some claims rejected
all claims allowed
3mo.
Reply (file supplementary IDS)
argue/amend
limit to allowed claims
SECOND OFFICE ACTION
file RCE, cont or C/P
at least some claims finally rejected (if not final, 3rd OA possible)
all claims allowed
3mo.
Reply
argue, offer amendments
limit to allowed claims
ADVISORY ACTION
INTERNAL REVIEW
within 3mo. of final rejection
all claims allowed
Notice of Appeal
2mo.
Appeal brief
EXAMINER'S ANSWER

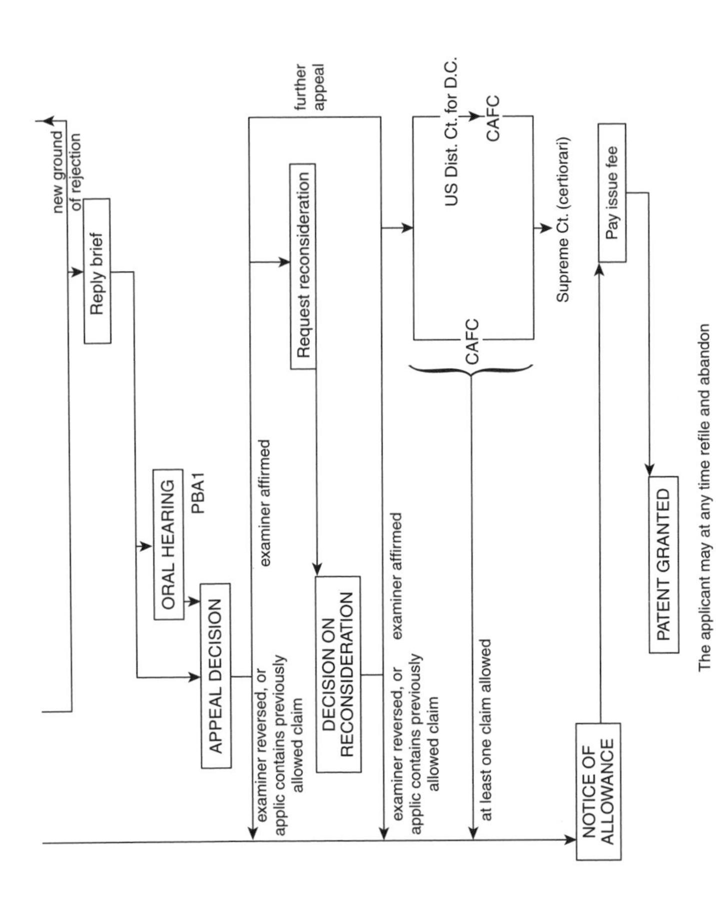

The applicant may at any time refile and abandon

FIG. 12. US Patent Office Grant Procedure

US005509859A

United States Patent [19]

Klees et al.

[11] **Patent Number: 5,509,859**

[45] **Date of Patent: Apr. 23, 1996**

[54] **LEASH WITH SOUND**

[75] Inventors: **Daniel J. Klees**, 224 W. Park St., Mundelein, Ill. 60060; **Terri Shepherd**, Mundelein, Ill.

[73] Assignee: **Daniel J. Klees**, Mundelein, Ill.

[21] Appl. No.: **496,601**

[22] Filed: **Jul. 29, 1995**

[51] **Int. Cl.**[6] **A63J 5/04**

[52] **U.S. Cl.** **472/64**; 446/397; 119/792

[58] **Field of Search** 446/297, 303, 446/397; 472/57, 64; 119/792

[56] **References Cited**

U.S. PATENT DOCUMENTS

1,652,382	12/1927	Swisher	446/397
1,667,125	4/1928	Majewicz	446/397
3,870,296	3/1975	Ellis	472/57
4,282,681	8/1981	McCaslin .	
5,145,447	9/1992	Goldfarb	446/408
5,316,515	5/1994	Hyman et al.	446/28

FOREIGN PATENT DOCUMENTS

791389	12/1980	U.S.S.R.	446/303

Primary Examiner—Mickey Yu
Attorney, Agent, or Firm—Dick And Harris

[57] **ABSTRACT**

A novelty item for creating the illusion of an imaginary pet including a hollow, elongated leash with a handle at one end and a collar and harness adjacent the other end. Housed within the handle, which is hollow, is a battery power source and an integrated circuit for producing a plurality of animal sounds. Also carried by the handle are an on/off switch and at least one selector switch for the sound circuitry. Mounted within the collar end of the leash is a micro speaker which is connected by wiring through the hollow leash to the circuitry in the handle.

8 Claims, 1 Drawing Sheet

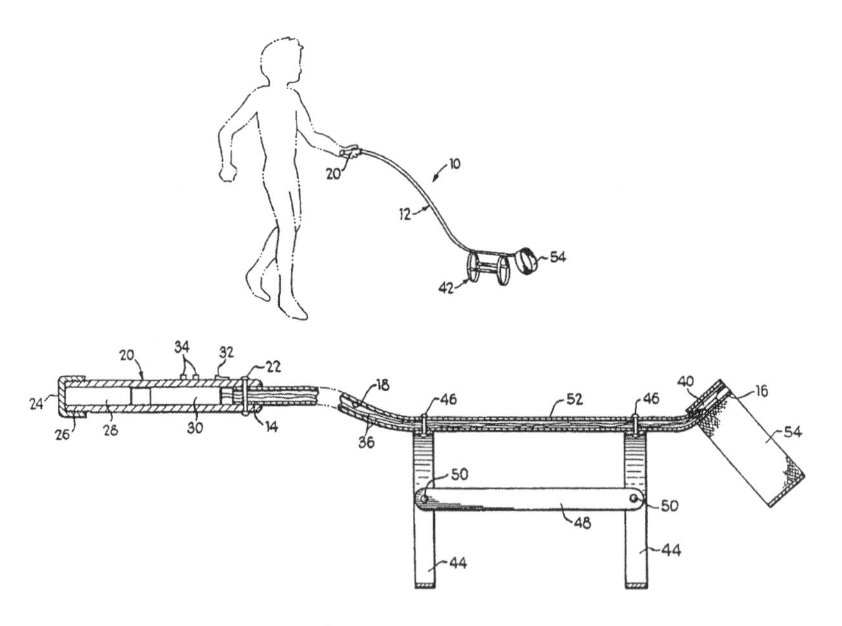

FIG. 13. Front page, US patent

catastrophe, but the patent happened to cover Prozac, Lilly's best-selling product, and, the next day, Lilly's share price dropped by a third.

The repeated filing of continuation and divisional applications, coupled with the fact that pending US-only applications are not published and that until recently the expiry date of a divisional could be much later than that of its parent, has enabled some US applicants to keep an application pending almost indefinitely and finally to have the patent granted with claims that had been inserted many years after filing specifically worded so as to cover exactly what a competitor was then doing. Such patents are referred to as 'submarine patents', because they surface without warning and can effectively torpedo the competitor's business. This tactic is no longer so attractive now that the term of such patents does not run from their date of grant and will be even less so if early publication is also applied to US-only applications. Furthermore, US courts are now applying stricter standards of sufficiency of disclosure,[17] which has already led to the invalidation of some submarine patents.

More recently, the CAFC considered an extreme example of a submarine patent invented (if that is the word) by Jerome Lemelson (see p. 439) and considered that it could be held unenforceable for 'prosecution laches'. The usual meaning of 'laches' is unreasonable delay in bringing suit, which may lead the court to deny relief; here, it was applied to unreasonable delay in prosecution of the patent application.[18] This pragmatic approach was welcomed by Lemelson's victims, but was sharply criticized in a dissenting opinion of Circuit Judge Newman as lacking any basis in the patent law. On remand to the District Court, not only did the judge (in Lemelson's home state of Nevada) hold the patent unenforceable, but also ruled that it was not infringed and was invalid for lack of enabling disclosure.[19] The USPTO may now base a rejection on prosecution laches and refuse to examine the case further if multiple refilings are made without serious attempts to advance prosecution. This practice has been upheld by the CAFC,[20] but it is not clear how it will be applied in cases less extreme than *Lemelson* and *Bogese*.

If the examiner, after finding some or all of the claims allowable, considers that another pending patent application or granted patent claims substantially the same invention, an interference may be declared to determine which applicant has priority of invention. The procedure involved is extremely complex and beyond the scope of this book.

Proposed Rule Changes

In January 2006, the USPTO published for public comment significant changes to the Rules relating to refilling, as well as to numbers of claims. The Final Rules were

17 See, e.g., *Genentech v. Novo Nordisk* 42 USPQ 2d 1001 (Fed Cir 1997).
18 *Symbol Technologies v. Lemelson* 61 USPQ 2d 1515 (Fed Cir 2002).
19 *Symbol Technologies v. Lemelson* 67 PTCJ 258 (Dist Ct Nev 2004).
20 *In re Bogese* 64 USPQ 2d 1448 (Fed Cir 2002).

published in August 2007[21] and a date of 1 November 2007 was set for their implementation. Final Rules 78 and 114 permit an applicant, as a matter of right, to file two continuation or cip applications, plus a single RCE, after an initial application. Any additional refiling would require the applicant to show why the amendment, argument, or evidence could not have been presented previously. Final Rule 75 proposed that if an application contains more than five independent claims, or more than 25 claims in all, the applicant must submit an examination support document (ESD, see Chapter 19), which would represent a severe burden upon applicants.

Glaxo Smithkline and an individual named Triantaphyllos Tafas filed complaints under the US Administrative Procedure Act of 1946, as amended, before the Federal District Court for the Eastern District of Virginia (where the USPTO is located) and sought injunctions against the implementation of the new Rules. The Court granted a preliminary injunction on 31 October 2007 and, by summary judgment of 1 April 2008, made the injunction permanent, on the basis that the changes in the Rules were substantive and outside the authority of the USPTO.[22] The USPTO appealed to the Federal Circuit and a bench decision on 20 March 2009 held that the Rules were procedural rather than substantive, but that Final Rule 78 was inconsistent with 35 USC 120. The USPTO decided to take no further steps until a new director had been appointed and, in the meantime, a petition for *en banc* rehearing was granted by the CAFC on 6 July 2009.

Finally, on 8 October 2009, David Kappos, the newly appointed director, rescinded the new rules and, jointly with Glaxo SmithKline, filed a motion at the Federal Circuit to dismiss the appeal and vacate the District Court decision. This was only partially successful: the appeal was dismissed, but the DC decision stands.[23]

21 72 Fed Reg 46,716 (21 August 2007).

22 *Tafas v. Dudas* 541 F Supp 2d 805, 814 (ED Va 2008).

23 *Tafas v. Kappos* (Fed Cir *en banc* 13 Nov 2009).

9

MAINTAINING A PATENT IN FORCE AND EXTENDING THE PATENT TERM

Thou shalt not kill, but needst not strive
Officiously to keep alive.

Arthur Hugh Clough, 'The Latest Decalogue' (1862)

Patent Term

Finally, after periods of time ranging from a few months to several years from the filing dates, the applicant receives granted patents in the countries in which he applied. These patents do not go on indefinitely, but, subject to any extension provisions, have a fixed legal term that, in the majority of countries, is now the 20 years from filing required by the TRIPs Agreement.[1] Previously, the patent term in the USA and in Canada was 17 years from the date of grant, but this was changed in Canada for patents filed after 1 October 1989, and in the USA, by the Uruguay Round Agreements Act (URAA) of 1994, so that, for US patents and applications filed before 8 June 1995, the term is now 20 years from filing or 17 years from grant, whichever is the longer; for patents granted on applications filed after that date, the

[1] Article 33, TRIPs.

term is 20 years from the first US filing date.[2] This means that, if a continuation or continuation-in-part (cip) application is filed after that date, the 20-year term does not run from the filing date of that application, but from that of the parent application. It used to be that refiling an application, by delaying grant, extended the patent term, but this is no longer the case.

Renewal Fees

For any non-expired US patent granted on an application filed before 12 December 1980, or for a Canadian patent with an application date before 1 October 1989, the patentee need do nothing more. No fees need be paid in order to keep the patent in force and it automatically runs for the full term unless positive action is taken to abandon it. For nearly all US and Canadian patents, and for patents in nearly all other countries in the world, it costs money to keep the patent in force. The renewal fee must be paid to the patent office of the country in question, normally upon an annual basis, although, in some countries, payments are made at less frequent intervals. US law, for example, provides for renewal fees payable three-and-a-half, seven-and-a-half, and eleven-and-a-half years after grant.[3] If a renewal fee is not paid within a stipulated period from the due date, the patent lapses. As a general rule, renewal fees are relatively low in the early years and rise more or less steeply towards the end of the patent term. Often the first few years are free; thus in the UK the first renewal fee falls due at the end of the fourth year from application.

The reasoning behind the requirement for renewal fees is twofold. Firstly, from a purely financial perspective, it helps to pay for the running of the patent office. In the US Patent and Trademark Office (USPTO), around a third of all patent revenues come from annuity fees. In the UK, the Patent Office is expected to pay for itself out of fees charged to its users. If this were done only by application, search, and examination fees, these would have to be set at such a high level that there would be a considerable disincentive to potential applicants, in particular the individual inventor would be unable to afford to apply for a patent. It seems reasonable that a considerable amount of the money should come from those who are making commercial use of their patents.

This brings us to the second reason, which is that the renewal fee system means that only patents that are of real or potential commercial value to their owners are maintained for more than a few years. In this way, inventions in which there is no great commercial interest become public property much earlier than would otherwise be the case. In the USA, recent statistics show that the first renewal fee was paid for 83 per cent of a sample of granted patents, the second renewal fee for 74 per cent, and the final renewal fee for 49 per cent.[4] In countries where the fees are

2 35 USC 154(a)(2).
3 35 USC 41(b).
4 *USPTO Performance and Accountability Report Fiscal Year 2008*, USPTO, 2009.

paid annually, the fraction of cases for which the final renewal fee is paid will be much less than half.

For granted European patents, renewal fees are collected by the national patent offices and are the same as for nationally granted patents. The European Patent Convention (EPC) provides, however, that the European Patent Office (EPO) shall receive a part of these renewal fees, so that the financing of the EPO is not dependent upon application, search, and examination fees alone.[5] This part may be up to 75 per cent, but the current level is now set by the Administrative Council at only 50 per cent.[6] In the early years, in which there was little or no income from renewals of European patents, the EPO received direct funding from the contracting states, but nowadays the situation is reversed. The national patent offices receive substantial sums from the renewal fees on European patents to which they have contributed nothing and it will be difficult to change this as long as the Administrative Council of the European Patent Organization is composed of representatives of the national patent offices. In January 2006, the Italian Parliament suddenly abolished all renewal fees for Italian national patents and European patents in Italy, but after protests from the EPO, reintroduced the fees one year later. The reintroduction was not retroactive, so no renewal fees needed to be paid for the year 2006.

In the USA, by contrast, it used to be government policy to subsidize heavily the operations of the USPTO, with the object of encouraging inventors. The application and issue fees, together with sundry fees that may be charged during prosecution, used to cover no more than a small part of the USPTO budget, the rest of which came from funds voted by Congress to the Department of Commerce. In practice, this policy resulted in the USPTO being constantly short of money and, because there were no renewal fees, many US patents remaining in force that had nothing more than nuisance value. In 1982, fees were drastically increased, with the intention that, within ten years, 100 per cent of actual costs would be covered by user fees, including renewal fees. But 'small entities' (that is, individuals, non-profit organizations, and small companies) pay only 50 per cent of the normal fees.[7] This increase in self-generated funding of the USPTO was expensive for applicants, but had a marked effect on its efficiency: additional examiners were recruited, backlogs decreased, and the average time from application to grant of a US patent was reduced.

In 1990, the US Omnibus Budget Reconciliation Act (OBRA) dedicated a portion of the USPTO fees to general deficit reduction: that is, to disappear into the black hole of the general US government finances. For many years, annual appropriations from Congress for the funding of the USPTO fell far short of the money actually collected in fees, the difference amounting to over $500 million up to 2004. The US patent applicant was no longer seen as someone to be encouraged and

[5] Article 39(1), EPC.
[6] OJ 1984, 296.
[7] 35 USC 41(h)(1).

supported, but rather as a cash cow to be milked to the maximum extent. Since 2005 the USPTO has been able to spend all or nearly all of the money that it collects in fees each year—a very welcome change. But in 2009 the recession caused a decline in patent filings and renewals, so that fee income would not meet the budgeted costs. As of writing, it is proposed to allow the USPTO to compensate for this by diverting surplus funds from trademark renewals to its patent operations.

It is a somewhat pointless exercise to discuss the absolute amounts of renewal fees in various countries, since fluctuations in exchange rates and inflation levels constantly alter the figures; in addition, if renewal fees are paid through local agents, their professional charges must be taken into account. For the sake of interest, however, renewal fees in the UK in 2009 ranged from £50 for the fifth year to £400 for the twentieth year. The costs in the UK are lower than in many European countries, and there is no particular correlation between cost and market size, strength of patent rights, or any other measure of value received.

The need to pay renewal fees to keep patents alive means that a patentee, whether a large company or an individual, should annually review the patents that are held and decide whether or not each patent should be kept in force. If the patent relates to a marketed product, it should be considered for each country involved what would be the commercial consequences of the patent being allowed to lapse; if no market product is involved, it should be asked whether the patent has any useful function as a 'defensive patent' to keep competitors away from an area of interest, or whether there is any prospect of outlicensing the patent. For a company with a large number of patents, very considerable sums can be saved by rooting out the patents that have outlived their usefulness and this, at least in the chemical and pharmaceutical industry, is preferable to trying to accumulate as many patents as possible. The patent policy of an electronics company may, however, be very different (see Chapter 22).

It is clear that a good record-keeping system is essential in order to ensure that renewal fees are paid when they fall due. Such a system must also be 'fail-safe'—that is, a patent should always be renewed in the absence of a positive decision to allow it to lapse. Patent agents in private practice will normally provide this service for their clients; companies with their own patent departments may do their own record-keeping. This type of system is ideally suited for operation by computer and there are a number of firms that specialize in the computerized management of patent renewal fees, either directly for industrial clients or on behalf of firms of patent agents.

However good a system may be, the possibility of error cannot be ruled out. For this reason, the Paris Convention protects patent holders by requiring that member states provide a grace period of at least six months for late payment of renewal fees and permits provisions for restoration of lapsed patents.[8] In the UK, the due date for

[8] Article 5*bis*, PC.

the payment of a renewal fee is now the last day of the month in which the anniversary of the filing date falls, in line with the procedure for maintenance fees in the EPO and in many EPC states.[9] The renewal fee for the next year can be paid during the three months before this due date. The fee is sent to the Patent Office, together with a standard form, receipt of which is acknowledged. If payment is not made by the due date, the patent technically lapses as of that date.[10] No later than six weeks after this (but usually well before this deadline), the Patent Office sends out a reminder that the fee is still unpaid.[11] The patentee can act on this reminder by paying the renewal fee at any time up to six months after the date on which it was due, but after the first month, an extra fee must be paid for each month the payment was late.

If payment is made within this six-month period, the patent continues in force as if it had never lapsed and action can still be taken against anyone who infringed the patent while the fee remained unpaid.[12] After this period, however, there is no automatic right to late payment of the renewal fee. The lapse of the patent is no longer a technicality, but a reality, and the ex-patentee must, within 13 months of the end of the six-month grace period, petition for restoration of the patent. In order to do this, he has to show that the failure to pay the renewal fee on time, or at least within the first six-month period, was unintentional.[13] This is a less strict requirement than the previous version of section 28(3)(a), which provided for restoration only if the patentee had taken 'all reasonable care' to pay the fee on time. Nevertheless, a mere assertion that the failure was unintentional is not enough; rather, some reasonable explanation must be given.[14] If the Patent Office allows restoration, a further additional fee must then be paid. What is more, the patent is no longer treated as if it had never lapsed. Anyone who began to do anything falling under the patent claims after the end of the grace period, or even made real preparations to do so, has the right, after the patent is restored, to go on doing what he began or prepared to do, without having to pay any royalty to the patentee.[15]

In the USA, also, there are provisions for late payment of a renewal fee during a six-month period, on payment of an extra fee. After this period, the USPTO may still accept a renewal fee if the failure to pay the fee on time was 'unavoidable'.[16] In most civil law countries, such as Germany and France, the patentee must show a higher degree of 'all due care' in order to re-establish rights after a missed

[9] Section 8(1), PA 2004; rr. 37 and 38, PR 2007.

[10] Section 25(3), PA 1977.

[11] Section 25(5), PA 1977; r. 39, PR 2007.

[12] Section 25(4), PA 1977.

[13] Section 28(3)(a), PA 1977, as amended.

[14] *Matshusita Electric Industrial Co. Ltd and ors v. UK Intellectual Property Office* [2008] RPC 35 (Pat Ct).

[15] Section 28A(4), PA 1977.

[16] 35 USC 41(b).

annuity payment. In any case, third parties have the certainty that, one year after the annuity fee was missed, restitution is no longer possible.[17]

An interesting situation occurred in France when the French Patent Office confirmed, in a letter, that a certain annuity fee for a complementary protection certificate (known by its French abbreviation, CCP) had not been paid and therefore the right had lapsed, whereas the holder of the CCP could provide evidence of actually having paid the annuity fees. A third party had in the meantime relied on the notice from the Patent Office and had launched a product that most likely would be held to fall under the patent/CCP claims. The first-instance court nevertheless found infringement of the CCP.

Extension of Term

Effect of Changes in Patent Term

A number of countries, including the USA, had to change their patent terms because of their adherence to the World Trade Organization (WTO) and therefore to the TRIPs Agreement. Japan and Korea had a patent term defined as the shorter of 20 years from filing or 15 years from publication for opposition, and this alternative had to be deleted. Similarly, within Europe, Austria had a term of 20 years from filing or 18 from publication on acceptance, and whereas in practice this was unlikely ever to give a term of less than 20 years from filing, the possibility existed and had to be removed. And, of course, Portugal had to change the term of its national patents from the fixed 15 years from grant (usually 17–18 years from filing). In all of these cases, the transitional provisions of TRIPs[18] required that the extended term also should apply to existing patents, provided that there was no obligation to restore patents that had already lapsed.[19] Thus Greece, although it had changed its patent term well before its adherence to TRIPs, was also obliged to extend the term of its old 15-year patents.

The USA, Japan, and Greece acted promptly to extend the term of existing patents by the deadline of 1 January 1996. Korea and Portugal did so only in the middle of 1996, so that some Korean and Portuguese patents expired during 1996, which would not have done so had the TRIPs obligations been met in good time. For example, in Portugal the patent for the drug enalapril expired on 9 April 1996, but should not have expired until 4 December 1999 if the TRIPs provisions had been in effect. The question of whether TRIPs Article 30 can be interpreted by national courts as having direct effect was referred to the European Court of Justice (ECJ),[20] which held that there was nothing in Community law to prevent national courts

[17] For example, §123 of the *Patentgesetz* (German Patent Act).

[18] Article 70.2, TRIPs.

[19] Article 70.3, TRIPs.

[20] C-431/05 *Merck Genéricos v. MSD* (ECJ, 11 September 2007).

from doing so. The Portuguese Supreme Court could therefore apply TRIPs directly, making Genéricos liable for infringement from 1996 to 1999.

Canada extended the term of its old patents only after being forced to do so by a WTO Dispute Panel decision in 2000. Other countries, such as Brazil and Pakistan, have not made any extension of the term of the patents granted under their pre-TRIPs laws. Argentina has adopted an intermediate solution: patents granted under the old law still have a term of 15 years from grant, but for those still in force on 1 January 2000, it is possible to make individual requests for extension.

Patent Term Adjustment in the USA

When the USA changed to a term of 20 years from filing, there were concerns that applicants could lose effective patent term (that is, the time from grant to expiry) because of delays in the granting process that were outside the control of the applicant. Accordingly, the URAA allowed the term of a US patent filed on or after 8 June 1995 to be extended to compensate for delays in issuance due to interference proceedings, secrecy orders, or successful appeals. The patent term adjustment (PTA) provisions of the American Inventors Protection Act (AIPA) of 1994, which apply to all patent applications filed on or after 29 May 2000, extended this principle to include situations in which the USPTO fails to take certain actions within specified time frames and in which the USPTO fails to issue a patent within three years of the filing date. In effect, the applicant receives a day-for-day extension of term for each failure or delay. For example, one such time frame is that there must be an initial action on the application within 14 months from the filing (or PCT national phase entry) date. But this first action may be no more than a restriction requirement.

Actions of the applicant have an effect on PTA: for example, filing a request for continued examination (RCE) cuts off any additional PTA due to failure to issue a patent within three years, but does not otherwise affect the PTA. The period of adjustment is reduced for any period during which the applicant failed to engage in reasonable efforts to advance the prosecution of the application, for example by failing to reply to an office action within the initial three-month period. The consequences of filing an appeal must be carefully considered. A successful appeal (that is, one in which all rejections of at least one claim are reversed) will result in an extension of patent term, whereas an unsuccessful appeal will result, if not in complete refusal, at least in a patent of reduced scope and shorter effective term.

Currently, about 75 per cent of US patents are granted with a PTA, the median length of extension being just over one year. In other words, the patent term in the USA is more likely to be 21 years from filing than 20 years. The principle on which the USPTO calculates PTAs is presently being challenged in the case of *Wyeth v. Dudas*, currently on appeal to the Court of Appeals for the Federal Circuit (CAFC). If the first-instance decision favourable to Wyeth is upheld,[21] average patent term

[21] *Wyeth v. Dudas* 88 USPQ 2d 1538 (DDC, 2008).

will become even longer. This is a highly specialized subject, on which advice from a US patent attorney should be obtained.

Extensions to Compensate for Regulatory Delays

Although the legal term of a patent is normally 20 years from filing, the *effective patent term*, defined as the length of time for which a product is marketed with the benefit of enforceable patent protection, may be very much shorter. Firstly, of course, there are no enforceable exclusive rights until the patent is granted, a process that normally takes from two to five years in most countries, so that between 15 and 18 years of effective patent term remain. In many fields, the development and marketing of a product will also take two to five years, so that often the product will be put on the market at much the same time that patents covering it are granted. In certain fields, however, development of a product necessarily takes very much longer than this, because the approval of regulatory authorities has to be obtained before marketing is allowed. This is particularly the case for the pharmaceutical industry, since no new drug can be approved without extensive clinical testing to prove that it is safe and effective, and this process may easily take eight to 12 years, or even more, from the filing date of the original patent application, leaving an effective patent term of only eight to 12 years instead of approximately 17, as is the case for most other products (see Chapter 22). It is not only pharmaceuticals, however, which are subject to regulatory approval for marketing, but also, for example, agrochemicals such as herbicides and insecticides, which may be sprayed upon food crops or could be hazardous to workers applying them.

The USA: The Hatch-Waxman Act

In the USA, where the Food and Drug Administration (FDA) was one of the most difficult health authorities in the world from which to obtain marketing approval, the delays in marketing a pharmaceutical product were even longer than elsewhere and the effective patent life for pharmaceuticals was correspondingly shorter. It was calculated that the average effective patent life for a drug in the USA fell from 13.6 years in 1966 to 9.5 years in 1979, and calculations such as this led to pressure from the research-based pharmaceutical industry for an extension of patent term as of right, to compensate for regulatory delays in the FDA. This proposal met with strong resistance from the manufacturers of generic (that is, non-patented) pharmaceuticals and, after considerable lobbying and political in-fighting, a compromise was worked out that granted patent extensions (but with certain important limitations) while making it easier for generics manufacturers to market a product after its patent protection expired. This compromise was embodied in the US Drug Price Competition and Patent Term Restoration Act of 1984, which is known as the Hatch-Waxman Act after the names of its sponsors in the House of Representatives and the Senate, and which is now incorporated into US patent law.[22]

[22] 35 USC 156.

Prior to enactment of the Hatch-Waxman Act, there were three types of new drug application (NDA) that could be approved by the FDA so that a drug could be marketed.

(a) A normal NDA required full animal and human testing of safety and efficacy, performed by the applicant.
(b) A 'paper NDA' had the same requirements, but these could be met by reliance on published data rather than original work—but the full range of studies required for NDA approval were rarely published in sufficient detail to enable a paper NDA to be filed.
(c) Finally, an abbreviated NDA (ANDA) required only equivalence and bioavailability studies to show that the product was the same as a product already marketed, but ANDAs were not available for drugs first marketed after 1961.

The 'drug price competition' provisions of the Hatch-Waxman Act have the effect of making ANDAs available for all new drugs once patent protection (including any extension) expires, provided however, that, irrespective of the patent situation, neither paper NDAs nor ANDAs can be granted for a new chemical entity for five years after the date of its first approval by the FDA. As a further benefit to the generics companies, the case of *Roche v. Bolar*[23] (see p. 190) was legislatively overruled to the extent that use of a patented invention 'reasonably related to the development and submission of information under a Federal law which regulates the manufacture, use or sale of drugs' is no longer patent infringement.[24] Testing can henceforth be carried out before patent expiry, with a view to coming on to the market as soon as possible afterwards.

The patent term extension provisions of the Hatch-Waxman Act provide that the patent may be extended for a period corresponding to one half of the investigative new drug (IND) clinical testing time, plus all of the NDA approval time, up to a maximum of five years, provided that the maximum patent term does not exceed 14 years from NDA approval date and provided that any such IND or NDA time prior to the grant of the patent is not taken into account.[25] The extension period can, however, be shortened if a third party is able to show that delays occurred that were attributable to the NDA applicant himself, and not to the FDA. The patentee must apply for extension of his patent within only 60 days of marketing approval being granted and must pay a fee. Application is made to the director of the USPTO, who passes the request on to the FDA for a determination of the regulatory period.

Only one patent is capable of being extended in connection with the first NDA approval; if, for example, there were separate product and use patents, then only one of these could be extended. A patent cannot be extended more than once, even if it covers two FDA-approved products, and no extension is allowed if the patent

[23] *Roche Products Inc. v. Bolar Pharmaceutical Co. Inc.* 221 USPQ 937 (Fed Cir 1984).
[24] 35 USC 271(a)(1).
[25] 35 USC 156(c).

has already expired: a patent, once dead, cannot be revived. A product patent when extended is limited in scope to the approved product for any approved use: that is, it will no longer cover other compounds within the original scope, nor any non-pharmaceutical use of the approved product.[26] Until recently, there was still controversy as to exactly what is meant by the 'approved product'. Normally, one specific salt form of the active substance is specified in the marketing approval, but it had been generally supposed that the scope of the extended patent covered all pharmaceutically useable salt and ester forms of the compound. Because generics competitors normally wished to market exactly the same product as that sold by the patentee, this assumption was never tested until 2002, when the Indian generics company Dr Reddy's Laboratories sought marketing approval for the maleate salt of Pfizer's product amlodipine, which Pfizer sold as the besylate salt. Pfizer sued for infringement, based on a patent extended under the Hatch-Waxman Act, and Dr Reddy argued that the patent covered only the besylate salt for which marketing approval had been granted. The New Jersey District Court agreed, but fortunately the CAFC reversed this decision.[27] Pfizer had argued that the extended patent should have the same scope of protection as the patent before extension, but this extreme position was not accepted by the CAFC. Nevertheless, the reasoning of the CAFC is open to criticism because it completely ignored an earlier contrary decision,[28] so perhaps we have not heard the last of this issue.

Japan: Patent Term Extensions

Under legislation that came into force in 1988, it is possible to obtain up to five years' patent term restoration in Japan, without any exemption of pre-expiry generics activity from patent infringement. The period of extension allowed is the time from patent grant or start of clinical testing, whichever is the later, to the date of marketing approval, with a maximum of five years. Separate extensions may be obtained in respect of different approved indications for the same product and more than one patent may be extended in respect of the same marketing authorization. In both the Japanese and the US systems, the date of patent grant is potentially critical, but whereas in the USA the patent would normally be granted well before the start of clinical testing, this was not the case under the Japanese deferred examination procedure. Whereas prior to 1988 pharmaceutical companies were happy to wait out the full seven-year deferred examination term in Japan, it then became important to ensure that patent grant was not delayed too long. Grant after clinical testing has begun shortens the possible extension period and if grant occurs after marketing approval, there is no extension at all. One therefore had to request examination on filing, or at least within three years of filing, in order to ensure no loss of extension.

[26] 35 USC 156(b).

[27] *Pfizer Inc. v. Dr. Reddy's Laboratories Ltd et al.* 67 USPQ 2d 1525 (DC NJ); 69 USPQ 2d 2016 (Fed Cir 2004).

[28] *Glaxo Operations UK Ltd v. Quigg* 13 USPQ 2d 1628 (Fed Cir 1990).

Now that the period within which examination must be requested has been reduced to three years, the problem no longer exists.

More importantly, as long as Japan had pre-grant opposition, a competitor could delay grant by filing opposition, thus reducing the extension available. Fortunately this is no longer possible.

Europe: Supplementary Protection Certificates

Following the enactment of the Hatch-Waxman Act in the USA and the extension provisions in Japan, there was considerable pressure for European countries to bring in similar measures. This was difficult in view of the fixed term of 20 years provided for European patents by the EPC[29] and, to avoid this problem, it was proposed to grant what would legally be a new form of intellectual property right, that is, the supplementary protection certificate (SPC), which would be granted so as to take effect upon expiry of the patent and would be of limited scope, covering only the marketed pharmaceutical product for which regulatory approval had been obtained.

The German government was unhappy about this legalistic method of avoiding the clear provisions of the EPC and at its instigation a diplomatic conference was held in December 1991 at which Article 63 was revised so as to allow member states to extend patent term or to 'grant corresponding protection which follows immediately on expiry' if 'the subject matter of the European patent is a product or a process of manufacturing a product or a use of a product which has to undergo an administrative authorization procedure required by law before it can be put on the market'. (Note that this provision is not limited to pharmaceutical products.)

On 18 June 1992, the EC Council issued a Regulation creating the SPC for pharmaceuticals[30] as of 1 January 1993. The Regulation was automatically binding upon all member states, but for those that did not at that time have product protection for pharmaceuticals (that is, Spain, Greece, and Portugal), it did not take effect until January 1998. France and Italy had already introduced certificates of complementary protection (CCP, which are similar to SPCs, but of longer duration) by national law; these provisions were replaced by the new Regulation, but French and Italian CCPs already granted retained their full duration. From 2004, however, the duration of Italian CCPs, which could be as long as 18 years, was progressively reduced.[31]

Although SPCs are governed by an EU Regulation, they are national rights that must be applied for on a country-by-country basis. In each country, there must be a basic patent protecting the active ingredient per se, its method of manufacture, or its use, as well as a marketing authorization to put the product on the market as a medicinal product. Application for an SPC is to be made in each country within

[29] Article 63, EPC.
[30] EC Regulation 1768/92.
[31] Decree No. 112/2002.

six months of either the date of marketing authorization in that country, or the date of patent grant, whichever is the later.[32] The marketing authorization may be issued by the regulatory authority of the country concerned, or centrally by the European Medicines Evaluation Agency (EMEA). On introduction of the Regulation, there were transitional provisions that allowed SPCs to be granted for products protected by a patent and which had received their first marketing authorization after 1 January 1985 (1982 for SPCs for Belgium and Italy; 1988 for Denmark, Germany, and Norway), as long as the application was made within six months of the entry into force of the Regulation.[33]

In some countries, it may not be possible to sell a product, even though a marketing authorization has been granted, until additional requirements, such as agreement on reimbursement price, have been met. Nevertheless, the six-month period runs not from the date on which the product can actually be sold, but from the date of issue of the formal marketing approval. This was decided by the ECJ in the *Hässle* case,[34] relating to the blockbuster drug omeprazole (Losec®), developed by Hässle (subsequently Astra and AstraZeneca). Omeprazole had received marketing authorizations in France on 15 April 1987 and in Luxembourg on 11 November 1987, both before the critical date of 1 January 1988, meaning that no SPC would be available in Denmark or in Germany. Agreement on pricing was reached in Luxembourg on 17 December 1987 (still too early) and in France not until 22 November 1989. Omeprazole was included in a published list of approved products in Luxembourg on 21 March 1988. In applying for an SPC in Germany, Hässle gave 21 March 1988 as the date of first marketing authorization in the European Community (EC), and the SPC was granted. Ratiopharm contested this in the German courts and a reference was made to the ECJ. The Court held that only the date of marketing approval under Directive 65/65[35] was relevant, so that the date of first marketing authorization was, indeed, 15 April 1987 and the subsequent dates did not count. Consequently, the German SPC was invalid.

That was not the end of the story, however. Hässle had arguments (albeit not very strong ones) to support its position and what it should have done was to present these arguments, together with the full facts, to the German Patent Office when applying for the SPC. Instead, it concealed the facts, misled the German Patent Office, and in subsequent litigation even took the position that it could not sell omeprazole in Luxembourg before publication of the list on 21 March 1988, whereas in fact it had put the product on the market ten days before that date. This led to an investigation by the European Commission, which imposed a fine of

[32] Article 7, EC Reg. 1768/92.

[33] Article 19, EC Reg. 1768/92.

[34] C-127/00 *Hässle v. Ratiopharm* [2004] ECR I-14781.

[35] Council Directive 65/65/EEC of 26 January 1965 on the approximation of provisions laid down by law, regulation or administrative action relating to medicinal products.

€60 million on Astra for abuse of a dominant position.[36] As of writing, an appeal to the European Court of First Instance (CFI) is still pending, and there may be further unpleasant consequences for the entire pharmaceutical industry (see Chapter 25).

The situation of old compounds marketed before Directive 65/65 came into effect is not clear. Merz had marketed a compound called memantine in Germany before any safety and efficacy testing was required, and was allowed to keep it on the market even after the implementation of Directive 65/65. Then, in 2002, Merz obtained an EU-wide marketing authorization for memantine in the treatment of Alzheimer's that did comply with Directive 65/65, and applied for an SPC on the basis of this marketing approval and a second-use patent. Synthon challenged the SPC in the Netherlands and the UK. The Dutch Patent Office considered that, since the ECJ had held that only authorizations granted under Directive 65/65 were relevant, the 2002 date could be accepted and the earlier ones ignored. In the English Patents Court,[37] Floyd J thought that the correct view was that the SPC Regulation did not apply at all to products first marketed before Directive 65/65 came into effect, because the purpose of the SPC was to compensate for delays caused by compliance with Directive 65/65. He did not reach a decision, however, and referred the question to the ECJ.[38]

An SPC, once granted, in effect lies dormant until the normal expiry of the basic patent, and then has a duration equal to the time between patent filing and the grant of the first marketing authorization in the Community reduced by five years, up to a maximum duration of five years.[39] The aim is to give an effective patent protection (from marketing authorization to SPC expiry) of 15 years. Unlike the situation in the USA and Japan, the date of patent grant is irrelevant, as is the duration of clinical testing as such. If first marketing approval is granted within five years of the patent filing date, no SPC will be granted; if marketing approval takes more than ten years from patent filing, the SPC will be capped at five years. If the period from filing to approval is between five and ten years, the duration of the SPC will be such as to give exactly 15 years' patent and SPC protection from the date of first marketing approval.

By the Agreement establishing the European Economic Area (EEA), the reference in the Regulation to 'the Community' was expanded to the EEA, which today consists of the European Union (EU), plus Norway, Iceland, and Liechtenstein. The date of first marketing approval in the EEA is therefore of critical importance for determining the duration of an SPC and this has been affected by the anomalous situation of Liechtenstein. Although Liechtenstein is a sovereign state, it is part of a customs and patent union with Switzerland. Both countries signed the EEA

[36] Commission Decision 2006/857/EC of 15 June 2005 relating to a proceeding under Article 82 of the EC Treaty and Article 54 of the EEA Agreement (Case COMP/A.37.507/F3—AstraZeneca).

[37] *Synthon BV v. Merz Pharma GmbH* [2009] EWHC 656 (PAT).

[38] Pending as C-195/09.

[39] Article 13, EC Reg. 1768/92.

Agreement in 1992, but Switzerland, after a referendum, did not ratify the Agreement, whereas Liechtenstein did. Accordingly, Liechtenstein became a member of the EEA (after some delay), while Switzerland remained outside. Liechtenstein did not have its own pharmaceutical regulatory authority, but recognized a Swiss marketing approval as being effective in Liechtenstein. The EU Commission and the majority of member states took the view that, because of its effect in Liechtenstein, a Swiss marketing approval may be regarded as a first marketing approval in the EEA and therefore starts the clock running for the SPC duration. This conclusion was also reached by the ECJ in two test cases decided in 2005.[40]

This decision had serious consequences for companies (not only Swiss ones) that obtained marketing approval for their products in Switzerland earlier than in any EU country. About 40 pharmaceutical products lost on average six months' patent protection in the entire EU, representing a loss to the industry of approximately €1 billion. The present situation is that a Swiss marketing authorization for an NCE extends to Liechtenstein only after a delay of 12 months, which can be extended upon request, so that for new products an earlier approval in Switzerland should no longer reduce the SPC duration. For the 40 products already affected, however, nothing can be done to recover the lost protection.

An SPC may be granted on the basis of a national or a European patent. Only one certificate may be granted to a single patentee in respect of a single product, although it is possible for two or more certificates for different products to be based upon the same patent. When more than one patent covering the same product is held by different parties, an SPC may be granted to both parties in respect of the same product. This is clearly stated in the later EC Regulation 1610/96 extending SPCs to plant protection products and the same principle applies also to pharmaceutical SPCs. In this situation, when only one of the parties holds a marketing authorization for the product, the ECJ decided that the body that granted the marketing authorization must supply a copy to the other party, so that his application for an SPC may proceed.[41] The ECJ also clarified that a second SPC can be granted even if a first SPC had already been granted for the same product, under the conditions that the patents on which the two SPCs are based are held by different entities and the later SPC application could not have been filed before the first SPC was granted because the basic patent for the second SPC was granted well after the marketing authorization of the product had been issued and therefore the six-month period for filing an SPC application only started from the later patent grant.[42] An additional uncertainty was added by a new codified form of the medicinal SPC Regulation,[43] which replaced and repealed Regulation 1768/92, and this new version does not include the above

40 C-207/03 and C-252/03 *Novartis/Millennium Pharmaceuticals* [2005] ECR I-3209.

41 C-181/95 *Biogen v. SmithKline Beecham* [1996] ECR I-00717.

42 C-482/07 *AHP Manufacturing BV v. Bureau voor de Industriële Eigendom*.

43 EC Regulation 469/2009 of 6 May 2009.

provision from Regulation 1610/96 on the possibility of two or more SPCs for the same product filed by different patent holders. Nevertheless, the reference to Regulation 1768/92 in the plant protection SPC has been replaced by a reference to Regulation 469/2009, so the position is presumably unchanged.

Application fees and annual maintenance fees are payable on SPCs. Questions of the scope of protection given by SPCs are dealt with in Chapter 13.

Most of the ten states that joined the EU in 2004 had already introduced SPCs on or before their date of accession. The transitional provisions were rather complex, however, and the deadlines for filing applications differed from country to country. Switzerland has its own national SPCs under provisions essentially identical to those of the EU Regulation and these also extend to Liechtenstein, which cannot issue its own SPCs under the Regulation.

Paediatric extensions in Europe

It has long been a problem for parents as well as doctors to know what dosage of a drug is the correct one to give to children. The old rule of thumb of halving the dose is often inappropriate and sometimes dangerous, and dosage regimes can only be properly established by clinical trials on young patients. Until recently, there has been no real incentive for pharmaceutical companies to carry out such tests, but this problem has been addressed by a Regulation on medicinal products for paediatric use,[44] which gives incentives in the form of an extension of SPC protection.

Any application for marketing authorization of a new product should now include the results of an agreed paediatric investigation plan (PIP), unless the drug is one that is clearly not suitable to be given to children (for example drugs for geriatric indications). For a drug that has already been approved, PIP studies may be filed subsequently. Once the results have been approved by the Paediatric Committee of the EMEA, an applicant holding an SPC or a patent qualifying for the grant of an SPC may obtain a six-month extension of the duration of the SPC, as long as this paediatric extension is applied for at least six months before expiry of the original SPC term (noting that, after five years from the entry into force of the Regulation, in January 2012, this period will be two years).[45]

Complications arise in the situation in which the time between the patent filing date and the marketing authorization is between four-and-a-half and five years. Normally, in this situation, no SPC would be applied for or granted, because it could give no extension of term. But if an SPC were to be granted having a negative or zero term, the effect of the six-month paediatric extension would be to give a real extension of patent protection. Thus, if the above period were four years and nine months, grant of an SPC with a term of minus three months would give a net extension of three months and grant of a zero-term SPC would allow the full six-month

[44] EC Regulation 1901/2006 of 12 December 2006.

[45] Article 7(4) and (5), EC Reg. 1768/92, as amended by Art. 52, EC Reg. 1901/2006.

extension to apply. As of writing, some patent offices have granted negative or zero-term SPCs, and some have refused to do so. Clearly, at some stage, the ECJ must give a ruling on how the Regulation is to be applied. It would be more equitable if the extension were to apply to a patent or to an SPC, whichever would be the later to expire.

Another requirement for an extension to be granted is that there must be marketing approval for the product in all member states. It has already been held in the UK that it is not sufficient to show that the PIP has been completed to the satisfaction of the Paediatric Committee, but that a first marketing approval incorporating this must be presented.[46] The same paediatric extension was allowed in the Netherlands, Sweden, Italy, France, and Germany, where the patent offices were more flexible and accepted late evidence. It has even been suggested by the EU Commission that corresponding amended marketing approvals for *all* member states must be presented whenever an application for paediatric extension of an SPC is made in any state, which might well make it impossible to meet the filing deadline.

Other Countries

Korea and Taiwan originally provided patent term extension only on the basis of bilateral agreements, but now have extension provisions in their laws similar to those of Japan. In Korea, however, unlike the situation in Europe or Japan, more than one patent may be extended for the same product, but the same patent may be extended only once. Australia has a system similar to that of the EEA, except that process patents (other than those for DNA technology) cannot be extended. Israel has an extremely complicated system based upon extensions in certain other countries, and Russia and some Commonwealth of Independent States (CIS) states have also introduced patent term extension.

Working Requirements

In the majority of countries, there is an obligation upon a patentee to exploit the invention within the country and, if he does not so 'work' the invention, his rights may be reduced, for example by compulsory licensing, or the patent may even be revoked. If a patent becomes liable to compulsory licensing, it means that the patentee no longer has a monopoly; any other person can obtain permission to use the invention, generally upon payment of a reasonable royalty to the patentee. Complying with working requirements may therefore be seen as one of the actions necessary to maintain a patent in force with full rights.

The Paris Convention places restrictions on the extent to which signatory countries can take action against a patentee for not working an invention. Compulsory licensing generally cannot be applied until three years after the grant of the patent,

[46] *In the Matter of EI Du Pont Nemours & Co.* [2009] EWHC 1112 (Ch).

so that the patentee has a breathing space in which he can make preparations to work the invention. The patent may be revoked for non-working only if compulsory licensing has failed, and then only after two years from grant of the first compulsory licence.[47] According to TRIPs, there must be no discrimination between products that are imported and those that are locally produced, so that importation from another WTO member must count as local working.[48] Measures against non-working should apply only if the patentee is refusing to supply the market in the country in question, whether by importation or otherwise.

The patent laws of most countries have more or less elaborate provisions for compulsory licensing of patents that are not worked, but, in the great majority of developed countries, these provisions are seldom if ever applied. As a matter of practical experience, only about 10 per cent of granted patents are ever actually worked; the majority of the others may be allowed to lapse by non-payment of renewal fees, but even those patents that are maintained in force and not worked are very rarely the subject of compulsory licence applications. The reasons for this are fairly clear, the chief one being that if it is not commercially worthwhile for the patentee to work the invention, it is not likely to be so for others either. Even if a patent covers what could be a good commercial product but is being maintained unworked for defensive reasons, a compulsory licence would only come in question if the parties were hostile and no agreement could be reached on a voluntary licence. In many cases, it would then cost more time and money to obtain a compulsory licence against the patentee's opposition than it would be worth.

In a few countries, of which the USA is the most important, there is no statutory obligation to work a patent and the patentee may, in theory, sit upon the invention, neither working it himself, nor being forced to grant a licence to anyone else. Nevertheless, the patentee may be forced to grant licences if he has been guilty of anti-trust violation or patent misuse (see p. 499). Some inventions can thus be effectively suppressed by a US patent, but only, of course, for a limited time. Allegations that are made from time to time about patents being used to permanently suppress inventions that might upset established industries are fictitious, because any patented invention must be published and must be available to the public when the patent expires.

Licences of Right

In the UK and in certain other countries, including Germany, France, and Italy, a patentee may voluntarily put himself into the position of someone whose patent is subject to compulsory licensing. The patentee does this by stating officially that anyone may as of right have a licence under his patent upon reasonable terms,

[47] Article 5A(3), Paris Convention.

[48] Article 27(1), TRIPs.

which will be set by the patent office or the courts if the parties cannot agree.[49] The patent may then be useful as a source of royalty income, but can no longer be used to exclude others from carrying out the invention. In return for accepting these restricted rights, the patentee can maintain the patent in force on payment of only half the normal amount of renewal fees.

In Germany, this action is irrevocable, but in the UK and other countries, the original position can be restored, if the patentee pays the balance of the renewal fees that would otherwise have been due in full and if the rights of any existing licensees are protected.[50] This 'licences of right' provision is also one that is not used as often as one might expect. If a patentee wishes to license the patent, he will normally try to do so by direct negotiation with one or more potential licensees rather than by throwing the patent open to all. This is because the most important thing that the patentee has to offer is the right to exclude others and a licensee will often not be interested unless he can obtain exclusive rights. This is not possible if a patent has been endorsed 'licences of right' and a licensee will generally not be willing to pay a high royalty rate for a licence that competitors may also have on the same terms.

[49] Section 46, PA 1977.
[50] Section 47, PA 1977.

10
ENFORCING PATENT RIGHTS

> No property is so uncertain as 'patent rights'; no property more speculative in character or held by a more precarious tenure. An applicant who goes into the patent office with claims expanded to correspond with his unbounded faith in the invention, may emerge therefrom with a shrivelled parchment which protects only that which any ingenious infringer can evade. Even this may be taken from him by the courts.
>
> *E. Bement & Sons v. La Dow* 66 Fed 185 (Cir Ct NY 1895)

What Constitutes Infringement

Once a patentee has obtained a granted patent and has done whatever is necessary to keep it in force, what can he do with it? As we have seen in Chapter 1, a patent does not give the patentee the right to practise the invention, but only to prevent others from doing so. Although this right is, in the UK, granted by the Crown, it is not up to the Crown to enforce it. Infringement of a British patent is not a crime for which one can be prosecuted, but a tort, that is, a civil wrong for which one can be sued in the civil courts. Essentially, the right given by a patent is the right to sue for infringement.

While this is true for all countries having an Anglo-Saxon legal system and also for many others, there are nevertheless a number of countries in which patent infringement *is* a criminal offence, either generally or in special circumstances, such as when there is deliberate and wilful infringement. Although criminal prosecutions for patent infringement are rare, the possibility may act as an additional deterrent (in Japan, for example, an infringer could, in theory at least, be sentenced to up to five years' hard labour), or may give procedural advantages to the patentee (thus, in Switzerland, a patentee alleging criminal infringement may be able to obtain a police search of the alleged infringer's premises to obtain evidence of infringement).

Infringement in the UK

Before the Patents Act 1977

The UK Patents Act 1977 is the first Patents Act to define what constitutes infringement. Previously, infringement was defined only by judicial interpretation of the wording of the patent grant (see Fig. 1 in Chapter 1), which gave the patentee the sole right to 'make, use, exercise or vend' the invention and forbade others to 'directly or indirectly make use of or put in practice the said invention'. For example, when the invention was a new chemical compound, the claim to that compound was considered to be infringed by making the compound, using the compound, or selling the compound. Because of the territorial limitations of the patent, any such activity had to be within the UK in order to be an infringement. Thus if the compound was made abroad and imported into the UK, the manufacture would not infringe the UK patent, but sale and use in the UK of the imported product would do so.

If the invention claimed was a method of making an article, including a chemical compound, then it was settled by early legal precedents that the patent was infringed not only by using the process of the invention, but also by using or selling the product of that process. This is frequently referred to as 'product-by-process protection', meaning that the unpatented product is protected by the process claim. This terminology may, however, cause confusion with the term 'product-by-process claim', in the sense of a product claim in which the product is defined in terms of the process used to prepare it. For the first meaning, the term 'derived product protection' is perhaps better. Furthermore, this reasoning was extended to situations in which the patent claimed a process for making a chemical compound used as an intermediate and it was held that sale in the UK of the unpatented final product infringed the patent. Because this principle was laid down in a series of cases involving intermediates for saccharin, it is commonly referred to as the '*Saccharin* doctrine'.[1]

If the claimed invention was a method of using an article or a compound, then the patent was infringed only by carrying out that method of use and not by making or selling the article used. A patent might, of course, contain claims of more than one type, but if any one claim of the patent was infringed, that sufficed to establish infringement of the patent.

In the course of time, a number of court decisions extended the meaning of the simple words used in the patent grant, to cover certain other situations in which commercial use was being made of the invention. For example, although mere possession of a patented article was not infringement, possession for trade purposes, such as with intention to resell, was considered to be 'vending'. The term 'use' was also held to cover situations in which a patented article was in position ready for use: for example a fire extinguisher was being used before it was actually operated to put out a fire.[2] Sales of a patented article as a 'kit of parts' to be assembled by the consumer could also constitute infringement.[3]

Case Law and Statute Law

This provides a good illustration of how British patent law, as many other branches of law in common law countries, is a blend of statute law and case law. Statute law is what is enacted by Parliament: for example the Patents Act 1977. Case law is a set of precedents derived from judicial decisions; these decisions may go back to the origins of English law, or may be judicial interpretations of Crown prerogative (such as the words of the patent grant), or of statutory enactments such as successive Patents Acts. What enables a reasonably consistent body of case law to be built up is the principle of *stare decisis*, by which a court is bound by an earlier decision of a higher court or a court of equal status to itself. Only since 1966 is the House of Lords not bound by its own precedents.

At first sight, this principle seems to rob judges of any independent power of decision, but this would be a gross exaggeration. The strict application of precedent applies only when the factual situation is the same, and this is rarely the case. The judge may always 'distinguish' the case at hand from an earlier one by pointing out differences in the facts of the case, and it is usually not too difficult to do so when he wishes to reach an opposite conclusion. Thus, there are some patent cases decided by the House of Lords that are generally thought to have been wrongly decided and which have never really been followed since. There may also be two distinct lines of cases taking opposite sides on a particular point, in which case, the judge must decide which set of precedents to follow. The Court of Appeal until

[1] For example, *Saccharin Corp. v. Anglo-Continental Chemical Works* (1900) 17 RPC 307 (HC).

[2] Hypothetical example given in *BUSM v. Simon Collier* (1910) 27 RPC 567 (HL).

[3] *Dunlop v. Moseley* (1904) 21 RPC 274 (CA).

recently considered itself strictly bound to follow its own earlier decisions, at least so long as there was a clear *ratio decidendi* on a point of law.[4] This principle was, however, relaxed in the case of *Actavis v. Merck*,[5] in which the Court followed decisions of the European Patent Office (EPO) Boards of Appeal rather than an earlier decision of the court in upholding the validity of a 'Swiss-type' claim based on a new dosage regimen (see p. 264). In the words of Lord Justice Jacob, 'this court is free but not bound to depart from the *ratio decidendi* of its own earlier decision if it is satisfied that the EPO Boards of Appeal have formed a settled view of European Patent law which is inconsistent with that earlier decision'.

What is clear, however, is that, if a point of law has been settled by the House of Lords, no lower tribunal, whether it be the UK Patent Office, the High Court, or the Court of Appeal, can simply decide differently if the factual situation is essentially the same. This means that the outcome of litigation is, in many cases, reasonably predictable, so that parties who are in dispute, for example a patentee and an alleged infringer, can make a reasonable estimate of their chances of success. If the precedents are clearly against one of the parties, that party will be well advised to settle out of court rather than spend large sums on litigation.

In countries having different legal systems, the system of precedent may not always apply. In Spain, it appears that only after the third concurring decision does a binding precedent exist, a principle first enunciated by Lewis Carroll.[6] In some countries, there is no precedent system whatsoever, so that the outcome of litigation is totally uncertain. In China, decisions of the Peoples' Courts are not binding upon the courts at the same or lower level, but a ruling on a point of law from the Supreme Peoples' Court is binding upon all lower courts. Such precedential rulings are rare, however.

In the USA, there is indeed a system of binding precedents, although it applies only to cases that have been officially designated as precedential. One difficulty there is that the US Federal District Courts are divided into 11 different circuits, each of which has a Circuit Court of Appeals, and a decision of a District Court or Circuit Court of Appeals is a precedent only within the same circuit. It is true that decisions of the Supreme Court are binding upon all lower courts, but the separate independent jurisdictions at circuit level have the result that the same point of law may be decided differently in different circuits. In non-patent litigation and previously also in patent cases, it can be of great importance to the litigants in which circuit the case will be tried, and a great deal of thought and effort goes into the question of where the action should be brought, and whether the jurisdiction of the court can be challenged and the action removed to a court in a different circuit. Such 'forum-shopping' tactics unnecessarily delay and complicate US litigation.

[4] *Young v. Bristol Airplane* [1944] KB 718 (CA).

[5] *Actavis UK Ltd v. Merck Ltd* [2008] EWCA (Civ) 444.

[6] 'What I tell you three times is true': Lewis Carroll, *The Hunting of the Snark*, 'Fit the First' (1876).

In patent litigation, however, the situation has now been much improved by giving the Court of Appeals for the Federal Circuit (CAFC) jurisdiction to hear appeals on patent cases from all Federal District Courts (see p. 23). Because the *stare decisis* principle is less strictly applied in the USA than in the UK, however, the CAFC does not consider itself bound by its own precedents, except that an *en banc* decision of the whole Court is binding upon any panel of the Court.

In countries in which there is the possibility of development of case law built up on judicial decisions, it sometimes happens that a body of case law relative to patents arises that is contrary to the original intentions of the legislature, or which has effects that the legislature wishes to change. When a new patent law is being considered, amendments may then be made with the intention of cancelling that part of the case law. As soon as the statute law is changed, the case law that interpreted the parts of the statute law that have been altered will normally cease to have any value as precedents. Thus, although the main object of the Patents Act 1977 was to adapt UK patent law to the European Patent Convention (EPC), the opportunity was taken to correct some minor points where it was felt that the case law was wrong or gave an unintended result.

More frequently, the legislature agrees with the case law developed by the judges and amends the statute law to make it say explicitly what the judges have held the previous law to mean. The case law is then codified as statute law and the earlier cases become superfluous as precedents, even though their principles are still followed. This is what has happened to the law of infringement in the UK. Previously, one had only the case law interpreting the cryptic wording of the patent grant; now, the Patents Act 1977 gives a statutory definition of what amounts to infringement. This definition now incorporates much of what had been decided in the earlier case law and also gives effect to the provisions on infringement that have been agreed as part of the Community Patent Convention (CPC), even though the CPC never entered into force. In some minor respects, it reverses previous case law and it also introduces one completely new type of infringement: contributory infringement.

Infringement Under the 1977 Act

It is now specified that importation of a patented product for commercial purposes constitutes infringement, as does keeping the product for commercial purposes.[7] In these respects, existing case law was written into the new statute law. The same rules as apply to patented products were made to apply to the direct product of a patented process. Derived product protection thus became for the first time part of the written law, but here the 1977 Act did not go so far as the old case law did. The use of the word 'direct' is clearly meant to do away with the *Saccharin* doctrine, so that sale of the final product of a multi-step chemical process is no longer an infringement of a patent for anything other than the last step of that process and the

[7] *Macdonald v. Graham* [1994] RPC 407 (CA).

sale of the end product will not infringe a patent covering the intermediate. The demise of the *Saccharin* doctrine has been confirmed by the Court of Appeal (see p. 451).[8]

A new aspect introduced in the 1977 Act was provision for what is usually referred to as 'contributory infringement', although it is not called that in the Act. Contributory infringement occurs under certain conditions when someone supplies any of the essential means for putting a patented process into effect and is as much an infringement of the patent as is carrying out the process itself. The conditions are that there is infringement only if the person supplied is not entitled (for example by a licence) to work the patented process and if the supplier knows, or should know, that the means are both suitable for putting, and intended to put, the invention into effect in the UK. If the means in question are a staple commercial product, there is no infringement unless the supply is made for the purpose of inducing infringement.

This definition, of which the above is only a summary, is a complex one, but the effect of the provision is that it should now be possible for a patentee who has invented a new use for a known substance to take action not only against unauthorized users, but also against persons selling the material for the new use. The patentee will generally not wish to sue the end users, who are, after all, customers or potential customers. The person whom the patentee wishes to sue is the rival manufacturer, which is using the patentee's invention to increase its own sales of the substance. If the substance has no other significant uses, it is enough if the seller merely knows that the substance is intended to be used for the patented use; if the substance is a commercial product with non-infringing uses, there will be infringement only if the seller sells the material with instructions to infringe.

Under the Patents Act 1977, certain rights are given as from the date of publication of an unexamined application. The applicant cannot sue anyone or obtain an injunction to stop the infringement until the patent has been granted, but may then claim damages as from the date of early publication, so long as the infringing act fell within the claims of the application as published, as well as the patent as granted. Similar provisions are found in the laws of most countries in which patent applications are published before examination.

It would obviously be unjust if a manufacturer could be stopped by someone else's patent from doing something that it had already been doing before the patent was first filed. Of course, if the manufacturer's activity gave public knowledge of the invention, there would be no problem, because the invention would no longer be novel and a later patent could not be valid. But if this activity was a process improvement that the manufacturer wished to keep as secret know-how, a subsequent patent would still be novel. The UK Patents Act 1949 got around this problem by making secret prior use a separate ground of invalidity, although not one that

[8] *Pioneer Electronics Capital Inc. v. Warner Music Manufacturing Europe GmbH* [1997] RPC 757 (CA).

could be brought in opposition proceedings. The 1977 Act approaches the problem from the other direction: the validity of the patent is not affected, but the person who began to use the invention before the priority date (or who made serious preparations to do so) has the right to continue what he was doing, or planning to do, after the patent is granted. This seems a fairer solution, because it leaves the patentee with rights that he can enforce against other parties who did not begin to use the invention until after the priority date.

As we have seen, unauthorized sale and use of a patented article constitutes infringement. What is the position, though, of a person who buys the article from the patentee? The answer is that the sale of the article by the patentee gives an implied licence to use and to resell it, and that this right is passed on to subsequent owners of the article, in the absence of a clear statement to the contrary.[9] The law in the UK still is that the patentee may sell the patented article subject to conditions (for example forbidding export to other countries), which, if properly notified, are binding on the purchaser and, if clearly marked on the goods, are binding also upon subsequent owners.[10] Such export prohibitions have essentially been abolished within the European Economic Area (EEA, see Chapter 25) and the other main purpose of conditional sale (resale price maintenance) has been abolished by UK statute law. Nevertheless, such conditions may be useful, for example, if a bulk chemical substance is sold to a non-EEA country for local formulation and resale, and the formulated product may be cheaply re-imported into the UK.

In most European countries, the principle of exhaustion of patent rights applies. Once a patentee has put the patented article on the market in the country in question, his rights in it are exhausted and he cannot impose any conditions on its subsequent sale or use. Many countries, including Japan[11] and, from 1 October 2009, China, extend this principle to sales anywhere in the world and do not allow national patents to be used to stop unauthorized 'parallel imports' of genuine goods put on the market in another country by the patentee or with his consent. This concept of worldwide exhaustion of patent rights by the first sale in any country runs counter to the national character of patent rights and can be very harmful to the patentee, particularly where artificial price differences exist between one country and another. Unfortunately this issue is deliberately avoided by the TRIPs Agreement.[12] This topic is discussed in more detail in Chapter 22.

The purchaser of an article that does not originate from the patentee or someone authorized by him may be liable for infringement by using it. In the UK, however, the purchaser has a measure of protection in the Sale of Goods Act 1979, which states that a seller gives an implied warranty of the right of 'quiet possession' of the

[9] *Incandescent Gas Light Co. v. Cantelo* (1895) 12 RPC 262 (QB).

[10] *National Phonograph Co. of Australia Ltd v. Menck* (1911) 28 RPC 229 (PC).

[11] *BBS Kraftfahrzeugtechnik AG v. Jap Auto Products* H-7(o) No. 1988 (Japanese Supreme Court, 1 July 1997).

[12] Article 6, TRIPs.

goods. If the buyer is sued for infringement, he can in turn sue the seller for breach of this implied warranty and should be able to recover any damages in that way.[13]

Infringement in the USA

In the USA, the patent law specifies that the patent shall grant to the patentee the right to exclude others from 'making, using, or selling the invention throughout the United States'. Where the invention is a process, however, it was until recently not infringement to use or sell the product of the process, a fact that greatly reduced the value of process claims in the USA. Thus a US patentee, whose patent covered a process, could not sue for infringement someone who carried out the process in another country and imported the product into the USA. The law was changed in this respect in 1988 and, since 23 February 1989, importation or sale of the product of a patented process has been infringement. Derived product protection is now a requirement of the TRIPs Agreement.

The patent owner may also be able to take action of a different kind against such imports by invoking the provisions of the US Tariff Act of 1930[14] against unfair competition. If it can be proved that the imports are products patented in the USA, or products of a process that would infringe a US patent if carried out in the USA, and that there is, or is being established, a relevant industry in the USA, then the imports may be stopped on application to the US International Trade Commission (ITC), which carries out its own investigation. This is much faster and cheaper than a normal infringement action, because, although the ITC is no longer legally required to give a decision within 12 months, it still aims to dispose of normal cases within this time frame. Also it avoids the need to sue the importer, who may be a customer or potential customer of the patentee. This procedure, which favours US patentees over their foreign competitors, might have been considered reasonable when there was no other way of enforcing a US process patent against imports of the product. Now that importation of the product is an infringement that can be dealt with in the normal courts, there is no remaining justification for the ITC jurisdiction, which is arguably contrary to the TRIPs Agreement, even though some procedural changes were made in 1994 to make things a little less favourable for the complainant.

Contributory Infringement

The principle of contributory infringement has long been recognized by case law in the USA but its operation has always been in conflict with another principle of US law, that of patent misuse. Patent misuse (see p. 498) may occur if the patentee attempts to extend his monopoly beyond the strict limits granted by law and has the

13 *Microbeads v. Vinhurst Road Markings* [1976] RPC 19 (CA).
14 19 USC 1337.

effect of making the patent unenforceable. It may be considered patent misuse to 'tie in', that is, to make a licence contingent upon the purchase of unpatented articles, and it has been argued that attempting to prevent contributory infringement is similarly trying to extend the patent monopoly beyond what is claimed, and hence to be patent misuse.

The US Patent Law of 1952 attempted to clarify the situation by defining contributory infringement for the first time and excluding from the scope of contributory infringement the sale of 'staple articles of commerce' for which there was at least one substantial use other than the patented use. Furthermore, the law explicitly stated that enforcing one's patent rights against contributory infringement did not constitute patent misuse. In spite of this seemingly clear statement of the law, the question remained in doubt until it was finally resolved by a decision of the Supreme Court in 1980.[15] The Court agreed that a patentee whose patent claimed the sole commercial use of an unpatented compound could stop others selling the compound for that use and was not obliged to grant a licence to other sellers (see Chapter 23). It should be noted that selling an article with instructions to use it to infringe a process patent is not treated as contributory infringement in the USA, but as inducement to infringe, which is considered to be a form of direct infringement.

Infringement in Relation to Drug Registration

As part of the compromise between research-based and generic pharmaceutical interests that culminated in the US Drug Price Competition and Patent Term Restoration Act of 1984 (see p. 170), generics companies were able to file abbreviated new drug applications (ANDAs) for drugs registered after 1961. To prevent sales before expiry of relevant valid patents, the innovator company lists its relevant patents in a Food and Drug Administration (FDA) publication known as the 'Orange Book', and an applicant for an ANDA must certify with the application:

(a) that the compound has not been patented;
(b) that any listed patents have expired;
(c) that listed patents exist, but that introduction will not take place before expiry; or
(d) that listed patents exist, but are invalid or would not be infringed by sale of the product.

Making the last type of certification (known as a 'Paragraph IV certification') is defined by statute to be an act of infringement and the patentee has 45 days in which to file suit if it wishes to challenge the ANDA. If the patentee does sue, approval of the ANDA is delayed for 30 months or until a court decision finding invalidity or non-infringement, but on ANDA approval, the first company to file a Paragraph IV certification may obtain a 180-day exclusivity period.

15 *Dawson Chemical Co. v. Rohm & Haas Co.* (1980) 206 USPQ 385 (Sup Ct).

This area of US law is, however, extremely complex and the above description must be regarded as an oversimplification.

Research Exemption

A question still under discussion in the USA is the extent to which experimental use of a patented product or process constitutes infringement. This was the subject of litigation in the case of *Roche v. Bolar*, decided in 1984 by the CAFC.[16] Bolar had been making preparations to introduce in the USA a generic version of a Roche pharmaceutical product as soon as Roche's patent expired, and had carried out clinical testing of its product while the patent was still in force, relying on earlier case law that held that experimental use did not amount to infringement. The CAFC held that the exception for experimental use must be construed narrowly: experimentation for pure speculative research was not infringement, but as soon as the experiments were directed to a clear commercial goal, then infringement occurred. The narrowness of the common law research exemption was made clear in the case of *Madey v. Duke University*,[17] in which it was held to cover only what was done 'for amusement, to satisfy idle curiosity, or for strictly philosophical inquiry'. Even use in a university for teaching purposes constituted infringement, because that was part of the legitimate business of the university and the non-profit status of the university was immaterial. As part of the US Drug Price Competition and Patent Term Restoration Act of 1984 (the Hatch-Waxman Act), Congress overruled the decision in *Roche v. Bolar*, specifying that it was not infringement to make, use, or sell a patented invention 'solely for uses reasonably related to the development and submission of information under a Federal law which regulates the manufacture, use, or sale of drugs'.[18] *Roche v. Bolar* is, however, overruled only in the pharmaceutical, veterinary, and medical device fields, and experimental testing of a patented agrochemical product, for example, would still constitute infringement, even though regulatory approval is required in order to sell the product.

Although it was clear that the Hatch-Waxman exemption covered all manufacture or use of material used in clinical testing, it was unclear to what extent it covered pharmaceutical research generally. In the case of *Integra v. Merck*,[19] the German company Merck AG argued that all activities involved in drug discovery and development were exempted, but the CAFC did not follow this argument, holding that preclinical research activities, as well as research the results of which were not actually submitted to the FDA, did not fall within the 'safe harbour' created by the Hatch-Waxman Act. To hold otherwise, the Court said, would be to destroy the value of all research tool patents. Judge Newman dissented. In her opinion, the Integra patent was not for a research tool and what Merck and its collaborators at

[16] *Roche Products Inc. v. Bolar Pharmaceutical Co. Inc.* 221 USPQ 937 (Fed Cir 1984).

[17] *Madey v. Duke University* 64 USPQ 2d 1737 (Fed Cir 2002).

[18] 35 USC 271(e)(1).

[19] *Integra Lifesciences Ltd et al. v. Merck KGaA et al.* 66 USPQ 2d 1865 (Fed Cir 2003).

the Scripps Research Institute had done was research to find new uses for the invention. There was, or should be, a common law research exemption for such activities and §271(e)(1) took over where that left off.

Certiorari was granted, and the Supreme Court largely adopted the position of Judge Newman and refused to limit the extent of the safe harbour.[20] The decision did not say anything about whether or not research tools used in connection with FDA submissions might fall under the Hatch-Waxman exemption, but the CAFC has since held that a device used for measuring the parameters of aerosol sprays in connection with FDA approval of nasal inhalers was not immunized from infringement.[21]

Sovereign Immunity

US universities were not happy with the decision in *Madey v. Duke*, but an earlier decision of the Supreme Court[22] had given certain universities carte blanche to infringe any patents that they pleased. The Supreme Court, in a five-to-four decision, reversed a decision of the CAFC and held that neither a state, nor an agency of a state, may be sued for patent infringement in a Federal court without its consent in view of the 11th Amendment to the US Constitution. A state university is considered an agency of the state, and thus cannot be sued (unless it consents). A law that had been passed by Congress (the US Patent Remedy Act of 1992) specifically to allow states to be sued for patent infringement was overturned as being unconstitutional.

In view of the fact that state universities such as the University of California are among the most prolific users of the patent system and do not hesitate to enforce their own patents, it is totally unjust that they should have immunity from suit themselves. Proposals have been made in Congress for legislation to bar state universities from suing other parties for infringement unless they waive their own immunity, but, so far, such a law has not been passed and might be held unconstitutional if it were. The Supreme Court had the opportunity to revisit the subject in the case of *Biomedical v. California*,[23] but, unfortunately, declined to grant certiorari.

A company that has contract research completed on its behalf at a state university is, however, liable for inducing infringement.[24] What the university is doing is still infringement, even if it cannot be sued for it itself.

Research Exemptions in Other Countries

In the UK, it is fairly safe to say that the activities of the accused infringers in both *Integra v. Merck* and *Madey v. Duke University* would have been considered to fall

[20] *Merck KGaa v. Integra Lifesciences* 545 US 193 (Sup Ct 2005).

[21] *Proveris Scientific Corp. v. Innovasystems Inc.* 536 F.3d 1258 (Fed Cir 2008).

[22] *Florida Prepaid Postsecondary Education Expense Board v. College Savings Bank* 51 USPQ 2d 1081 (Sup Ct 1999).

[23] *Biomedical Patent Management Corp. v. State of California* 505 F.3d 1328 (Fed Cir 2007).

[24] *Syrrx Inc. v Oculus Pharmaceuticals Inc.* 64 USPQ 2d 1222 (D Del 2002).

within the statutory research exemption,[25] excluding from infringement acts done 'for experimental purposes relating to the subject-matter of the invention'. Similar provisions exist in the laws of most European Union (EU) countries, with wording originally derived from the CPC.

Decisions similar to *Roche v. Bolar* have been reached by courts in a number of countries, including the UK and Germany, in litigation between Monsanto and Stauffer.[26] Stauffer planned to introduce a product very similar to Monsanto's patented herbicide Roundup® and made very extensive field trials of its substance. The courts held consistently that the field trials were just as much an infringement as commercial sales would have been.

There have, however, been contrary decisions in a number of countries and more countries are writing so-called '*Bolar* exemptions'[27] into their laws. In Germany, the first step was the case reported as *Klinische Versuche I*,[28] in which the finding of non-infringement was justified by the *Bundesgerichtshof* (BGH, or Federal Supreme Court) on the basis that the testing was to discover new uses of the drug. This was extended by a subsequent case in which there was no new use, but it was considered not to be infringement to seek to remove uncertainties about the effects and tolerability of a pharmaceutical composition containing the patented substance.[29] In Japan, the decision of the Tokyo District Court in *Otsuka*[30] gave a sweeping refusal to consider testing as infringement, based on public policy considerations.

In 1998, the Australian patent law was amended to allow extensions of term of pharmaceutical patents, but it was enacted that not only in the extended period, but also in the main term of the patent, testing for commercial purposes should not be considered as infringement from the day on which the application for extension is granted, which will usually be long before the extension starts. The law in Canada allowed even manufacturing and stockpiling of a patented product prior to patent expiry, and the compatibility of this law with TRIPs was the subject of World Trade Organization (WTO) dispute proceedings between the EU and Canada. In August 2000, the Dispute Panel decided against the stockpiling provisions, but held that the basic '*Bolar* exemption' was not contrary to TRIPs.

The pharmaceutical industry's fight against *Bolar* is unfortunately a lost cause. More and more countries, most recently China, are incorporating such provisions into their laws and it has now become standard throughout Europe as part of a deal in which the duration of regulatory data protection (the time during which a generic competitor cannot copy or refer to the data submitted by the innovator to obtain marketing approval) was extended (see p. 268).[31]

25 Section 60(5)(b), PA 1977.

26 *Monsanto v. Stauffer* [1985] FSR 55 (Pat Ct).

27 Wrongly so called, because *Roche v. Bolar* decided that such use *was* infringement.

28 X ZR 99/92 *Klinische Versuch I* (OJ 1997, 588); [1997] RPC 623 (BGH).

29 X ZR 68/94 *Klinische Versuch II* (OJ 1997, 589) (BGH).

30 *Otsuka/Procaterol Hydrochloride* (Tokyo Dist Ct) 7430/1966.

31 Article 2.6, Directive 2004/27/EC.

Procedure in the UK

One of the innovations introduced by the Patents Act 1977 is that an infringement action in the UK may be heard by the UK Patent Office,[32] provided that both parties agree. The procedure has seldom been used, probably because the Patent Office has no power to grant an injunction, which is usually the most important remedy, and because the Patent Office has a tradition of giving the applicant or patentee the benefit of any doubt, making it unattractive for the defendant. If the case is of any importance, any decision of the Patent Office would in any case be appealed to the courts, so even though the Patent Office procedure is cheap, it saves time and money to go to the courts in the first place. The Patents Act 2004 provides a new procedure whereby the Patent Office may, on the request of any party, give a non-binding opinion on whether a particular activity would infringe a particular patent. It is not yet clear how much use will be made of this possibility.

Accordingly, nearly all infringement actions are still heard by the courts. Until recently, this meant in the first instance the Patents Court set up under the 1977 Act as a part of the Chancery Division of the High Court to hear all patent matters, including appeals from the Patent Office. The judges of this Court are specialist patent judges, who have a considerable amount of technical, as well as legal, knowledge. If the technology involved is particularly complex, the judge may sit together with a scientific adviser, although this provision has never been used. There have, however, been cases in which a scientific adviser has been used to instruct the judge in the technical subject matter of the invention prior to trial. For this purpose, it may be more important that the scientist be a good teacher who can explain matters clearly than that he be a leading expert in the field. A leading expert was, however, used to teach the judges of the House of Lords the basics of recombinant DNA technology before they heard *Kirin-Amgen v. TKT* (see p. 290).[33] Now, the Patents County Court (PCC) is an alternative first-instance forum (see below).

Infringement proceedings are now regulated by the Civil Procedure Rules (CPR) 1998,[34] which came into effect on 26 April 1999, and apply to all civil litigation in England and Wales. The 'overriding objective' of the CPR is to deal with cases justly, which includes equality of the parties, economy, proportionality, and expedition. More powers are given to the judges to manage the case so as to prevent undue delay, the extent of discovery of documents is reduced, and there is more use of written evidence presented to the judge before trial. The old practice of highly paid barristers reading out evidence in court for days on end no longer exists.

An infringement action under English High Court procedure now begins by the 'claimant' (formerly known as the 'plaintiff') serving the defendant with a claim form (rather than a writ). The claim form must indicate which claims of the patent

[32] Section 61(3), PA 1977.

[33] *Kirin-Amgen Inc. v. Transkaryotic Therapies Inc.* [2005] RPC 9 (HL).

[34] SI 1998/3132.

are alleged to be infringed and give at least one example of each type of alleged infringement. The defendant, if he wishes to defend the action, must file a statement of defence and, if he makes a counterclaim that the patent is invalid, must give the full grounds of invalidity on which he intends to rely. It is possible to add to these later, but late submission of grounds may lead to additional liability for costs. The claimant has the opportunity to reply to the allegations of invalidity at this stage.

Within 14 days after all statements of case have been filed, the claimant must request a case management conference (CMC) at which the judge will set a timetable for the case, and decide what expert evidence is required and what should be the scope of disclosure. Each party may at this stage request the other side to admit relevant facts, give further particulars, or answer written questions (requests for information). At this stage, 'disclosure' (formerly, 'discovery') is carried out: that is, each party must deliver to the other a list of all relevant documents within the scope of disclosure ordered by the judge. The other party is entitled to inspect these documents and use them in evidence if it wishes. Certain documents may be privileged, for example a lawyer's or patent agent's advice to his client, and these may be listed separately and are not made available to the other party. If it is proposed to carry out experiments to resolve technical issues, these should be notified to the other side.

Finally, the case comes on for trial, which consists of oral arguments before the judge (no jury is involved in British patent cases), and examination and cross-examination of witnesses on either side. These may include expert witnesses who have been asked by one side or the other to give opinions on technical matters or to carry out experiments or trials. Finally, the barristers on each side sum up their arguments. The judge may give his decision on the spot, but more usually will reserve judgment and give a decision two or three weeks later. The losing party may ask the first-instance court for leave to appeal to the Court of Appeal, and if this is refused, may ask the Court of Appeal directly. In such cases, a committee of the Court of Appeal considers whether the party has any reasonable prospect of success on appeal and, if not, leave to appeal is refused. This procedure was recently challenged on the basis that the TRIPs Agreement requires judicial review of any decision to revoke a patent,[35] but it was held that the preliminary review by the committee met this requirement and that there was no obligation to allow a full appeal in every case.[36] A party that loses on appeal may ask for leave to appeal further. Up to October 2009, the court of final instance was the House of Lords (strictly speaking, the Judicial Committee of the House of Lords, consisting only of senior judges, known as 'Law Lords'); after that date, it is to a new body set up by the Constitutional Reform Act 2005, that is, the Supreme Court.

[35] Article 32, TRIPs,

[36] *Pozzoli v. BDMO* [2007] EWCA Civ 588.

Patent litigation in England is an expensive matter: the litigant will normally employ a solicitor and a patent agent, as well as at least one barrister; if a Queen's Counsel (QC) is briefed, as would be usual for an important case, then junior counsel must also be retained. Costs are chiefly a factor of how many days (or weeks) the hearing lasts and, despite the progress that has been made by the courts in reducing the length of patent cases by increased use of written evidence, costs remain high.

Since April 2003, a new 'streamlined procedure' has been available for patent litigation if both parties agree. Under this procedure, all evidence is given in writing, there is no disclosure of documents, there are no experiments, and there is only limited cross-examination. The trial duration is fixed and will not normally be more than one day. It seems doubtful, however, that this will be much used in practice.

One consequence of the high cost of patent actions is that a small company may own a valid patent, but not be able to afford to enforce it, particularly against an infringer who has greater financial resources. Attempts have been made from time to time to launch insurance schemes under which a patentee pays a fixed annual premium for each insured patent; if the patentee wishes to sue an infringer and obtains counsel's opinion that he has a reasonable case, the insurance underwriters will pay the legal costs of the infringement action. Such schemes have not proved very successful. Naturally, insurers want to pay only for those cases that have a high probability of success, but clear cases are normally settled out of court and it is the doubtful ones, the results of which cannot be predicted, which go to trial.

The procedure described above is that of England. It should not be forgotten that Scotland has a separate legal system based on civil law rather than common law, and that patent actions may also be brought in Edinburgh before the Outer House of the Court of Session, from which appeal lies to the Inner House of the Court of Session, and from there (with leave) to the Supreme Court. The procedure is similar, but the nomenclature is different. The litigant will now be the 'pursuer' (claimant) or 'defender' (defendant); the pursuer will employ 'writers to the signet' and 'advocates' instead of solicitors and barristers, and may if successful get an 'interdict' instead of an injunction. Indeed it is generally quicker and easier to obtain an interim interdict in Scotland than the corresponding remedy in England, and such an interdict need not be limited to Scotland, but may be granted to cover the entire UK. If the case proceeds to a full trial, however, this will take longer than in England.

The Copyright, Designs and Patents Act 1988 set up the PCC as an alternative designed to make the procedure faster and cheaper. The idea was that the PCC would make more use of written pleadings, more along the lines of the procedure before the EPO Boards of Appeal (the judge first appointed to the PCC was a former member of the EPO Enlarged Board of Appeal). In practice, this Court has not been used as much as had been expected. A number of its early decisions were overturned by the Court of Appeal and the recent steps to expedite the hearing of High Court patent actions means that there is little difference in speed between the two alternatives, so that the extra cost of High Court proceedings may be considered by some litigants as justified by a higher reliability of the decision. A mediation service in patent disputes offered by the PCC has also met with little or no response. Even with the appointment of a new judge and the possibility of using new streamlined procedures, the use of the PCC has remained low. Proposals to remedy this include mandatory use of streamlined procedures and a cap on costs.[37]

Remedies

If the decision of the court is that the patent is valid and has been infringed, then the plaintiff may be granted the following remedies:

(a) an injunction;
(b) damages (or an account of profits);
(c) delivery up or destruction of infringing goods;
(d) a declaration of validity; and/or
(e) costs.

In most cases, the injunction is the most important remedy of all. It is a direct order from the court to the defendant to refrain from future acts of infringement; disobedience of the order is contempt of court, which the court can punish by fine or imprisonment. Although the injunction normally ends when the patent expires, an injunction was, in one case, granted to stop sales for a period of 12 months after patent expiry to compensate for pre-expiry infringing activity.[38] This is distinct from an interlocutory injunction, which is an order from the court issued before the trial to stop acts alleged to be infringements until the case has been settled. Interlocutory injunctions in patents cases are now more easy to obtain than they used to be and can be extremely useful to the patentee (see Chapter 23). But if an interlocutory injunction is granted and the patent is subsequently found to be invalid or not infringed, the patentee will have to pay compensation to the defendant for the time during which his activities were halted by the injunction.

Damages may be awarded by the court in respect of past acts of infringement (but not any that occurred more than six years before the writ was issued). Damages and account of profits are two alternative ways of reckoning the payment to be made by the infringer to the patentee: damages are based on the loss sustained by the patentee as a result of the infringement; an account of profits is based on the profits made by the infringer. Of the two, damages are by far the most common. At the least, the successful patentee should get the equivalent of a fair royalty plus interest from the infringer. This, of course, is no real deterrent to infringement, because the infringer, if he loses the case, may pay no more than he would have done had he taken a licence in the first place. Post-expiry damages to compensate for pre-expiry infringement have been granted by the Court of Appeal.[39]

[37] See, e.g., G. Morgan and R. Price, 'Providing UK patent litigants with a real alternative' (2009) 211 Patent World 31.
[38] *Dyson v. Hoover* [2002] RPC 465 (CA).
[39] *Gerber Garment Technology Inc. v. Lectra Systems Ltd* [1995] RPC 383 (CA).

There are, however, certain restrictions on the payment of damages. Damages are not awarded against an 'innocent infringer' who can prove that he was not aware of, and had no reason to be aware of, the existence of the patent. It is practically impossible for a large company, particularly one with its own patent department, to claim to be an innocent infringer. The patentee can also make it difficult for imitators to claim to be innocent infringers by marking the patented articles that he makes with the patent number. It is not enough merely to mark 'patented' if the number is not given. It is an offence punishable by a fine to represent falsely that an article offered for sale is patented[40] or that a patent application on the article has been filed.[41]

The courts also have discretion to refuse damages in respect of a period for which a renewal fee was paid late and may reduce or refuse damages if the patent was amended after publication, or was found to be only partially valid, unless the patent was originally framed 'in good faith and with reasonable skill and knowledge'. Damages can be refused if the patent was originally drafted deliberately far too broadly and had to be cut back in scope later.

Delivery up of infringing goods applies only to articles currently in the possession of the infringer; often it is not asked for, but if the infringer has a large inventory or if the infringing device is a machine capable of making large quantities of articles, this form of relief can be well worth having. Goods delivered up do not become the property of the patentee; all that he can do is destroy them or have them rendered non-infringing.

A declaration of validity of the patent is a useful deterrent to future infringers and may affect the costs in any future action.

The party that wins an infringement action is normally awarded costs, which means that the other party must contribute towards the expenses that he has incurred in the litigation. The party awarded costs must not expect everything to be paid; costs are assessed by an official called the 'Taxing Master' according to a fixed scale and normally will amount to less than half of the winning party's actual out-of-pocket expenses. If a certificate of validity has been granted in earlier proceedings, however, and the patentee successfully sues a second infringer, then he may be awarded costs on a higher scale, which might actually approximate to his real outlay.

Who May Sue

We have referred to the person suing for infringement as the patentee, but this may not always be the case. The original patentee may have sold the patent to another party; that party, the assignee, will then have all of the rights associated with the

[40] Section 110, PA 1977.
[41] Section 111, PA 1977.

patent, including the right to sue infringers. The patentee may also have granted licences under the patent.

We shall discuss licensing more fully in Chapter 24, but it should be noted here that there are basically three types of licence, as follows.

(a) An *exclusive* licence gives the right to operate under the patent to the licensee to the exclusion of all others, including the patentee himself.
(b) A *sole* or *semi-exclusive* licence is one under which the licensee is assured that no more licences will be granted, but the patentee retains the right to work the invention himself.
(c) A *non-exclusive* licence leaves the patentee free to grant any number of further licences.

British patent law gives to an exclusive licensee, but not to a sole or non-exclusive licensee, the same right to sue for infringement as that of an assignee. There is one catch, however: an assignment or exclusive licence is supposed to be registered at the Patent Office; under previous Patents Acts, this requirement was often ignored, but under the 1977 Act, an assignee or exclusive licensee who does not register within six months of the agreement being signed loses the right to claim damages for any infringement before the date on which he does register.[42]

There may of course be more than one patentee, because a patent may be granted to any number of co-patentees jointly. When there is more than one patentee, each co-patentee may sue for infringement without the consent of the others, but is not free to grant licences without the agreement of the other co-owners of the patent.[43]

Threats

A patentee must be careful not to threaten others rashly with an infringement action, because it is possible for any person aggrieved by such conduct to retaliate with a suit for unjustifiable threats. This seems a somewhat anomalous provision of the patent law: after all, if I threaten to sue someone for trespass, breach of contract, breach of copyright, or indeed any civil tort other than infringement of a patent or a registered design, the person whom I threaten has no statutory remedy. The reason that patents were singled out was to put a stop to the once-common practice of a manufacturer threatening its competitors' customers with an infringement action. The customers, knowing nothing of patents and anxious to avoid any legal problems, were often frightened away from the competitors' goods, even if no infringement was actually present.

To avoid this abuse of patent rights, not only the customers actually threatened, but also the manufacturer whose goods were alleged to infringe could sue the person making the threats (whether or not he was the patentee) and if successful could

[42] Section 68, PA 1977.
[43] Section 36(3), PA 1977.

get an injunction and damages. To succeed, they had to show either that the acts were not infringements, or that the patent was not valid. This situation has been improved slightly by the UK Patents Act 1977, which in effect rules out a threats action if the person threatened is a manufacturer rather than a customer.[44] Nevertheless, the patentee normally has to start with the retailer marketing the infringing goods and may have no way of knowing the identity of the manufacturer. Thus the possibility of threats actions makes it difficult for a patentee to deal in any straightforward way with an infringement situation, because any warning letter that did more than notify the infringer that a patent existed could be held to be a threat. Even if an infringement action was planned, a warning letter gave the infringer the chance to lodge a threats action, thereby gaining certain procedural advantages and putting the patentee on the defensive. Furthermore, the fact that not only the patentee, but also a solicitor or patent agent who sends a warning letter on his behalf, may be personally liable in a threats action can easily disrupt the attorney–client relationship. The Patents Act 2004 improves matters somewhat by enabling the patentee to write to a retailer requesting information about the manufacturer or supplier of the goods, and to write a warning letter to the company representing the highest identifiable point in the supply chain, without risking a threats action.[45]

Procedure in the USA

In the USA, there is a dual system of state and Federal courts, but whereas state courts may hear certain cases related to intellectual property, for example those involving contract law or trade secrets, the Federal courts have sole jurisdiction in matters of patent law. Patent infringement actions are heard in the first instance by the Federal District Court in the district in which the alleged infringer resides or has his registered office, or in which he has a regular place of business *and* is infringing. There will generally be the possibility of bringing suit in more than one district, and the one chosen will depend on factors such as the pro- or anti-patent attitude of the courts, geographical convenience, and the backlog of cases before the court. The plaintiff will usually want a rapid decision and may want to seek out a district court with a small docket of cases; for a number of years, the US District Court for the Eastern District of Virginia has been known as the 'rocket docket' for this reason. More recently, the Eastern District of Texas has become popular as a venue for patent litigation because of its reputation as being patent-friendly: patent owners have won 77 per cent of trials. This court frequently accepted jurisdiction at the request of the plaintiff even where the defendant had little or no connection with Eastern Texas, but in some cases, the venue has been successfully challenged in either the Federal Appeals Court for the Fifth Circuit, or, more recently, in the CAFC.[46]

[44] Section 70(4), PA 1977.

[45] Section 70(5), PA 1977, as amended by s. 12, PA 2004.

[46] *In re TS Tech USA* 551 F3d 1315 (Fed Cir 2008).

The validity of the patent can, as in the UK, be contested by the defendant and a further complication is that, where there is an actual controversy, the infringer can take the initial step by bringing an action for declaratory judgment of invalidity, which can be brought in the Federal District Court in which the patent owner resides, or, for a foreign patent owner who has not appointed an agent to receive service of process, in the US District Court for the District of Columbia. An exception to the normal jurisdiction occurs when the alleged infringement is a sale to the US government. The seller cannot be sued for infringement and the patentee must sue the government in a special court known as the US Court of Federal Claims.

The procedure in US courts is somewhat similar to that in England, starting in the same way with formal pleadings. Compared to the English system, however, the trial itself is shorter and the pretrial proceedings far longer. It is not unusual for a party, in response to a request for production of documents, to make available for inspection and copying by the other party hundreds of thousands of pages of documents. The powers of the court to order discovery are exercised more extensively and there is also much taking of evidence on oath (depositions) before the trial begins. After evidence is taken, one side may move for summary judgment in its favour, which may be granted if there is no unresolved issue of fact, and a decision can be given on legal points alone—known as 'judgment as a matter of law' (JMOL). The court may also grant a preliminary injunction before trial. This used to be rare unless the patent had been held valid in previous suits, but is now becoming more common. In one case,[47] however, in which the District Court had granted a preliminary injunction and the defendant appealed to the CAFC, the CAFC not only vacated the injunction, but also declared the patent invalid without there ever having been a full trial of the action.

The judge may require a pretrial conference with the lawyers of both sides to try to narrow the issues or reach an out-of-court settlement. If the validity of the patent is being challenged on the basis of prior art not considered by the US Patent and Trademark Office (USPTO), the judge may stay the infringement action until the validity has been re-examined by the USPTO, using the reissue or re-examination procedure described in the next chapter.

As well as purely legal defences, such as non-infringement and patent invalidity, the defendant may rely on so-called 'equitable defences', which are based on the old principle of equity that a plaintiff seeking relief must come into court with 'clean hands'. Thus, the defendant may allege that the plaintiff has 'unclean hands' because of patent misuse, inequitable conduct (fraud on the USPTO), undue delay in bringing suit, and any other reason that the fertile brain of his attorney can come up with.

The trial itself may be heard before a judge alone, or with a jury, whereby the jury can make findings of fact, but the judge rules on questions of law. In recent years,

[47] *Genentech v. Novo Nordisk* 42 USPQ 2d 1001 (Fed Cir 1997).

the number of jury trials has increased, largely because juries appear to be more sympathetic to the plaintiff than do judges. Between 1995 and 2008, the success rate of plaintiffs in a jury trial was 79 per cent as compared with 44 per cent in a bench trial (that is, with no jury).[48] But the role of the jury in a patent trial has been reduced by the 1996 Supreme Court decision in *Markman*,[49] which held that the construction of patent claims is a question of law rather than fact and so must be decided by the judge alone. As a result of this, there will often be a pretrial '*Markman* hearing', at which questions of claim construction will be decided before a jury is even selected. In many cases, the question of infringement turns on the issue of claim construction, so the *Markman* hearing may be determinative. The idea that a jury of laypersons is competent to decide on complex technical issues is one that seems absurd to many British observers, but jury trials of commercial civil cases were abolished only relatively recently in the UK[50] and, because the right to a jury trial in most civil cases involving more than US$20 is enshrined in the Seventh Amendment to the US Constitution, it is not something that is easy to change.

The trial itself will be relatively short by English standards and the judge will not normally be a specialist in patent law. If the patentee is successful, the remedies that he obtains are essentially the same as in England, except that costs do not include attorney's fees, which are awarded only in exceptional cases. An injunction to restrain further infringement used to be granted almost automatically in patent cases, but this has changed since the opinion of the Supreme Court in the *eBay* case.[51] In this case, eBay was found to have infringed a valid patent, but the District Court refused to grant an injunction because the facts that the patentee company had granted a number of licences and was not practising the invention itself indicated that it would not suffer irreparable harm if an injunction were not granted. On appeal to the CAFC, it was held that there was a general rule in patent cases that, if validity and infringement were found, a permanent injunction must necessarily follow, except in very special circumstances. The Supreme Court held that both lower courts were wrong: there was a well-established rule setting out a four-point test for the grant of injunctive relief, which applied also for patent cases, and this test had to be applied on the merits of each individual case. The four points are that:

(a) the plaintiff has suffered an irreparable injury;
(b) monetary damages are inadequate to compensate for this;
(c) considering the balance of hardships, a remedy in equity is warranted; and
(d) the public interest would not be disserved by a permanent injunction.

48 Federal Trade Commission (FTC) Conference, 'The evolving intellectual property (IP) marketplace', 11–12 February 2009.

49 *Markman v. Westview Instruments Inc.* 38 USPQ 2d 1461 (Sup Ct 1996).

50 A.P. Herbert, 'Why is a jury?', *Wigs at Work*, Penguin, 1933; 1966.

51 *eBay Inc. v. MercExchange LLC* 126 SC 1837 (2006).

Damages must be at least equivalent to a reasonable royalty and are usually reckoned in terms of the patentee's lost profits, in which case they may be very substantial indeed. Furthermore, damages awarded may be tripled if the court finds that there has been 'wilful infringement'. Until recently, this meant that there was an affirmative duty to exercise due care to determine whether or not one was infringing a valid patent by one's proposed activities,[52] and this normally meant that one had to get proper independent legal opinion that the patent was likely to be held invalid or not infringed. A recent *en banc* decision of the CAFC[53] considered that 'wilful' infringement required more than mere negligence and set the higher standard of 'recklessness', overruling *Underwater Devices*.

Damages may be awarded only if the infringer has been notified of the infringement[54] and marking the product with the relevant patent number is considered to give such notice. As in the UK, however, false marking is a crime, with a penalty of up to US$500 for 'each offence'[55] and, if the court considers each marked article to be a separate offence, this can rapidly become expensive. The law provides that a third party may sue for false marking, and if successful, may receive half of the penalty payment (the other half going to the government). This encouraged a Washington patent attorney to sue Solo Cup for false marking in respect of millions of paper cups; as of writing, the case is still pending. Most pharmaceutical companies do not mark their products with patent numbers: potential imitators are well aware of the relevant patents; an injunction is in any case a more important remedy than damages; and the administrative effort and risk of a false marking suit outweigh the benefits.

The consequences of losing a patent infringement action in the USA can be very serious, as Eastman Kodak found out to its cost when at the end of 1985 it was held to have infringed patents owned by Polaroid by the sale of its instant cameras and film.[56] An injunction against further sales meant that Kodak had to shut down its instant camera business, and had to pay damages and interest totaling over US$900 million.[57] (It has been suggested that the subsequent demise of Polaroid was due to lack of healthy competition from Kodak, but the advent of digital cameras is a more likely explanation.) From 1995 to 2008, there have been 22 cases in which damages of over US$100 million have been awarded, with six such awards in 2006 alone. The average amount of damages awarded in jury trials is an order of magnitude higher than in bench trials, the current record being the damages of US$1.6 billion, awarded to Centocor against Abbott Laboratories in 2009 by a jury in the Eastern

[52] *Underwater Devices Inc. v. Morrison-Knudson Co.* 185 F.3d 1259 (Fed Cir 1983).

[53] *In re Seagate Technology* 497 F3d 1360 (Fed Cir 2007, *en banc*).

[54] 35 USC 287(a).

[55] 35 USC 292.

[56] *Polaroid Inc. v. Eastman Kodak Co.* 228 USPQ 305 (D Mass 1985); upheld 229 USPQ 561 (Fed Cir 1986).

[57] US$909,457,567, to be precise: *Polaroid Inc. v. Eastman Kodak Co.* 16 USPQ 2d 1481 (D Mass 1990).

District of Texas, another reason for the popularity of that court among patent owners. The cost of an infringement suit in the USA is very high, normally ranging from US$100,000 to well over US$1 million for the first instance alone. Most of the costs are incurred during the pretrial proceedings.

If the patent is found invalid, it is not formally revoked, as in England, but merely declared unenforceable. In the past, such a decision was binding only as between the parties, but the position now is that collateral estoppel exists, meaning that the patentee is barred from asserting the patent against any other infringer. Thus, an unfavourable decision in a Federal District Court that is notoriously anti-patent will, unless successfully appealed, prevent the patentee from enforcing that patent in a court in which he would have had a good chance of success.

The CAFC, which also hears appeals from the USPTO and from the US Claims Court, has, indeed, done away with the problem of conflicting precedents in different circuits in patent cases, but because many of its decisions are non-precedential and because most cases are not heard by the whole Court of 12 judges, but normally by three-judge panels, total consistency cannot be expected. Indeed, there have already been instances of the same point being decided in opposite ways by two different panels of the CAFC within a period of two or three months. It is possible to petition the CAFC for a rehearing *en banc*, that is, by the full court, and this is occasionally granted. In 2008, looking at cases decided on the merits, 55 per cent of patent cases arising from the district courts were affirmed, 24 per cent were affirmed in part, and 19 per cent were reversed.

Although the US Supreme Court is not an appellate court in the normal sense, it may agree to review cases decided in the CAFC, by granting a petition for a writ of certiorari. It normally does so even more rarely than the House of Lords hears patent cases in England, but in the last five years, there have been an unusually high number of patent cases heard by the Supreme Court, so that it is sometimes said that, because Congress is incompetent to reform US patent law, the Court is doing the job instead.

Patent litigation in the USA is very much a growth industry. In 1987, 500 US patents were litigated; in 1991, 1,000; in 1999, 2,000; and in 2007, no fewer than 4,000. In view of the high legal costs, the patent-friendly attitude of juries and most judges, the enormous sums that may be awarded in damages, and the devastating effects of an injunction, the mere threat of litigation is often enough to intimidate an accused infringer, even if no infringement exists or the asserted patent is invalid.

Procedure in Continental Europe

Courts in continental Europe have a different type of procedure from the Anglo-Saxon system of the UK and the USA. Most of the evidence is written, rather than oral, and the role of the judge is seen more as that of an investigator than that of a referee. In some countries, such as France, the Netherlands, and Switzerland, the same court can hear both the infringement action and a counter-claim of invalidity, but in the others, the question of validity is treated as a separate issue that is heard in a different court or is referred back to the patent office.

In Germany, for example, infringement actions are brought first in one of a selected group of state courts (*Landgericht*), which have a certain degree of specialization in patent matters. In practice, most cases are heard in Düsseldorf or Munich. Appeal lies to the higher state court (*Oberlandesgericht*) and from there (on questions of law) to the BGH. The validity of the patent cannot be considered by these courts, however; only the *Bundespatentgericht* (BPatG, or Federal Patents Court) in Munich can do so.

The proceedings consist of alternate exchanges of written briefs and short hearings. Often, a neutral expert is appointed by the court to give a report on the technical aspects of the case. There is essentially no cross-examination of witnesses as there is in England. Interlocutory injunctions are unusual, but may be granted by the court in clear cases.

Whereas in England the judgment of the court of first instance is normally suspended if the case is appealed, in Germany the plaintiff who wins in the *Landgericht* may compel enforcement of its judgment by depositing a bank guarantee sufficient to compensate the defendant should the defendant win on appeal. A judgment of the *Oberlandesgericht* is enforceable at once without the need to deposit a bond, and if the plaintiff loses on appeal to the BGH, he is liable only to pay back any damages that he has received.

Under the old law in the Netherlands, there was a very strong presumption that a Dutch patent was valid; indeed whereas in most countries, if a patent is invalidated, it is treated as if it had always been invalid (*ex tunc*), in the Netherlands, a patent granted under the old law is invalid only as from the date of the decision (*ex nunc*), so that damages may be awarded for infringements made before the determination of invalidity. The same patent-friendly attitude has been evident in the willingness of Dutch patent judges to grant injunctions against infringement not only of Dutch patents, but also of corresponding patents in other EU countries, applying the principles of the Brussels Convention on Jurisdiction and Enforcement. (Such 'cross-border injunctions' will be discussed further in Chapter 23.) Another consequence of this attitude is that even unexamined Dutch registration patents had a good chance of being enforced and were often asserted aggressively against alleged infringers, despite their high likelihood of invalidity.[58]

There are a number of countries, even within Europe, in which it is extremely difficult to enforce a patent even against a clear infringer. To the extent that these countries did not have product protection for chemical compounds at the relevant date, enforcement is doubly difficult in the chemical field. Spain used to be a particularly bad example of this. Under Spanish law prior to 1986, holding a Spanish patent gave a positive right to work the claimed invention irrespective of possible

[58] D. van Engelen, 'Beware the wolf' (2004) 158 Patent World 26.

domination by an earlier patent. Thus when a foreign patentee tried to enforce his Spanish patent for a process of preparation of a new compound, he would find that the infringer could not be stopped because it had a patent for a process differing only in minor details from the patentee's own. This part of the Spanish law was repealed on Spain's accession to the European Economic Community (EEC), but it is still a feature of the law of certain Latin American countries, deriving from that of Spain.

One of the main needs of industry is the provision of rapid, inexpensive, and reliable enforcement of patent rights within Europe. The EU Commission has attempted to harmonize IP enforcement in Europe by means of Directive 2004/48,[59] now implemented by all EU countries, but this does little more than set out general principles, for example relating to damages and costs, that differ only in detail from those already in the TRIPs Agreement. Nevertheless, the requirement of the Directive that EU states should provide court-ordered procedures for securing evidence has led to a significant change in practice in Germany (see p. 454). The original draft of the Directive required member states to provide criminal penalties for all kinds of IP infringement, but this is now the subject of a parallel Directive that has not yet been finalized.

Improvement of patent litigation in Europe remains one of the main issues to be resolved in connection with any future Community patent (CP). The original proposal whereby a CP could be revoked for all of Europe by a non-specialist national court sitting in Palermo or Salonika, with the damage correctable only upon appeal, was unacceptable. The next proposal of the Commission was a European Union Patent Court, which at first instance would have national and regional divisions as well as a central division; at second instance, it would have a single appeals court for the whole of the EU. The first-instance national division courts would have three judges, two from the country in which the court is located and one from a different country, whereas the appellate court would have multinational panels of five judges. It remains unclear whether or not the courts would have power to decide on issues of both infringement and validity. It is also hard to see how these courts could attract a sufficient number of experienced judges without leaving the national courts empty, nor how the proposed complex language procedures could be made to work.

An alternative proposal, the European Patent Litigation Agreement (EPLA), has been under discussion for ten years. This would allow a decision of a central European patent court on infringement and validity to have effect in all states party to the EPLA. Unlike the proposed CP Regulation, this would be an optional agreement, open to all contracting parties to the EPC. The EPLA was opposed by the EU Commission and Parliament, which took the position that EU member states would

[59] Directive 2004/48/EC of the European Parliament and of the Council of 29 April 2004 on the enforcement of intellectual property rights.

violate the Treaty of Rome by entering into such an agreement. As of writing, the EPLA is dead and a compromise solution that would open the EU proposal to non-EU members of the EPC is under discussion.

Procedure in Asia

Japan

In Japan, patent infringement cases are heard by the district courts, of which Tokyo and Osaka have designated chambers specializing in IP matters. The case is heard by a panel of three judges, who normally have no technical training, but may be assisted by technical advisers from the Japanese Patent Office (JPO). Appeal lies to the IP chamber of the Tokyo High Court and, from there, on matters of law only, to the Supreme Court. It is not usually possible to stay judgment pending an appeal. The Tokyo High Court also has exclusive jurisdiction to hear appeals from the trial division of the JPO.

There is no discovery procedure under Japanese law. There is no single trial as such, but rather a series of court hearings, usually at intervals of one or two months. A decision is finally reached after a period of at least two years from the filing of the original complaint. Interlocutory injunctions are rarely granted in patent cases. Damages used to be based on normal royalty rates, which were no real deterrent to an infringer; damages may now be set at higher levels and the record amount of 8.4 billion yen (over US$60 million) was awarded in 2002 in respect of infringing *pachinko* slot machines.[60]

In Japan, like Germany, the court hearing the infringement action is supposed to assume that the patent is valid, validity being contested separately before the JPO. In one particularly clear case,[61] the Japanese Supreme Court held that an infringement court could refuse to enforce a patent if it was so clearly invalid that an infringement action could be considered an abuse. Since then, lower courts have held patents invalid on a number of occasions, but these have generally been for simple mechanical inventions. There have so far been no pharmaceutical or biochemical cases in which the courts have made use of this power.

China

China has a dual system of patent enforcement: either judicial, through the People's Courts; or administrative, through the Intellectual Property Bureaus (IPBs) set up in each province and some major cities. People's Courts exist at four levels: the basic 'intermediate' and higher People's Courts; and the Supreme Court. The system is a unitary one, with no distinction corresponding to the US state and federal courts.

[60] *Aruze Corp. v Sammy Corp.* (Tokyo District Court, 19 March 2002).
[61] *Fujitsu v. Texas Instruments* (Japanese Sup Ct, 11 April 2000).

There are normally only two instances in a patent case. Most cases start at one of the 52 Intermediate People's Courts that have jurisdiction over patent cases, with a final appeal to the Higher People's Court. A particularly important case might start at the Higher People's Court and be appealed to the Supreme Court. Rarely, the Supreme Court may retry a case that has already gone through two instances.

The patentee may sue in the domicile of the infringer or that in which infringement occurs. Because it is not a good idea to sue in the infringer's home territory, most patent infringement actions are brought in the Intermediate People's Court of Beijing. The procedure is similar to that of Japan, but strict formal requirements must be met: for example written evidence (even prior art documents) may be ignored unless legalized and notarized.

Although China has a poor reputation for IP enforcement, this is mainly a problem in the area of trademark and copyright infringement. Patents can usually be successfully enforced in China.

11

INVALIDITY AND AMENDMENT OF GRANTED PATENTS

A considerable portion of all the patents granted are worthless and void, as conflicting with, and infringing upon one another, or upon public rights not subject to patent privileges; arising either from a want of due attention to the specifications of claim, or from the ignorance of the patentees of the state of the arts and manufactures, and of the inventions made in other countries, and even in our own.

Report to US Senate by Senator John Ruggles (28 April 1836)

Introduction

The fact that a patent has been granted is no guarantee that the patent is valid, and the patent laws of many countries state this explicitly. The chance that a patent will be held to be valid if challenged, in other words, the presumption of validity of the patent, is essentially a function of the completeness of the search performed, the strictness of examination of the patent in the patent office, the legal and technical competence of the patent examiner, the chance of intervention by third parties during the patent office proceedings, and the pro- or anti-patent attitude of the national courts. Whereas patents granted by the European Patent Office (EPO), Japanese Patent Office (JPO), or US Patent and Trademark Office (USPTO) are subject to a relatively rigorous substantive examination, patents in some other countries, such as South Africa, Belgium, or even Switzerland, are granted after a merely formal examination, without any investigation as to whether the claimed invention meets the patentability criteria. Some patent offices, for example those of France or Italy, do not perform a substantive examination, but do at least publish the granted patent together with a report that is based on a patentability search. Individual granted patents may have a high presumption of validity if they have already been unsuccessfully opposed or challenged in the courts.

Grounds of Invalidity in the UK

The validity of a British patent may be challenged at any time during its term by an application for revocation made to the UK Patent Office or to the Patents Court. The grounds on which the patent can be revoked are the same in each case—that is, that:[1]

(a) the invention is not a patentable invention;
(b) the patent was granted to a person not entitled;
(c) the specification does not disclose the invention clearly and completely enough for it to be performed by a person skilled in the art;
(d) new matter was added to the disclosure after the filing date; and
(e) the scope of protection was extended (after grant of the patent) by an amendment that should not have been allowed.

We have already looked at what constitutes a patentable invention (see Chapter 4), and found that a patentable invention is one that is capable of industrial application, novel, and involves an inventive step. The first of these grounds of attack therefore includes all allegations that the patent is old or is obvious in view of the state of the art, as well as the objection that the type of invention is one that is not capable of industrial application (for example a mere nucleic acid sequence without any indication of a specific and credible use), or which is specifically excluded from patentability (for example a plant variety).

The second ground is one that can only be used by a person who alleges that he, and not the patentee, was entitled to the grant of the patent. Normally an application

[1] Section 72(1), PA 1977.

for revocation on this ground must be made within two years of the grant of the patent, unless it can be shown that there was deliberate fraud by the patentee.

Ground (c) is an attack based on insufficiency of disclosure: the patent specification must give sufficient description for a person skilled in the art to perform the invention without having to do undue experimentation.

The last two grounds, which were introduced for the first time in the Patents Act 1977, are based on the rule that no amendment, whether before or after grant, may add new matter to the specification, or, after grant, extend the scope of protection beyond that as originally filed. Any patent in which this has been done is invalid. The first of these in particular must constantly be kept in mind when amending the specification during prosecution. Neither of these grounds should arise if the Patent Office does its job properly, because amendments introducing new matter during prosecution should not be allowed any more than broadening amendments after grant. The difference is that whereas if the Patent Office mistakenly allows a single application to be granted for more than one invention, or allows claims to be granted that are too broad or ambiguous, the validity of the patent is generally not affected, Patent Office mistakes in allowing amendments can lead to invalidity (see p. 223 for EPO practice.)

Revocation Proceedings

Before the Courts

A revocation action may be filed at the Patents Court or the Patents County Court (PCC) at any time during the life of the patent. The person seeking revocation does not have to show, as he would in the USA, that he is likely to be sued under the patent, and he may even be a 'straw man' acting on behalf of another party.[2] The validity of the patent may also be challenged as a counterclaim in infringement proceedings and in certain other proceedings: for example a threats action, or an action for a declaration of non-infringement. In all cases, the grounds that can be relied upon are the same five as can be used in revocation proceedings.

If a patent is revoked as the result of a final court decision, it is treated as always having been invalid. If, in earlier litigation, damages for infringement had been awarded against another party, however, that party cannot avoid paying damages even though the patent is now considered to have been invalid all along. In 1996 Coflexip SA sued Stolt Offshore MS Ltd for infringement of a patent relating to laying pipe from a ship; in 2000 the Court of Appeal found the patent valid and infringed. Later, a company called Rockwater attacked the patent using prior art that was not relied on by Stolt and was successful at first instance—one of the few cases of a patent being revoked after having been found valid by the Court of Appeal. Stolt asked for the ongoing enquiry into damages to be stayed, on the basis

[2] *Cairnstores Ltd v. AB Hassle* [2002] FSR 35 (Pat Ct).

that Coflexip would be unjustly enriched by collecting damages for an invalid patent. However, the Court held that, as between Stolt and Coflexip, the matter was *res judicata* and Stolt could not rely on the later success of another party.[3] Finally, the Court of Appeal reversed the Patents Court, finding the patent valid and infringed for the second time,[4] so the question of damages on an invalid patent became moot.

The same question arose again in a case between Unilin Beheer and Berry Floor. Unilin's European patent (GB) was asserted against Berry Floor before the PCC, which decided that the patent was valid and infringed. The Court of Appeal confirmed, even though an opposition against the same patent was still pending at the EPO.[5] Following refusal of the House of Lords to hear the case, the first-instance PCC had to decide on the awards of damages and costs, and ordered an interim payment with the liberty to request a repayment. The PCC was of the opinion that there would be no *res judicata* effect in the event that in the opposition proceedings, the patent would finally be revoked or restricted such that the alleged infringing product would no longer be covered by the patent. This decision was again appealed to the Court of Appeal, which overruled the PCC, stating that validity and infringement were *res judicata* between the involved parties. Costs and damages awarded by a final court decision in England cannot be subject to repayment, even if the patent were to be finally revoked or restricted in the EPO opposition procedure so that the basis for the payment would retroactively cease to exist.[6] That situation did not occur in this particular case because the parties settled and the opposition was withdrawn before the EPO Boards of Appeal issued a decision.

The situation in England is in contrast to that in other European jurisdictions, such as France, where the *Cour de Cassation* decided in a similar case that damages paid as a result of a French court decision establishing validity and infringement of a patent have to be paid back to the infringer if the patent that was the basis for the decision is later finally revoked in an opposition procedure.[7]

Germany, where the courts, due to the split jurisdiction on patent validity and infringement (see p. 204), are much more familiar with such situations of diverging views on validity of a particular patent by different bodies, even has a legal mechanism by which damages or costs that had been first awarded to a patentee have to be paid back if the patent finally is revoked, even after the court decision awarding the costs or damages becomes final. Since the factual or legal basis for the damages or cost claim is retroactively removed, an infringer can initiate an action for restitution of the earlier decision[8] with the same court, which will then reopen the proceedings

[3] *Coflexip SA v. Stolt Offshore MS Ltd* [2004] FSR 7, 118 (Pat Ct).

[4] *Rockwater v. Technip France SA (formerly Coflexip SA)* [2004] EWCA Civ 381 (CA).

[5] T 1040/04 (OJ 2006, 597).

[6] *Unilin Beheer BV v. Berry Floor NV* [2007] EWCA Civ 364.

[7] *Normalu, Scherer v. Newmat* (Cour de Cassation, 12 June 2007).

[8] Known as *Restitutionsklage*, under §580, No. 6 of the *Zivilprozessordnung* (ZPO, or German Code of Civil Procedure).

and set the earlier decision aside. Another mechanism to prevent the enforcement of a German court decision awarding costs or damages based on a patent that later was revoked or restricted is an action to avert enforcement.[9] Because such an action will not change the earlier decision itself, but prevents its enforcement, it is useful only if the awarded damages or costs had not yet been actually paid to the patentee.

Before the Patent Office

Under the 1949 Act, validity could be considered by the British Patent Office only in opposition or belated opposition proceedings, only for a limited time, and only on limited grounds. It was the rule that any doubt should be resolved in favour of the patentee, because the courts could always consider validity more thoroughly later. As a result, there was little point in filing an opposition as a serious attack upon a patent and the few oppositions that were filed were often mere delaying tactics to hold up the patent grant. Under the Patents Act 1977, grant cannot be delayed in this way, and the Patent Office can consider validity at any time and on the same grounds as the courts, so there is no reason why the patentee should have the benefit of the doubt. It was supposed that such proceedings before the Patent Office would become widely used, but this has not been the case. In any case of real importance, the losing party would always appeal from the Comptroller to the courts, so it makes more sense to go to the courts right away. Under the Patents Act 2004, the Patent Office may, on request of any party (including the patentee), give a non-binding opinion on the validity of the patent, a procedure similar to that for opinions on infringement (see p. 193).[10]

If someone unsuccessfully applies to the Patent Office for revocation of a patent, he may appeal, but cannot simply try his luck again by making a separate application to the courts, unless the courts give special permission. Anyone accused of infringement may always contest the validity of the patent by a counterclaim in the infringement action, however, even if he has already applied for revocation before the Patent Office or the courts without success.[11]

It should also be mentioned that the Patent Office may of its own accord revoke a patent in two rather special circumstances: firstly, when the patent is found to be anticipated by an earlier application on the 'whole contents' principle; and secondly, when both a British patent and a European patent (UK) are granted to the same patentee for the same invention.[12] In the latter case, the national patent is revoked and the European patent remains in force. If for any reason the patentee wishes to keep the national UK patent rather than the European patent (UK), he must abandon the European patent, or at least drop the designation of the UK, before the British

[9] *Vollstreckungsabwehrklage*, under ZPO §767.

[10] New ss. 74A and 74B, PA 1977, introduced by s. 13, PA 2004.

[11] Section 72(5), PA 1977.

[12] Section 73, PA 1977.

national patent is granted. If the two granted patents have coexisted for any period, the national patent must be revoked.[13]

Amendment of British Patents

To Cure Partial Invalidity

As a result of revocation proceedings or a validity challenge in an infringement action, the patent may be found wholly valid, partially valid (that is, some claims valid and some not), or wholly invalid. If it contains valid claims, the court may enforce these without requiring that the patent be amended, or may require amendment to delete the invalid claims and possibly make consequential amendments.[14] If the patent contains no valid claims, or none that would be infringed, but does contain validly patentable subject matter within the scope of an invalid claim, the court may allow the patentee to amend the specification, but this is at the discretion of the court, and may be opposed by the other party (or indeed by any third party).[15] Finally, if the patent is wholly invalid and no allowable amendment would cure this, the patent must be revoked. An amendment may be requested under section 75(1) of the Patents Act 1977 at any time during the proceedings, not only when a validity determination has already been made.

Amendment post-grant was always discretionary under UK patent law and amendment could be refused if claims were over-broad, if the patentee knew that the claims were invalid, but delayed in seeking amendment, or if he had acted inequitably in some other way. The fact that amendment was at the discretion of the court and not as of right often made it very difficult for the patentee to make a limiting amendment if the patent was held partially invalid. In *inter partes* proceedings, the patentee's patenting strategy and entire conduct could be called into question, and if amendment was refused, the consequence was frequently total invalidity of the patent.

The discretionary nature of amendments after grant, confirmed by the Court of Appeal in 2000,[16] and 2001,[17] has been changed by the Patents Act 2004. This leaves unaltered the wording that the courts or the Comptroller *may* allow amendment, but adds the subsection: 'In considering whether or not to allow an amendment proposed under this section, (the court or) the Comptroller shall have regard to any relevant principles applicable under the European Patent Convention.'[18] Since the European Patent Convention (EPC) 2000 now provides the patent proprietor with the possibility of central limitation of a European patent as a matter of right,

[13] *Albright & Wilson's Patent* [1994] SRIS O/142/94.
[14] Section 63 (1) and (3), PA 1977.
[15] Section 75(1) and (2), PA 1977.
[16] *Kimberley Clark v. Proctor & Gamble* [2000] RPC 422 (CA).
[17] *Oxford Gene Technology v. Affymetrix Inc. (No. 2)* [2001] RPC 18 (CA).
[18] Sections 27(6) and 75(5), PA 1977, added by s. 2, PA 2004.

it seems clear that if the 'relevant principles' are to be followed, amendment must be as of right within the UK as well. The courts or the Comptroller may, however, refuse to award damages for infringement of a partially valid patent if it appears that the specification was not drafted in good faith, and with reasonable skill and knowledge, or if the infringement proceedings were not brought in good faith.

Voluntary Amendment Before the Patent Office

In addition to amendment of British patents in the course of litigation, a patentee may, of his own accord, apply to the Patent Office for permission to amend the patent specification (which includes the claims). If the Patent Office finds the proposed amendment acceptable, it is advertised in the *Official Journal (Patents)* and any other party may oppose the allowance of the amendment. If there is no opposition, or if opposition is unsuccessful, the amendment will be made and the patent will have effect as if it had originally been granted in the amended form. A patentee may not apply for amendment in this way if proceedings are pending in which the validity of the patent is being put in issue.

Allowable Amendments

Under the 1949 Act, an allowable amendment had to fall under one of the three categories of disclaimer, explanation, or correction. Under the 1977 Act, this is no longer the case, because the Act says nothing about types of amendment. In principle, any amendment to a granted patent is permissible as long as (unless it is the correction of an obvious mistake) it does not introduce new matter or extend the scope of the claims. Whereas under the 1949 Act an amendment once allowed could not be called in question, allowance of an amendment subsequently held to add new matter may invalidate the patent.

Under the 1977 Act, any amendment by disclaimer should in principle be allowable, provided that the new narrower scope was already described in the specification. It is not clear, however, whether *any* deletion would constitute an allowable amendment because it may be considered, paradoxically enough, that deletion amounts to adding new matter if it results in the emphasis of the disclosure being substantially different from what it was before.[19] The same prohibition about addition of new matter applies to amendments during prosecution, although during prosecution the scope of the claims may be enlarged by amendment as long as the basis for such broader claims was already in the specification.

Explanatory amendments to clarify or remove ambiguities from the specification run a particular danger of being held to constitute new matter, so that applicants are well advised not to amend in this way without very good reason. And it is quite clear that invalidity due to insufficiency of disclosure cannot be cured by amendment because any amendment that makes an insufficient disclosure sufficient must

[19] *Raychem's Application* [1986] RPC 547 (Pat Ct).

necessarily add new matter, which is itself a ground of invalidity. As far as corrections are concerned, this is covered by a separate section of the Act[20] dealing not only with clerical errors, but also with false translations and mistakes generally. No mistake can be corrected unless it is obvious what the correction should be, but corrections of mistakes do not fall within the prohibitions on adding new matter or extending the scope. Note that corrections may also be made by the Patent Office on its own initiative, or on the request of a party other than the patentee.

Anyone who reads a patent should be alive to the possibility that the specification may have been amended. If a significant amendment has been made, the specification will normally be republished with an indication that it is an amended version; the reader may, however, have an old copy, perhaps photocopied from a library copy of the original version. If the matter is important, for example if possible infringement of the patent is being considered, a new copy should be obtained directly from the Patent Office website. Better still, the Register of Patents, or the file of the patent may be consulted, either online or in hard copy, to check whether an amendment has been made, as well as such matters as whether the patent is still in force, or has been assigned or exclusively licensed.

Invalidity and Amendment in the USA

US law has no exact counterpart to a revocation action in which anyone can attack the validity of a patent before the court. Validity can be challenged in the courts only by a counterclaim in an infringement action or by a suit for declaratory judgment of invalidity; the latter can only be filed by a party that can establish what is called a 'case of actual controversy'. Declaratory judgments are a general feature of US law, not limited to patent issues.[21] Until 2007, the US courts applied a relatively strict two-step test in order to determine whether there was a case of actual controversy.[22] According to that test, the plaintiff in a declaratory judgment action needed to show:

(a) that there was a reasonable apprehension of imminent suit by the patentee, which could be a direct threat to sue or, for example, a newspaper article containing a general threat to sue imitators;[23] and
(b) that he conducted potentially infringing activities or at least made serious preparations to do so.

In 2007, the US Supreme Court handed down its decision in *Medimmune v. Genentech*, which abandoned the two-step 'reasonable apprehension test' for

[20] Section 117, PA 1977.

[21] See US Declaratory Judgment Act of 1934 (28 USC 2201(a)); this concept is based on Art. III of the US Constitution.

[22] See, e.g., *Teva v. Pfizer* 395 F3d 1324 (Fed Cir 2005, certiorari denied).

[23] *Dr Reddy's Laboratories Ltd v. aaiPharma Inc.* 64 PTCJ 555 (SDNY 2002).

finding a case of actual controversy and replaced it by an analysis of 'whether the facts alleged, under all the circumstances, show that there is a substantial controversy, between parties having adverse legal interests, of sufficient immediacy and reality to warrant the issuance of a declaratory judgment'.[24] Up to that time, a licensee in good standing had no possibility of challenging the licensed patent by a declaratory judgment action, because there could be no apprehension of an infringement suit as long as royalties were paid. Medimmune was licensed under patents and patent applications of Genentech for the production of monoclonal antibodies by gene technology (see p. 293), but when all other patents had expired and a new 'co-expression' patent issued, Genentech demanded royalty payments for another 17 years. Medimmune considered the new patent to be invalid and not infringed, but if it stopped payment of royalties, it made itself potentially liable for triple damages for wilful infringement, not to mention an injunction against sales of its product. The Court held that, in these circumstances, a declaratory judgment action could be filed.

Following that decision, it has been held that enforcement of related patents can give rise to declaratory judgment action in respect of the non-asserted patents. In one case, for example, a patentee had listed five patents covering a certain drug in the Orange Book (see p. 189), but initiated patent litigation on only one of the listed patents after another party had filed an abbreviated new drug application (ANDA) alleging that all five listed patents were either not infringed, or were invalid or unenforceable. According to the Court of Appeals for the Federal Circuit (CAFC), the ANDA filing constituted a single act of infringement against all five patents and the fact that the patentee did not choose to sue on all of them at once did not prevent declaratory judgment actions in relation to the four non-asserted patents.[25] Even a covenant not to sue may not be sufficient to avoid a declaratory judgment action jurisdiction, if the patentee would still be able to derive some advantages out of the patent: for example that the US Food and Drug Administration (FDA) would not issue a generic marketing authorization before the patent would expire or be declared invalid, non-infringed, or unenforceable.[26]

In another case, the CAFC confirmed declaratory judgment jurisdiction for three new configurations of a device for filling catalyst into chemical reactor tubes for which another configuration was already the subject of ongoing patent infringement proceedings, because the alleged infringer had made serious steps to bring those three new configurations to the market and, taking into account all circumstances, it could be reasonably expected that the patentee would also start infringement proceedings against the new three configurations.[27]

[24] *Medimmune v. Genentech* 549 US 118 (Sup Ct 2007).

[25] *Teva v. Novartis* 482 F.3d 1330 (Fed Cir 2007).

[26] *Caraco v. Forest Labs* 527 F.3d 1278 (Fed Cir 2008).

[27] *CAT Tech v. Tubemaster* 528 F.3d 871 (Fed Cir 2008).

Re-Examination

Similarly, there are as yet no *inter partes* opposition or revocation proceedings before the USPTO in which a patent can be attacked on all of the grounds available before the court. Since 1980 any person could, at any time during the life of a US patent, cite new prior art to the USPTO and, upon payment of a fee, request re-examination of the patent, but the requester did not become a party to the proceedings and this was not a very attractive procedure for that reason.

As part of the American Inventors Protection Act (AIPA) of 1999 (see p. 23), *inter partes* re-examination was made available for all patents filed on or after 29 November 1999, but this still has some disadvantages for the party requesting re-examination. It is expensive, slow (typically taking up to five years), it can be based only upon written prior art (and not, for example, on evidence of prior use or sale), and the USPTO may simply decide that no new question of patentability exists, in which case, the proceedings stop there (although most of the fee is then refunded).

If a new question of patentability exists, re-examination is opened and the examiner issues an initial action on the merits, which may be a rejection or an allowance. If there is a rejection, the patentee can respond on the merits and may amend the claims, and the requester may comment upon each response of the patentee. The procedure is rapid, only 30 days being allowed for each response and comment. Finally, there will be an office action closing prosecution, which will either reject or allow each claim. Appeal lies to the Board of Patent Appeals and Interferences, and from there, to the CAFC. Originally, only the patent owner could appeal to the courts, but since 2002 both parties have this right. Any parallel litigation on the patent is usually stayed during the re-examination proceedings, but this is not mandatory.

Although this is an *inter partes* proceeding, the parties are still not given equal treatment as they are in an EPO opposition. The re-examination is primarily a dialogue between the examiner and the patent owner, with the requester commenting from the sidelines. Furthermore, there is no formal or informal oral hearing at any stage at which the parties could confront each other directly.

The main disadvantage for the party requesting re-examination, however, is that he is estopped in subsequent litigation from contesting validity of the patent on any prior art grounds that were raised or could have been raised in the re-examination procedure. This means that, except in the unlikely case of totally new prior art being found, the requester could argue only on grounds such as prior use or sale, lack of written description, and similar grounds that cannot be used in re-examination.

A third party may still request the old *ex parte* re-examination, in which the requester is not a party to the proceedings. This is cheaper and the extent of subsequent estoppel is less. Over the last few years, the number of requests for *ex parte* re-examination has increased substantially, reaching almost 700 per year in 2007–08. Although any party requesting re-examination needs to show that there is a substantial new issue of patentability,[28] the chances that the director of the USPTO will find that criterion to be met are quite good. In 2007–08, around 94 per cent of all *ex parte* re-examination requests were admitted.[29] But only 13 per cent of all *ex parte* re-examination proceedings requested by third parties led to a final rejection of all claims, as compared with 73 per cent for *inter partes* re-examinations. It is a long shot to attack a granted US patent by way of *ex parte* re-examination unless one has a printed publication that destroys the novelty of the patent—and if one has that, one can simply ignore the patent and save the fee. In practice, re-examination is also used by the patentee, particularly in the situation in which new prior art has been found after the patent has been granted and the patentee wishes the USPTO to confirm that the granted claims are still patentable over this prior art. If it is clear that the new art anticipates some of the claims and that amendment will be needed, it would be more usual to apply for reissue (see below). Despite all of its disadvantages, re-examination may be an attractive addition or alternative to court proceedings, since the USPTO is not bound by a validity decision of a district court. Indeed, the USPTO had to admit a request for re-examination in a case in which not only was there no new prior art cited, but the validity of the patent had already been upheld in a court decision based on the same cited prior art. It was found sufficient that the known prior art was used together with new arguments against the validity of the patent.[30]

Grounds of Invalidity

The statutory grounds on which a patent may be found invalid by a US court are essentially the same as those on which the USPTO could have refused grant. Those are that the invention is not patentable because it is not novel, obvious, or useful, or that the specification does not meet the requirements that it must describe the invention, enable any person skilled in the art to make and to use the invention, set out the best mode contemplated by the inventor of carrying out the invention, and contain claims that particularly point out and distinctly claim the invention. Even an incorrect dependency of claims may be a ground of invalidity.[31] It is not, however, a ground of invalidity that the application on which the patent was granted had lapsed and was wrongly revived by the USPTO.[32]

The description requirements for the specification will be considered more fully in Chapter 17. In summary, they may be considered as four separate requirements of 'written description', 'how to make', 'how to use', and 'best mode'. The first three are considered separately in relation to each claim of the specification and

[28] 35 USC §302ff.

[29] *USPTO Performance and Accountability Report Fiscal Year 2008*, Table 13A, p. 127.

[30] *In Re Swanson* 540 F3d 1368 (Fed Cir 2008).

[31] See *Pfizer v. Ranbaxy* 79 USPQ 2nd 1583 (Fed Cir 2006).

[32] *Aristocrat Technologies Australia v. International Game Technologies* 521 F3d 1328 (Fed Cir 2008, certiorari requested).

are objective requirements; the last refers to the disclosure in general and is subjective, because it requires inclusion of what the inventor considered to be the best mode at the time of filing. The written description requirement is more than only a requirement that the claims be fairly based on the description; it amounts to a requirement that the applicant was, at the filing date, actually in possession of the invention, and this has important consequences for biotech inventions (see p. 288). The 'how to make' requirement corresponds to the British sufficiency requirement and the 'how to use' requirement necessitates a utility statement, particularly in pharmaceutical cases.

Independent from invalidity, a US patent may also be held unenforceable in an infringement action if it has been obtained by inequitable conduct, particularly by concealing from the USPTO relevant prior art known to the applicant during prosecution, or by giving false or misleading results in a showing to overcome an obviousness rejection. This ground has at times been overplayed by litigants, and an allegation of inequitable conduct (formerly known as 'fraud on the patent office') became a standard plea brought forward in practically every patent case and often applied to trivial acts of omission which could at worst be regarded as negligence. After a period in which the courts gave a more limited interpretation to 'inequitable conduct', there have been many cases in the last few years in which the Federal Circuit has upheld findings of unenforceability for this reason, usually over the spirited dissent of Judge Newman. In 2003 she described this as 'the new plague';[33] in 2006 she complained that 'if the inventor provided selected references, he was accused of inequitable conduct in the selection; and if he provided an entire search report, he was accused of burying the significant references'.[34]

It is a fact, however, that the USPTO places a very heavy burden upon the applicant, and his agents, to disclose all relevant matter during the prosecution of the application and in any subsequent re-examination, and not to conceal anything that could be relevant to patentability. There is a breach of the duty of disclosure only if the information withheld is material, material information being facts that a reasonable examiner would regard as important in determining patentability. One difficulty with this definition is deciding what a hypothetical reasonable examiner would think, when so many real examiners often seem to be unreasonable. If this breach is due to inadvertence, mistake, or a minor act of negligence, the situation may be remedied, but if it is due to deliberate intent to deceive or conceal, bad faith, or gross negligence, the patent is incurably unenforceable. What is more, not only is that patent unenforceable, but even related patents may be infected and held unenforceable, and any attempt to enforce a patent obtained by inequitable conduct can make the patentee liable under the anti-trust laws for triple damages (see p. 499).

[33] *Hoffmann-La Roche Inc. v. Promega Corp.* 66 USPQ 2d 1385 (Fed Cir 2003).

[34] *Ferring BV & Aventis Pharmaceuticals v. Barr Laboratories* 437 F3d 1181 (Fed Cir 2006).

Reissue

If a US patent is wholly or partially invalid, or defective because of error without deceptive intent, for example if it is partially anticipated by prior art that was found only after grant of the patent, the patentee may correct matters by way of a reissue application. This differs procedurally from an application to amend a British patent, because it consists of a new patent application incorporating any proposed amendments, coupled with an offer to surrender the original patent. The whole patent, including any unamended claims, is subject to a new examination in the light of any additional prior art that may be found. If the reissue application is granted, a new patent is issued, having a five-figure number prefixed with the letters 'Re' and expiring on the same date as the original patent. The surrender of the original patent takes effect when the reissue patent is granted.

A reissue application may not contain any new matter, but it is permissible for it to contain claims of broader scope than those of the original patent, provided that the application for reissue was filed within two years of the grant of the original and provided that the new claims had not been deliberately cancelled during prosecution. If a reissue patent issues with broadened scope, it cannot be asserted against someone who started to use the invention before the grant of the reissue unless he is infringing a valid claim of the reissue patent that was also present in the original.

Reissue applications are notified and the files are open to the public. Any member of the public may protest the reissue, and submit additional prior art and comments, but the protester, as in the case of re-examination, is not a full party to the proceedings. It is suggested by many attorneys that a patentee wishing to reissue before suing for infringement should invite the infringer to intervene as a protester in the reissue proceedings. This puts the infringer in a dilemma: if he declines to intervene, that refusal may be (at least subconsciously) held against him by the courts when he subsequently counterclaims for invalidity in the infringement action; if he does intervene, he does so at a disadvantage, because he is not a full party to the proceedings and has no opportunity to obtain useful evidence by means of discovery.

One thing that reissue and re-examination procedures cannot do is cure a patent that really has been obtained by inequitable conduct. If the 'new' prior art is relevant, and if it was, in fact, known to the patentee during prosecution and deliberately concealed from the USPTO, then, if the facts come out, no reissue will be granted and the original patent will become unenforceable. Reissue applications are scrutinized carefully for any suspicion of inequitable conduct; if there is any such 'dirty linen', reissue cannot be used to launder it.

Reissue is not the only procedure available to amend or correct a US patent: individual claims may simply be deleted by disclaimer without any new examination of the remaining claims. Any mistakes made by the USPTO, which are usually minor typographical and clerical errors, but may be more substantial than this, can be corrected by means of a certificate of correction attached to copies of the specification.

Invalidity and Amendment in the European Patent Office

Opposition

The validity of a granted European patent may be challenged in the EPO by way of opposition proceedings, but once a European patent has survived opposition in its original or in amended form, or when the opposition period has expired without opposition having been filed, the validity of the European patent can only be judged by the national courts. The relevant national laws have largely been harmonized with the EPC along the same lines as the UK Patents Act 1977. Previously, amendment of European patents after grant and outside of the opposition period could be made only under national law, but the EPC 2000 now provides a central procedure for revocation or limitation of a European patent at the request of the patent owner.

Any person may file a notice of opposition against a European patent within nine months of the publication of the mention of grant. Previously, the real identity of the opponent had to be known, that is, 'straw man' oppositions were not permissible.[35] It was, however, difficult to challenge an opponent who was suspected of acting on behalf of another party;[36] in 1999, the Enlarged Board of Appeal (EBA) reversed its earlier decision and held, in two consolidated cases,[37] that it was permissible to lodge an opposition on behalf of another party, as long as there was no abuse of process, for example an evasion of the rules about representation. Opposition differs from reissue proceedings in the USA in two respects: firstly, the proceedings are truly *inter partes* and the opponent is fully involved; secondly, in the EPC, it is not permissible for the patentee to oppose his own patent. It was previously held by the EBA[38] that 'any person' must include the patent owner, but this decision was overturned by a later EBA decision,[39] which stressed that opposition proceedings were *inter partes* and not a continuation of prosecution.

The admissible grounds, given in EPC Article 100, are:

(a) that the invention is not patentable under Articles 52–57;
(b) that there is insufficient disclosure of the invention; and
(c) that new matter has been included by amendment after filing.

The grounds provided by Article 100(a) include not only lack of novelty, lack of inventive step, and lack of industrial applicability, but also that the invention is not patentable subject matter under Articles 52(2) and 53.

[35] T 219/86 (OJ 1988, 254) *Naming of opponent.*
[36] T 590/93 (OJ 1995, 337) *KOGYO GIJUTSUIN/Photosensitive resins.*
[37] G 3/97 (OJ 1999, 245) *INDUPAK*; G 4/97 (OJ 1999, 270) *GENENTECH.*
[38] G 1/84 (OJ 1985, 299) *MOBIL OIL/Opposition by proprietor.*
[39] G 9/93 (OJ 1994, 891) *PEUGEOT/Opposition by patent proprietor.*

Insufficiency of Disclosure and Breadth of Claim

It is stated in EPC Article 83 that a European patent application must disclose the invention in a manner that is sufficiently clear and complete for it to be carried out by a person skilled in the art, and this requirement also applies to a European patent and bases a ground of opposition. If the requirement is not met, the patent is invalid. Note that the requirement applies to the whole patent rather than to any particular claim. The necessary disclosure may be found in the description, claims, or drawings, but not in the abstract.

The disclosure must be reproducible without undue burden by the skilled person using his own general knowledge. This means that although some experimental work may still be needed to reproduce the invention, this must not be excessive and in particular must not require any inventive step. The question of what constitutes undue burden is one that has been discussed many times in the Boards of Appeal and will depend upon the facts of the particular case. A general problem from the point of view of the opponent is that it is always very difficult to prove a negative: for example that the examples of the opposed patent cannot be repeated. Good evidence from an independent expert would be more valuable than the opponent's own trials.

For microbiological inventions, deposit of the microorganism will normally be required as a condition of sufficiency. For inventions relating to monoclonal antibodies (see p. 291), the antibody can now be characterized by its amino acid sequence, but before this was possible, it was always necessary to deposit the hybridoma cell line that produced it, the production of the hybridoma itself being an inherently random and non-reproducible process. If the deposited hybridoma became non-viable in storage, the deposit could be replaced,[40] but if it was viable, but could not be made to produce an antibody having the described characteristics, the patent was insufficient.[41]

EPC Article 84 requires that the claims of the patent be supported by the description, but non-compliance with Article 84, unlike Article 83, is not a ground of opposition. This has caused difficulties to parties wishing to attack a patent primarily on the ground that the claims are unduly broad in view of the patentee's actual invention. Such an attack can usually only be made on the basis that it amounts to the same thing as insufficiency, that is, that the description does not teach how to carry out the invention over the whole scope claimed.

Particularly in the relatively early days of biotechnology, many European patents were granted containing very broad claims. Sometimes a real pioneering invention of general applicability had been made and broad claims were justified; more often, the examiner had been persuaded to grant claims that gave the patentee more than was deserved. An example of the first type was Genentech's patent for polypeptide

[40] Rule 34, IR.

[41] T 418/89 (OJ 1993, 20) *ORTHO/Monoclonal antibody*.

expression, an important improvement on the basic Cohen/Boyer work (see p. 283). When this patent was unsuccessfully opposed, it was held that in this case a single example of one specific gene expressed in one host cell was enough to establish sufficiency over the entire scope[42] because the skilled person could apply the procedure to any gene and any host cell. Unfortunately this statement was applied out of context in a number of subsequent cases and it became, for a time, the view of the EPO that in all cases a single reproducible example made any patent sufficient, no matter how broad its scope. Two later cases,[43] however, re-emphasized the basic point that the protection given by a patent must correspond to the technical contribution to the art, however, and that the description must enable the invention to be performed over the whole scope of the claim. This approach was specifically endorsed by the English Court of Appeal[44] and in the same case the House of Lords[45] held the patent invalid essentially on the ground of excessive breadth of claims (see p. 289).

Addition of New Matter

The criteria for what constitutes new subject matter are the same as are applied during prosecution. If this ground of opposition is established, the entire patent is invalid unless the added subject matter is removed by amendment. Except in the special case of conflict with Article 123(3), which is described below, this will always be possible, but the removal may result in the patent being invalid for other grounds (for example insufficiency of disclosure).

Opposition Procedure

The notice of opposition will be held inadmissible if, by the end of the opposition period, it does not include a reasoned statement setting out the extent of opposition (that is, which claims are attacked), the grounds of opposition, and the arguments applied.

The opposition is examined by a three-person Opposition Division (OD), although, as in examination, one member of the OD will take prime responsibility. This will often be the same examiner who dealt with the application as part of the Examining Division. The patentee will be invited to comment and all communications are transmitted to all parties (there may of course be a number of different opponents). The OD may give a preliminary opinion and invite all parties to comment on this. Oral proceedings are usual and must be held if any party so requests. Parties who did not specifically request oral proceedings are also entitled to appear

42 T 292/85 (OJ 1989, 275) *GENENTECH/Polypeptide expression.*

43 T 409/91 (OJ 1994, 653) *EXXON/Fuel oils*; T 435/91 (OJ 1995, 188) *UNILEVER/Detergents.*

44 *Biogen v. Medeva* [1995] RPC 68 (CA).

45 *Biogen v. Medeva* [1997] RPC 1 (HL).

The OD or Board of Appeal, in considering amendments offered by the patentee, is required by Article 101(3) to consider whether the amended patent meets all of the requirements of the Convention. This has been held to mean that whereas, for example, the clarity of claims required by Article 84 cannot be an issue in opposition proceedings in respect of the claims as granted, any claim amended during the opposition must comply with the Article.[57] It is irrelevant in opposition proceedings, however, whether the European patent, before or after amendment, has unity of invention.[58]

Central Limitation and Revocation Proceedings

Before the EPC 2000 revision came into effect on 13 December 2007, it was not possible for a patent proprietor to limit his granted European patent in a central procedure. If the patentee wanted to limit the patent, for example because new relevant prior art had been identified after the grant, the only possibility was to check, for each EPC country in which the patent had been validated, whether it was possible to limit the claims under the particular national law. As we have seen above, it was possible to limit the claims in the UK before the Patent Office under certain circumstances. But some countries did not provide for any voluntary limitation, so that a patentee had the disadvantage of being forced to start patent enforcement proceedings based on claims that he knew were too broad, in the hope that the judge would allow a limitation in order to render the claims valid. The lack of any central limitation procedure had led some patentees to file opposition against their own patents, until this practice was banned by the EBA (see above). If the patentee wanted to have the patent revoked during ongoing opposition proceedings, one way in which to do so was to notify the opposition division that the patentee no longer approved the text in which the patent was granted and did not submit any amended text. The OD then had to revoke the patent. A request by the proprietor to revoke the patent was considered to have an equivalent effect.[59]

Since 13 December 2007, there is now a much more elegant way to limit centrally any claims, that is, by way of the new central limitation procedure.[60] At the same time, the possibility was introduced for the patent proprietor to have the patent centrally revoked, independently of any ongoing opposition procedure. One important difference between requesting revocation of a patent and merely abandoning it, for example by not paying the annuity fees, is that a revocation takes away any effects of the patent from the beginning (*ex tunc*), whereas in the case of an abandonment the patent just ceases to be in force from the time of the abandonment (*ex nunc*).

A central limitation can be requested by the proprietor at any time after grant of a European patent and may be requested even if the patent is no longer in force.

[57] T 301/87 (OJ 1990, 335) *BIOGEN/Alpha-interferon.*

[58] G 1/91 (OJ 1992, 253) *SIEMENS/Unity.*

[59] See Legal Advice L11/82 (OJ 1982, 57).

[60] Article 105a–105c, EPC; rr. 90–96, IR.

PART III

PATENTABILITY OF INVENTIONS IN SPECIFIC TECHNICAL FIELDS

12

CHEMICAL INVENTIONS

> There stood a hill not far, whose grisly top
> Belched fire and rolling smoke, the rest entire
> Shone with a glossy scurf, undoubted sign
> That in his womb lay hid metallic ore,
> The work of sulphur.
>
> John Milton, *Paradise Lost* (1667)

Novel Compounds

There are different categories of invention in the chemical field: for example new compounds, new compositions, new manufacturing processes, and new uses. The most straightforward case is that of a new chemical compound of known structure, which has been synthesized in a research laboratory. A novel compound cannot be a patentable invention unless it is industrially applicable. In university laboratories, hundreds of thousands of new compounds may be made every year, but the great majority of these are only of theoretical interest. It is not enough to make a compound patentable that it is useful in the elucidation of some problem of reaction mechanism or that it has an interesting ultraviolet (uv) absorption spectrum, but if, for example, the latter property were to indicate that the compound would be useful as a uv stabilizer in plastics, then the compound could be patentable.

Compounds may be patentable even if they have no uses except as intermediates in the preparation of other compounds. The rule is that if the end products are industrially applicable, then so are the intermediates, and that this applies not only to the immediate precursor of the final product, but also to the products of earlier steps in the reaction sequence. Such intermediates can therefore be patented as long as they meet the other criteria of being novel and unobvious.

Normally, of course, the invention will not consist of a single compound, but will encompass a group of compounds having some structural features in common and which all have the same end use. It will be the task of the inventor in the research laboratory to synthesize sufficient compounds to form an idea of which compounds will work and which will not, and that of the patent agent to decide in consultation with the inventor what the scope of the claimed invention should be, taking into account not only the inventor's findings, but also the prior art.

Obviousness

A compound may be new and useful, but still not be patentable, because it is so close to the prior art that there is no inventive step involved in making it, or in other words, that it is obvious. In considering how close a compound is to a compound described in the prior art, one must consider not merely the structural formulae of the compounds, but also the compounds themselves, including their properties. Thus, suppose the invention is a certain group of brominated aromatic compounds useful as flame retardants and the closest prior art is a compound that would fall within the scope except that it is chlorinated instead of brominated. If this chloro-compound were to be described in an academic publication suggesting no use for it, or if it were to have a use quite different from that as a flame retardant, then the invention should be patentable in spite of the very close structural similarity, simply because it would not be obvious that the bromo-compounds would be useful as flame retardants.

If, however, the prior-art chloro-compound were itself known to have flame-retardant properties, then the bromo-analogues would be considered likely to share these properties. In order for the bromo-compounds to be patentable, it would be necessary for them to be surprisingly better flame retardants than the chloro-compound, at least in some respects. If it were known that in similar types of compound brominated derivatives generally were better flame retardants than the corresponding chlorinated compounds, then an improvement over the prior-art compound would be expected and would not suffice to make the bromo-compounds patentable unless the difference was very large. Alternatively, it might be enough for patentability if the bromo-compounds, although no better than the prior art on most substrates, could flameproof one substrate for which the chloro-analogue was ineffective.

It is thus particularly difficult to patent compounds that while new are very closely structurally related to known compounds. Among the closest structural relationships are (in decreasing order): salts of acids and bases; geometrical isomers; positional isomers; homologues (for example within the alkyl series); and adjacent halogen compounds, as in the example above. If the structural similarities are less close, then the new compounds may be patentable and non-obvious irrespective of whether they have improved properties. In the UK, for example, a case in 1970 decided that it would not have been obvious to substitute -CF_3 for -Cl at the 2-position of a phenothiazine ring system, even though the prior-art compounds (for example chloropromazine) and the novel compounds (for example trifluoroperazine) were both tranquillizers.[1] After trifluoroperazine itself was known, however, the substitution of -CF_3 for -Cl in similar types of compound would be obvious to try and the results would be patentable only if surprising advantages were found.

There is no general requirement that an invention in order to be patentable must be better than what has gone before. Germany at one time did require an invention to show 'technical progress', but this requirement has now been abolished. It is now required only that a surprising improvement in properties may be evidence of the presence of an inventive step, if the closeness of the prior art should cast doubt on this.

A considerable degree of uncertainty exists in situations in which a novel compound that is structurally close to the prior art has some advantageous properties that are predictable and some that are not. On the one hand, it can be argued that the unpredictable advantages confer patentability; on the other, it can be said that the fact that certain advantages were predictable made it obvious to prepare the new compound and that determination of further advantages is then merely discovery of the properties of an unpatentable substance. The latter view was taken by the Technical Board of Appeal (TBA) in the European Patent Office (EPO), where an invention relating to sepulchrate complexes was held obvious over a publication by the inventor of similar complexes.[2] The publication enabled the skilled man to arrive at the claimed complexes without inventive effort and their predicted advantageous properties would give him an incentive to do so. The further unexpected property that was found was 'not relevant to the issue of patentability'. In another

[1] *Olin Matheson v. Biorex* [1970] RPC 157 (Ch D).

[2] T 154/82 (unpublished) *AUSTRALIAN NAT. UNIVERSITY/Metal complexes.*

TBA decision, a novel ortho-substituted benzamide compound was held to be unpatentable due to lack of inventive step over a disclosure of the para-substituted benzamide isomer, because the patentee could not show that a skilled person would not have expected the claimed compound to work in a similar way as the earlier published para-substituted isomer.[3]

This approach does not appear to be followed in the USA, where the properties of the compound are considered as a whole to determine whether surprising advantages exist and the presence of a predictable advantage would not necessarily negate patentability. The significance of the properties with respect to which the claimed compound exhibited surprising advantages would be weighed against the significance of the properties for which it did not. Even in Europe, the principle of the cited sepulchrate complexes and substituted benzamide cases may not be generally applicable, and the result will depend upon the facts of the particular case.

Selection Inventions

Particular problems may arise in patenting compounds that, although individually new, fall within an earlier disclosure of a broader group of compounds. The invention then can only be the selection of a particular compound or relatively small group of compounds from the larger group previously disclosed in broad terms. As an example, suppose C_1-C_{20} monoalkylphenol ethoxylates had been broadly described as being useful non-ionic surfactants, but the only ones that were specifically identified were those in which the alkyl group was methyl, ethyl, or a higher branched-chain alkyl. Then, we should be able to protect as an invention the C_8-C_{12} straight-chain monoalkylphenol ethoxylates, provided that this subgroup possesses some advantage over the generality of the broad group of compounds previously disclosed.

It may well be asked how this differs from the normal situation of an invention with close prior art; indeed it could be said that a selection invention *is* the normal situation, since all classes of compounds are already known and in that sense any new compound is a selection from some previously described group such as 'steroids' or 'azo-dyestuffs'.

I.G. Farben *Rules*

There was, however, an important early case in the UK (*I.G. Farben's Patents*)[4] in which special rules for selection inventions were formulated by the judge. Apart from the basic requirement that the compounds must be novel, even though

[3] T 998/04 (unpublished) *ROMARK LABORATORIES/Tizoxanide.*

[4] *I.G. Farbenindustrie's Patents* (1930) 47 RPC 239 (Ch D).

selected from a previously disclosed group, the three '*I.G. Farben* rules' were as follows.

(a) There must be some substantial advantage to be secured by the use of the selected members.
(b) All of the selected members must possess the advantage (although a few exceptions would not invalidate the patent).
(c) The selection must be in respect of a property that can fairly be said to be peculiar to the selected group.

If we look at these requirements in the light of present-day practice, we see that the first two are no more than the normal rules whereby any compound that is prima facie obvious over a prior art disclosure may be shown to be non-obvious by possessing a surprising advantage. As regards the third of the *I.G. Farben* rules, it is by no means clear what logical basis supports it. If, out of a previously disclosed large group of compounds, a smaller 'group A' can be identified having a non-obvious advantage, then the compounds in group A should be patentable as a selection invention. If, subsequently, a second 'group B' is identified, also having that property, then group B may or may not be patentable (it may, of course, be obvious in view of A), but why should this affect the validity of the patent claiming A?

It must be remembered that *I.G. Farben* was decided at a time when the distinction between novelty and inventive step was not as clear as it is today. The UK law on selection inventions was restated in *Dr Reddy v. Eli Lilly*,[5] in which Floyd J concluded essentially that there were no special rules for selection inventions, and that the normal rules of novelty and inventive step applied. Novelty depends upon the degree to which the disclosure of the prior art group is 'individualized', and the troublesome House of Lords decision in *Du Pont v. Akzo*[6] was explained away on this basis (see below).

Certainly in the USA a disclosure from which the invention is a selection is not treated any differently from other prior art. In fact, in the USA, if the disclosure from which the invention is selected is very broad, it is not even necessary to show any advantages. For example, the Court of Appeals for the Federal Circuit (CAFC) held that the prior-art disclosure of 'substituted ammonium salts' of a particular compound was too broad to render the claimed 2-(2'-hydroxyethoxy) ethylammonium salt prima facie obvious, where the closest disclosed salts were the 2-hydroxyethylammonium and di-2-hydroxyethylammonium salts.[7]

[5] *Dr Reddy's Laboratories v. Eli Lilly* [2008] UKHC 2345 (Pat).
[6] *Du Pont v. Akzo* [1982] FSR 303 (HL).
[7] *In re Jones et al.* 21 USPQ 2d 1941 (Fed Cir 1992).

Selection Inventions in the European Patent Office

In the leading EPO case on selection inventions,[8] the prior art had listed 20 specific starting materials and five different reaction conditions. But because each specific starting material always gave the same product whichever reaction conditions were used, a selection of one starting material and one set of reaction conditions was not regarded as novel. Nevertheless, it was stated that if the prior art had given two different lists of two types of starting material that had to be combined to give the end products, then selection of one (or more) specific end products would be a novel selection invention. The same reasoning applies when a generic formula has two variable substitution points[9] and possible substituents are listed for each, and a later case[10] applied the same conclusion to claims covering mixtures of compounds.

The approach of the EPO is more sensible than that taken at much the same time by the English Court of Appeal and the House of Lords, in the case of *du Pont v. Akzo*.[11] In 1950, ICI had obtained a patent for polyesters of terephthalic acid and certain glycols, in which polyalkylene oxide units were incorporated into the chain in order to improve dyeability of fibre made from the material. Although the specific examples all used ethylene glycol, four other glycols including 1,4-butanediol were mentioned by name as possible reactants. In 1972, Du Pont filed an application based on its discovery that polyesters of this type using 1,4-butanediol, because the glycol had improved mechanical properties making it particularly suitable, for example, for the manufacture of hydraulic hose. Instead of claiming this new use of the material, however, Du Pont claimed the polymer per se. The application was allowed and was opposed by Akzo. In the UK Patent Office, the hearing officer held that the claim was anticipated by the ICI disclosure, since the starting material was named, even though the final product was not specifically disclosed. This decision was overturned on appeal, and further appeals to the Court of Appeal and the House of Lords were lost, so that the Du Pont patent was upheld. One would have expected a selection of one out of five named possibilities to be sufficiently individualized as to constitute an anticipation.

Selection from a Small Class

One problem with establishing the novelty of a selection invention is that, when the earlier disclosed class of compounds is small, it may be argued that even a general disclosure of the class is equivalent to a specific disclosure of each of its members, thereby rendering every member of the class no longer novel. In England, a disclosure of a process involving an 'alkali metal' was held to be a disclosure of the process with lithium.[12] In the EPO, it was held[13] that a description of N-alkylation

[8] T 12/81 (OJ 1982, 296) *BAYER/Diastereomers*.
[9] T 7/86 (OJ 1988, 381) *DRACO/Xanthines*.
[10] T 401/94 (unpublished) *ELF ATOCHEM/Novelty—selection invention*.
[11] *Du Pont v. Akzo* [1982] FSR 303 (HL).
[12] *General Mills (Miller's) Application* [1972] RPC 709 (PAT).
[13] T 181/82 (OJ 1984, 401) *CIBA-GEIGY/Spiro compounds*.

of a specific compound with a 'C_{1-4} alkyl bromide' amounted to a specific disclosure of the N-methyl compound, because C_1 was mentioned as the lower end of the range and C_1 could only be methyl. The implication is that none of the other seven N-alkyl compounds were specifically disclosed, which leads to the somewhat absurd conclusion that if the specification were to have used the term 'an alkyl bromide having less than five carbon atoms', there would not have been disclosure of the N-methyl compound, although, to the chemist, the two expressions have precisely the same meaning.

In Germany an extreme view on this point has been applied until recently, holding that, even in a relatively large generic group of compounds, disclosure of the group is, to the skilled chemist, fully equivalent to a disclosure of each compound within the group.[14] Selection inventions, in the normal sense of the word, have been regarded as unpatentable in Germany and some selection compound patents granted by the EPO for Germany had been revoked by the Bundespatentgericht (BPatG, or Federal Patents Court). But the Federal Supreme Court clarified in a landmark decision in late 2008 that not all compounds embraced by a generic formula are automatically regarded as individually disclosed. Therefore, later claims to a subgroup of compounds falling under the previously disclosed generic formula can be found novel, if the skilled person would not have had compounds of the subgroup 'in his hands' from reading the disclosure of the generic formula.[15]

Selection from a Patent Disclosure

Another argument that has been used against the concept of selection inventions is applied when the earlier disclosure is a patent, as usually is the case. It can then be argued that once that patent has lapsed its entire scope should be in the public domain and that a patent for a selection invention unjustifiably prolongs the monopoly, particularly if the same patentee owns both the patents. Some courts have held patents for selection inventions to be invalid on this basis. This approach is extreme and would, if applied logically, have the absurd effect of invalidating all improvement patents, of which chemical selection patents are only a special case. If the earlier disclosure is broad and general, it may very well be a meritorious invention to find that a small subgroup has particularly good properties. In consideration for the patent grant, the public is being given information that it did not have before, and which it might be very difficult to find out by trial and error.

Optical Isomers

Perhaps the closest that a new compound can be to the prior art is the situation in which the new compound is an optically active enantiomer of a compound previously known only in racemic form. This may be regarded as an extreme form of a

[14] *Fluoran* [1988] GRUR 447 (BGH), under the former German Patent Law of 1968.
[15] X ZR 89/07 *Olanzapin*, Mitt. 3/2009, 119 (BGH).

selection invention and it has been argued that the optically active form cannot be regarded as novel if the racemate is known, since the racemate could be considered as an equimolar mixture of the R- and S-forms. The EPO and most other patent offices consider, however, that isolated optical isomers of known racemates may be considered as novel per se[16] as long as the individual enantiomers have not been explicitly disclosed[17] and that the patentability of the optical isomers is rather a question of inventive step.

This view is in agreement with case law in the UK[18] and USA.[19] Inventive step may, however, be difficult to establish. It is obvious from the presence of an asymmetric centre in the molecule that optically active forms can exist and it is usually obvious that they can be isolated by one or other of the standard methods of resolution. The only way in which an optical isomer of a known compound can be patentable is if it has surprisingly superior properties as compared with the racemate, if it has a use that the racemate did not have, or if it can be shown that no separation processes for the enantiomers from the racemate were available in the prior art.

At one time, it was relatively easy to obtain patents for an optical isomer of a pharmaceutical compound on showing superior pharmacological activity. Nowadays, it is recognized that it is normal for one optical isomer to have much higher activity than the other, so that superior activity for at least one of the isomers as compared to the racemate is only to be expected. For example, the patent for amoxicillin, an antibiotic that is an optical isomer of a known racemate, was upheld in the UK only to the extent that claims to an oral composition containing amoxicillin were held valid, it being shown that amoxicillin had particularly high activity on oral administration, whereas the racemate did not have this high activity.[20] In France and Germany, however, the amoxicillin patent was held to be invalid. The EPO Boards of Appeal have held that when the racemate is known to be active, no inventive step is normally involved in selecting which of the two possible enantiomers has the higher activity.[21]

Nevertheless, in the UK, Lundbeck's patent for escitalopram, the (+) enantiomer of the known drug citalopram, was held valid by the Patents Court and the decision was upheld by the Court of Appeal.[22] In this case, the separation of the enatiomers was unusually difficult and required an inventive step. It was argued that, because the invention resided in the method of separation of the enatiomers, the patent was insufficient if it covered escalitopram however it was made (known as '*Biogen* insufficiency', see p. 289). But Lord Hoffmann, sitting in the Court of Appeal, took

16 T 296/87 (OJ 1990, 195) *HOECHST/Enantiomers*.

17 T 658/91 (unpublished) *ELF SANOFI/Novelty*.

18 *ICI (Howe's) Application* [1977] RPC 121 (PAT).

19 *In re May and Eddy* 197 USPQ 601 (CCPA 1978).

20 *Beecham Group Ltd's Application* [1980] RPC 261 (CA).

21 See, e.g., T 857/04 (unpublished) *DEGUSSA/Verwendung von R - (+) alpha-Liponsäure*.

22 *Lundbeck v. Generics (UK) and ors* [2008] EWCA Civ 311.

the opportunity to clarify his judgment in *Biogen* and explained that this did not apply to a product per se claim such as Lundbeck's. In contrast, the Dutch High Court considered additional evidence not presented to the Patents Court and held that the same patent (and the supplementary protection certificate (SPC) based upon it) were novel, but invalid for lack of inventive step. In Australia, the corresponding patent was held valid and infringed, but the full Federal Court held that it should not have been extended, being merely for a 'purified form' of citalopram, which had already been registered as a pharmaceutical.

In a similar UK case,[23] the patent for levofloxacin, an enantiomer of ofloxacin, was found valid and it was specifically held that the earlier SPC granted for the racemic drug did not prevent the grant of an SPC for the enatiomer.

Grades of Purity

What may be regarded as an even more extreme form of selection is to attempt to claim a known compound in a specified high level of purity: for example a specific optical isomer containing no more than 1 per cent of the enantiomer, or a particular compound at least 99.5 per cent pure. The general rule of the EPO is that a document disclosing the preparation of any low molecular weight compound is taken to make the compound available to the public in all desired grades of purity and that a claim to a specific purity level lacks novelty.[24] This general rule applies unless the applicant can show that all prior attempts to obtain the substance in the claimed level of purity using conventional methods had failed. Claims to 'substantially pure' compound also lack clarity.[25] It would seem more logical to hold that, whereas a compound in a previously unknown state of purity is novel, purification to any desired level normally lacks inventive step. Note that T 990/96 applies specifically to low molecular weight compounds, so different criteria might apply, for example, to polymers or purified proteins. By expressly distinguishing from the ruling in T 990/96, a process for preparing certain polycarbonates differing from the prior art only in that starting compounds of a certain high purity level were used was held novel and inventive by an EPO Board of Appeal. The patentee provided evidence that the level of purity of the starting compounds had a significant influence on the properties of the final polymer product, and the Board saw a fundamental difference between the purity requirements presumed to exist for the isolation of a final product as referred to in T 990/96 and those for the starting compounds of a preparative process in which the skilled person would normally prefer to use less pure, cheaper, raw materials.[26]

[23] *Generics (UK) v. Daiichi Pharmaceutical Co. Ltd and anor* [2008] EWHC 2413 (Pat).
[24] T 990/96 (OJ 1998, 489) *NOVARTIS/Erythro compounds.*
[25] T 728/98 (OJ 2001, 319) *ALBANY/Pure terfenadine.*
[26] T 786/00 (unpublished) *GENERAL ELECTRIC COMPANY/Polycarbonates.*

Overlapping Groups

The distinction between a selection invention and a normal prior-art situation becomes blurred when the 'selection' is in fact a large class of compounds that overlaps with a class of compounds known in the prior art, even though no specific compound disclosed in the prior art falls within the claimed scope. In the EPO, it was held[27] that in the latter case the claimed invention was not novel, at least where the claimed compounds formed a major part of the previously disclosed class. Similar considerations apply to claims defining an invention in terms of one or more numerical parameters when the claimed range overlaps with a range disclosed in the prior art.[28]

Disclaimers

The opposite situation to that of a selection invention occurs when the invention is a group of compounds and one compound, or a very small group of compounds, within the group is part of the prior art. If this situation is known before the patent application is filed, the claims can be drafted so as to take account of the prior art and a suitable basis for the claims written into the specification. If the prior art is discovered only during prosecution, it may in certain circumstances be appropriate to insert into the claim a disclaimer to the prior art compound, in the form (for example):

A compound of formula I in which X is $C_{(1\text{-}6)}$ alkyl and Y is $C_{(1\text{-}6)}$ alkyl, phenyl or benzyl, *provided that when X is methyl Y is other than phenyl.*

In other words, the compound in which X is methyl and Y is phenyl is in the prior art, and is disclaimed. For many years, it was standard practice in the EPO that such a disclaimer, in the correct circumstances, was not regarded as adding new matter even if there was no basis for the disclaimer in the specification as filed.[29] A disclaimer could, however, only be used to give novelty and could not be used to establish inventive step. The EPO did not allow a claim to be amended by a disclaimer lacking basis in the specification as filed if the purpose was to establish inventive step rather than novelty.[30] Furthermore, even if such a disclaimer was made for the purpose of avoiding anticipation, it could not impart inventiveness to an obvious claim.[31]

Thus, in the example given above, suppose that the compounds of formula I were useful as antihypertensive agents. If the compound in which X is methyl and Y is

[27] T 124/87 (OJ 1989, 491) *DU PONT/Copolymers.*

[28] For example, T 666/89 (OJ 1993, 495) *UNILEVER/Washing composition.*

[29] For example, T 433/86 (unpublished) *ICI/Diisocyanates.*

[30] T 170/87 (OJ 1989, 441) *SULZER/Hot-gas cooler*; T 597/92 (OJ 1996, 135) *BLASCHIM/Rearrangement reactions.*

[31] T 710/92 (unpublished) *KANEGAFUCHI/Polyolefin.*

phenyl was known from a prior publication only for an unrelated use (for example as a dyestuff), then the prior art does not base any attack on the inventive step of the remaining scope once the novelty is restored by disclaimer and the disclaimer will render the claim valid. But if the prior art compound was also known for this antihypertensive use, then the rest of the scope, especially, for example, the closest compound in which X is ethyl and Y is phenyl, is prima facie obvious and the disclaimer will not change this.

In this situation, the only possibility is to cut back to a narrower scope for which there is already basis in the specification. Here, for example, a scope such as 'a compound of formula I in which X and Y are each, independently, $C_{(1\text{-}4)}$ alkyl' could be valid if the group were shown to have unexpected advantages.

Disclaimers have been particularly useful in the European system when the prior art is not a prior publication, but an unpublished application of earlier priority date that is prior art under Article 54(3) of the European Patent Convention (EPC). In this situation, the prior art could be used only to base a novelty attack and inventive step is not considered. Therefore, a disclaimer of one or more specific compounds could be made without danger even when the compounds have the same utility as those claimed.

In 2002, a TBA held[32] that this practice was incompatible with the stricter approach to priority and basis of claims taken in G 2/98 (see p. 88). This issue was referred to the EBA by another Board and the EBA essentially restored the status quo. A disclaimer does not constitute new matter when it is used to establish novelty where inventive step is not an issue: for example in the case of Article 54(3) prior art, or an 'accidental anticipation' in which the prior art related to a different use or gave no utility at all. The catch is that if it subsequently turns out that inventive step is relevant, for example if the application loses its priority date and a piece of prior art under Article 54(3) becomes Article 54(2) prior art, the disclaimer will become retroactively inadmissible, with likely fatal consequences for the application.

Over-Broad Disclaimers

In many cases, especially when numerical ranges are involved, the extent of the disclaimer necessary to avoid anticipation may not be clear. The applicant may be tempted to disclaim a broader part of the scope than strictly necessary, in the hope that the resulting amended claim will be more likely to withstand an obviousness attack, and the examiner may well (wrongly) allow such an over-broad disclaimer.

This can, however, give rise to a very dangerous situation. If the patent is contested in opposition proceedings, it can be argued that the unsupported disclaimer, if going beyond what was strictly required to avoid anticipation, amounts to the addition of new matter, a ground of opposition under EPC Article 100(c). The patentee

[32] T 323/97 (OJ 2002, 463) *UNILEVER/Disclaimer*.

is then in the situation described above (p. 225): if the over-broad disclaimer is indeed new matter, he cannot correct it by restricting the disclaimer to disclaim a smaller part of the scope because this would broaden the scope of the claim, which, in opposition proceedings, is forbidden by EPC Article 123(3).

Compounds of Unknown Structure

A compound may still be patentable even if its chemical structure is uncertain or unknown. The difficulty lies in defining the compound in such a way that it can be claimed unambiguously. There are two main ways of doing this: firstly, by defining the compound in terms of its properties; and secondly, by defining how the compound is made.

Fingerprint Claims

If one is claiming only a single compound, for example a new antibiotic obtained from a mutant strain of a micro-organism, it may well be possible to obtain the compound in pure form and define it in terms of properties that may be:

(a) physical—for example melting-point, infrared spectrum, nuclear magnetic resonance (nmr) spectrum, optical rotation, uv spectrum, mass spectrum, or crystal form;
(b) physico-chemical—for example solubility in various solvents, molecular weight, or elementary analysis;
(c) chemical—for example its action with various reagents; or
(d) biological—for example its effect upon various bacteria.

The greater the number of such characterizing properties that can be found, the better: the compound can then be claimed in terms of these properties in a so-called 'fingerprint claim'. This is preferable to making a guess at an uncertain structure, because, if the guess is wrong, the mistake cannot normally be corrected later. In one case in the UK, the patentee was allowed to replace a compound per se claim based on an incorrect structural formula by a fingerprint claim characterizing the compound by the melting points of a number of derivatives, although he was not allowed to substitute the correct formula.[33] This was rather a special case in that the patent was claiming a single, well-characterized compound. Normally, chemical patents claim a group of compounds and would not contain enough data to put together a fingerprint claim for any one compound in the group.

The need for such fingerprint claims is less now that techniques for structure determination have improved, but if a full structure determination may take some months, it may be advantageous to file a priority application containing such claims and to add a normal structure-based claim at the foreign filing stage.

33 *Egyt's Patent* [1981] RPC 99 (Pat Ct).

Although all available data should be mentioned in the description, only the totally reliable data should be given in the claims. Including questionable data that may later turn out to be false will weaken the patent.

Product-by-Process Claims

If it is not possible to define the product adequately in terms of its properties, or if it is desired to claim a group of compounds, the product may be defined as the product of reacting certain reagents under certain conditions: that is, by a product-by-process claim. In the EPO, a claim such as 'the product obtained by reacting A with B' would be construed as a product claim covering the product as such however it was made, and is patentable only if the compound itself is patentable and if there is no alternative way of defining the compound.[34] If the compound is known, the product-by-process claim adds nothing to the protection given by a process claim and for this reason is not allowed as a separate claim. In the UK, the above claim would be construed not to cover the same product when prepared by a different process, and if product per se protection were required, it would be necessary to use wording such as '*obtainable* by reacting A with B'.

For example, in a case in which a bromo-compound is formed by removing hydrogen bromide from a dibromide and it cannot be determined on which of two carbon atoms the remaining bromine atom is located, the claim could read:

The compound of formula [compound structure showing generically two possibilities] wherein Br is in the 9 or 10 position *obtainable by* dehydrohalogenation of the compound of formula [. . .] with alcoholic potassium hydroxide.

Ideally, one should (in the UK) add a dependent claim using the words 'obtained by', which should, by the rules of claim construction (see p. 442), ensure that the 'obtainable by' claim is construed more broadly. But many patent offices simply refuse to accept claims drafted in 'obtainable by' form.

In the USA, there have been conflicting cases on the interpretation of product-by-process claims, but the situation has now been clarified, albeit in an unsatisfactory way, by an *en banc* decision of the CAFC.[35] This held that a claim to a crystal form, characterized as 'obtainable by' a certain crystallization process, did not cover the product however it was made, but was infringed only if that specific process was used. Judge Newman and three other judges dissented, and would have adopted the European construction of such claims.

Polymeric Compounds

Strictly speaking, all polymers are mixtures of various molecular species, so that a precise structural description is usually not possible. If the polymer is completely

[34] T 150/82 (OJ 1984, 309) *IFF/Claim categories.*

[35] *Abbott Laboratories and Astellas Pharma Inc. v. Sandoz Inc. and ors* (Fed Cir 18 May 2009).

new, in the sense that it is obtained from monomers never previously polymerized, or by modifying functional groups on a polymer backbone in a new way, the structure can normally be defined in terms of the structural formula of its repeating unit.

Given that this basic formula of the polymer is already known, it is still possible to obtain new and patentable polymers by various kinds of modification, provided that these give useful and non-obvious results. Perhaps the simplest of these would be control of the molecular weight to give products of a specific average molecular weight or molecular-weight distribution. But the correlation between polymer properties and molecular weight is fairly well understood, so it is difficult to find non-obvious advantages of a particular molecular weight range.

The nature of the end-groups on a polymer chain may modify the properties of the polymer very considerably, even though the end-groups may represent only a very small fraction of the total structure. In the same way, the presence of quite small quantities of comonomers may give a copolymer with surprising advantages over the original homopolymer.

The chemistry of polyoxymethylene, $CH_3\text{-}O\text{-}(\text{-}CH_2\text{-}O\text{-})_n\text{-}H$, illustrates both of these points very well. Under suitable conditions, pure formaldehyde can be polymerized into a high-molecular-weight polymer with excellent mechanical properties. The unmodified polymer is quite useless, however, because it has no thermal stability: on heating, the polymer chain ‘unzips’ and depolymerizes back to formaldehyde.

One approach to this problem was to modify the end groups by converting hydroxy end-groups to methoxy. The resulting polymer, $CH_3\text{-}O\text{-}(\text{-}CH_2\text{-}O\text{-})_n\text{-}CH_3$, had sufficient thermal stability to be useful and was a patentable new product. It still had the problem that if the end-group should be cleaved off, the remainder of the chain would rapidly unzip, so that, above a certain temperature, depolymerization would set in very rapidly.

The problem was solved by the introduction of a small amount of ethylene oxide as comonomer, giving a polymer that, after end-capping, had the structure:

$$CH_3\text{-}O\text{-}(\text{-}CH_2\text{-}O\text{-})_n\text{-}CH_2\text{-}CH_2\text{-}O\text{-}(\text{-}CH_2\text{-}O\text{-})_m\text{-}[\ldots]\text{-}CH_3$$

The ethylene oxide units in the chain acted as ‘zipper jammers’, so that if depolymerization should set in, it would stop at the first ethylene oxide unit and not continue along the whole chain. Here, again, a patentable new product had been invented.

If larger amounts of comonomer units are present, their distribution in the chain may give rise to different products, such as random, regular, or block copolymers, and if the polymer chain contains optically active centres, the sterochemistry of the polymer may be used to define new products. For example, isotactic polypropylene is an important product, patentable over ordinary atactic polypropylene, and in the USA, there was a conflict that lasted for over 30 years on the question of who had the rights to this invention.

Cross-linking polymer chains gives products differing markedly from uncross-linked polymers and, again, such products may be patentable, although there are increasing difficulties in defining the product once cross-linking occurs.

New Salt Forms

Many organic compounds may exist in the form of salts: for example metal or organic base salts of an acidic compound, or salts formed between a basic compound and an inorganic or organic acid. In the field of pharmaceuticals, there are relatively small groups of anions and of cations used to make pharmaceutically acceptable salts with basic and acidic active substances, and it is difficult to obtain a patent for a new salt form unless it can be shown to have unexpected advantages over known salts. In 2007, the CAFC overturned the District Court and held Pfizer's patent for amlodipine besylate (benzenesulphonate) invalid for obviousness over Pfizer's earlier patent disclosing the maleate and other salts.[36] The besylate was prima facie obvious because, although salt properties were not predictable, the person skilled in the art would have a reasonable expectation that the besylate would be suitable. Such improved properties as there were were not enough to overcome the finding of prima facie obviousness.

New Physical Forms

New physical forms of known compounds may also be patentable chemical inventions. For example, certain pigments may exist in different allotropic solid forms and it sometimes happens that a particular new solid form may have improved stability or better colouring properties than the form in which the pigment was previously known. Even mere reduction in particle size of a known compound may be a patentable invention. The drug griseofulvin was known to be an effective agent against fungal infections of the skin, but could only be used locally because the compound was so insoluble that if it was taken orally none of it was resorbed into the bloodstream. It was found that micronized griseofulvin could be used orally and was patentable per se.

In the pharmaceutical field, patents relating to new crystal forms of known substances may have great economic importance. If, some years after a patent application is filed for a new chemical entity, a new crystalline form is found that has some advantage, such as improved stability or solubility, then this form can itself be patented. If it is the new form that is subsequently marketed and approved, an imitator must normally copy this form in order to obtain registration and it will be of no help to him that the substance itself, in the original form, is patent-free. In this way, Glaxo Wellcome was able effectively to prolong the patent life of its blockbuster

[36] *Pfizer v. Apotex* 480 F.3d 1348 (Fed Cir 2007).

anti-ulcer drug ranitidine (Zantac®). The validity of such patents can in theory be attacked if it can be shown that the method of preparation of the substance that was disclosed in the original patent would in fact give the 'new' crystal form, but this can be very difficult to prove in practice, particularly if it can be argued that the experimental conditions used did not rule out the possibility of 'seeding' by minute crystals of the new form.

An important case relating to crystalline forms was that of *Synthon v. SmithKline Beecham*,[37] which dealt with two conflicting applications originally claiming paroxetine methansulphonate (PMS) as a new salt form. This salt was, however, mentioned in the prior art and so both applications were amended to claim the crystal form of the salt. The question was whether Synthon's earlier unpublished application destroyed the novelty of SmithKline Beecham's later filing. The situation was complicated because Synthon had given incorrect X-ray diffraction data and had chosen an unsuitable solvent that made crystallization difficult. In the House of Lords, Lord Hoffman restated the general principle that anticipation requires a prior disclosure that, if performed, would infringe the patent. Here, evidence showed that the substance was monomorphic, so that the incorrect data were irrelevant; furthermore, the disclosure was enabling, because the skilled person would try a different solvent if the first were not to work.

Servier's attempt to enforce its patent for the α-form of the t-butylamine salt of perindopril failed miserably when the Patents Court and the Court of Appeal both held that this crystal form would clearly be obtained by carrying out the process disclosed in the basic patent for the substance. Lord Justice Jacob was particularly scathing, saying: 'It is the sort of patent that can give the patent system a bad name.'[38] And it is, of course, the sort that gives ammunition to persons complaining about 'evergreening' (see p. 428).

New Synthetic Processes

Another major category of chemical inventions is that of new processes for the preparation of known compounds. These may be completely new and applicable to a wide range of end products, in the way that, for example, reduction using boron hydrides or alkylation using Grignard reagents were at the time that they were invented.

Such a process is patentable, because it will be easy to specify at least one industrially applicable end product that can be made using it; furthermore, such processes may involve the use of novel reagents that could themselves be patentable as new compounds. But the majority of such advances in general synthetic chemistry

[37] *Synthon v. SmithKline Beecham* [2005] UKHL 59.
[38] *Servier v. Apotex* [2008] EWHC Civ 445.

are made in university laboratories and are published in the scientific literature rather than patented.

In industrial laboratories, research on new synthetic methods is generally applied to particular commercially important compounds. Such methods may range from an entirely new synthetic route representing the first commercially feasible method of producing a whole new group of compounds to a minor improvement in the established process for a single product.

In deciding whether or not to seek patent protection for such an invention, one must balance the relative merits of obtaining patent protection and maintaining the new process as secret know-how. A patent costs money and should be applied for only if a commercial benefit is expected from it. If the process of the invention is such that it cannot be determined from the end product or other evidence (such as trace quantities of characteristic by-products) whether or not a competitor is using it, then any patent rights will be unenforceable. The patent will then have value only to the extent that the competitors are ethical enough to respect patents that they know they could infringe with impunity. What is worse, the publication of the patent application informs competitors, ethical or otherwise, how to carry out the invention.

Keeping a process invention as secret know-how will be feasible only if the invention cannot be deduced from the end product, but if it is feasible, it does have certain advantages: firstly, it costs nothing over and above the normal overheads of maintaining business security; secondly, it gives nothing away to competitors; and thirdly, the effective period of monopoly can in theory be prolonged indefinitely and is not limited to the term of patent protection. But if the secret is lost other than by theft, or if someone else makes the same invention independently, the original inventor can do nothing about it.

If someone else does make the same invention and patents it, there is a risk of being liable for infringement of the new patent. In most European countries and in Japan, if someone has been using the invention, or has made serious and effective preparations to use it, before the priority date of the patent, that person has the right to continue to use it. However, it must be possible to document this prior use and the permitted use usually cannot be expanded, for example by building another factory for the process. In the USA, there is no right of prior use (except in relation to the specific class of business method patents, see p. 331), and the earlier invention would not invalidate the later patent because the first inventor concealed the invention.

On balance, it is preferable in many cases not to patent process improvement inventions, since such patents are extremely difficult to enforce. But if it is intended to keep a process as a trade secret, documentary evidence must be kept of when preparations to use the process started and when actual use began. It must also be made clear to employees that it is a secret. If this is not done, an employee who leaves and joins a competitor may not feel under any particular obligation of confidentiality with regard to the process, and it is in this way that trade secrets may most easily be lost.

Analogy Processes

In the UK and the EPO, it has always been the rule that if a group of compounds is new and inventive, then not only are claims to the compounds per se patentable, but so also are claims to the process for the preparation of the compounds, even if the starting materials are known and the process itself is known as a method for making similar compounds. Similarly, if an intermediate compound is novel and inventive, then a process for making known end products from the intermediate will be patentable, even if the process is known for similar starting materials. In the USA, however, such 'analogy process' claims were held to be unpatentable unless they were inventive in themselves.[39] This may not appear to be particularly important, because, if the product itself is patented, a claim to a process for making it adds little additional protection. But the biotech company Amgen found itself seriously disadvantaged, because although it had a patent for recombinant host cells expressing erythropoietin (EPO), a known substance, it was unable to obtain a claim to the conventional process for obtaining EPO by culturing those cells. When Chugai made such cells in Japan and imported the product into the USA, the importation could not be stopped because the imported product was not the product of a patented process.[40] As a result of lobbying from the biotech industry, *Durden* was legislatively overruled by the Biotechnology Patent Protection Act of 1993, which amended the patent law so that a biotechnological process is not to be regarded as obvious if it uses or results in a novel and unobvious composition of matter.[41] This legislation has the serious defect that it has effect only for one specific field of technology, which is arguably contrary to the provisions of the TRIPs Agreement. It has, however, effectively been rendered moot by subsequent CAFC decisions that have overruled *Durden* and which are equally applicable to all technical fields.[42]

New Compositions and Mixtures

A new composition may sometimes be claimed when the invention is really a new compound or a new use of an old compound. For example, if a new pigment is useful in paints, one could broadly claim a paint composition containing the pigment and a pharmaceutical composition claim may be a suitable way in which to claim the invention that a known compound has a pharmaceutical use. Such functional definitions of compositions are only feasible if it is clear, for example, what is meant by a 'paint': would it include a nail varnish or an ink? The borderline between pharmaceutical and cosmetic compositions is also somewhat unclear.

[39] *In re Durden* 226 USPQ 359 (Fed Cir 1985).

[40] *In the Matter of Certain Recombinant Erythropoietin* 10 USPQ 2d 1906 (ITC 1989); *Amgen v. US International Trade Commission* 14 USPQ 2d 1016 (Fed Cir 1990).

[41] 35 USC 103(b).

[42] For example, *In re Ochiai* 37 USPQ 2d 1127 (Fed Cir 1995); *In re Brouwer* 37 USPQ 2d 1663 (Fed Cir 1995).

If the composition itself is claimed, it must of course really be new and one does not make an old composition new merely by giving it a new name or supplying a new use for it. Thus, for example, if a compound of formula X was known to be a uv stabilizer and had been used in paint compositions for that purpose, then the invention that X could be used as a corrosion inhibitor could not be protected by means of a claim to 'an anticorrosive paint containing a compound of formula X'. Only if anticorrosive paints differed in composition from paints in which X had been used and these differences were expressed in the claim would the claimed invention be novel.

This approach is that of Anglo-Saxon countries and is different from that of Germany, where a claim to an anticorrosive *Mittel* (means) could be considered novel even if the composition were to be the same as that previously used for a different purpose. The EPO has also adopted a different approach, as an extension of the practice of claiming a first pharmaceutical use of a known compound by means of a 'product for use as' claim. This will be discussed in the next chapter.

There are many inventions in relation to which the invention itself lies in the preparation of a composition consisting of two or more components, which is particularly useful for some industrial application. The general term 'composition' covers both simple mixtures, such as a new plant protection agent formulation of known ingredients, and cases in which there is some physico-chemical interaction between the components: for example glass compositions, which may be described in terms of their content of various metal oxides, silica, etc.

There are some fields, for example those of detergents, cosmetics, or textile finishing agents, in which most commercial products are simple mixtures of various components, each of which is known to have been used in other similar products. To get any kind of patent protection for these mixtures is not easy: firstly, because of the difficulty of defining the composition precisely; and secondly, because of the large amount of close prior art. The difficulty of definition is not helped by the fact that many of the individual components are themselves mixtures, or are commercial products sold under a trade name, the actual structure or composition of which is not always known. The close prior art means that most such mixtures will be prima facie obvious and will have to have some sort of advantage in order to be patentable.

New Uses and New Application Processes

This broad category of chemical invention includes the discovery of a specific new use for a known compound that may or may not have some existing use, as well as all new methods of treating substances or articles with known compounds: for example an improved dyeing process using known dyes and auxiliaries in a new way.

Such inventions may come from research and development laboratories in which old compounds are screened for possible new activities, but they often arise out of non-research activities, such as product development, customer service, or even marketing. Many patentable inventions coming from such sources may be overlooked, because they are not recognized as such either by the inventor or by management, and yet an invention made in the course of solving a particular problem for a customer, for example, will have more chance of being commercially interesting than many more speculative and less practical inventions arising in the laboratory.

Inventions of new uses for old compounds may be claimed as the process of using the compound, reciting one or more actual process steps, for example 'A method of preventing corrosion of metals by applying to the metal a compound of formula I'. or, alternatively, as a use claim such as 'The use of a compound of formula I as an anti-corrosion agent for metals'. The EPC does not distinguish between the two and regards them as equivalent.[43] Special considerations arise when the new use is a pharmaceutical use and these are discussed in the next chapter.

For uses, just as for compositions themselves, the UK approach is that, for a new use to be patentable, the corresponding process steps themselves must have some new technical feature and that 'mere novelty of purpose' is not enough. This approach is not followed in the EPO, in the light of two 1989 EBA decisions.[44] In one of these, the invention was that a substance previously known as a plant growth regulator was also a fungicide; in the other, a compound known as a motor oil additive to inhibit rust also acted to improve lubrication properties. In both of these, it could be argued that there was inherent lack of novelty, because the technical embodiment of the new use was in each case identical with that of the known use. Plants were sprayed with the same solution of the same compound and if this killed fungi on the plant, it must equally have done so when it was previously being used as a growth regulator. Similarly the improved lubrication effect must have been present, even if unrecognized, when the same compound was put in the same oil. Nonetheless the EBA decided that the new result was to be regarded as a functional technical feature that could give novelty to the claims and the patents were granted.

The conclusion that there is in effect no such thing as inherent lack of novelty because, like secret prior use, it does not make the knowledge of the new technical effect available to the public is a far-reaching one, which has since been followed, not without reluctance, by the House of Lords (see p. 253). What is not resolved is the question of what scope of protection will be given by such claims and how infringement will be judged. Presumably someone who sold the old product with instructions to use it in order to obtain the new effect would be an infringer,[45] or at least a

[43] G 5/83 (OJ 1985, 64) *EISAI/Second medical indication.*

[44] G 2/88 (OJ 1990, 93) *MOBIL OIL/Friction reducing additive*; G 6/88 (OJ 1990 114) *BAYER/Plant growth regulating agent.*

[45] In Germany, someone who prepared a compound to be used for the new claimed purpose, e.g., by mentioning the purpose in a product leaflet is regarded as a direct infringer: see ZR 29/88 X *Geschlitzte Abdeckfolie* [1990] GRUR 505 (Federal Supreme Court).

contributory infringer. But if the product does not contain any indication for the new use, even a commercial end-user could perhaps avoid infringement by alleging that he was using the product for the old, known, effect. It is an unsatisfactory situation if infringement cannot be judged objectively, but depends upon the intention of the alleged infringer.

A number of developing countries have specifically excluded new uses from protection. This was done, for example, by Decision 486 of the Andean Community, covering Bolivia, Colombia, Ecuador, Peru, and Venezuela. When, in 2000, Peru granted a new use patent to Pfizer for the most famous use of Viagra® (sildenafil was first patented as an anti-angina agent), this was challenged in the courts and, in December 2001, the Andean Court of Justice gave a decision subsequently ratified by the Peruvian Administrative Court that the patent should not have been granted. India has similar provisions in the Patents (Third Amendment) Act 2005. According to section 3(d) of the Indian Patents Act 1970, new uses, crystalline forms, and even derivatives of a known compound are only patentable if the patentee can show an improved efficacy. The argument is that, once the compound is known, finding out new uses for it is the 'mere discovery' of its properties and so not patentable. The real motivation is a fear that use patents are being used to prolong artificially the duration of patent protection for pharmaceuticals. In fact, it is a rather unusual situation if a compound is marketed only for a later-invented indication. Normally the innovator company will be selling the product for an indication disclosed in the original patent and the existence of later use patents will not prevent any generic product for the earlier first use. In terms of promoting innovation, it is important that new indications of known pharmaceutical compounds can be protected by patents, because a new indication can be of equal value to patients suffering from a certain disease as a new drug with a new active ingredient that would be clearly patentable.

Pharmaceutical companies are sometimes, however, so concerned that others might be able to hinder their activities by filing new use patents that they often disclose every conceivable use in the original patent, thus making it difficult or impossible for anyone, including themselves, to get such patents.

13
PHARMACEUTICAL INVENTIONS

> . . . medical science is as yet very imperfectly differentiated from common cure-mongering witchcraft . . . one practitioner prescribing six or seven scheduled poisons for so familiar a disease as enteric fever where another will not tolerate drugs at all . . .
>
> George Bernard Shaw, 'Preface', *The Doctor's Dilemma* (1906)

New Chemical Entities

In the previous chapter, we outlined the various types of invention that may arise in the chemical field. Where the field of application of the invention is the field of pharmaceuticals, however, some special considerations apply and these will be considered in this chapter.

Many of these special problems arise from the provision in the UK Patents Act 1977 and corresponding provisions in the European Patent Convention (EPC) excluding from patentability 'methods for treatment of the human or animal body by surgery or therapy and diagnostic methods practiced on the human or animal body',[1] although it is clearly stated that this does not prevent a substance or composition from being patentable for use in any of these methods.[2] Under the EPC, such methods were excluded by being defined as not industrially applicable; the new approach introduced by EPC 2000 is more logical and also clarifies the patentability of further medical uses (see p. 262). Novel compounds that have a pharmaceutical utility are patentable per se in all countries that have implemented the TRIPs Agreement.[3]

Prodrugs and Active Metabolites

It is found that, when a pharmacologically active compound is administered to the human or animal body, some of the compound may be excreted unchanged, but some or all of it may instead be subject to more or less complex chemical changes resulting in a series of metabolites that are excreted or broken down further. It frequently happens that one or more of the metabolites is also active, and it is not uncommon to find that the activity of the compound that is administered is entirely due to an active metabolite. If a compound is known to be pharmacologically active, it is regarded as a drug; a compound that itself is inactive, but which is hydrolysed or otherwise metabolized in the body to form the active drug, is considered to be a 'prodrug'. If a compound owes its activity to the fact that it is metabolized to another compound, it may be merely a matter of historical accident whether the compounds related in this way are considered to be prodrug and drug, or drug and active metabolite.

If the real active substance is invented first, a subsequently invented compound metabolized to it (a prodrug) may have some pharmacokinetic advantage. Often only the development of a prodrug makes it possible to select a different way of administration to the patient. For example, the cephalosporin antibiotic cefuroxim can only be administered by injection, whereas the later-developed prodrug cefuroxime axetil makes oral administration possible. Another example of a drug/prodrug pair are the antivirals acyclovir and its prodrug valacyclovir. Both active ingredients are on the market in orally administered dosage forms (acyclovir is additionally available in topical forms), but the later-developed prodrug valacyclovir has a much higher bioavailability due to a quicker intestinal resorption rate. Patents with specific claims to the prodrug valacyclovir have been granted.[4]

When drafting patent applications for drug substances, if there is the possibility that prodrugs could later be developed, it should be considered whether it is worth including language in the claims and specification already directed to prodrugs.

[1] Section 4A(1), PA 1977; Art. 53(c), EPC 2000.

[2] Section 4A(2), PA 1977; Art. 53(c), second clause, EPC 2000.

[3] Article 27.1, TRIPs.

[4] For example, EP308065B1.

Such language may help later on in a situation in which someone else develops a prodrug in order to evade the patent protection for the drug substance itself. A court may be more easily convinced to regard any sale of the prodrug as an infringement, even if it literally falls outside the compound claims of the patent. But any, even speculative, language about prodrugs in the basic patent for the drug will make it more difficult to establish patentability of a particular prodrug that may be developed later.

In the UK, it was held that sales of hetacillin, an acetone adduct of ampicillin that was immediately hydrolysed in the body to ampicillin, infringed the ampicillin patent because it was 'ampicillin in disguise'.[5] Here, of course, the prodrug in question was considered to be a deliberate attempt to capitalize on ampicillin while evading the letter of the patent claims and it is not at all clear that this case would be followed if the person selling the prodrug were to have acted in good faith without realizing that activity was due to a patented metabolite, or if the prodrug had any activity of its own.

Patents for Prodrugs

Following the hetacillin case, Beechams succeeded in obtaining in the UK (under the Patents Act 1949) a patent claiming a novel cephalosporin, its salts, and 'pharmaceutically acceptable bioprecursors thereof'. A claim such as this appears to be too broad and may lack sufficiency of disclosure, because it will require an undue amount of experimental work to determine whether or not a given compound could fall under the vague definition of 'bioprecursor'. It is nevertheless possible to draft allowable claims to drugs that literally cover prodrugs: for example one can claim 'physiologically hydrolysable and acceptable esters' of alcohols or acids. A typical description of such esters is as follows:

> By physiologically hydrolysable and acceptable ester as used herein is meant an ester in which the hydroxy group is esterified and which is hydrolysable under physiological conditions to yield an acid which is itself physiologically tolerable at dosages to be administered. The term is thus to be understood as defining regular prodrug forms. Examples of such esters include for example acetates, as well as benzoates, of the compounds of the invention.

The situation is different when a compound is patented as a drug and subsequently found to be only a precursor of the real active substance, that is, the active metabolite. This should not affect the intrinsic patentability of the substance: if a pharmacological result is obtained by giving the substance, it is immaterial by what process the result is obtained. It is never necessary to explain in a patent how an invention works; the fact that it does work, whether directly or through an intermediary compound, is sufficient. In such a case, however, if claims to the first substance do

[5] *Beecham v. Bristol Laboratories* [1977] FSR 215 (HL).

not literally include the second, they cannot be construed to cover it indirectly, that is, sales of the metabolite would not necessarily infringe the patent for the drug.

Patents for Active Metabolites

The active metabolite, if novel and inventive, can of course be patented separately and then the question arises whether sales of the original drug substance infringe the patent claiming the active metabolite. By analogy with the 'prodrug' scenario described above, one might think that they should, but this leads to an unacceptable conclusion. Merrell Dow attempted to do just this in respect of the drug terfenadine, which had been sold for many years as an antihistamine and the patents on which had expired. Merrell Dow had, however, found that the activity was due to an active metabolite and it patented this compound separately. When a generic competitor began to sell terfenadine, Merrell Dow sued it for infringement under the metabolite patent.

This was clearly a blatant attempt to extend patent protection far beyond the statutory period. Patients had been taking terfenadine, at least in clinical trials, before the active metabolite was patented; the patent on terfenadine had expired. Yet supplying terfenadine for the same act, that is, to be swallowed by a patient, was now alleged to be infringement. In the UK, the case came before the House of Lords in 1995.[6]

It was absolutely clear that this misuse of the patent system could not be allowed; the difficulty lay in finding a good legal argument to stop it. In the USA, a District Court, faced with the same problem, held that the patent was valid, but not infringed,[7] which seems difficult to reconcile with case law on contributory infringement and prodrugs. In Germany also, the courts found that there was no infringement.[8] The House of Lords clearly wanted to find the patent invalid, but was faced with the difficulty that the Enlarged Board of Appeal (EBA) of the European Patent Office (EPO) had held in effect that there was no such thing as 'inherent lack of novelty' (see p. 248).

By analogy, even though at least some members of the public had been making the active metabolite by swallowing terfenadine before the date of the patent, they had not been aware of the fact and thus this prior use could not be novelty-destroying. The House of Lords accepted this argument, but held that the disclosure of the terfenadine patent specification itself, although it did not mention the active metabolite, made available to the public the invention of the acid metabolite because it 'enabled the public to work the invention by making the active metabolite in their livers', that is, by taking terfenadine. Accordingly the patent was invalid to the

[6] *Merrell Dow v. Norton* [1996] RPC 76 (HL).

[7] *Marion Merrell Dow Inc. v. Baker Norton Pharmaceuticals Inc.* 41 USPQ 2d 1127 (SD Fl 1996); [1998] FSR 158.

[8] *Terfenadine* [1988] FSR 145 (Munich *Oberlandesgericht*).

The other two methods of drug discovery exploit the fact that it is now possible to do initial screening assays very rapidly and with very small amounts of many compounds, the entire process being fully automated. Tens of thousands or sometimes millions of different compound–ligand interactions can be tested in such high-throughput screening systems within a couple of weeks or days. The problem is to find enough test compounds to put into the assays. One way is to screen large groups of samples randomly, which may be compounds from non-pharmaceutical sources, for example compounds originally synthesized as photographic chemicals, or may be from natural sources. In the latter case, the samples need not be pure compounds; rather, they may be plant extracts, soil samples or fermentation broths from microorganisms. If activity is found, the active compound can be isolated and identified subsequently.

The other way is to use the technique of combinatorial chemistry, in which thousands of different compounds may be synthesized simultaneously by a combination of different starting materials, reaction steps, and reagents to produce a 'library' of compounds that can then be screened. Clearly there is no point in producing a random mixture of all of the compounds together: in order to get meaningful results from the screening, it must be possible to identify which compounds are the active ones. A number of different techniques have been used for this, many of which involve carrying out the reactions on a solid surface, for example a plastic bead, and 'tagging' the surface in some way so as to identify the sequence of reactions and reagents to which the original compound on the bead was subjected, and hence the structure of the final compound on the surface. In other methods, the initial screen may only identify that one compound in a mixture of, say, ten compounds of known structure has the desired activity and a second step may then be used to select the individual active compound. A combinatorial library may be designed to be as diverse as possible, or may be targeted to focus on a narrower group of compounds.

It is an interesting problem to what extent patent protection can be obtained for a compound library as such, as distinct from the group of compounds contained in the library. Although it is known what structures will be present in this group and it may be possible to draft a generic formula covering them, simply claiming the compounds will usually have the problems that some of the compounds may be old and that the great majority will be inactive. It may be possible to claim the library as an array of compounds defined by the way in which it was synthesized, including any tagging technology that was used. As long as it is an array or library of compounds that is claimed and the claim does not cover individual compounds, there should be no novelty problem if some of the compounds are known. Normally a group of compounds would not be considered as industrially applicable if their only use was to be the subject of tests to find a real utility for one or more of them. But compound libraries are commercially important and may be sold or licensed for large sums of money, indicating that the pharmaceutical industry certainly thinks that they are useful. This ought to be enough to establish utility in the patent sense. If the claims cover the individual compounds, each compound being attached to a solid support, the problem of novelty should be overcome, but here utility will be more of a problem. No patents claiming compound libraries seem to have come before the courts so far.

Pharmaceutical Compositions

Novel pharmaceutical compositions may be of three distinct types:

(a) combination preparations comprising two or more known pharmaceutically active ingredients;
(b) new drug delivery systems or galenic forms (for example a new kind of tablet giving a controlled rate of release of drug when swallowed); and
(c) compositions comprising a compound not previously used as a drug, together with any conventional pharmaceutical carrier or excipient.

Combination Preparations

As in other chemical fields, simple mixtures of known pharmaceuticals can be patented only if they are novel and if inventive step can be shown by some improved property. For example, patentability can be established by demonstrating that the person skilled in the art would not have combined these ingredients, or that the combination is synergistic, that is, it has a 'superadditive effect' that could not be expected.

Classical synergism between two drugs can be extremely difficult, if not impossible, to prove, largely because it is practically impossible to predict what would be expected if there were no synergistic effect. In fact, synergism in the strict sense can be mathematically proved only by comparing dose-versus-response curves for the two components separately, as well as for the combination. This can be a long and difficult exercise even for a simple quantifiable measurement such as blood pressure; when the effect that has to be measured is the rate of incidence of some infrequent occurrence, for example the percentage of patients suffering heart attacks in a given time, and particularly where the results have to be obtained from clinical studies on humans because no suitable animal tests exist, then rigid proof of classical synergism becomes a practical impossibility.

If synergism can be demonstrated, this will help to make the combination patentable, but even in the absence of classical synergism, other advantageous and unobvious results should be usable to establish the presence of an inventive step. Suppose compounds A and B each have the same pharmaceutical effect at a dose of 100 mg, but that, at this dosage, each gives some undesirable side effects; suppose further that a combination of 50 mg A and 50 mg B gives exactly the same desired effect, but with reduced side effects. There is no evidence for synergism between A and B, but nevertheless the advantage of lower side effects, assuming that it was not predictable, should be sufficient for patentability of the combination.

Drug Delivery Systems
In compositions of this type, the invention lies not in the pharmaceutically active material, but in the other constituents that enable it to be administered in a particular way. Very often, such inventions are applicable to a very wide range of drugs. Because it is becoming more and more difficult and expensive to find and develop new drugs, more effort is being put into finding ways of delivering existing drugs more effectively, and inventions of this type are increasing in frequency and in commercial importance.

For example, it is often desirable to have low but more-or-less steady concentrations of a drug in the body over a relatively long period of time. Alternatives to the frequent swallowing of tablets at regular intervals include, for example, adhesive patches from which a drug is absorbed slowly through the skin, or depot injections of fine particles of a biodegradable polymer that releases a drug over a period of weeks or months. Many hormones, such as insulin, are peptides that are destroyed by the digestive juices if taken orally and which normally have to be given by injection; alternative methods have been developed, such as nasal sprays, from which the peptide can be absorbed through the mucous membrane of the nose. In the case of inhalable insulin, however, a first product that was brought to the market in 2006 was withdrawn from the market for lack of commercial interest in 2008.[12]

Conventional Pharmaceutical Compositions
Where a compound is already known for non-pharmaceutical purposes and can no longer be claimed per se, the invention that it has a pharmaceutical use may conveniently be claimed by claiming 'a pharmaceutical composition comprising a compound of formula . . . in association with a pharmaceutically acceptable diluent or carrier', which would include all forms in which the compound could be administered, from a complex drug delivery system to a simple tablet, or even a solution in sterile water for injection. This claim may lack novelty if the compound was known to exist in solution and, in the USA, examiners may insist on limitation to solid dilutents or carriers for this reason. This is, however, a special case of the question of 'first pharmaceutical use' protection, which is dealt with more fully below.

First Pharmaceutical Use

A compound made, tested, and found to be useful as a pharmaceutical may, when a search is carried out, be found to be novel. There is then no problem in patenting the substance itself (assuming that it is not obvious). It may be, however, that the search reveals the substance to be already known: for example it may be mentioned in an expired patent describing photographic sensitizers, or as a softener in polymers. In both cases, product protection is ruled out and an alternative solution has to be found.

12 Exubera®; see the European Medicines Evaluation Agency (EMEA) Public Statement 557896/2008 of 10 November 2008.

As already mentioned, one possibility is to claim broadly pharmaceutical compositions containing the active ingredient. Such claims have the advantage that they are not limited in respect of any specific pharmaceutical indication; they would equally well cover the compound in a cough syrup or in a haemorrhoidal suppository. An alternative approach is to claim the use of the compound as a pharmaceutical. Here there are difficulties to be overcome, because the use of a substance as a pharmaceutical is equivalent to a method of medical treatment, which as we have seen is specifically excluded from patentability by the EPC and by the Patents Act 1977.

But both the EPC and the Patents Act 1977 provide that a known substance or composition may be patentable for use in a method of treatment of the human or animal body by surgery or therapy, or in a diagnostic method, provided that its use for any such method is not within the state of the art.[13] Under the original version of the EPC, the Technical Board of Appeal (TBA) of the EPO had already held[14] that claims in the form 'Compounds of formula . . . for use as an active therapeutic substance' were allowable and that, by analogy with pharmaceutical composition claims, such claims should cover all therapeutic uses of the substances and not only use for the specific indication that was disclosed. Of course, non-medicinal uses of the claimed compound are not protected by such a patent.

In spite of the long-standing rule in British patent law that a statement of purpose cannot add novelty to an old composition, the UK Patent Office now follows the practice of the EPO in this respect. Pharmaceutical uses form a statutory exception to this general rule, which still applies in other areas.

Second Pharmaceutical Use

From discussion of the protection available for the invention that a known chemical compound may be used as a pharmaceutical follows naturally the question of what patent protection may be obtained for the invention that a compound, already known to have one or more pharmaceutical uses, has a new pharmaceutical utility unrelated to any earlier known use.

A priori, there is no fundamental reason why an invention of this type should be less capable of patent protection than any other. The amount of work involved in making the invention, the potential benefit to the public, and the potential commercial importance may all be as great as for the invention of a new chemical entity having pharmaceutical utility. And, certainly, it is possible to obtain a patent for the dyeing of nylon with a dyestuff previously known to be useful only for dyeing wool. So if new uses of dyestuffs can be patented (given that they are novel and unobvious), why not new uses of pharmaceuticals?

[13] Article 54(4), EPC 2000; s. 4A(3), PA 1977.

[14] T 128/82 (OJ 1984, 164) *HOFFMANN-LA ROCHE/Pyrrolidone derivatives.*

Methods of Medical Treatment

The answer, as far as many countries are concerned, lies in the form of wording of the claims. When the only novel feature was the new utility, the claim had to have wording that was essentially equivalent to 'A method of treating disease Y by administering compound X', or 'The use of compound X in treating disease Y'. The problem is that such claims are claims to the medical treatment of humans and, as such, were denied patent protection in many countries either by specific wording in the law or by legal precedent.

The rationale for this exclusion from patentability has never been clearly stated, but it seems to derive from the idea that a doctor must be free to treat his patient as he sees fit, without having to worry about being sued for patent infringement. This is a perfectly valid point, even though the likelihood of doctors being sued by pharmaceutical companies is remote, but it could equally well have been dealt with by allowing claims to methods of medical treatment, while specifically providing that treatment of a patient by a medical professional would not be an infringement of such a claim. In exactly the same way, the UK Patents Act 1977, like many other national patent laws, already provides that a claim to a pharmaceutical substance or composition is not infringed by a pharmacist making up an individual prescription written by a doctor or dentist.[15]

In the USA, claims to medical treatment of humans have been allowed for a long time. A typical claim of this type relating to a new use of a pharmaceutical would read 'A method of treatment of disease Y comprising the administration, to a human in need of such treatment, of an effective dose of compound X'. Claims to surgical procedures are also patentable and this caused controversy when a US surgeon patented a new type of incision for eye operations,[16] demanded royalty payments from hospitals carrying out this technique, and in 1993 sued a clinic in Vermont for patent infringement. The doctor concerned was roundly denounced by the American Medical Association (AMA), which considered his actions unethical. The issue did not stay with the AMA, but led to the introduction of a Bill in Congress that would have banned this type of patent. The pharmaceutical and biotech industry quickly pointed out that the Bill initially proposed would prevent not only the patenting of surgical procedures, but also of new uses of pharmaceuticals, and a compromise proposal that tackled the problem in terms of the remedies available to the patentee and not to issues of patentability was adopted.

The new subsection of the US patent law, 35 USC 287(c), which was tacked on to an Appropriations Bill in September 1996, exempts from infringement performance of a medical activity by a medical practitioner and a 'related health care entity' (for example the hospital where the doctor works). 'Medical activity' does not include the use of patented drugs or equipment, nor patented uses of drugs,

15 Section 60(5)(c), PA 1977.

16 Pallin USP 5.080,111, 14 January 1992.

nor biotechnological processes, so that in practice only surgical procedures are protected from infringement suits. Despite the strange way in which it was enacted, the resulting change in the law is certainly preferable to excluding categories of invention from patentability and this is an approach that Europe should also consider.

In Germany, the courts did not feel quite so restricted by what the law actually said. In the case of Bayer's *Hydropyridine* application, the invention was the use of a known cardiovascular agent (Nimodipine) to treat cerebral disorders. The application was refused by the German Patent Office and an appeal to the Bundespatentgericht (BPatG, or Federal Patent Court) was unsuccessful. On a further appeal to the Federal Supreme Court in 1983, however, the Court held[17] that the German law, which also states methods of medical treatment to be unpatentable,[18] did *not* preclude the patenting of new uses of known pharmaceuticals. A claim of the form 'Use of compound X for the treatment of disease Y' has therefore been accepted in Germany.

'Swiss-Type' Claims

The Swiss Patent Office was asked for its views in the light of this German decision and gave the opinion[19] that, although claims of the German type would be forbidden by Swiss patent law, claims of the form 'Use of compound X for the preparation of an agent for the treatment of disease Y' should be acceptable.

The same application by Bayer that was successful in Germany was also filed in the EPO, where its rejection by the Examining Division was appealed to the Board of Appeals and referred to the EBA. After long deliberation, the EBA decided in *Bayer* and other related cases[20] that the German form of claim was a claim to a method of medical treatment, and was not patentable, but that the Swiss-type claim would be granted by the EPO. The EBA was perhaps glad to be able to adopt a compromise position originally suggested by a small neutral country, and thus avoid offending the UK and France by uncritically adopting the German view, or offending Germany by rejecting it totally.

The Swiss form of claim suffers from the logical objection that it lacks novelty, because it claims the use of the compound for preparation of a medicament and normally the medicament itself will be the same as that already used for the first pharmaceutical indication. Accordingly, there was concern that patents granted with such claims by the EPO could be held invalid by national courts in countries such as the UK.

These fears were allayed by an unexpected decision of the UK Patents Court, in which both patent judges sitting together decided that a British national patent

[17] X ZB 4/83 *Hydropyridine* IIC 2/84 215 (BGH).

[18] German Patent Law, §2a(1).

[19] Legal Advice from Swiss Federal Intellectual Property Office, 30 May 1984 (OJ 1984, 581).

[20] G 5/83 (OJ 1985, 64) *EISAI/Second medical indication.*

by the EPO as being essentially a method of treatment because the novel feature is the method of administering the drug, which is typically an activity carried out by a doctor. But in 2004, a TBA found[26] that a dosage regimen should not be regarded as a method of treatment, but should be patentable subject matter as a further indication, provided that it is formulated in the proper further medical use claim format. Additionally, it was discussed that a new way of using a known drug for a known indication can potentially be a very valuable invention for patients, because it may represent a major improvement over the known treatment (for example by reducing side effects). For the patentee, investigations to come up with new improved dosage regimens often require a substantial amount of investment; without the possibility of patent protection, there would probably be less investment into finding better ways to administer known drugs for known indications. Although the EPO Examining Division is, at the moment, following this decision, and is granting patents on novel and inventive dosage regimens, the question has now been referred to the EBA,[27] where the case is pending.[28]

National courts have also dealt with the question of patentability of dosage regimens. In one early case relating to the anti-cancer drug taxol, a Dutch court considered that the claim was really to a method of medical treatment and refused an injunction.[29] The corresponding *Bristol-Myers Squibb* case in the UK had the same result (see above), but in 2008, in the case of *Actavis v. Merck*, the Court of Appeal followed the EPO *Genentech* decision rather than its own BMS precedent (see p. 184) and held valid a Swiss-type claim for the use of finasteride for treatment of male baldness in a dose of 0.05–1.0 mg (the prior art showing administration of 5 mg for the same purpose). The reference to the EBA (see above) was made before judgment was given in *Actavis v. Merck* and it would be rather embarrassing if the EBA were to overrule *Genentech*, but the Court of Appeal stated that a late appeal to the House of Lords could still be made after the EBA gives its decision.

In Germany, the Federal Patent Court had allowed a dosage regimen patent[30] following the logical line of the Federal Supreme Court (*Bundesgerichtshof*, BGH) ruling on the German-type claim format that held that a claim to a method of use of a known product covers not only the direct process of using the product, but also covers acts that put the product in a position to carry out the claimed use (*sinnfällige Herrichtung*), for example by marketing the product together with written instructions for use (see above). Acts by which the product is individually prepared for the special claimed use are clearly carried out in an industrial environment, and therefore the whole use has industrial applicability and is in principle patentable

26 T 1020/03 (OJ 2007, 204) *GENENTECH/Method of administration of IGF-1*.

27 T 1319/04 (OJ 2009, 36) *KOS LIFESCIENCES/Dosage regimen*.

28 G 2/08 *KOS LIFESCIENCES/Dosage regimen* (EBA pending).

29 *Bristol-Myers Squibb v. Yew Tree* (1997) 2255 Scrip 9 (District Court, The Hague).

30 *Knochenzellenpräparat* [1996] GRUR 868 (BPG).

subject matter. But the Supreme Court found in 2006 that dosage regimen features could be taken into consideration when examining novelty and inventive step requirements only if the claim expressly covered steps to prepare a composition that is suitable for such dosage regimen.[31] For Germany, it is therefore advisable for patent applicants for dosage regimens to include in their European applications claims in a format such as 'Use of Compound X in the manufacture of a medicament for the treatment of disease Y wherein the medicament is prepared to be administered in a dosage scheme Z'.

At best, a patent containing a Swiss-type claim may be used to prevent a competitor from actively promoting the compound for the new use, including by advertisements and package inserts, but it cannot prevent doctors from prescribing for the patented new use a generic product that is already on the market for an earlier indication that is no longer patent protected. This situation is often called 'off-label' use, because the actual medicinal use is outside of what is prescribed on its product label.

Scope of Supplementary Protection Certificates

We discussed, in Chapter 9, the provisions in Europe for the grant of supplementary protection certificates (SPCs) as a means of extending patent term to compensate for regulatory delays. The SPC does not extend the entire scope of the patent on which it is based, but is limited to the product covered by the marketing authorization and for any medicinal use of the product that has been authorized before the expiry of the certificate.[32] Thus sales of the product for non-medicinal uses do not infringe, but the SPC would be infringed by sales of a medicinal product by a third party even if that party were to have a marketing authorization for a different indication. Apart from this, the SPC confers the same rights as the basic patent, and is subject to the same limitations and obligations,[33] so that existing licences under the basic patent would continue under the SPC. The scope of protection given by a US patent during the extension period under the Drug Price Competition and Patent Term Restoration Act of 1984 (the Hatch-Waxman Act) is essentially the same.

A major area of uncertainty relating to SPCs has been the question of what is the 'product' that is protected by the SPC, that is, is it the active ingredient in all possible salt forms and formulations, or is it only the specific salt form and formulation for which the marketing authorization was obtained? As far as salt forms are concerned, Regulation 1768/92[34] leaves this unclear, but the wording of Regulation

[31] X ZR 236/01 *Carvedilol II* (BGH, 19 December 2006).

[32] Article 4, EC Reg. 1768/92.

[33] Article 5, EC Reg. 1768/92.

[34] Council Regulation (EEC) No. 1768/92 of 18 June 1992 concerning the creation of a supplementary protection certificate for medicinal products.

1610/96[35] extending the grant of SPCs to plant protection products makes it clear that where the basic patent covers salts and esters, then so will the SPC, even though the marketing authorization is for one specific form.[36] This was also supposed to give guidance for the interpretation of the medicinal product SPC Regulation.[37]

The matter was settled by the European Court of Justice (ECJ) in a case[38] relating to Farmitalia's compound idarubacin. The German Patent Office wanted to limit the SPC to the hydrochloride form of idarubacin, because this was the form covered by the marketing authorization. The ECJ held that where a product in the form referred to in the marketing authorization is protected by a basic patent in force, the SPC is capable of covering the product, as a medicinal product, in any of the forms enjoying the protection of the basic patent. It is not possible to evade the SPC by using another therapeutically equivalent salt (or ester) form.

SPCs will not normally be granted for galenic formulations of an active ingredient with inactive excipients, which is just as well, since the grant of such SPCs would necessarily imply, contrary to the above ECJ decision, that the scope of protection given by the SPC in respect of the basic patent was limited to the specific formulation originally approved and would not be infringed by the same active ingredient in a different formulation. But if it can be argued that the other material has an additional effect, an SPC may be obtainable. For example, a British SPC[39] was granted for itraconazole as an inclusion complex with hydroxy-β-cyclodextrin for 'pulse dosage'.

In 2006, the ECJ decided the *Massachusetts Institute of Technology (MIT)* case on SPCs for galenic formulations.[40] MIT's patent claimed an old cytotoxic drug, carmustine, in a polymeric matrix of polifeprosan, an inactive excipient that enables carmustine to be used in the form of a cranial implant. SPCs were allowed in the UK and France, but not in Germany. On appeal, the BGH made a reference to the ECJ. Although the opinion of AG Leger was favourable to the applicant, the decision of the Court gave a strict interpretation of 'active ingredient': such an ingredient must have an effect of its own on the human or animal body. The SPC was refused. Shortly thereafter, a case referred by the English Patents Court reached the same conclusion.[41]

Administration devices may not normally be the subject of SPCs, because they are not the 'active ingredient or combination of ingredients of a medicinal product'.[42]

35 Council Regulation (EC) No. 1610/96 23 July 1996 concerning the creation of a supplementary protection certificate for plant protection products.

36 Recital 13, EC Reg. 1610/96.

37 Council Regulation (EEC) No. 1768/92 of 18 June 1992 concerning the creation of a supplementary protection certificate for medicinal products.

38 C-392/97 *Farmitalia Carlo Erba Srl's SPC Application* [2000] RPC 580.

39 SPC/GB/96/047 *Janssen.*

40 C-431/04 *Massachusetts Institute of Technology* (ECJ, 4 May 2006).

41 C-202/05 *Yissum Research and Development Company of the Hebrew University of Jerusalem. v. Comptroller General of Patents* (ECJ 2008).

42 *AB Draco's SPC Application* [1996] RPC 417(Pat Ct).

It is possible to get an SPC for a combination of two active ingredients, even if these have already been the subject of separate SPCs, but the composition must be specifically covered in the patent on which the SPC is based, that is, the combination must be claimed, or at least there must be basis in the specification for such a claim. It is not enough that the patent would be infringed by sale of the combination, because it covers one (or even both) of the ingredients. This was decided in the UK Patents Court on a case involving several SPC applications of Takeda and the judge refused to make a referral to the ECJ, holding that the matter was *acte claire*. Nevertheless, in the *Gilead* case,[43] the Patents Court allowed an SPC application for a combination of tenofovir and emtricitabine based on a patent relating only to tenofovir, on the somewhat flimsy basis that the last claim of the patent covered a composition containing tenofovir 'and optionally other therapeutic ingredients'.

The *Takeda* decision was challenged by Astellas in a case[44] in which the basic patent contained no combination claim and no mention of the second ingredient. Although the judge had some sympathy with the view that the 'infringement test' should apply, he preferred to leave the decision on this to the Court of Appeal.

Other Exclusivity Periods

In addition to patent and SPC protection, pharmaceutical products may benefit from other forms of exclusivity that mainly have to do with the marketing authorizations that are necessary before a pharmaceutical product can be put onto the market. It is important to understand that those other exclusivities have different purposes and are generally independent from patent protection, because patents are granted for innovations, whereas those other forms of exclusivity are designed to protect certain investments, such as creation of clinical test data, or to give incentives in certain areas in which other measures have shown not to be sufficient to promote the development of new appropriate drugs. Such an area is, for example, pharmaceutical therapies for very rare diseases, often called 'orphan indications'. As we will see, some countries have introduced a special type of protection for orphan drugs. Another such area is the development of dosage regimes specifically for children, which may give rise to so-called 'paediatric exclusivity'. In Europe, this is given in the form of an additional period of SPC protection and is therefore described in Chapter 9. The system in the USA will be described below.

Generic Drugs

The purpose of patents, SPCs, and other forms of exclusivity is to delay the introduction of competing products, generally referred to as 'generic drugs'. We should be clear what is meant by this term.

[43] *Gilead's SPC Application* [2008] EWHC 1902 (Pat).
[44] *Astella's SPC Application* [2009] EWHC 1916 (Pat).

A generic drug is defined as one having the same active ingredients as an earlier-approved drug, often called the 'originator' or 'reference' drug, and the applicant for a marketing authorization for a generic drug needs only to show that its product produces the same effects in patients as the originator drug and has no additional side effects. Usually, this is shown by 'bioequivalence studies', which are small-scale clinical trials on volunteers in which it is shown that the generic drug achieves the same blood plasma levels of the active ingredient as the originator drug. Pharmacokinetic properties such as the maximum blood concentration Cmax, the time until Cmax is reached (Tmax) and the total amount of active ingredient in the plasma over a given time, expressed as 'area under the curve' (AUC), are important parameters to establish bioequivalency. To recognize bioequivalency, it is usually required that the 90 per cent confidence intervals of the natural logarithmic ratios of Cmax and AUC between the generic and originator product lies between 80 per cent and 125 per cent. Thus, it can sometimes be quite useful for an originator to include patent claims claiming certain pharmacokinetic parameters, provided that they are new and inventive.

To show safety and efficacy of the active ingredient, the concept of a generic drug requires that the generic marketing authorization refers to the corresponding data of the originator's product marketing authorization. Those data are in principle owned by the sponsor of the originator drug. Thus, the generic applicant would need the consent of the originator marketing authorization holder, which often would not be voluntarily given.

Regulatory Data Exclusivity

In order to obtain a marketing authorization, a large amount of pre-clinical and clinical testing needs to be performed to generate the data on which the drug regulatory health authorities examine whether a medicinal product is safe and effective. In the case of veterinary drugs for farm animals used to produce food, additional safety tests are necessary to show that any possible residues are not harmful for the end consumers. These generated data are principally the property of the sponsor of all of those trials and investigations, who could in theory prevent any third party from using them. But it is in the public interest that the data may also be used by generic applicants who refer to them to provide evidence that the generic products are equally as safe and effective as the reference product. In order to allow such a cross-reference, but to protect the substantial investment that has been made into the creation of the data, policy considerations therefore led to the concept that the submitted data should be protected for a limited period of time from being referred to by any third party. This concept is called 'regulatory data exclusivity'.[45]

45 Sometimes also called 'regulatory data protection'; not to be confused with protection of private or personal data.

Data exclusivity runs from the actual grant of the marketing authorization, and usually has a duration of between three and 11 years, depending on the country and the type of marketing authorization. For example, in the USA, all new small-molecule drugs approved by the Food and Drug Administration (FDA), that is, new chemical entities (NCEs), benefit from five years of data exclusivity, whereas new indications, dosage forms, and ways of administration get three years of data exclusivity. For all drugs with a new active ingredient or a new combination of active ingredients authorized in the European Union (EU) for the first time after October/November 2005, there is generally a uniform period of ten years of data exclusivity, which may be extended to 11 years if during the first eight years an additional indication of substantial clinical benefit is authorized (8+2+1 formula).[46] A generic applicant can submit his marketing authorization referring to the originator's data at the earliest after eight years from the start of the data exclusivity, but will not get an actual marketing authorization before expiry of the ten-year (or, if extended, 11-year) period.

Apart from this additional year, however, the EU does not provide for any data exclusivity for new indications, new dosage forms, or ways of administration developed by the same marketing authorization holder or persons related to him. If a new indication of an authorized medicinal product is independently developed by a different entity, not related to the entity that developed the earlier indication of the same active ingredient, and is submitted to the European health authorities as a full stand-alone dossier to get marketing authorization, it seems that a full ten-year data exclusivity period for the product with the new indication would be recognized by the European health authorities. But if the same work of developing the new indication is done by the person who already holds the marketing authorization for the earlier indication, or a person related to that holder, then there will be no new separate data exclusivity period, because the new indication will be regarded as an extension of the earlier marketing authorization dossier.[47]

Although in line with the concept of a 'global marketing authorization', within which all different forms of a product containing the same active ingredient are regarded as belonging to the same family of marketing authorizations independently of regulatory pathways, this practice of different treatment depending on who is the applicant for the later-filed marketing authorization is unfair, because the development of a new indication of a known drug will often require a comparable amount of investment to that of a medicament having a new active ingredient, and

[46] Directive 2001/83/EC of 6 November 2001 on the Community Code relating to medicinal products for human use, Art. 10; EC Regulation No. 726/2004 of 31 March 2004 laying down Community procedures for the authorization and supervision of medicinal products for human and veterinary use and establishing a European Medicines Agency, reg. 14.

[47] See EU Commission ENTR/F2/BL D(2002), 'Notice to applicants', Rev. 3 (2005), Vol. 2A, Ch. 1, point 7.2.

it is hoped that either the legislator or the courts will correct this flaw in the present legislation.

TRIPs provides for regulatory data exclusivity, but does not specify any duration for it.[48] Japan and Canada provide for eight years of data exclusivity for human drugs with new active ingredients, whereas China, South Korea, and Turkey have six years, and Australia, New Zealand, Mexico, Colombia, and Argentina provide for five years of data exclusivity. Some important countries, for example Russia, India, and Brazil, still do not provide for any effective data exclusivity, sometimes reasoning that referencing the originator's data is not 'unfair commercial use' according to TRIPs and sometimes based on the theory that referencing is not necessary to approve a generic product, because the health authorities know that the data exist.

In many cases, for a new drug, the period of data exclusivity will be shorter than the term of patent protection and will therefore bring no additional benefit. But if the drug lacks patent protection, or if the data is for a new indication of an old drug, data exclusivity can be extremely important.

Orphan Drug Exclusivity

For the pharmaceutical industry, it was for a long time not very attractive to invest in development of new drugs against diseases that affect only a small number of patients, because of the small potential market for the new drug. In order to increase incentives to develop new pharmaceuticals for such rare diseases, often also called 'orphan diseases', orphan drug exclusivity was introduced as a new type of exclusivity in some industrialized countries.

The USA was the first country to create particular incentives for orphan drugs through its Orphan Drug Act of 1983.[49] Those incentives include fee reductions and technical assistance on the regulatory procedure, tax deductions of up to 50 per cent of certain development costs, and a seven-year period of orphan drug exclusivity. During that seven-year period after issuance of the first marketing authorization in the USA, the FDA is not allowed to issue another marketing authorization in the same orphan indication for the same active ingredient, or for similar products with the same clinical performance and the same 'active moiety', that is, another member of the same therapeutic class.

If, however, clinical superiority can be shown by a later applicant for a similar product, the later product is not regarded as the 'same' and will be allowed even during the orphan drug exclusivity period. For example, the interferon beta product Betaseron® got FDA approval as an orphan drug against multiple sclerosis in 1993. A few years later, another company submitted another interferon beta product,

[48] Article 39.6, TRIPs, says that member states should ensure that regulatory data are kept confidential and shall be protected from unfair commercial use. Of course, it depends on what is considered to be 'unfair'.

[49] 21 USC §360aa–ee, the Orphan Drug Act and related Orphan Drug Regulation 21 CFR §316.

Avonex®, with the FDA and got also a marketing authorization as an orphan drug in 1996, during the seven-year orphan drug exclusivity period for Betaseron®. Because it could be shown by clinical trial data that Avonex® caused far less swelling at injection sites than Betaseron®, the two products were not regarded as the same under the orphan drug legislation. Other orphan or non-orphan indications for the same active ingredient may be allowed during the orphan drug exclusivity period.[50]

Despite their different regulatory pathways, the seven-year orphan drug exclusivity is available for new human drugs, vaccines, biological, and antibiotics. For drugs and antibiotics, the orphan drug exclusivity has to be listed in the Orange Book (see p. 189) and, like the relevant patent, can be subject to a six-month extension if the FDA grants a paediatric marketing authorization following successful clinical trials in children. In order to get orphan drug status, a marketing authorization applicant needs to obtain from the FDA a designation as an orphan drug before submitting the marketing authorization dossier. There are two alternative criteria for an orphan drug designation: a prevalence of fewer than 200,000 patients affected by the disease in the USA, which corresponds to approximately 0.07 per cent of the population; or if the applicant can show that there is no reasonable expectation that the cost of developing and making available a drug in the USA for such disease will be recovered from US sales of the drug. So far, all orphan drug designations have only used the first criterion.

The EU introduced, in 2000, an Orphan Drug Regulation that is applicable in all EU member states.[51] The concepts are similar to those of the US legislation, and incentives also include free technical assistance and fee reductions for the regulatory proceedings. But the term of the orphan drug exclusivity in the EU is ten years from the issuance of the marketing authorization of the product for the rare indication.

The criteria in the EU for an orphan drug designation are that the disease must be potentially life-threatening or leading to chronically debilitating conditions, and either that it is established that no more than 0.05 per cent of the EU population is affected, or it can be established that there would be no sufficient profitability without the incentives provided by the legislation. Additionally, it needs to be shown that there are no other therapeutic means, including non-pharmacological options, to control or treat the disease sufficiently in the EU, or that patients will have a significant benefit from the new drug compared with the situation at the time of the orphan drug application.

[50] *Sigma-Tau v. Schwetz* 288 F3d 141 (CA 4th Cir 2002)

[51] EC Regulation No. 141/2000 of 16 December 1999 on orphan medicinal products; EC Regulation No. 847/2000 of 27 April 2000 laying down the provisions for implementation of the criteria for designation of a medicinal product as an orphan medicinal product and definitions of the concepts 'similar medicinal product' and'clinical superiority'.

In the USA, orphan drug exclusivity can be revoked only if the marketing authorization holder cannot sufficiently supply the US market, so that even if the prevalence increases to more than 200,000 affected patients in the USA during the orphan drug exclusivity period, there will be no reduction or lapse of the period.[52] By contrast, in Europe, the ten years of orphan drug exclusivity may be reduced to six years, if it can be shown at an optional review started after five years from marketing authorization that the drug no longer meets the criteria for orphan drug registration. Such a review can be requested by an EU member state, and the European Medicines Evaluation Agency (EMEA) will examine whether the criteria on which the orphan designation originally had been granted are still met and, if not, whether other reasons exist to maintain the orphan drug status. Increases in prevalence because of prolonged patient life that can be attributed to the application of the orphan drug itself will not be taken into account. Only if no justification can be established for maintaining the orphan drug status will the EU Commission shorten the ten-year exclusivity period to six years.[53] Up to mid-2009, no such review procedure has been initiated.

Since January 2007, there is a possibility to extend the ten-year orphan drug exclusivity to 12 years if pediatric studies are carried out according to an EMEA-agreed pediatric investigation plan (PIP) and the results, whether negative or positive, are accordingly reflected in the product information (see Chapter 9).

Once an orphan drug marketing authorization has been granted by the EMEA,[54] the orphan drug exclusivity will prevent another medicinal product from being authorized or even accepted for examination by health authorities in respect of the same therapeutic indication, a similar molecular structure, and the same mechanism of action. Thus orphan drug exclusivity protects not only against generic applications referring to the data of the earlier approved products, but also prevents marketing authorization grants for independently developed products for the same indication with a similar structure. Although a guideline was adopted setting out what would be regarded as a 'similar structure', for example if the international non-proprietary names (INNs) of the two compared products have the same prefix or substem, such as propanolol and atenolol,[55] it will always be a case-by-case determination whether a later-developed product will be regarded as 'similar' under the orphan drug legislation. For example, the drugs Revatio® (INN sildenafil) and Volibris® (INN ambrisentan) have been held by the EMEA to be not of similar

[52] That is the reason why a number of HIV treatments were issued in the early 1990s as orphan drugs, although, today, HIV is scarcely an orphan disease in the USA.

[53] For details on the procedure of the review and the criteria, see Guidelines on aspects of the application of Art. 8(2) of Regulation 141/2000, Communication from the EU Commission of 17 September 2008, C(2008) 4051 final.

[54] The centralized approval procedure according to Regulation 726/2004 is mandatory for orphan designated drugs.

[55] Guideline on aspects of the application of Art. 9(1) and (3) of Regulation 141/2000, Communication from the EU Commission of 19 September 2008, C(2008) 4077 final.

structure to Tracleer® (INN bosentan) in treating pulmonary hypertension. Additionally, even if a similar structure and mode of action can be established, a later new medicinal product may be approved within the orphan drug exclusivity period of the earlier product if the later product can be shown to be clinically superior over the earlier drug.

Japan introduced in 1993 the possibility of orphan drug designations, providing as the main incentive an extended regulatory re-examination period of ten years compared to the regular data exclusivity period of eight years (see above). A few other countries, such as Australia and Singapore, have orphan drug provisions in place, however, which do not extend any exclusivity period for the products. Switzerland is considering the introduction of an orphan drug exclusivity period of ten years.

Integrated View on Exclusivity

It should not be forgotten that orphan drug exclusivity periods in the USA and EU are independent from terms of regulatory data exclusivity and patents. Thus, if a new active ingredient is approved for the first time in an orphan indication in the EU, then the ten-year orphan drug exclusivity runs in parallel with the ten-year regulatory data exclusivity period according to the 8+2+1 formula. If, for example, three years after the first orphan indication, a second orphan indication were to be approved, then a new ten-year orphan drug exclusivity would start with regard to the second orphan indication, whereas the ten-year regulatory data exclusivity could be extended to 11 years if substantial clinical benefit could be shown (which would be likely if the second indication were also to get orphan status). But any calculation of an SPC expiry will be started by the very first approval date of the first indication. This last factor is similarly true for the USA, where any patent term extension would also be calculated based on the first marketing authorization of the active ingredient, and needs to be factored in for strategic planning of the order of marketing authorizations for orphan and non-orphan indications for the same active ingredient.

Commercially, it may be more logical to get the non-orphan indication authorized first and the orphan indication later, because otherwise the protection time for the higher sales volume of the non-orphan indication would be shorter. There are, of course, other factors, for example establishing market entrance or ethical considerations, which may outweigh the commercial advantage of having, first, a non-orphan and, later, the orphan indications.

This example shows the importance of looking at the protection of a pharmaceutical in an integrated way, not only looking at the patent and perhaps extension status, but also taking into account regulatory data exclusivity, orphan drug exclusivity, and pediatric exclusivity periods. Furthermore, the possibility of parallel trade and influences of different exclusivity expiry dates in various countries on the pricing of a pharmaceutical product needs to be taken into account to build up a complete picture of the protection of a drug.

14

BIOTECHNOLOGICAL INVENTIONS

The Microbe is so very small
You cannot make him out at all . . .
But Scientists, who ought to know
Assure us that they must be so . . .
Oh! let us never, never doubt
What nobody is sure about!

Hillaire Belloc, 'The Microbe' (1912)

What is Biotechnology?

Classical biotechnology may be defined loosely as the production of useful products by living microorganisms and as such it has been with us for a long time.

The production of ethanol from yeast cells is as old as history, and over 80 years ago, the production of various industrial chemicals, such as acetic acid and acetone, by fermentation processes was well known. Indeed, even the word 'biotechnology' is not recently coined. In 1920, a Bureau of Bio-Technology was established in Leeds, and published a journal dealing with fermentation technology and related topics. Back in 1873, Louis Pasteur obtained a US patent[1] claiming 'Yeast, free from organic germs of disease, as an article of manufacture', an early case of a patent for living organisms.

More recently, the antibiotics industry was based upon the isolation of products from selected strains of microorganisms and although the majority of antibiotics are now produced synthetically, many are still made from microorganisms either found in nature or artificially mutated. Not only antibiotics, but also other drugs, for example the immunosuppressant cyclosporin, are produced by fermentation of a microorganism.

What may be described as modern biotechnology, as distinct from the classical fermentation technology, began in the 1970s with the two basic techniques of recombinant deoxyribonucleic acid (DNA) technology and hybridoma technology. In the first of these (also referred to as 'gene splicing' or 'genetic engineering'), genetic material from an external source is inserted into a cell in such a way that it causes the production of a desired protein by the cell; in the second, different types of immune cell are fused together to form a hybrid cell line producing monoclonal antibodies. More recently, the techniques of genetic engineering have been applied to higher organisms to produce transgenic animals and plants, and even to humans (gene therapy), to replace missing or defective genes coding for a protein required by the body, for example, or to introduce genes into cancer cells that will render them easier to kill. The high speed at which DNA fragments can be sequenced has led to the completion ahead of schedule of the Human Genome Project and the identification of many human genes, and this science of genomics can be used to find genes that could make useful protein products, which could be applicable in gene therapy and which might be useful in the elucidation of disease mechanisms, in diagnostic kits, and in screening for new drugs. There are also many research tools and techniques making use of biotechnological processes.

Patents and Biotechnology

Biotechnology has based a whole new industry and patent protection for biotechnological inventions is of immense commercial importance. But patent law and practice have had serious difficulties in keeping up with the rapid scientific progress in this field, and issues such as inventive step, sufficiency of disclosure,

[1] USP 141,072.

and permissible breadth of claims have proved troublesome. There has been much litigation of biotech patents and courts have found it difficult in such a rapidly moving field to determine what was the general knowledge of the skilled person at the time that the invention was made. A procedure to find and clone a specific gene, and to express it in a suitable host, may have been a breakthrough at the time that it was first done and purely routine work not many years later.

There is also the problem of opposition by special interest groups against anything to do with genetic engineering and particularly against the existence of patents in this area. We shall look at this in more detail in the following chapter, but some basic points may be mentioned here. It has been suggested, for example, that the products of biotechnology cannot be patented because they are natural products, or even that the patent system is inherently unsuitable for protecting inventions of this type.

The first of these suggestions is incorrect. As we have seen (p. 254), natural products are, in principle, patentable provided that the product is technically useful and that the claim is worded or interpreted in such a way that it does not cover the product in its natural environment. As to the second, there is no reason why the basic requirements for patentability should not apply in the field of biotechnology as in any other. The difficulty is that the inherent complexity of living systems is such that it becomes more difficult to ensure that these requirements are met where living organisms are involved, particularly the requirement of a sufficient and reproducible disclosure.

Microbiological Inventions

Patentability of Microorganisms

Microbiological inventions generally involve the use of a new strain of microorganism to produce a new compound or to produce a known compound more efficiently (for example in higher yield or purity). The new organism may have been found in nature (for example by screening of soil samples), or may have been produced in the laboratory by artificially induced random mutation or by more specific techniques, such as genetic engineering.

If the microorganism produces a novel product, such as a new antibiotic, then the novel product may be claimed as any other new chemical compound can, subject to the requirements of a sufficient description being given. If the end product is already known, process protection is available, but this protection is weak and it would be preferable to patent the new microorganism itself.

Most patent laws do not deal specifically with the question of whether or not a new living strain of microorganism is itself patentable, but the UK Patents Act 1977 and the European Patent Convention (EPC) do not exclude the possibility. Plant and animal varieties are excluded from protection, as is any biological process for their production, but not excluded is a microbiological process or the product of such a process—which may, of course, be a microorganism. Both the British Patent Office and the European Patent Office (EPO) grant patents for microorganisms as such. The TRIPs Agreement[2] makes it obligatory for all World Trade Organization (WTO) members, after the end of any applicable transition period, to grant patents for microorganisms. If the microorganism is one that occurs in nature, it will be necessary to claim it in the form of an isolated strain, in order to avoid possible novelty objections. It must be remembered that the term 'microorganism' is interpreted broadly so as to include not only bacteria and fungi, but also viruses, and animal and plant cells.

In the USA, in spite of the precedent of the Pasteur patent mentioned above, it had become the practice of the US Patent and Trademark Office (USPTO) to refuse claims to living systems as not being patentable subject matter. In 1980, however, the Supreme Court decided (by a five-to-four majority), in the famous *Chakrabarty* case,[3] that a new strain of bacteria produced artificially (by bacterial recombination, not genetic engineering) was a patentable invention. Although Chakrabarty's bacteria did not produce a useful product, they had the useful property that they could feed on, and so disperse, oil slicks. Since the product that would be sold would be the bacterial strain itself, it was particularly important in this case to obtain a per se claim to the microorganism. The *Chakrabarty* decision aroused great public interest at the time, but in all of the fuss, it went unremarked that the British Patent Office had already granted the corresponding British patent in 1976.

Description and Deposition Requirements

Whether the claimed invention is a new microorganism itself or a new product obtained from it, the patent will be invalid unless it gives a disclosure of the invention that is sufficient to enable it to be reproduced. In normal pharmaceutical cases, a reproducible disclosure presents no problems. But it is practically impossible to define a strain of microorganism unambiguously by a written description, and even if a complete description were possible, this would not necessarily put the public in possession of the invention when the patent expired. Anyone who wished to carry out the process of the invention would first have to catch his bacterium; he could perhaps tell when he had got the right one, but to get it, by search in nature or random mutation, might take years or might take forever.

The approach that has been developed to meet this problem is that of deposition of the strain in a recognized culture collection, which will maintain the strain in a viable condition and make samples of it available to the public. In the USA, the Court of Customs and Patent Appeals (CCPA) held, in 1970,[4] that such a deposit would suffice to meet the disclosure requirements of the US patent law. The deposition had to be made on or before the US filing date, but no access to the deposited

[2] Article 27.3(b).TRIPs.

[3] *Diamond v. Chakrabarty* 206 USPQ 193 (Sup Ct 1980).

[4] *In re Argedoulis* 168 USPQ 99 (CCPA 1970).

strain need be allowed until the patent was granted, whereupon it had to be made available unconditionally to the public. In 1985, however, the US Court of Appeals for the Federal Circuit (CAFC) held that it was *not* essential that the deposit be made by the date of filing of the application, as long as the applicant had the strain and could make it available to the USPTO upon request.[5] Deposit could be made at any time during the pendency of the application and addition to the specification of information about the deposit did not constitute new matter. The requirement that as of the date of grant the strain must be publicly available from a recognized depository remained unchanged.

The Budapest Treaty

The majority of developed countries have now adopted the solution of requiring deposit of strains and the Budapest Treaty of 1977, which came into force in 1980 and which, as of 2008, had been ratified by seventy countries and is also followed by the EPO, establishes a list of international depository authorities (IDAs) and provides that a single deposit made at any of these will suffice for all signatory states. The Budapest Treaty is mainly concerned with formalities relating to the deposit and maintenance of the culture. These provide for the possibility of re-deposit if a strain becomes non-viable on storage and for a minimum storage period of 30 years from the original deposit.

A serious problem arises, however, from the fact that most countries now have early publication of patent applications 18 months from the priority date and consider that, as part of the publication, the deposited strain must be made available from this time. In other words, the applicant must make the means for carrying out his invention available to the public, including any competitors, before there is any assurance that he will actually obtain any patent protection. The traditional concept of patent protection as exchange for disclosure has thereby been distorted so as to require, before any protection exists, not only disclosure, but also what has aptly been described as a 'pocket factory handed over to the imitator on a silver plate'.

It is true that, in most countries, a person who obtained a sample of the deposited strain after early publication had to give an undertaking that he would use the strain only for experimental purposes and would not give it to any third party, but the enforceability of such undertakings remains highly questionable. In the EPO, the 'expert solution' has been adopted, whereby a deposit is still acceptable if the depositor stipulates that, up to the grant of a European patent, the strain can be made available not to a competitor directly, but only to an independent expert, who could carry out experiments on behalf of a third party, but not pass the strain on to the party for whom he was acting.[6] In many national patent offices, however, the rules

[5] *Ex p Lundak* 227 USPQ 90 (Fed Cir 1985).
[6] Rule 32, IR; EPC 2000.

still insist that any third party may get his hands on the deposited strain at any time after early publication and, as long as the application is filed on at least one such country, the fact that the EPO applies the 'expert solution' does not help much. In the USA, the rule is that when an application is early published 18 months from the priority date, the applicant does not have to make any deposited material available to the public, a logical consequence in view of *Lundak*.

Another serious problem relating to deposits of microorganisms is that most states not only require deposits to be made, but also require a written description. In Germany, it was held[7] that a deposit may be essential to comply with the general rule of sufficiency of the description as a whole, but that a deposit alone is not enough to support a product per se claim to the microorganism.

Understandably, industry would like to see a uniform practice applied in all developed countries, ideally one in which a deposited strain would be recognized as sufficient description, but the strain would not be released until the patentee had enforceable rights. Failing such additional safeguards, a company inventing a process for making a known substance by using a new strain of microorganism should consider keeping the new process as a trade secret instead of trying to patent it. Patenting will require deposit of the new strain, and if anyone obtains this and uses it, it will be difficult to prove that the patent is being infringed. The best form of protection in this case is to keep the new strain safely in one's own hands.

Apart from microorganisms themselves, the deposition of biological material is becoming less important. At one time, it was necessary to deposit vectors containing DNA, or hybridomas producing monoclonal antibodies, because reliable DNA sequence information was not available; now, it is sufficient for this type of invention to file a sequence identifier giving the complete sequence of the DNA or protein in question.

Recombinant DNA Technology

Scientific Background

Genetic information is carried in the cell by molecules of DNA. As elucidated by Watson and Crick,[8] the molecule is a linear polymer in the form of a twin-stranded double helix, each strand of which has a backbone of deoxyribose phosphate, with one of four bases attached to each deoxyribose unit. The four bases are adenine (A), cytosine (C), guanine (G), and thymine (T), and the two strands are held together by hydrogen bonding between pairs of complementary bases, an adenine being always paired with a thymine and a cytosine with a guanine.

[7] *Bäckerhefe* (1975) 6 IIC 207 (BGH).

[8] J.D. Watson and F.H.C. Crick [1953] 171 Nature 737.

The sequence of bases along one strand (the 'coding', or 'sense', strand) of the double DNA chain encodes the information for the synthesis of proteins within the cell, each of the possible 64 triplets ('codons') of the four bases coding for one specific amino acid, or for a signal to start or to stop the coded sequence. Because there are only 20 amino acids used in protein structure, the code is redundant: that is, there is generally more than one triplet coding for the same amino acid. The other strand (the 'non-coding', or 'antisense', stand) is complementary to the coding strand.

The mechanism whereby this code controls the formation of protein is a two-stage one. In the first stage ('transcription'), the twin strands separate, and the coding strand of the DNA acts as the template for the building up of a complementary and chemically related molecule: ribonucleic acid (RNA). In RNA, ribose units replace deoxyribose and the base uracil replaces thymine. This single-strand 'messenger RNA' molecule separates from the DNA and is transported to a cell structure known as a 'ribosome', which, in the 'translation' stage, assembles amino acids in a sequence corresponding to that of the base triplets on the messenger RNA, which in turn corresponds to that of the original DNA. The resulting protein is the 'gene product' of the gene: that is, of the DNA sequence coding for its production.

In simple cells such as bacteria, which have no distinct nucleus ('prokaryotes'), the genetic material is normally a single closed loop of DNA carrying all of the genes required to code for all of the proteins produced in the cell. In the more complex cells known as 'eukaryotes', for example yeast cells and cells of higher organisms, there is a separate nucleus that contains the genetic material in the form of a number of chromosomes consisting of DNA associated with protein.

In eukaryotes, in contrast to prokaryotes, genes are generally not found to exist as one continuous sequence, but consist of sequences called 'exons', which code for parts of the gene product, separated by regions called 'introns', which are not transcribed into m-RNA and which do not contribute to the protein. Furthermore, individual genes are usually separated by long non-coding sequences of DNA, which may amount to more than 90 per cent of the total genome. A sequence of one or more structural genes each coding for a specific protein are under the control of 'promoter' and 'terminator' DNA sequences, which do not code for protein, but which control the starting and stopping of transcription into m-RNA.

The chromosomes of a complex eukaryotic cell, such as a mammalian cell, contain all of the genetic information necessary for the entire organism, but in any one type of cell only a very small amount of this information will be used. For example, a human islet cell is specialized to produce insulin, so that the gene for insulin, which is present in all human cells, is 'expressed' particularly in cells of this type. In all cells, there is a more or less complex system of regulators and promoters that determines which of the many genes present will be expressed, and hence which proteins will be produced by the cell. These regulators and promoters may have the effect that the same DNA sequence may be read in more than one way, giving rise to more than one protein product for the same gene. Some mechanism such as this is required to account for the observation that the number of genes found in the human genome is much smaller than expected and does not seem to be large enough to account for the number of proteins in the human body, even taking post-translational modifications into account.

Gene-Splicing Techniques DNA can be manipulated in specific ways by enzymes that break or rejoin the DNA molecule. A number of enzymes called 'restriction endonucleases' are known that cause the DNA to be cut only at specific base sequences. If two separate DNA molecules are cut by the same restriction endonuclease, they can recombine with each other just as easily as with their original partners, to give 'recombinant' DNA. Enzymes called 'DNA ligases' are also available to join DNA molecules together.

Transformation and Expression The above methods may be used to insert a DNA sequence coding for a desired protein into a 'vector', which can then be introduced into a host cell. It is no use simply to insert the DNA sequence alone into the cell, because even if it could express the gene product, it would not replicate within the cell and would be lost when the cell divided as part of its normal growth. A suitable vector has the property of being able to replicate itself within a host cell, thus perpetuating any gene associated with it.

For bacterial host cells such as E. coli, the most common vectors are 'plasmids', which are circular molecules of DNA that can be transferred from one bacterium to another and which, in nature, are associated with the transmission of antibiotic resistance between bacteria. A suitable plasmid can be cut at one or more restriction sites, a foreign gene inserted, and the circle re-closed. The plasmid can then be inserted into the host cell, for example by incubating in the presence of calcium ions; the cell containing the plasmid is referred to as a 'transformant'. The plasmid will still normally have an intact region conferring antibiotic resistance, so that transformants containing the plasmid can be isolated from non-transformed E. coli cells by their ability to grow in a culture medium containing antibiotic. Other types of vector include 'phages', 'cosmids', and 'yeast artificial chromosomes' (YAKs), which are more suitable for the expression of large genes.

The gene to be inserted into the vector may be obtained by one of two main methods. If m-RNA corresponding to the gene product can be isolated, DNA can be produced from it by 'reverse transcription'. This type of DNA, which consists only of coding regions without introns, is described as c-DNA (the 'c' standing for complementary). Alternatively, the total DNA of a cell known to contain the desired gene can be cut into fragments using restriction enzymes and all of the selected fragments inserted into vectors. This method (known as 'shot-gun cloning') has the disadvantage that the gene fragments will normally contain introns and such genes might not be expressed directly in a prokaryotic host such as E. coli. This point was the central issue in the case of *Biogen v. Medeva*,[9] discussed below. But it is also

[9] *Biogen v. Medeva* [1997] RPC 1 (HL).

possible to use eukaryotic cells as hosts, and such cells may be able to excise introns and also to produce the protein in the same final form as it is produced in the human body. For this purpose, mammalian cell cultures such as Chinese hamster ovary (CHO) cell lines are most suitable, but these are difficult to grow in large quantities, whereas yeast cells can be grown very easily.

The selection of DNA fragments containing the desired gene is made much easier by the probe technique. The DNA is converted to single-strand form and mixed with a synthetic radioactively labelled short single strand of DNA (the probe), which is complementary to a sequence of DNA known to be present in the desired gene. If this target sequence is present, the probe hybridizes with the DNA fragment containing it: that is, the complementary bases pair with each other to form a short length of two-strand DNA, which can be picked out by autoradiography. The technique of polymerase chain reaction (PCR), which in principle enables a single DNA molecule to be copied exponentially, then enables the production of the selected DNA molecule in any desired quantity.

Isolation of the Gene Product Even if the desired protein is successfully expressed by the transformant, there remains the problem of isolating it in pure form. Yeast cells and some types of bacteria may secrete the product into the culture medium, but concentration of protein from this very dilute solution is by no means easy. In E. coli, the protein normally remains in the cell, which must be disrupted and the desired protein must then be separated from other cell proteins. The scaling up of such processes to commercial levels also presents serious problems.

The protein that is expressed by a transformed host cell containing the correct gene for the desired natural product may not correspond exactly to the natural material, even if it contains the correct amino-acid sequence. For one thing, it may contain a 'leader sequence' of one or more additional amino acids at the N-terminal end of the molecule. Such leader sequences are frequently expressed by the natural gene *in vivo*, but are then split off by enzymes in the cell to give the mature form of the protein. If the transformed host cell lacks the necessary enzyme system, the leader sequence will not be removed in the recombinant product. Such molecules are new compounds that may be patented, even if the natural protein had previously been isolated, but generally they will be less desirable for therapeutic use than the natural product itself.

The natural protein may be 'glycosylated': that is, carry one or more sugar units attached to the amino acid chain. A recombinant protein obtained from a prokaryotic cell will not be glycosylated. In many cases, glycosylation does not seem to be important, and the unglycosylated form has the same activity as the natural protein. In some cases, however, the glycosylated form is more active and it may be necessary to use a eukaryotic host in the hope of obtaining a correctly glycosylated product.

It is not only the chemical structure of the protein molecule that is important, but also its physical structure, or shape.

(a) The 'primary structure' of a protein is its amino acid sequence, represented as a simple linear structure and including any disulphide bridges linking cysteine residues in different parts of the molecule.
(b) 'Secondary structure' refers to the spatial arrangement of amino acid residues that are near to one another in the linear sequence and which frequently build regular structures such as α-helices, β-sheets, etc.
(c) 'Tertiary structure' refers to the overall shape of the protein molecule, for example in a crystal or in an aqueous environment.

Changes in the environment, for example the ionic strength of the medium, or binding to a ligand molecule, may cause large changes to the tertiary structure, and in some cases, a change in primary structure of only a single amino acid may cause a profound change in tertiary structure. An example is the point mutation in the haemoglobin molecule that changes the tertiary structure of desoxyhaemoglobin, causing sickle cell disease.

If the protein can be crystallized, which is often difficult and sometimes impossible, computerized X-ray crystallography can now rapidly determine the crystal coordinates, and from this, the tertiary structure of the molecule. It may then be possible to calculate the shape of a putative low molecular weight ligand molecule that would fit into a receptor part of the protein molecule.

Patentable Inventions in Recombinant DNA Technology

General Techniques There are two basic types of patentable invention in this field. The first relates to techniques and methods that are generally applicable to the production of a wide range of gene products; the second relates to specific products, such as proteins and DNA sequences. The patenting of gene sequences other than those coding for a known protein is discussed in the next chapter.

The basic invention of gene-splicing techniques was made by Cohen at Stanford and Boyer at the University of California. Stanford University filed patent applications to cover its invention, but because publication in a scientific journal[10] had already taken place before filing, patents could be granted only in the USA. The basic Cohen/Boyer patent[11] was issued in December 1980 and claimed a method of producing a protein by expression of a gene inserted into any unicellular host, thus covering the great majority of all genetic engineering processes. The patent expired in December 1997, having earned hundreds of millions of dollars in royalty payments for the two universities (see Chapter 22).

[10] Cohen et al., [1973] 70 Proceedings of the National Academy of Science 3240.
[11] US Patent 4,237,224.

Other universities and companies also own patents that are generally applicable to a wide range of products: some of these are pure process patents; others claim novel vector systems, such as plasmids that give particularly good replication in host cells, or, for example, promoter systems that can regulate inserted genes so as to give high expression rates of product.

Specific Gene Products and DNA Sequences The other main type of invention in the field of recombinant DNA technology relates to the production of a specific protein product by a transformed microorganism. The product may, on the filing date of the patent, be: one the structure (amino-acid sequence) of which is already known; one that has been isolated in pure state, but the structure of which is not yet elucidated; or even one known only by its activity in some impure mixture. In the last of these cases, the product can be claimed per se as a new compound characterized by its structure (which will generally be known once the gene has been obtained and sequenced). The gene itself, or at least the c-DNA coding for the protein, can also be claimed.

Of course, if the product has previously been obtained in pure state, a per se claim is no longer possible, but the invention can be claimed in a variety of ways having the effect of covering the product whenever made by recombinant DNA techniques. In the EPO, for example, the patentee could claim the isolated gene for the product, a vector containing the gene, the host cell transformed with the vector, the process for obtaining any of these, and finally the process for obtaining the end product, which would be infringed by sale of the product obtained by that process. It may be possible to claim the unglycosylated protein per se even when the natural glycosylated form is known. In the UK, Genentech was able to claim 'human tissue plasminogen activator as produced by recombinant DNA technology,[12] although such broad claims would probably not now be granted.

Examples of products of previously known structure that have been obtained by genetic engineering methods include human insulin and tissue plasminogen activator (tPA), which dissolves blood clots and is useful in the treatment of myocardial infarction. Products the structures of which were unknown or only partly known when they were first cloned include hormones (for example human growth hormone); lymphokines (immune system mediators secreted by white blood cells), for example interferons and interleukin-2); and blood-clotting factors (for example factor VIII, the hereditary deficiency of which is the cause of haemophilia). The protein may also be a receptor molecule that is normally bound to the surface of a cell and instigates changes in the cell metabolism when it binds the appropriate ligand. Receptors are particularly useful as targets in assays designed to find potential biological activity of small molecules.

[12] UK Patent 2,119,804

Inventive Step

It is questionable whether the inventive step necessary for a valid patent is present if the inventor merely uses standard methods to produce an obviously desirable product. Certainly, in the early days of genetic engineering, the work involved to obtain a product was enormous and there was no guarantee of success, so that the presence of an inventive step could readily be established. But the rate of progress in this field is so extraordinarily rapid that what was once revolutionary very quickly becomes standard practice. The fact that the state of the art changes so dramatically within the time during which a patent application is pending makes it very difficult to judge the invention in the light of what was the state of the art at the filing date, and different tribunals can easily reach different conclusions. Thus, in the EPO, Genentech's European patent claiming a recombinant DNA process for the preparation of tPA was upheld by the Board of Appeal (albeit in restricted scope) in the face of a determined opposition alleging lack of inventive step,[13] whereas the corresponding British patent, the main claim of which was quoted above, was held invalid by the Court of Appeal[14] on the basis that the substance was known, it was obviously desirable to produce it by recombinant DNA technology, and that the methods used were standard ones.

In the USA, it was originally considered by the USPTO that the gene coding for a protein of known amino acid sequence was prima facie obvious and unpatentable, but in 1993, the CAFC held otherwise,[15] considering that the redundancy of the genetic code meant that there were over 10^{36} distinct DNA sequences coding for the protein in question (insulin-like growth factor) and that the prior art did not suggest any particular one of these sequences. Two years later, the Federal Circuit, in the case of *Deuel*,[16] reversed the Board's conclusion that a prior-art reference teaching a method of gene cloning, together with a reference disclosing a partial amino acid sequence of a protein, rendered DNA molecules encoding the protein obvious. The court also stated that 'obvious to try' is not the test for obviousness.

This latter conclusion of *Deuel* was disapproved by the Supreme Court in *KSR v. Teleflex*[17] (see p. 75), with the result that when a similar biotech case came before the Federal Circuit in 2009,[18] the Court affirmed a Board finding of obviousness and returned to the 'obvious to try' jurisprudence of *O'Farrell*.[19] In *Kubin*, the patent application claimed a human gene coding for a protein designated NAIL (natural killer cell activation inducing ligand). The protein was known to be identical with one previously described in the literature and the specification stated that the DNA

[13] T 923/92 (OJ 1996, 564) *GENENTECH/Human t-PA (EP 93619)*.
[14] *Genentech v. Wellcome* [1989] RPC 147 (CA).
[15] *In re Bell* 26 USPQ 2d 1529 (Fed Cir 1993).
[16] *In re Deuel* 51 F.3d 1552 (Fed Cir 1995).
[17] *KSR v, Teleflex* 550 US 398 (Sup Ct 2007).
[18] *In re Kubin* 561 F.3d 1351 (Fed Cir 2009).
[19] *In re O'Farrell* 853 F.2d 894 (Fed Cir 1988).

sequence could be obtained by 'conventional methodologies known to one of skill in the art', referring to a standard work on cloning technologies (Sambrook).[20] The CAFC confirmed the finding of obviousness, pointing out that the applicant could not rely on Sambrook to establish sufficiency while ignoring its implications for lack of inventive step.

Many commentators have considered the finding in *Kubin* to be too harsh and even to spell the end of patenting in DNA technology. In fact it seems to be doing no more than applying the same standards of inventive step to biotech inventions as already apply to other technical fields. In any event, this type of 'classical' biotech invention, in which one first has a protein and then tries to find the gene coding for it, is already history; now, the more usual situation is that one has a DNA sequence obtained from the Human Genome Project, which, from homology, appears to be a gene, and the problem is to find the function of the protein that it codes.

Breadth of Claims

In any natural protein, there are some regions in which it is possible to change one or two amino acids without affecting the function of the protein, and other regions in which any change in the exact amino-acid sequence will alter or destroy the activity. Thus, although porcine and bovine insulin differ slightly from human insulin, they have essentially the same activity in humans. In view of such possibilities, a claim to a recombinant protein product defined by one specific amino-acid sequence is likely to give a scope of protection that is too narrow. Patentees have attempted to solve this problem by drafting claims to proteins that have a certain degree of homology with the defined amino-acid sequence, or which may have a certain number of possible amino-acid deletions, additions or substitutions (some of which have been so broad as to claim practically all possible proteins). But such a claim must necessarily cover a large number of useless products, in view of the fact that a change of one amino acid may cause complete loss of activity, and is likely to be invalid for this reason. A better claim combines such possibility of structural variation with a requirement that the product must have a certain defined activity, but even this, unless there is a clear correlation between structure and activity, may be held not to meet the written description requirement. In *Kubin*, there was a claim to 'An isolated nucleic acid molecule comprising a polynucleotide encoding a polypeptide at least 80% identical to amino acids 22-221 of SEQ ID NO:2, wherein the polypeptide binds CD48'. Although this claim is of the type suggested above, it was held by the CAFC to be invalid. An alternative approach is that the courts should interpret a claim to a specific protein structure as covering also minor variations such as might be expected to occur in nature and which do not alter the properties of the claimed product, but one can hardly rely on this type of

[20] J. Sambrook et al., *Molecular Cloning: A Laboratory Manual* (2nd edn, 1989) pp. 43–84.

interpretation in a country, such as the UK, with a tradition of literal claim construction.

By analogy, claims to DNA sequences may be placed in four categories of increasing breadth:

(a) a 'picture' claim to one specific DNA sequence;
(b) including other DNA sequences coding for the same protein (genetic code redundancy);
(c) including DNA sequences coding for modified proteins having the same function;
(d) including DNA sequences coding for significantly modified proteins, some of which may not be functional, or including non-coding DNA sequences.

In the USA, it is not only the literal wording of the claim granted by the USPTO that is important, but also the extent to which the courts will broaden the wording by application of the doctrine of equivalents, or restrict it, for example by consideration of the written description requirement. In the litigation between Genentech, Wellcome, and Genetics Institute (GI) over the clot-dissolving drug tPA, GI had developed a modification of tPA called FE1X, which had 15 per cent fewer amino acids than did natural tPA, as well as several minor changes and which as a result stayed active in the blood ten times longer than the natural product. Genentech's claims literally covered only the natural sequence and naturally occurring variants, but before the District Court, a jury found infringement by equivalence. The Federal Circuit reversed the decision on appeal,[21] on the basis that there was no evidence that FE1X functioned in the same way as natural tPA.

Sufficiency, Enablement, and the Written Disclosure Requirement

The US courts have frequently found broad claims to proteins and DNA sequences to be invalid for lack of enabling disclosure. Thus, for example, in the litigation between Amgen and Chugai over erythropoietin (epo), a claim of category (c) above, claiming any purified and isolated DNA sequence 'encoding a polypeptide having an amino acid sequence sufficiently duplicative of that of erythropoietin to allow possession of (biological properties of epo)' was held invalid for lack of an adequate disclosure of how to make other DNA species in this broad genus.[22] The CAFC made it clear, however, that broad generic claims could be valid if they corresponded to the disclosure of the invention.

The situation frequently arises in which a patent application is written at an early stage at which only part of the work has been done, but it is possible to guess how it may be continued. In such cases, broad speculative claims may be drafted and may be allowed by the patent office. Both the USPTO and the EPO have, at times,

[21] *Genentech Inc. v. Wellcome Foundation Ltd* 31 USPQ 2d 1161 (Fed Cir 1994).
[22] *Amgen Inc. v. Chugai Pharmaceutical Co.* 927 F2d 1200 (Fed Cir 1991).

granted biotech patents with claims of excessive width. In the USA, such a claim, once granted, has, at least until recently, been very difficult to attack, but in the infringement litigation between the University of California and Eli Lilly,[23] the CAFC dealt a serious blow to broad claiming. The University inventors had isolated and sequenced only the gene for rat insulin, and the patent claimed genes for all mammalian insulin, especially human. The description was *sufficient* in that the reader would be able to isolate human insulin gene using the method described in the patent. But it did not meet the 'written description' requirement of 35 USC 112 because the patentee never obtained the human insulin gene and did not describe its structure.

The same issue was central to the later case of *Enzo v. Gen-Probe*. Enzo's claims were directed to nucleic acid probes that preferentially hybridize to the DNA of Neisseria gonorrhoeae rather than Neisseria meningitides. Deposits of three such probes were made at the American Type Culture Collection (ATCC). The CAFC initially held[24] that claims directed to nucleic acid probes described solely by their function of selective hybridization do not satisfy the written description requirement, notwithstanding the fact that ATCC deposits had been made.

But a few months later, on a petition for rehearing, the same three-judge panel reversed itself[25] and held that a deposit was sufficient to meet the written description requirement, at least for the sequences that were actually deposited. The case was remanded to the District Court to determine whether the deposits that were made satisfied the written description requirement for the claims that were broader in scope. The decision was in line with *Eli Lilly* in holding that claims directed to materials of biological origin that describe the materials merely by their function do not satisfy the written description requirement of §112 even if the precise claim language appears in the specification. Something more is needed: for example a structure, a partial structure, or a deposit.

Over the dissent of three judges, the Court declined to grant an *en banc* rehearing. The dissenters took the position that §112 did not contain a written description requirement that was separate from the enablement requirement, and wanted the Court totally to revisit the written description requirement and to overrule *Eli Lilly*. So far this has not happened, but the debate continues, most recently in another case involving Eli Lilly.[26] In this case, the CAFC found a 'mechanism patent' invalid for lack of written description, and a separate concurring judgment by Judge Linn contended that the patent was invalid for lack of enablement and that no separate written description requirement was necessary. In August 2009, the Court granted a petition to rehear the case *en banc*, on the specific questions of whether 35 USC

[23] *University of California v. Eli Lilly* 43 USPQ 2d 1398 (Fed Cir 1997).
[24] *Enzo Biochem Inc. v. Gen-Probe Inc. et al,* 62 USPQ 2d 1289 (Fed Cir 2002).
[25] *Enzo Biochem Inc. v. Gen-Probe Inc. et al.* 63 USPQ 2d 1609 and 1618 (Fed Cir 2002).
[26] *Ariad Pharmaceuticals Inc. v. Eli Lilly & Co.* 560 F.3d 1360 (Fed Cir 2007, en banc)

112(1) contains a separate written description requirement distinct from enablement, and, if so, what is its scope and purpose. The decision is eagerly awaited.

In the EPO (as discussed at p. 222), it has been difficult to attack over-broad claims because violation of EPC Article 84 is not a ground of opposition; for some time, it was difficult to allege insufficiency of disclosure because the *Genentech* case[27] was interpreted to mean that in all cases a single working example established sufficiency for the entire scope, no matter how broad. Later cases now agree that the description has to be sufficient over the whole claimed scope.

For early-stage work in a rapidly moving field, obtaining a valid claim to priority is often a critical issue, because there will frequently be publications during the priority year that, if they were prior art, would invalidate the claims. This was shown in the UK in the case of *Biogen v. Medeva*,[28] which illustrated the difficulty of upholding the validity of broad biotech claims. The patent claimed a recombinant DNA molecule characterized by a DNA sequence for a polypeptide displaying hepatitis B virus (HBV) antigen specificity, and covered the genes for both core and surface antigen, although only one of these antigens had been made at the priority date.

In the EPO, the patent was opposed, and the Board of Appeal held that it was sufficient in that it enabled the reader to make both core and surface antigens. The patent was upheld, whereas in the UK, the European patent (GB) was held invalid even though the House of Lords agreed that the specification enabled the production of both core and surface antigens by the single method described. This was the method of shotgun cloning, which was used because the structure of the antigen protein was not known. It was not clear at the time that this could work on a large gene containing introns and its successful use was considered inventive. But between the priority date and the European filing date, the structures of the antigens were published. Thereafter anyone could make the genes by known methods, and Biogen accepted that the claims were obvious as of the filing date and were invalid unless entitled to the priority date.

The House of Lords said that it was not, because the claims covered every way of achieving a result, while showing only one way of doing so (see p. 382). This seems to contradict the normal law for chemical inventions, under which a novel compound may be validly claimed per se if one method of making it is given and it is immaterial if a much better method is invented six months later. But the distinction is that, here, the claim was to a *recombinant* DNA, which imports into the claim the way in which the product is made.

Lord Hoffmann, sitting in the Court of Appeal in *Lundbeck v. Generics* (see p. 236), revisited this point and clearly stated that the concept of '*Biogen* insufficiency' does not apply to a compound per se claim.

[27] T 292/85 (OJ 1989, 275) *GENENTECH/Polypeptide expression*.
[28] *Biogen v. Medeva* [1997] RPC 1 (HL).

The *Biogen* case gave rise to a debate about the nature of sufficiency in the UK, just as *Eli Lilly* did for the written description requirement in the USA. A major piece of biotech litigation in both countries has been the infringement actions by Amgen and others against Transkaryotic Therapies (TKT) and its licensees. The Amgen patent related to recombinant erythropoietin (epo), a naturally occurring hormone that stimulates the production of red blood cells. The substance itself was known and the DNA sequence of the epo gene was also known, although not previously recognized as such. Claim 1 of the patent claimed a defined DNA sequence for expressing in a host cell a polypeptide with the biological characteristics of epo. Claim 26 was to a polypeptide product of the expression in a eukaryotic host cell of a DNA sequence according to claim 1.

Amgen described only the classic recombinant DNA method of producing the protein by inserting and expressing exogenous DNA in a host cell, while TKT used a quite different method in which a specific regulatory sequence was inserted into a cell and this stimulated expression of the natural endogenous epo gene that was already there. There were thus two related issues: whether the claims, properly construed, covered the protein prepared by this method; and whether the specification met the sufficiency requirement.

In the Patents Court,[29] Neuberger J held that, although TKT did not literally infringe claim 26, its process was an immaterial variant of what was claimed. On sufficiency, the judge distinguished between 'classic insufficiency', which arose when the teaching of the patent did not support what it purported to deliver, and '*Biogen* insufficiency', in which a claim was broader than that enabled by the teaching of the patent. Although certain claims were held invalid for classic insufficiency, this did not extend to claims 1 and 26. In the result, claims 1 and 26 were valid, and claim 26 was infringed by TKT.

The Court of Appeal held[30] that it was wrong to consider two different types of insufficiency. Section 72(1)(c) of the UK Patents Act 1977 provides only one ground of invalidity based on insufficiency. It had been held that the patent lacked sufficient teaching for the performance of the claim because the skilled person would not be able to determine whether or not his product was within the claim. The issue was therefore really one of clarity 'dressed up to look like insufficiency'. Lack of clarity was not a ground for revocation and all claims were held valid. Claim 26 was a product-by-process claim and would be infringed only by epo products actually made by the specified process, which involved using an exogenous DNA sequence. TKT's process was not an immaterial variant, because it did not work in the same way as the Amgen process. The final outcome was that Amgen's patent was held to be wholly valid, but not infringed by TKT.

29 *Kirin-Amgen v. Roche Diagnostics/Kirin -Amgen v. Transkaryotic Therapies* [2002] RPC 1.

30 *Kirin-Amgen Inc. v. Transkaryotic Therapies Inc.* [2004] RPC 31 (CA).

In a further appeal to the House of Lords, the main judgment given by Lord Hoffmann gave a detailed analysis of the principles of claim construction (see Chapter 23). The final result was that the patent was not infringed and the relevant claims were invalid for lack of novelty or insufficiency. The problem basically was that Amgen had invented a process for making epo, but was essentially trying to claim the compound itself, however it was made.

In the USA, there was similar litigation about the use of TKT's technology to produce α-galactosidase. Genzyme sued TKT for infringement of its patent for the 'classical' recombinant DNA production of this product, but it was held that the limitation that the DNA coding for the product be 'chromosomally integrated' into the host cell was not met and there was no infringement.[31]

Monoclonal Antibody Technology

Scientific Background

Another basic area of biotechnology is based upon the workings of the immune system rather than upon molecular genetics. The major role in the immune system is played by the white blood cells, or 'lymphocytes'. These cells originate as 'hematopoietic stem cells' in the bone marrow, and then differentiate and mature either in the bone marrow to 'B-lymphocytes' (B cells), or in the thymus gland to 'T-lymphocytes' (T cells). The main task of the B cells is the production of antibodies. When a foreign substance, such as a foreign protein molecule, enters the body, this antigen activates those B cells that carry on their surface receptors that can fit the antigen just as a key fits a lock. The activated B cell undergoes rapid division and develops into a 'clone' of identical 'plasma cells', all of which secrete antibody molecules that have the same specificity to the antigen as did the original B cell.

The antibodies, or 'immunoglobulin' (Ig) molecules, are complex proteins having the approximate shape of the letter Y, and contain, at the ends of the branches of the Y, binding sites specific for a particular antigen. On reacting with an antigen molecule, the antibodies form a cross-linked insoluble structure, thereby effectively removing the antigen from circulation. If the antigen is located on the surface of a foreign cell such as bacterium, the antibodies bind to the surface of the cell, and this renders the cell liable to destruction by macrophages and other components of the immune system.

Apart from any direct therapeutic uses, the high specificity of the antibody–antigen reaction makes antibodies, in principle, a very powerful tool in diagnosis and in research: for example in the identification of antigen sites on cell surfaces. But the great difficulty was that it was never possible to obtain the desired antibody in a pure state. Then, in 1975, Milstein and Köhler succeeded in fusing together a mouse myeloma cell from a malignant B cell and a normal B cell from the spleen of

[31] *Genzyme v. Transkaryotic Therapies Inc.* 68 USPQ 2d 1596 (Fed Cir 2003).

a mouse. The resulting hybrid cell line, which they called a 'hybridoma', combined the properties of both parent cells. It produced the antibody associated with the normal B cell, but like the myeloma parent, it would grow in culture. In effect, the B cell had been immortalized. By choice of a myeloma that did not produce antibodies of its own, only the antibody from the normal B cell was produced. Using spleen cells from a mouse immunized with sheep erythrocytes (red blood cells), 'monoclonal antibodies' (MAbs) against sheep erythrocytes were obtained.[32]

Although the importance of this work was not immediately recognized, after two or three years, it was seen that the process was of general applicability. To obtain MAbs against any desired antigen, a mouse or rat was immunized with the antigen. After a few days, lymphocytes were recovered from the spleen of the animal and fused with cells from a suitable myeloma line, individual hybridoma cells were isolated by dilution, and clones grown from them. Clones that produce the desired MAb could then be selected and the selected hybridoma line be cultured to give the MAb in any desired quantity.

Initially, the amino-acid sequence of an antibody molecule could not be determined with the available techniques and a MAb could be characterized only by deposition of the hybridoma producing it. Later work not only allowed the sequence, but also the full three-dimensional structure to be determined, and also made it possible to produce monoclonal antibodies by recombinant DNA technology in host cells other than hybridomas. One problem with the therapeutic use of the original form of monoclonal antibodies is that they are mouse proteins, and if repeated administration is necessary, the patient's immune system may react against them, reducing their effectiveness or even causing harmful allergic reaction. One way in which to reduce this problem is to produce by rec-DNA methods 'chimeric' MAbs in which the variable regions (the arms of the Y) remain murine, but the constant regions (the base of the Y) are replaced with the constant regions of a human antibody. A further step is to replace all but the actual hypervariable regions that give the specificity, to give a 'humanized' antibody.

More advanced techniques can now cause fragments from human antibody genes to be expressed on the surface of a carrier such as a bacteriophage. This 'phage display' technique enables fragments coding for hypervariable regions of desired specificity to be selected and incorporated into genes that can be expressed to give fully human MAbs, thus making the technology for the production of chimeric and humanized antibodies obsolete by the time that the first of these products came on the market. Phage display can also be used as a screening technique to find large or small molecules that bind to a particular structure: for example one corresponding to a receptor or its ligand.

32 G. Kohler and C. Milstein [1975] 256 Nature 495–7.

Patentability

Whereas Cohen and Boyer obtained patents at least in the USA, the basic work of Köhler and Milstein was not patented. The work was sponsored by the Medical Research Council (MRC), and the National Research and Development Corporation (NRDC), which was responsible for the commercial exploitation of MRC research, has been blamed by some for failing to protect the invention. This may be unjust, because it seems that the NRDC was informed only when it was already too late; in any event, the importance of the work was not immediately apparent.

Many patents have been granted claiming any MAbs directed to particular antigens, or classes of antigen. In the UK, the Patents Court upheld the Patent Office in rejecting for obviousness an early patent application by the Wistar Institute,[33] which claimed broadly monoclonal antibodies to any viral antigens. Corresponding applications were, however, allowed in the USA and Japan. Patents have been granted and upheld, both in the USA and by the EPO, claiming any MAbs to specific known antigens (for example to alpha-interferon) or to certain groups of cells (for example certain groups of human T cells), but their validity is questionable. Since 1977, when the general applicability of the hybridoma technique was recognized, it can be argued that there was no longer any invention in producing MAbs to any previously known antigen, in the absence of special difficulties that had to be overcome. If the antigen was unknown at the time, inventive step may not be an issue, but there may be problems of sufficiency of disclosure. Nevertheless, in the USA, the CAFC[34] has endorsed the proposition in the USPTO Guidelines that a claim to 'any antibody which is capable of binding to antigen X' would have sufficient support in a written description that disclosed the fully characterized antigen. In one case before the EPO Board of Appeal,[35] however, the patent contained broad claims to antibodies reacting with certain human blood cells, but not others, but there was only a single example and it was shown not to fall within these claims. Not only were the broad claims invalid, but a claim limited to the deposited hybridoma was also rejected, since the written description did not correspond to what was deposited.

Chimeric antibodies were broadly patented by Genentech (the 'Cabilly' patent) and the technique of recombinant expression of antibodies by Celltech (the 'Boss' patent), so that anyone wishing to produce recombinant chimeric antibodies would require licences under both patents. In 2000 and 2001, the EPO Technical Board of Appeal (TBA) restricted the claims of both patents considerably: the Boss patent was limited to expression in yeast cells;[36] and the Cabilly patent was limited to non-glycosylated chimeric mouse–human antibodies.[37] In the USA, there was an

[33] *Wistar's Application* [1983] RPC 255(Pat Ct).

[34] *Noelle v. Lederman et al,* 69 USPQ 2d 1508 (Fed Cir 2004).

[35] T 418/89 (OJ 1993, 20) *ORTHO/Monoclonal antibody.*

[36] T 0400/97 (unpublished) *CELLTECH/Immunoglobulins.*

[37] T 1212/97 (unpublished) *GENENTECH/Immunoglobulin preparations.*

interference that resulted in a new version of the Cabilly patent being granted that would not expire until 2018.

Not only are monoclonal antibody reagents very powerful diagnostic tools, but therapeutic MAbs are also becoming increasingly commercially important. There are now over 20 therapeutic antibodies approved by the US Food and Drug Administration (FDA), some of which have annual sales of US$1.5 billion, the whole market being worth over US$30 billion in 2009. These commercial products include murine, chimeric, humanized, and fully human MAbs. In addition, over 150 therapeutic MAb products are in development. Nevertheless, there has recently been litigation on one therapeutic MAb: Genentech's product Herceptin®, a humanized antibody used in the treatment of breast cancer. Chiron sued Genentech on the basis of a patent dating from applications made in 1984, 1985, and 1986, disclosing murine antibodies against the relevant breast cancer antigen, but claiming chimeric and humanized as well as murine MAbs. It was held that Chiron's patent was not entitled to these earlier filing dates, because these applications did not meet the written description requirement in respect of chimeric and humanized antibodies, the relevant technology not being known at that time. Consequently the patent was invalid over an intervening publication.[38]

More Recent Technologies

Genomics

The science of genomics correlates genetic information with biochemical pathways and specifically with disease mechanisms. Some diseases, for example cystic fibrosis, haemophilia, or sickle cell disease, are caused by defects in a single gene, but usually the picture is much more complex and disease states may be associated either with mutations in a number of different genes, or with over- or under-expression of normal genes. Thus if, in a certain disease, a particular gene is over-expressed, then the protein product of that gene may be a target for intervention by a drug molecule that would interfere with its function, for example by blocking a receptor for that protein.

Information on levels of gene expression in diseased and healthy tissue may, for example, be obtained using the very powerful technique of DNA chips, in which an array of tens of thousands of oligonucleotides of known structure, corresponding to partial structures of genes, is produced on a small glass plate. Labelled cDNA obtained from tissue RNA is incubated with the chip and hybridization with the probe DNA is detected. In this way, the expression levels of large numbers of genes can be determined in a single experiment.

[38] *Chiron Corp. v Genentech Inc.* 70 USPQ 2d 1321 (Fed Cir 2004).

Proteomics

Proteomics takes the process a step further by analysing directly the levels of protein molecules produced in the cell by gene expression on a large scale (the 'proteome') and how this varies as a function of time. Proteomics also takes account of post-translational modifications, such as phosphorylation events, which play a critical role in protein function. The term is also applied to the elucidation of protein tertiary structures to enable the design of drug molecules that would interact with the protein molecule.

Antisense Technology

When a disease state may be caused by or associated with the expression of a particular gene, it may be desirable to prevent the gene from being expressed. One way in which this might be done is to add to the cell a short fragment of 'antisense' DNA, which is complementary to a part of the coding strand of that gene. When the strands separate, this fragment will bind to the coding strand and block the transcription of that strand to mRNA. A fragment of 20 bases in length will normally be completely specific for one particular gene and therefore not interfere with the expression of other genes.

This sounds very simple in theory, but in practice there are two major difficulties. One is the low stability of single-stranded DNA *in vivo*, which makes normal DNA unsuitable for use as a drug. This problem may be overcome by chemical modification of the backbone of the DNA chain, for example replacing the phosphate groups by groups that are hydrolysed less easily. Such chemical modifications may of course be patented in the same way as normal chemical inventions, irrespective of the actual sequence of the nucleotide residues attached to the backbone. The other problem is to find a suitable delivery system that will allow the antisense drug to penetrate the cell wall and enter the cell nucleus.

RNA Interference

As a method of inhibiting gene expression, an alternative to antisense DNA is to use short sections of double-stranded RNA (short interfering RNA, or siRNA). These have a well-defined structure: a short (usually 21-nt) double strand of RNA, with 2-nt 3' overhangs on either end. siRNAs can be introduced into cells by various transfection methods to bring about the specific knockdown of a gene of interest, which may make it unnecessary in many cases to produce a transgenic mouse lacking the target gene (known as 'knock-out mice', see p. 305). Essentially any gene of which the sequence is known can be targeted. This has made siRNAs an important tool for gene function and drug target validation studies, and they also have potential as therapeutic agents.

Stem and Progenitor Cells

In recent years, it has been possible to study cell differentiation and identify stem cells that give rise to a range of different, but related, tissue types. This has been done largely with the aid of monoclonal antibodies that recognize specific molecules on cell surfaces and allow separation of the cells by the fluorescence-activated cell sorting (FACS) technique. In this way, the development of the various types of blood cell (the hematopoietic system) can be traced back to the hematopoietic stem cell (HSC), which can give rise to all other types. Similarly, mesenchymal stem cells (MSCs), which can differentiate into all types of connective tissue, such as cartilage, bone, and tendon, have been isolated from bone marrow. Such cells are patentable (if the normal criteria are met) as long as the claims do not cover the cells when in their natural state in the human body (see next chapter).

Such stem cells can have direct therapeutic uses: for example HSCs may be administered to cancer patients whose hematopoietic system has been damaged by chemotherapy or radiotherapy, and MSCs may be useful, for example, to regenerate cartilage damaged by injury or arthritis. Initially, at least, such treatment would be autologous, that is, the cells would be harvested from the patient, isolated, expanded in culture, and then returned to the same patient. The cells may also act as suitable vehicles for delivering proteins in the context of gene therapy, that is, the cells may be genetically modified before being returned (see below).

Embryonic stem cells, which may give rise to any types of human tissue, present ethical problems that are discussed in the next chapter. A possible alternative to embryonic stem cells with less ethical concerns is the use of adult stem cells derived from certain types of already differentiated tissue, although adult stem cells often have less potential to differentiate than pluripotent, or even totipotent, embryonic stem cells.

Cell and Gene Therapy

If a disease state is caused by one specific gene being defective, so that a necessary protein is not expressed in normal amounts, or at all, it would clearly be desirable to somehow insert the normal gene into the patient so that the gene product can be expressed. One could imagine somehow modifying the gametes that are the precursors of the ova or sperm cells, so that the desirable change would be inherited in the next generation, but this 'germ-line gene therapy' is considered unacceptable for ethical reasons (see next chapter) and attention is concentrated on 'somatic cell gene therapy', which affects only non-reproductive cells.

There are, however, a number of possible types of somatic cell gene therapy. One may be characterized as *ex vivo*: that is, cells are removed from the body, genetically modified, and returned (cell therapy). In another form, *in vivo* gene therapy, the intention is to modify body cells without the need to remove them first. The first reported human trial of *ex vivo* gene therapy was carried out on a child suffering from a rare form of immunodeficiency caused by the lack of a specific protein. Lymphocytes from the child's blood were isolated, transformed *in vitro* with a vector containing the normal gene, and returned to the patient. Near-normal concentrations of the missing protein were found in the blood. But because the lymphocytes had a half-life in the body of only a few months, the treatment had to be repeated at regular intervals.

In theory, if the transformed cells were HSCs instead of differentiated cells such as lymphocytes, repeated treatment might not be necessary, but this would depend upon the replacement gene being inserted into the chromosomal DNA so that it would not be lost upon cell division. Retro- and lentiviruses (such as HIV) replicate by inserting their own DNA into the genome of the host, and retroviral vectors, modified so as to be non-pathogenic and incapable of reproduction, have therefore been used for this purpose. For example, cystic fibrosis is a disease caused by a genetic defect leading to the lack of an enzyme that breaks down mucus in the lungs. Attempts have been made to 'infect' cystic fibrosis patients with integrating viral vectors carrying the natural gene, but results have been disappointing.

Another type of gene therapy has the aim not of restoring a missing function to otherwise normal somatic cells, but rather of disrupting the functioning of abnormal cells such as cancer cells, either killing them directly or making them more likely to be killed by other agents. Here the problem is one of how to target the vector specifically to the cancer cells. The first treatment of this type was approved in China in October 2003, and uses an adenoviral vector to target cells of head and neck squamous cell carcinoma. In the USA, the FDA gave clearance in January 2009 for the first clinical trial with a treatment based on human embryonic stem cells for acute spinal cord injury. As of mid-2009, however, the trials were suspended pending FDA review of new non-clinical animal study data submitted by the applicant.

Although, in Europe, gene therapy methods, being methods of medical treatment, are not patentable, the vectors and constructs used may be so. In addition, *ex vivo* process steps will be patentable as long as the last step of administering the transformed cells to the patient is not claimed. In the USA, of course, the claims may cover the entire process.

Pharmacogenomics

Pharmacogenomics is the study of how an individual's genome affects the body's response to drugs. It is well known that both the efficacy and the side effects of drugs may vary considerably from patient to patient, and it is now possible in many cases to correlate these differences with nucleic acid or protein 'biomarkers'. These may be single nucleotide polymorphisms (SNPs) in the chromosomal DNA, in which a single nucleotide is substituted for another. SNPs occur about every 100–300 bases along the whole genome, and while the majority of these may have no effect whatsoever, some turn out to be correlated with drug response. In future, analysis of a patient's genome, for example by using DNA chip technology, will be able to indicate which drug, at what dosage, would give the best response with the lowest side effects. RNA or protein expression patterns may also be efficiently analysed with chip technology and can provide clinically relevant biomarkers for diagnosis, drug efficacy, or side effects.

This potential for personalized medicine presents both risks and opportunities for the pharmaceutical industry, and may lead to a complete change in its

At the time of writing (mid-2009), there are several Bills in the legislative process that would establish a regulatory pathway for biosimilars in the USA. One point that is quite hotly debated is the period of regulatory data exclusivity: some generic manufacturers would like to see the same five-year period as for SCE data exclusivity; some originator manufacturers argue for up to 14 years of data exclusivity, because of the allegedly higher cost of investment in the development of biological drugs. Two Committees of the Senate have voted for a 12-year data exclusivity period, whereas the White House seems to advocate a seven-year period. The first interchangeable biosimilar is supposed to benefit from a semi-exclusivity period of one to two years according to various proposals.

It is hoped that the US legislators will also find the right balance in order to open a pathway for safe and affordable biosimilars, while maintaining sufficient incentives for research and development into innovative biological products.

15

PATENTING OF GENES, PLANTS, AND ANIMALS

All imbalances will be gone, longevity assured. We will have a race of men . . . test-tube-bred . . . incubator-born . . . superb and sublime.

Edward Albee, *Who's Afraid of Virginia Woolf*? (1962)

There will be no patents on monsters while I am Commissioner.

US Patent Commissioner Bruce Lehman (1998)

Patents and Life

In the previous chapter, we discussed from a purely scientific and legal point of view the patenting of biotechnological inventions, including DNA sequences and microorganisms. In this chapter, we shall go on to consider the patenting of DNA sequences that are complete human genes or fragments thereof, as well as the patenting of higher organisms, including plants and animals. Here we can no longer stay at the objective level of scientific findings and patent laws, but must consider wider issues of ethics, social policy, and politics.

Many persons, for a wide variety of reasons, are opposed to the granting of patents for living organisms; the same persons often consider that someone who patents a human gene somehow acquires control of human life, or even of an individual from whose tissue the gene was isolated. Where the invention can be useful in the diagnosis or treatment of disease, the majority of the public may accept the need for such patents, but opposition to genetic modification of food crops and to patents on the resulting transgenic plants is more widespread.

Human Genes

Many of the DNA sequences that have been patented for years are human genes, because they code for a human protein, although they are normally c-DNA sequences, which omit any introns that may be present in the natural gene. But the emphasis has always been upon the protein, with the DNA or the gene seen simply as a means for the production of the product of interest, and, with the sole exception of the *Relaxin* case in the European Patent Office (EPO) (see below), no objections have been raised on ethical or moral grounds to any of these patents.

The situation becomes somewhat different when the focus of attention switches to the gene, particularly if it is discovered before the protein it encodes and before the specific function of the protein is known. Many companies and universities are engaged in research to identify genes that are associated with specific diseases, and will wish to patent any such gene that is found. There are some diseases, for example haemophilia or cystic fibrosis, which are directly attributable to a defect in a single gene. A far larger group of diseases involve a number of different genes, and result from a complex interaction between genetic and environmental factors. Thus certain diseases, for example hypertension, Alzheimer's disease, and certain types of cancer, have been shown to be associated with specific genes in the sense that people carrying a particular variant of that gene have a greater than average chance of developing the disease. Whether they will ever have the disease will depend upon a great many factors, including eating habits and exposure to carcinogens, many of which factors are as yet unrecognized.

Finding such disease-related genes often results from a combination of biotechnology with classical genetics, which may involve studies of large families with a high prevalence of the disease, or studies of genetically homogenous populations, such as Icelanders or residents of remote Canadian valleys, where there is a high chance that a disease-related gene in the population arose from a single common ancestor. It helps if the population kept good written records of their family trees so that familial relationships can be clearly established. The Mormon Church, for its own religious reasons, has accumulated what is probably the world's most extensive collection of genealogical data and access to these data helped Myriad Genetics of Salt Lake City to identify the BRCA1 gene associated with an increased risk of breast cancer.

Disease-related genes are potentially useful for screening to identify the persons who are at risk, as well as to base studies of the mechanism of the disease process. There are ethical issues involved in screening, however. It may be ethically correct to inform individuals that they have an increased risk of heart disease if, for example, by stopping smoking, eating a low-fat diet, or taking medicament to lower blood pressure, they can take effective counter-measures; it may be questioned whether it is acceptable to inform those that have an increased risk of Alzheimer's disease if, at the current state of medical knowledge, there is nothing that can be done about it. Furthermore, such genetic data on individuals may get into the hands of insurance companies and employers, with the danger of discrimination against persons carrying 'bad' genes. This problem has been addressed in the USA by the passage of the Genetic Information Non-discrimination Act in 2008, which makes such discrimination illegal. It does not, however, mention use of genetic information by law enforcement agencies and, in some states of the USA and in some other countries, DNA databases are kept based on samples compulsorily collected from convicted criminals, even non-violent ones, or even from suspects who were never charged or convicted. It is not difficult to imagine a scenario in which a certain gene is found to be associated with violent behaviour and all persons who carry the gene are made liable to police surveillance.[1] These are the areas in which the real ethical problems are to be found and arguments about whether or not one should patent a gene are completely irrelevant to these issues.

The Human Genome Project

A different way of looking for human genes is that of sequencing all or part of the entire human genome, finding which sequences correspond to expressed genes, correlating gene expression with cell type and disease state, and gradually building up a picture of the function of the genes that are found. This is the basic approach of the Human Genome Project, set up in 1988 by the US National Institutes of Health (NIH). The Human Genome Organization (HUGO) correlated sequencing at a number of academic centres and estimated that it would be possible to sequence the entire human genome by 2005. In fact, the project was completed well ahead of schedule in April 2003. Naturally, there is more interest in sequencing genes than the large amount of non-coding DNA in the genome, and the technique was developed of isolating messenger RNA from various tissues and obtaining from this a mixture of fragments of c-DNA (typically 200–300 bp long), known as expressed sequence tags (ESTs). Although by the mid-1990s only a small amount of the total genome had been sequenced, ESTs could be used as probes to isolate and identify full-length genes, and by comparing ESTs from different tissue types and from diseased and healthy tissues, it was possible to draw some conclusions about tissue-specific gene expression and about which genes were disease-related.

[1] The XYY chromosome abnormality has already been associated with criminality by some researchers.

Patenting of ESTs

In early 1993, the NIH filed a US patent application claiming 2,421 ESTs for which the structure and function of the full gene was unknown. This set off a bitter controversy within the scientific community and the patent application was later withdrawn. Although the NIH had retired from the patenting scene, HGS, Incyte, and other firms continued to file US patent applications claiming thousands, or tens of thousands, of ESTs in each application. In 1996, the US Patent and Trademark Office (USPTO) agreed that up to ten sequences could be searched and examined in the same application; meanwhile, the others would remain unpublished, and at any time, a divisional application for another ten could be filed. The aim was clearly the submarine patent strategy of waiting until someone else found an interesting gene and then obtaining grant of a patent with a claim that it was hoped would cover not only the EST, but also the gene itself.

In the EPO, patent applications for ESTs without a known function were seldom filed and almost never granted. A European application claiming 1,000 ESTs would attract an invitation to pay 999 additional search fees, which made the strategy impractical. As late as 1998, the USPTO took the position not only that ESTs were patentable subject matter, but also that a claim to 'A purified and isolated DNA composition comprising (the EST sequence)' would normally meet the requirements for enablement and written description. Such a claim of course reads on to the gene itself, which was exactly the result that the applicants were hoping to achieve. In 2002, however, the USPTO introduced new guidelines providing that a DNA sequence must have a stated utility that was 'substantial, specific and credible' in order to be patentable. The Federal Circuit finally, in 2005, had the opportunity to rule on an EST patent owned by Monsanto, and upheld the rejection of the claims by the Board for lack of utility and lack of enablement.[2] In this, the Court followed the reasoning of the Supreme Court in *Brenner v. Manson*,[3] in which it was held that a novel process for making compounds with no known use was not patentable. The situation of an EST that might be used to find a gene of no known utility was considered as equivalent to this. This puts a welcome end to the EST patent story.

Transgenic Animals

It is possible by the use of various techniques, including microinjection, to introduce extraneous genetic material into a fertilized mammalian ovum, insert the ovum into a pseudopregnant female, and obtain offspring in which the genetic material has become incorporated into the genome. By combining this process with classical breeding steps, such as back-crossing, it is possible to obtain a strain of animal that stably transmits the new gene to subsequent generations, which

[2] *In re Fischer* 421 F.3d 1365 (Fed Cir 2005).
[3] *Brenner v. Manson* 383 US 519 (Sup Ct 1966).

will display the corresponding phenotype according to the laws of Mendelian genetics.

Such transgenic animals have two main uses at present: one is as animal models for research; another is as a source of useful materials. In one type of genetic modification, the effect is not to add a functional gene, but to render inoperative an existing gene. Thus it is possible to create a transgenic 'knock-out mouse' in which almost any given gene is 'knocked out' and its gene product is not expressed. This will, in many cases, have a fatal effect, but often the knock-out mouse will be viable and sometimes perfectly healthy. If it is hypothesized that a defect in a particular human gene causes or predisposes to a disease, it can be tested whether the removal of the corresponding murine gene produces a similar disease state. A particularly effective technology for producing knock-out mice is the 'cre-lox' technique invented at Harvard University. The patents were assigned to Du Pont, and subsequently acquired by Monsanto, then by Bristol-Myers Squibb. Du Pont originally imposed very restrictive licensing conditions, but in the face of pressure from the NIH relaxed these for academic users of the technology.

Adding a gene to a mouse can also cause or predispose to a disease, and the diseased animal can then be used as a model for possible treatment. This was the situation for the Harvard OncoMouse®, the European patent[4] for which has given rise to more controversy than any other. The OncoMouse carries a gene that makes it very likely to develop cancer by a particular age; it can therefore be used as a model to test anti-cancer drugs in place of normal mice in which cancer has been induced by the use of carcinogens. Use of the OncoMouse gives more reliable and reproducible results, and can therefore reduce the overall number of test animals used.

If it is desired to produce commercial quantities of a human protein, one can arrange for the expression of the corresponding gene not in a unicellular host such as E. coli or yeast, but in a transgenic animal. Here, it is of little use if the gene expression takes place randomly throughout the organism and the aim is to produce localized expression, for example by combining the gene with a promoter that normally controls expression of the gene for one of the proteins found in milk, such as casein or a whey protein, so that the transgenic mammal will secrete the desired protein in its milk. Since the amount of milk obtainable from a mouse is somewhat limited and since cows are slow to breed, the species mainly used are sheep, goats, pigs, and sometimes rabbits. This approach (known as 'pharming') can work well for proteins such as human serum albumin and alpha-antitrypsin, which are required in large quantities and which do not have high biological activity in the host animal. But if a sheep were to be modified so as to secrete a highly active cytokine in its milk, enough of the product would get into the animal's circulation to have serious and possibly fatal effects.[5]

[4] EP 169 672B.

[5] 'Mary breeds transgenic sheep. A foreign gene she slips in/And from the milk this new Bo-Peep gets alpha antitrypsin./But if she used a cytokine to do her transgene cut on,/Her sheep, instead of doing fine, would be transgenic mutton.'(Anon.)

A further use for transgenic animals is as a source of organs for transplantation. Many thousands of patients whose lives could be saved by organ transplants die because not enough donor organs are available.[6] A few unsuccessful attempts have been made to transplant primate organs into humans, but the ideal donor is the pig, the circulatory system of which is remarkably similar to that of a human.[7] If a normal pig organ is transplanted into a human, it undergoes hyperacute rejection and becomes necrotic within a short time, due to a process in which elements of the human complement system recognize surface molecules on the pig cells as foreign. If a pig is transformed with a human gene so that the cells in its organs express a human surface protein, it is expected that the hyperacute rejection would be blocked, and that medium- and long-term rejection would be no more of a problem than with human–human transplantation. There are safety issues involved, since one must be very sure, for example, that porcine retroviruses could not cause disease in humans as a result of xenotransplantation, but the ethical issue is no different from that involved in eating a ham sandwich. It is less so, in fact, since both Jewish and Islamic law accept the use of pig tissue to save human life. However, this approach, which seemed very promising in the late 1990s, has not lived up to its promise and no successful transplants of transgenic pig hearts to humans have been carried out.

Transgenic Plants

The transformation of plant cells poses challenges that are not found for animal cells. Plant cells typically have not only a semipermeable membrane, but also a hard external cell wall, and external genetic material must somehow be introduced through this barrier. Furthermore, the interior of most plant cells largely comprises an air-filled space rather than an aqueous environment through which vectors and large molecules can move freely. Nevertheless, techniques have been developed that allow transformation events to occur; some verge on the bizarre, such as biolistic transformation, in which DNA molecules are placed on the surface of micronized glass beads that are then physically shot into the plant cell. Once transformation has occurred, conventional breeding techniques enable production of plants the seeds of which will pass the desired phenotype to further generations.

The long-term aim is to generate transformed plants that would have not only the desirable characteristics that are normally attainable by conventional breeding programmes, but also extra advantages such as high yield, growth in arid conditions, additional nutritional quality, and others that will be necessary in order to be able to feed an estimated world population of over 10 billion in 2050 with less arable land than at present. At present, however, not enough is yet known about the mechanism of action of plant genes and the changes introduced by genetic engineering have so

6 A consequence of the increased use of seat belts and airbags.

7 Not only the circulatory system: see George Orwell, *Animal Farm* (1945).

far been rather specific in nature. For example, one early product was a tomato in which the gene producing the enzyme that causes softening of the picked fruit had been suppressed. This meant that the tomato, instead of being picked while green and hard, and ripened artificially to give a typically tasteless product, could be harvested when vine-ripened without going soft on storage. The product was introduced by Calgene Inc. in 1994 under the name Flavr Savr®, but a combination of patent conflicts,[8] technical problems,[9] poor marketing, and lack of customer acceptance led to it being withdrawn from the market.

Other transformed plants have characteristics relevant to herbicide and insect resistance. For example, a crop plant may have a gene conferring herbicide resistance so that weeds can be killed by spraying while the crop plant is growing. Bt maize contains a gene from Bacillus thuringensis that produces a toxin harmful to insects, but not to mammals, so that the larva of the corn borer is killed by eating the plant and spraying with chemical insecticides may be reduced or avoided.

Patenting of Animals, Plants, and Human Cells

In the USA

Following the *Chakrabarty* case, the first US patent on a multicellular organism was granted in 1987 and covered polyploid oysters, produced by a technique that, like that in *Chakrabarty*, did not involve genetic engineering. Then in 1988 a US patent was issued on the Harvard OncoMouse®. Some US politicians called for a law imposing a moratorium upon animal patenting and, although no such law was enacted, there was a de facto moratorium of a four-year period up to 1993 during which no further animal patents were issued by the USPTO. Subsequently it has been made clear that any life form is patentable provided that human technical intervention is required in its production. The one exception is that human beings cannot be patented because this would be a violation of the Thirteenth Amendment to the Constitution, which prohibits slavery and hence any property rights in individual human beings.

In the European Patent Office

The situation in the EPO is less clear-cut because of the prohibition in the European Patent Convention (EPC) against the patenting of plant and animal varieties.[10] Whereas there is agreement as to what constitutes a plant variety,[11] it is totally unclear what constitutes an animal variety, and the terms used in the French and German texts (*races animales* and *Tierarten*, respectively) are not any more helpful.

[8] *Enzo Biochem Inc. v. Calgene Inc.* 29 USPQ 2d 1679 (E Cal 1994).

[9] For example, picking machines used for hard green tomatoes did not work with soft ripe ones.

[10] Article 53(b) EPC.

[11] See r. 26(4), IR.

The history of the Harvard OncoMouse® application in the EPO is a depressing saga, which continued from the filing date in 1985 up to July 2004. The Examining Division initially refused the application on the basis that it claimed an animal variety, contrary to EPC Article 53(b). The refusal was appealed and the Board of Appeal considered both parts of Article 53 in a decision[12] that held that Article 53(b) prevented the patenting of certain categories of animals, but not of animals as such, but returned the application to the Examining Division with instructions to consider Article 53(a) and weigh the possible suffering of the animal against the possible benefit to humanity. Why the Board thought that this should be the function of the EPO rather than that of the national authorities with responsibilies for regulating animal experiments is not clear.

The Examining Division went through this exercise as best it could and granted the patent,[13] whereupon it was opposed by a motley group of individuals and organizations, mainly related to churches and animal rights groups. After long delays, oral proceedings were held in Munich in 1995. These were turned into a media circus by the opponents, whereupon the Opposition Division lost its nerve, reverted to written proceedings, and put the whole case on ice until the Biotechnology Patenting Directive (see below) should be decided. Second oral proceedings were held in 2001 and a written decision was issued early in 2003, which upheld the patent, with claims limited to species such as rodents suitable for use as experimental animals, elephants, for example, being regarded as unsuitable. An appeal was filed in May 2003 and a final decision limiting the scope further (to mice instead of rodents) was given by the Board of Appeal in July 2004, just one year before the patent was due to expire.

The exclusion of plant varieties from patent protection was construed in a limited way in a case decided in 1984. In this case,[14] which related to chemically treated seeds, not transgenic plants, it was held that if the claim covered plant genera without claiming individual specific varieties, it did not contravene Article 53(b). In a later case,[15] however, in which transgenic plants were claimed, the decision was essentially that a transformed plant, having a characteristic phenotype stably transmitted, was by definition a plant variety and the prohibition of Article 53(b) was not avoided merely by using broader claim language. Claims to plant cells were granted, but not claims to plants or seeds. It was also held that although the first step in the process, that is, the transformation of the plant cell, was 'microbiological', this did not make the final plant the product of a microbiological process.

This decision was strongly attacked by the seeds industry and the president of the EPO referred the matter to the Enlarged Board of Appeal (EBA) on the basis that

[12] T 19/90 (OJ 1990, 476) *HARVARD/Transgenic Mouse.*

[13] Reported at OJ 1992, 588.

[14] T 49/83 (OJ 1984, 112) *CIBA-GEIGY/Propagating material.*

[15] T 356/93 (OJ 1995, 545) *PLANT GENETIC SYSTEMS/Plant cells.*

there were conflicting Board of Appeal decisions on the same point of law.[16] The EBA held the referral inadmissible because it did not consider that there was indeed any conflict between the cited decisions[17] and refused to decide the point.

Subsequently, the same Technical Board of Appeal (TBA) that decided the *Plant Genetics Systems* case referred to the EBA a series of questions on the interpretation of Article 53(b) arising out of a specific appeal. This referral could not be evaded and the EBA finally decided[18] that transgenic plants were patentable subject matter. The EBA did not, as some had expected, decide the issue on the basis that, at the time the EPC was written, the only way to make plant or animal varieties was by traditional breeding methods that are non-technical in nature and not reproducible. Transgenic plants or animals are made by a technical, reproducible process and therefore should fall outside the exclusion. Instead, the rationale of the decision was that, at the time that the EPC was ratified, most contracting states were members of the International Convention for the Protection of New Varieties of Plants (known by its French abbreviation, UPOV) in a version (since superseded) that prohibited dual protection of plants through UPOV and through patents. Therefore, plant varieties were excluded from patent protection only to the extent that they could be protected by plant breeders' rights under UPOV, which transgenic plants generally could not. A claim that does not individually claim specific plant varieties is not unpatentable merely because it may embrace plant varieties. This decision is perfectly logical as applied to plants, but cannot easily be applied to animals, because there is nothing corresponding to UPOV for animal varieties.

Article 53(b) also excludes from patentability 'essentially biological processes for the production of plants or animals' and Rule 26(5) states that a process is 'essentially biological' if it consists entirely of natural phenomena, such as crossing or selection. In a case[19] relating to a process for the production of broccoli with improved properties (increased level of certain cancer-preventing ingredients), the claimed process contained steps of crossing and selection, but also the use of molecular markers in a selection step. The question of whether the presence of such a single technical step allowed the claim to escape the exclusion was referred to the EBA, where it is currently pending as G 2/07, together with a similar case relating to tomatoes.[20]

The scope of protection given by patents relating to transgenic plants was considered in two cases in which Monsanto alleged infringement of its patents by importation of soy meal from soybeans grown in Argentina using its RoundUp Ready® technology. The patent claimed a DNA sequence inserted into the transgenic plants and a process for producing the plants. Small amounts of the relevant DNA could be

[16] Article 112(1)(b), EPC.

[17] G 3/95 (OJ 1996, 169).

[18] G 1/98 (OJ 2000, 111) *NOVARTIS/Transgenic plants.*

[19] T 83/05 (OJ 2007, 644) *PLANT BIOSCIENCE/Broccoli.*

[20] T 1242/06 (OJ 2008, 523) *STATE OF ISRAEL/Tomatoes* (pending as G 1/08).

detected in the soy flour. In the UK,[21] it was held that, although the patent was valid, it was not infringed because the DNA was not 'isolated' as required by the claim and because the flour was not the direct product of the process (see p. 452). In the Netherlands,[22] the court considered Article 9 of the Biotech Patenting Directive (see below), which states that the protection given by a patent for genetic information (DNA) extends to all material incorporating the genetic information and in which it performs its function. Even if DNA is present in the soy flour, it certainly does not perform its function there. The court referred to the European Court of Justice (ECJ) the question of whether this provision of the Directive limited the per se protection that otherwise might be given by Dutch law.

The Biotechnology Patenting Directive

For many years, the European Commission planned to harmonize the national laws of European Union (EU) member states on the question of the patent protection that should be given to biotechnological inventions, including genes, plants, and animals. Although, when the discussion was started in 1988, the status quo was that living organisms and DNA were patentable, or at least not stated to be unpatentable, in all member states, there was a fear that individual states might change their law so as to ban such patents. It was also felt that the situation in Europe, as compared with the USA, was uncertain enough to cause disadvantage to European research. Accordingly in 1989 the Commission adopted a draft Directive on the legal protection of biotechnological inventions (more usually referred to as the Biotechnology Patenting Directive (BPD)), which if adopted would enforce certain minimum standards of protection upon all member states and prevent any subsequent weakening of protection in this area.

This first version was rejected by the European Parliament in 1995, mainly because the Parliament considered that there were insufficient ethical controls on the patenting of the human body and its parts. After this rejection, some industry representatives considered that the whole idea should be allowed to die, because it was feared that any version to which the Parliament would agree would be less favourable than the status quo. The Commission persevered, however, and a new version, which, even after the incorporation of the numerous amendments proposed by the Parliament, was considered acceptable by industry, was adopted by the Council as its common position and was finally approved by the Parliament on 12 May 1998, entering into force on 6 July 1998 as Directive 98/44. Member states had until July 2000 in which to implement the Directive. The Netherlands challenged the Directive before the ECJ, which upheld the Directive in October 2001.[23] Nevertheless, not only the Netherlands, but also a number of other member states, including Germany, France, and Austria, have taken many more years to implement

21 *Monsanto Technology LLC v. Cargill International SA* [2007] EWHC 2257 (Pat).

22 *Monsanto Technology v. Cefetra* (The Hague District Court, 24 September 2008).

23 C-277/98.

the Directive and the result often runs counter to what the Directive says (see below).

Applying the BPD to the grant of European patents has proceeded more smoothly. There was the problem that the EPO is not an organ of the EU and the BPD is not directly applicable to the EPC. Nevertheless, it was clearly unacceptable that the EPC should be in conflict with the BPD and the question was how the areas in which there is conflict should be resolved. Although the cleanest way from the purely legal standpoint would have been to amend the EPC by Diplomatic Conference, this was undesirable for practical reasons: it would have been a long process requiring unanimity of EPC member states, including states that were refusing to implement the BPD and states that were not members of the EU; and other changes to the EPC would also be proposed that might cloud the issue. Instead of this, the Administrative Council adopted new Implementing Regulations incorporating the wording of the BPD.[24]

Important points in the Directive that are relevant to the patenting of genes, plants and animals include Article 5, which reads:

1. The human body, at the various stages of its formation and development, and the simple discovery of one of its elements, including the sequence or partial sequence of a gene, cannot constitute a patentable invention.
2. An element isolated from the human body or otherwise produced by means of a technical process, including the sequence or partial sequence of a gene, may constitute a patentable invention, even if the structure of that element is identical to that of a natural element.
3. The industrial application of a sequence or a partial sequence of a gene must be disclosed in a patent application.

and Article 4(2):

> Inventions which concern plants or animals shall be patentable if the feasibility of the invention is not confined to a particular plant or variety.

Thus, it seems clear that the BPD allows patenting of isolated genes. The intention is presumably not to allow patenting of functionless sequences such as ESTs, but this depends on the interpretation of 'industrial application' in Article 5(3). It is also clear that a claim should not be rejected merely because its scope includes plant varieties.

The ethical issues that caused the first version of the BPD to fail have been addressed in the newer version. Thus it is stressed that the human body, at any stage of its formation or development, including germ cells, cannot be patented,[25] and there is a non-exhaustive list of 'immoral' inventions that are non-patentable, including processes for the cloning of human beings, processes for germ-line modification of humans, use of human embryos for industrial or commercial purposes,

[24] Rules 26–29, IR.

[25] Recital 16, BPD.

and processes for modifying the genetic identity of animals that are likely to cause them suffering without any substantial medical benefit to man or animal.[26] These represent no more than a list of the hot issues at the time that the BPD was approved and give little guidance for the morality of future technologies not thought of in 1998.

Thus germ cell gene therapy is regarded as unethical and the BPD provides that it is unpatentable. It is not easy to see why this should be so: after all, if a couple may pass on to their offspring haemophilia or sickle cell disease, and if it were possible to eliminate this possibility by germ-line gene therapy, is there any convincing moral reason why this should not be done?[27] The real reason for this exclusion is the fear that germ-line intervention could be used not to prevent disease, but for eugenic reasons: a valid ethical concern, but one that has nothing at all to do with patents. A more pragmatic reason not to attempt germ-line gene therapy at present is that we do not know enough about human reproduction and human genomics to be sure that it could be done safely.

Use Limitations on Gene Patents

One problem that has arisen in the debates on implementation of the BPD is that it is felt strongly in academic circles that patents for genes should be use-limited, that is, that the patent for a gene would not be infringed by the use of the gene for a purpose other than the one disclosed in the patent application. This position is supported, for example, by the 2002 report of the Nuffield Council on Bioethics,[28] which was less concerned with ethical issues than with telling patent offices how they should examine patent applications for DNA.

As justification for this approach, it is sometimes suggested that genes are not merely chemical substances, but are primarily carriers of information. It is also argued that the same gene may code for more than one protein product (see p. 280), and a patent for a gene based on its coding for one protein should not cover the production of a different one. The problem with this is that it is impossible to distinguish this situation logically from that of a low molecular weight compound that may have a number of different and unrelated uses, but for which it has been settled patent law for decades that absolute product protection covering all possible uses is available.

Nevertheless, such use limitations have appeared in a number of national laws purporting to implement the BPD. For example, the provisions of the Italian law now include that the function of an element isolated from the human body must always be an integer of a product claim, that the origin of any biological material must be indicated in the application as filed, and that a statement on the consent to

26 Article 6(2), BPD.

27 See A.L. Caplan, in G.J. Annas and S. Elias (eds), *Gene Mapping*, Oxford University Press, 1992, ch. 7.

28 'The ethics of patenting DNA' [2002] 20 Nat. Biotechnol. 862.

obtain human biological material and the use of such biological material must be inserted in the application. Furthermore, there is to be an additional examination of biotech patents/applications by the 'National Committee for Biosecurity and Biotechnology'. The criteria that this body will apply are unclear.

Similarly in France, Article 12 of the revised Bioethics Law reads 'An element isolated from the human body or otherwise produced by means of a technical process, including the sequence or partial sequence of a gene cannot constitute a patentable invention', which is in clear contradiction to Article 5(2) of the Directive. The same article also limits the patent to the specific disclosed use of gene sequences and would not protect the sequences per se.

Germany has also adopted a use limitation on patents for genes (not for DNA sequences in general). Switzerland chose a slightly different solution and introduced in 2008 a new provision whereby gene sequences that are derived from naturally occurring sequences still obtain absolute per se protection, but one in which the protection is limited to those portions of the sequence for which a concrete function is expressly disclosed in the patent application.[29]

These limitations on gene claims are at least in part a response to the controversy about the Myriad patents for breast cancer genes. Myriad Genetics have patents claiming gene sequences (BRCA1 and BRCA2) associated with an increased risk of breast cancer, and their aggressive use of these patents to establish a monopoly in breast cancer screening in Europe caused much controversy.

For example, EP 705,902 claims 'an isolated nucleic acid which comprises a coding sequence for the BRCA1 polypeptide defined by the amino acid sequence set forth in SEQ ID NO:2, or an amino acid sequence with at least 95% identity to the amino acid sequence of SEQ ID NO:2', which gives absolute product protection to this class of DNA molecules, for any use. But abolishing per se protection for genes would not help much, because insertion of the diagnostic use in the claim would still give an effective monopoly.

Another Myriad patent, EP 699,754, claims:

> A method for diagnosing a predisposition for breast and ovarian cancer in human subject which comprises determining in a tissue sample of said subject whether there is a germline alteration in the sequence of the BRCA1 gene coding for a BRCA1 polypeptide defined by the amino acid sequence set forth in SEQ ID NO:2, or a sequence with at least 95% identity to that sequence, said alteration being indicative of a predisposition to said cancer.

This use claim gives Myriad full protection and is just as difficult to 'design around' as a gene substance patent, so that banning absolute product protection for genes would not solve the 'Myriad problem'.

A number of the Myriad patents have been limited or held invalid by the EPO, on classical patent law grounds, and one wonders if the corresponding US method claim would be upheld in view of *Bilski* (see p. 331).

[29] Article 8c of Swiss Patent Law, as amended.

Human Cell Lines

Another area of controversy is that of the patenting of human cell lines. The BPD clearly states that cells isolated from a human body may be patentable, but problems arise when the cells are human embryonic stem cells (hESCs) derived from early-stage embryos ('blastocysts'). Human ESCs were claimed in a US patent of Wisconsin Alumni Research Foundation (WARF) [30] and although, in view of the *Eli Lilly* case (see p. 288), it must be questioned whether the patent met the written description requirement, the patent was granted and upheld in re-examination. Many more US patents claim methods for the culture or selection of hESCs or their differentiation into lineage-specific cells lines, but, in Europe, there was a major controversy over one such patent granted to Edinburgh University and generally referred to simply as the 'Edinburgh Patent'. The problem arose initially because the patent covered animal cloning processes without disclaiming human cloning. But in opposition proceedings, the Opposition Division went much further than this, and held essentially that any invention (for example that of a differentiated cell line) deriving from the use of hESCs was excluded from patentability because it involved the destruction of a blastocyst and was therefore 'industrial or commercial use of a human embryo'.

This reasoning was followed by the EBA in the case relating to the European counterpart of the above WARF US patent.[31] It can perhaps be understood that the EBA felt itself compelled by the wording of Rule 28(c) and the BPD to reach this conclusion, but many people in Europe find it absurd that a procedure that is legal in the majority of European countries, which destroys only blastocysts that would in any event be discarded, and which has the potential to save human lives (real human beings, not a few cells in a dish) is regarded by the highest patent authority in Europe as 'immoral' and unpatentable—and that just at a time when the USA is relaxing its ideological opposition to stem cell research. Furthermore, the claimed cells can now be obtained from other sources that do not require destruction of blastocysts, but because everything must be looked at as of the filing or priority date and at that time no such alternative was available, the patent could not be granted. It seems questionable whether this general principle of patent law should properly be applied to EPC Article 53(a), which talks about the commercial exploitation of the patent, not about allegedly immoral actions taken before the patent was even applied for.

The same conclusion was reached by the *Bundespatentgericht* (BPatG, or Federal Patent Court) in Germany, which revoked part of a patent of Professor Brüstle directed to a process to prepare purified cells with certain properties wherein the first step was 'cultivating [human] embryonic stem cells'. The BPatG did not address the issue that human embryonic stem cells may be legally derived from

[30] USP 5,843,780 and 6,200,806.

[31] G 2/06 (OJ 2009, 306) *WISCONSIN ALUMNI RESEARCH FOUNDATION/Stem Cells*.

other sources, such as existing embryonic cell lines, and that such research is substantially funded and promoted by public institutions. The patentee appealed and the case is pending before the BGH, which referred to the ECJ two questions on the interpretation of the BPD. Courts in Sweden and the UK have, however, granted patents corresponding to the WARF European application.

Morality Issues

Ordre Public *and Morality*

It is easy to say that patents have to do with economics, not morality, and therefore that moral issues are irrelevant to patent law, but this position is difficult to maintain in view of the fact that EPC Article 53(a) specifically raises the issue of morality as a possible ground for denying the grant of a patent for what would otherwise be a patentable invention. It prohibits the grant of European patents for inventions the publication or exploitation of which would be contrary to *ordre public*, or morality, irrespective of whether or not the invention is patentable under Article 52.

The concept in Continental law of *ordre public* is one that is not found in English common law, which is perhaps why it is the only foreign-language phrase used as such in quotation marks in the English version of the EPC. It has been suggested that the difference between *ordre public* and morality is that adultery in private may or may not be considered immoral, but if you practise it in the street and frighten the horses, that is contrary to *ordre public*. But a breach of *ordre public* means more than what English law would call 'disturbance of the peace'; under German law, it would mean a violation of a basic constitutional right, such as the right to life, personal freedom, human dignity, and freedom from bodily harm. *Ordre public* means the proper order of society.

As already mentioned, the EPO guidelines give an anti-personnel mine as an example of the type of invention that should be excluded from patentability under this Article. Clearly, publication of a patent for such an Article would be undesirable for reasons of *ordre public*; equally clearly, its use would be immoral in most circumstances. It is worth noting that Article 53(a) does not itself prohibit publication of the application, but in practice the Receiving Section will refuse to publish an application falling under this category. Some 20 years ago, the British press reported that patents for chemical warfare agents, which had been applied for by the Ministry of Defence (it is hard to imagine for what purpose) and had been granted as 'secret patents',[32] had, years later, appeared on the shelves of the Patent Office library, enabling any competent chemist to make lethal nerve gases using readily available materials. The patents were subsequently removed, but under the EPC, this would have been a good case for refusing both publication and grant on *ordre public* grounds.

[32] Section 18(2), PA 1949.

In the UK, under the Patents Act 1949, an application could be refused if its use was contrary to law or morality,[33] although if both legal and illegal uses were possible, the illegal use could be disclaimed. There are some old reported cases relating to slot machines that could be used for (illegal) gambling. For example, the defendants in one infringement action[34] claimed that the patent was invalid for this reason, but their argument was weakened by the fact that they were selling such machines themselves. As Eve J said:

> if it be the duty of the Court to stop . . . the corruption of the youth of this country by letting them gamble in these instruments for pennies or sweets or bad cigars, it seems to me that that is an argument in favour of my granting the injunction rather than allowing two sets of corruptors of youth to be placing on the market these injurious instruments.

The judge was making an excellent point often forgotten by those who want to refuse patents for inventions they do not like, namely, that if there is no enforceable patent, everyone is free to exploit the invention.

On the other hand, Article 53(a) of the EPC specifically states that the exception to patentability does not necessarily apply even if exploitation of the invention is illegal in all contracting states. As the Guidelines remark,[35] it could be manufactured for export to countries in which use would be legal. Since patentability need not be denied even if the use of the invention is illegal in all member states; it should logically follow that if the intended use is legal everywhere, patentability cannot be denied.

It is when we move on to morality that we find ourselves on much shakier ground. Concepts of what is contrary to morality change with time and place, and are also variable within a single country at a single time. In the UK, in the 1930s, no patents were granted for contraceptives (although probably not so much because they were considered immoral as because they were considered embarrassing). Instead of invoking any specific section of the Act dealing with immoral inventions, patents were simply refused, as an exercise of the royal prerogative to grant or refuse patents at will. In one case,[36] contraceptives were described as 'articles which it would not be fitting to sell as being protected by Royal Letters Patent' and Sir Thomas Inskip, the Solicitor General, said: 'I decline to be any party to the grant of a patent for this class of article.'

One shudders to think of how Sir Thomas would have reacted to the discussion by an EPO Board of Appeal on the question of whether the fact that a prostitute might apply a contraceptive before having sex with a client constituted industrial applicability.[37] Prevailing standards of morality may also differ from one region of

[33] Section 10(1)(b), PA 1949.

[34] *Pressers and Moody v. Haydon* (1909) RPC 58 (Ch D).

[35] Guidelines C IV 4.2.

[36] *A & H's Application* (1927) 44 RPC 298 (SG).

[37] T 74/93 (OJ 1995, 712) *BTG/Contraceptive method.*

Europe to another; without making any value judgements, one can safely say that some behaviour may be considered immoral in Sicily, but not in Scandinavia, and vice versa.

Can Patent Office Examiners be Judges of Morality?

Patent office examiners are, in this respect, simply members of the public, with no special aptitude or training. As long as the criteria for applying EPC Article 53(a) are as stated in the Guidelines, namely, 'whether it is probable that the public in general would regard the invention as so abhorrent that the grant of patent rights would be inconceivable', then examiners, as members of the public in general, are as well fitted to judge this as anyone else. But if, as they are required to by the Board of Appeal in the OncoMouse® case and by the BPD, examiners are supposed to carry out sophisticated balancing of subjective moral values as part of the examination procedure, then they are wholly incapable of such a task. Are EPO examiners to have courses in moral philosophy or theology? And if so, of what variety?

The basic principle should apply that an invention is not immoral only because some people do not like it, but only if it is deeply offensive to the great majority of the population. Unfortunately we have allowed vocal minorities to determine what constitutes morality and the EPO is now dancing to their tune. The EPO decisions on the Edinburgh patent and the WARF case are perfect illustrations of the folly of making patent office examiners and Board of Appeal members into judges of morality, and of treating members of the European Parliament as experts on ethics.

Patents for Life: From Bacteria to the OncoMouse®

Many people believe that the US patent granted to Chakrabarty for oil-eating bacteria, the grant of which was upheld by the US Supreme Court, was the first patent to be granted for living matter. In fact, not only was the corresponding British patent granted without any fuss some time before the US Supreme Court case, but as we have seen (p. 275), Pasteur had also already been granted, in 1873, US Patent 141,072 containing a per se claim to live yeast. Mice are different, however, and many people who would set a trap if they had a mouse in their kitchen are extremely upset if a mouse is patented. This, it is claimed, 'degrades animals to the level of property'—as if there were no such thing as a cattle market or a pet shop.

Similar arguments are applied to the patenting of human genes and cell lines, and the idea is spread about that by patenting a cell or a piece of DNA derived from an individual, one somehow has patented that person. The prime example is John Moore, an individual who was suffering from a rare form of cancer and whose spleen was removed as part of the life-saving treatment that he received at the University of California hospital. His doctor succeeded in establishing an immortalized cell line from the spleen cells, which was found to overproduce certain cytokines that had not previously been isolated. The University patented the cell line, the genes, and the cytokines, and licensed the patents to a US biotech company, which sublicensed them to a Swiss pharmaceutical company. Then the doctor gave interviews to the press about his inventions and how much money they were worth. Mr Moore read the articles and then sued his doctor, the University, and both the US and the Swiss companies, alleging that the cell line was his property.

The case went through the California state courts up to the State Supreme Court, which held[38] that Moore had no property right in his cells once they were removed from his body and that he had no cause of action against the University or the companies. He did have a claim against his doctor, but only because the doctor had failed to inform him that commercial, as well as purely experimental, use might be made of his cells, and this claim was settled out of court.[39]

Moore, whose only cause for complaint was that the doctor who had saved his life had not also made him rich, then went around telling everyone how he had been patented and how terrible he felt about it.[40] If indeed the fact that a cell line derived from him had been patented were to give the patentee any control over him, you would think that someone would have made him keep quiet. The fact that he is far from quiet demonstrates the irrationality of the argument that was advanced by the Green Party of the European Parliament in its opposition to the patent on recombinant human relaxin in the EPO,[41] namely, that human life was being patented and that this amounted to slavery.

It is, indeed, hard to imagine how such absurd arguments could ever have been seriously put forward, and we must be grateful to the Opposition Division in that case for the clarity and vigour with which the arguments of the opponents were demolished:

> DNA is not 'life', but a chemical substance which carries genetic information . . . The patenting of a single human gene has nothing to do with the patenting of human life. Even if every gene in the human genome were cloned (and possibly patented), it would be impossible to reconstitute a human being from the sum of its genes.

Attempts are also made to mobilize religious opinion to fight against 'patents on life'. It is sometimes said that, by patenting living organisms, man is usurping God's role as Creator. It is also sometimes said, in jest, that God neglected to file patents for His work, but if He had done so on the date of Creation (4004 BC by Archbishop Usher's chronology), the patents would have already expired in 3984 BC. Patents do not last forever, although this also tends to be forgotten.

In 1998, Jeremy Rifkin announced that, together with a scientist named Stewart Newman, he had filed a US patent application claiming 'manimals', or chimeras between man and animal. It was not suggested that any such creatures had been

[38] *Moore v. Regents, U. California* 793 P.2d 479 (Sup Ct Cal 1990).
[39] Reportedly, for US$200,000, which is considerably better than nothing.
[40] 'Like being raped': quoted by Beth Burrows, Third World Network Features, 1996.
[41] *Howard Florey/Relaxin* [1995] EPOR 541 (EPO Opposition Division).

produced, but the 'inventors' claimed to have made an enabling written description of processes for their production. The object of this exercise was to provoke the sort of instinctive reaction given by the US Commissioner (quoted at the head of this chapter), and to use this as a basis to reopen the whole issue of patenting transgenic animals and, indeed, any living organisms. This piece of mischief turned out to last no longer than any other of Rifkin's campaigns, but it did indeed raise some interesting questions. Given that the genome of a human is 98 per cent identical with that of a chimpanzee, how many genes does it take really to make the difference?

Why are Patents Being Targeted?

This is a question that is hard to answer. If it is felt that it is immoral to make transgenic animals or human embryonic stem cell lines, then surely it would be more logical to concentrate on campaigning for legislation to stop or restrict such experiments, rather than trying to lock the door after the horse has bolted by stopping patents for work that has already been done?[42]

There are perhaps three reasons.

(a) It is easier and cheaper to oppose a patent than to mount a lobbying campaign for a piece of legislation, especially when any number of opponents are allowed to oppose for the cost of one opposition fee.[43]

(b) Patents appear to be a potent symbol of commercial exploitation and they take on a symbolic importance that bears no relationship to the limited rights actually granted by a patent. It is either not known, or ignored, that patents can only be used to stop others and do not give positive rights of exploitation. In 2002, the University of Missouri obtained a US patent[44] that was condemned by various non-governmental organizations (NGOs) because it covered, inter alia, human cloning. The response of the University was: 'Yes, we know it does, and if anyone tries to do human cloning, we will use our patent to stop it.' This seems a perfectly logical answer: invalidate the patent and anybody can do it.

(c) Finally, a patent is seen as a seal of official approval and, by attacking the patent, one indirectly attacks government policy. This shows a complete misunderstanding of the nature of patent rights.

A further point that needs to be made is that there is a disturbing trend, as shown in the WARF decision, to consider patents as invalid under EPC Article 53(a) not for anything to do with the *use* of the invention, but because of what the inventor did before filing the application. As an example, if tissue samples are taken from a

[42] For a thoughtful and balanced discussion of this topic, see G. Laurie, 'Patenting stem cells of human origin' [2004] EIPR 59.

[43] G 3/99 (OJ 2003, 347).

[44] USP 6,211,429.

human patient, prior informed consent must be obtained. Some people argue that, if this is not done, any patent on an invention made using such samples violates Article 53(a).

For many NGOs, the logic is clear: all pharmaceutical companies make money out of human suffering, therefore they are immoral, and therefore all patents that they obtain are contrary to morality and therefore invalid.

16

SOFTWARE-RELATED AND BUSINESS METHOD INVENTIONS

Dogbert: My patent for no-click shopping was granted. I'm sure some whiners will say it's an obvious idea. You'd better click something or I'll have to ship you some books.

Scott Adams, Dilbert cartoon strip (2000)

Relevance of Software and Business Method Inventions to Chemistry and Biotechnology

Discussion of software-related inventions may appear to be a digression in a book primarily concerned with the technical fields of chemistry, pharmaceuticals, and biotechnology. Nevertheless, the use of computers in research in these fields is continuously increasing. Computers and associated software are used in molecular modelling (see, for example, the *Fujitsu* case described below), rational drug design, the design and production of compound libraries by combinatorial chemistry, the carrying out and evaluation of high-throughput assays, and the sequencing of DNA fragments and their correlation with known sequences. Inventions that relate to improvements in any of these techniques may be in whole or in part software-related, and it is important to be aware of what may and may not be patented in this area. Similarly, inventions of business methods have certain similarities with inventions of diagnostic or pharmacogenomic methods, especially where these are computer-implemented. There is still a general perception that computer programs are not patentable, but although this is what the law in Europe appears to say, it is not in fact the case. In order to protect inventions of this type, it is, however, essential to use a patent attorney who is thoroughly familiar with what can be patented and how software patents should be drafted. A patent attorney who specializes in chemistry or biotechnology would be wise not to attempt it.

Patenting of Software-Related Inventions in the European Patent Office

Provisions of the European Patent Convention

Within Europe, national patent laws and the European Patent Convention (EPC) specifically prohibit the patenting of computer programs and computer programs have been seen as the equivalent of literary works, which are protectable only by copyright. The availability of copyright protection for computer programs is mandated in the European Union (EU) by the Directive on the legal protection of computer programs[1] and, for all World Trade Organization (WTO) members, by the TRIPs Agreement.[2] Patent protection, if additionally available, would clearly be superior to copyright, since it would allow broader protection for the basic concepts of the program and would protect against later independent development, whereas in order to enforce copyright protection, it is necessary to prove actual copying. Even proof of copying may not be sufficient, since the EU Directive allows copying of programs in the context of reverse engineering, so that competitors can design non-infringing compatible systems and not be locked in to the system of one supplier.

The relevant provisions of the EPC are to be found in Article 52(2), which gives a non-exhaustive list of subject matter that shall not be regarded as inventions:

(a) discoveries, scientific theories and mathematical methods;
(b) aesthetic creations;
(c) schemes, rules and methods for performing mental acts, playing games or doing business, and programs for computers;
(d) presentations of information.

One reason why it has proved difficult for courts and European Patent Office (EPO) Boards of Appeal to interpret the scope of these provisions is that they are not expressed, as those of Article 53 are, as being exceptions to patentability, so the normal rule of construction that exceptions are to be construed narrowly does not apply. There also seems to be no overriding principle binding them together, except perhaps that they relate to subject matter that is primarily abstract or non-technical, whereas a patentable invention must relate to a technical field,[3] and the claims must

[1] Directive 250/91.
[2] Article 10.1, TRIPs.
[3] Rule 42(1)(a), IR.

be defined in terms of the technical features of the invention.[4] Thus not only are computer programs stated to be unpatentable, but so are many of the things that are a part of computer programs (mathematical methods) and many of the applications to which they are put (including methods for performing rapidly calculations that a human mind could perform slowly, business systems, presentations of information such as screen displays, and word processing).

One would therefore expect that there would be no software patents in Europe. On the contrary, many thousands of such patents have been granted by the EPO. To understand why this is so, it is necessary to read Article 52(2) together with Article 52(3), which provides that the provisions of Article 52(2) exclude patentability of the subject matter referred to only to the extent to which the patent application relates to such subject matter as such. Thus computer programs *as such* are unpatentable, but inventions consisting of a computer program plus something additional may be patentable. The question of what additional subject matter must be present in order to confer patentability is not addressed in the EPC or the Implementing Regulations, and it has been the task of the Boards of Appeal to deal with this problem.

The general approach taken by the Boards of Appeal has been that patentability should depend upon there being a technical contribution to the art. Unfortunately, 'technical' is a word that is used in the Regulations without ever being defined[5] and it has proved difficult, in many cases, to determine whether an invention that incorporates a computer program, but is more than a computer program as such, may be regarded as 'technical', or falls into one or another of the other 'non-technical' exclusions of Article 52(2): for example methods for performing mental acts, business methods, or presentations of information.

Even where a technical contribution to the art can be found, there are two possible approaches to the question of patentability that were applied in early case law. One is to consider that if the invention as a whole makes such a contribution, then it is patentable, and it is immaterial whether or not the novel and inventive feature is to be found only in the subject matter excluded from patentability under Article 52(2). This is sometimes referred to as the 'whole claim approach'. The other is to take the position that if the contribution of the invention to the art, that is, what makes it novel and inventive, lies in excluded subject matter, then there is no invention at all (the so-called 'contribution approach').[6] Under the first approach, a conventional computer-controlled machine tool operated more efficiently because of a new and inventive computer program would be a patentable invention; under the other, it would not.

[4] Rule 43(1), IR.

[5] 'The word "technical" is not a solution. It is merely a restatement of the problem in different and more imprecise language': *CFPH's Application* [2006] RPC 59, *per* Mr Peter Prescott, QC, sitting as deputy judge.

[6] Newman, 'The patentability of computer-related inventions in Europe' (1997) 12 EIPR 703.

A strict application of the contribution approach would make most software-related inventions unpatentable and, as we shall see, this is not the approach generally adopted by the Boards of Appeal. But the whole claim approach, if adopted without reservation, would seem to suggest that a computer program should always be patentable if it is associated with something concrete, for example simply by being recorded on a conventional magnetic disc, which would make the exclusion of Article 52(2) essentially meaningless.

The original EPO Guidelines for Examination mandated the contribution approach, but this was changed as long ago as 1985 and further changes were made in the light of Board of Appeal decisions. The present wording of the Guidelines (as of April 2009) reads:

> a computer program may be considered as an invention within the meaning of Art. 52(1) if the program has the potential to bring about, when running on a computer, a further technical effect which goes beyond the normal physical interactions between the program and the computer. A patent may be granted on such a claim if all the requirements of the EPC are met.[7]

Somewhat balancing this apparent generosity is the fact that the EPO has become much more strict on the question of whether or not the invention solves a *technical* problem. The Guidelines (2003 version) now state:

> Features which cannot be seen to make any contribution, either independently or in combination with other features, to the technical character of an invention are not relevant for assessing inventive step. Such a situation can occur for instance if a feature only contributes to the solution of a non-technical problem, for instance a problem in a field excluded from patentability.[8]

In fact, since 2002, there has been a clear hardening of the EPO's attitude to this issue, so that cases that would once have proceeded without objection are now being refused on this basis. This is particularly a problem for inventions in relation to which the products are simply knowledge or refined data.

The Guidelines specify that, for inventions in the computer field, the disclosure of the invention should not be a program listing in a programming language, but should rather be written in normal language, so that it can be understood by skilled readers who are not programming specialists.[9] Short excerpts of programs written in programming language can, however, be used to illustrate a specific embodiment of the invention, and flow charts representing the operation of the program or routines within the program are a useful aid to understanding the invention and to meeting the sufficiency requirement.

[7] Guidelines C IV 2.3.6.
[8] Guidelines C IV 11.7.2.
[9] Guidelines C II 4.15.

Boards of Appeal Case Law

The first leading case on computer-related inventions[10] concerned a method for the digital enhancement of images, involving storing the image digitally in a computer, altering the data by means of a mathematical function, and displaying the new data as an enhanced image. The Examining Division had refused the application on the ground that the allegedly inventive features were a mathematical method and a computer program, both of which were excluded from patentability. This refusal was the consequence of applying the second of the approaches to patentability outlined above and was consistent with the original version of the Guidelines, which was still current at that time.

On appeal, the Board took the other approach, and considered that a mathematical method as such operated upon numbers and produced results in numerical form without producing any technical effect, whereas a technical process was one that produced a change in a physical entity. The critical point was that an image stored as an electrical signal was considered to be a physical entity, so that altering such an image by enhancement was to be considered a technical process and therefore patentable. This decision was consistent with the new version of the Guidelines, which had by then come into effect (although the Boards of Appeal are not bound by the Guidelines).

A similar conclusion was reached in a case in which the invention related to computer control of an X-ray apparatus in order to make the apparatus function more efficiently.[11] The invention was held not to be excluded from patentability by Article 52(2) even though the apparatus was known and the only novel feature was the computer program.

A method for displaying an error message on a computer screen was held to be patentable because it related to a technical problem and was not merely presentation of information.[12] But a method for the display of characters in a particular font was not a technical method.[13] In the latter case, the Board considered that a program that does no more than data processing is not patentable if the data do not represent operating parameters of a device, nor affect its physical or technical functioning. In a series of decisions relating to word processing, the line was generally taken that, in so far as the data to be processed had purely linguistic significance, patentability was excluded, because the process was essentially a method for performing a mental act. Thus a method of editing a document to detect and replace linguistic expressions exceeding a pre-set understandability level was excluded.[14] A method for converting printer instructions within a document to make the document compatible with a different word processing system was, however, not excluded from

[10] T 208/84 (OJ 1987, 14) *VICOM/Computer related invention.*
[11] T 26/86 (OJ 1988, 19) *KOCH & STERZEL/X-ray apparatus.*
[12] T 115/85 (OJ 1990, 30) *IBM/Computer related invention.*
[13] T 158/88 (OJ 1991, 566) *SIEMENS/Character form.*
[14] T 38/86 (OJ 1990, 384) *IBM/Text processing (understandability).*

patentability because the data in question were technical in nature, being significant to the computer, rather than linguistically significant to the human reader.[15]

The word-processing cases in which patentability was denied on the basis of the 'mental act' exclusion appear to have adopted the first of the two approaches outlined above: that is, instead of looking at the invention as a whole, the Boards have focused on the novel feature and denied patentability where this feature can be characterized as a method for performing a mental act, even where conventional technical elements are also present. This may be considered as an unduly restrictive approach. A less restrictive approach, however, would lead to the conclusion that, because electrical signals may be regarded as technical entities[16] and because a computer program always manipulates such signals, a computer programmed with a new program must always be patentable if there is an inventive step. This conclusion had been specifically rejected by the Boards of Appeal on numerous occasions.

The situation changed, however, with a decision of the Technical Board of Appeal (TBA)[17] that a computer program was only to be regarded as a computer program 'as such' when it lacked technical character. If there was a predetermined technical effect (other than that of how electrons moved within the computer), the program could be claimed as 'a computer program product' either 'directly loadable into the internal memory of a digital computer', or 'stored on a computer usable medium', or the program could even be claimed by itself, consistent with the confirmed patentability of a claim to 'a signal' (see above). At the same time, the Board rejected the 'contribution approach', stating that the 'further technical effect' could be something already known in the prior art. The end result is similar to that in the USA, where such claims are referred to as *Beauregard* claims, because they were allowed in that case.[18] Unlike the US Patent and Trademark Office (USPTO), however, the EPO will allow claims to a program per se, without the need to recite the presence of a data carrier.

This approach was taken further by a decision holding that a claim to a programmed computer is not claiming a computer program as such.[19] This and other decisions relating to business methods (see below) are inconsistent with *Vicom*, with UK case law, and, to some extent, with each other. As a result, the president of the EPO, at the instigation of the English Court of Appeal, referred a number of questions to the Enlarged Board of Appeal (EBA) in October 2008.[20] Concerns have, however, been expressed that the EBA may consider the referral inadmissible, because it does not clearly point out conflicting decisions of the Boards.

[15] T 110/90 (OJ 1994, 557) *IBM/Editable document form*.
[16] As, e.g., in T 163/85 (OJ 1990, 379) *BBC/Colour television signal* (patentable).
[17] T 1173/97 (OJ 1999 609) *IBM/Computer program product*.
[18] *In re Beauregard* 35 USPQ 2d 1383 (Fed Cir 1995)—see p. 330.
[19] T 0424/03 (unpublished) *MICROSOFT/Data Transfer*.
[20] As of writing, pending as G 3/08.

In the EPC revision conference of November 2000, it was proposed that the exclusion of 'programs for computers' should be removed, but because the Commission was then conducting a public consultation exercise on computer-implemented inventions, it was thought that deletion of the exclusion would pre-judge the issue and this proposal was not adopted. The wording of EPC 2000 thus remains unchanged.

Patenting of Software-Related Inventions in the UK

Wording practically identical to EPC Article 52(2) is to be found in the UK Patents Act 1977.[21]

Case law in the UK

A leading case in the UK is *Fujitsu's Application*, in which the invention was a computer-based method of modelling crystal structure, in particular, of combining two known structures so as to display the structure of a hypothetical combined product. The application was refused by the Patent Office, and the refusal was appealed unsuccessfully to the Patents Court[22] and the Court of Appeal.[23] Although each of these courts took a different approach, the conclusion was the same, and the result was that the practice in the UK became significantly more restrictive than in the EPO.

In the Patents Court, Laddie J started by agreeing with the EPO that a program recorded on a disc or stored in a computer was not patentable. This finding was in line with *Gale's Application*,[24] in which a ROM chip carrying an algorithm enabling a computer to calculate square roots more rapidly was held to be a computer program as such. A computer under the control of a program could, however, be patentable, but it depended on what it was that the computer was doing. If all that was being done, as a matter of substance, was something excluded by section 1(2) of the 1977 Act, then it was also unpatentable. In this case, the activity was essentially the performance of a mental act.

The Court of Appeal regarded the method as an exact computerized equivalent of the known process in which plastic models of the crystal structures were constructed and compared. The computerized process was faster and more accurate, but these were advantages inherent in computerization and did not amount to a technical contribution.

Although the Court used the same terminology as the Boards of Appeal in discussing technical contribution, it applied the term in a much more restrictive manner. It appears that the real objection to Fujitsu's application was that the courts

21 Section 1(2), PA 1977.

22 *Fujitsu's Application* [1996] RPC 511 (Pat Ct).

23 *Fujitsu's Application* [1997] RPC 610 (CA).

24 *Gale's Application* [1991] RPC 191.

considered it to be obvious, but in this field obviousness was equated with lack of technical contribution, leading to an exclusion under section 1(2) rather than to a determination of lack of inventive step under section 3. The difficulty may be that the term 'contribution' implies the question 'contribution to what?', to which the answer can only be 'to the state of the art'. But any evaluation of what constitutes a contribution to the state of the art is properly to be considered as a determination of the presence or absence of an inventive step, and not as something relevant to section 1(2). Perhaps it would be better if the more neutral term 'technical effect' were to be used.

The next major decision in the UK was that of the Court of Appeal in two cases heard together in 2006 because of the similarity of the inventions, although the parties were different.[25] The patent under which Aerotel was suing Telco related to a telephone calling system in which calls are routed through a special exchange in which the caller has established credit to pay for the call. A claim to the system was considered to be a new physical combination of hardware not excluded by EPC Article 52(2) and the associated method claim was a narrower claim to the use of the system, also not excluded.

Macrossan's application, the refusal of which by the Patent Office was on appeal, was for an automated method of acquiring the documents necessary to incorporate a company, which the Court of Appeal held to be both a computer program as such and also a method of doing business as such. In *Aerotel/Macrossan*, the Court, following its own precedent in *Merrill Lynch* (see below), adopted a four-point structured enquiry:

(a) properly construe the claim;
(b) identify the actual contribution;
(c) ask whether it falls wholly within the excluded subject matter; and
(d) check whether the actual or alleged contribution is actually technical in nature.

The UK Patent Office interpreted this decision to mean that claims to computer programs were always excluded from patentability, although claims to a computerized method and to a programmed computer were potentially patentable. A number of corresponding refusals of such claims were appealed together[26] and, in January 2008, Mr Justice Kitchin ruled that this was not what *Aerotel/Macrossan* had said; indeed, the patentability of computer program product claims was not an issue in that case. Application of the four steps of *Aerotel* would determine whether a computer program made a technical contribution and thus whether it could be patentable; in what form it was claimed was immaterial. The cases were remitted to the

[25] *Aerotel v. Telco and ors/Macrossan's Application* [2007] RPC 7.
[26] *Astron Clinica and ors v. Comptroller* [2008] EWHC 85 (Pat).

Patent Office for reconsideration. The Patent Office decided not to appeal to the Court of Appeal and changed its Practice Notice in accordance with this decision.

Later, in 2008, the *Symbian* case[27] was heard by the Court of Appeal. The application, which had been refused by the Patent Office, related to the patentability of a purely software-implemented invention having no clear effect outside the internal workings of the hardware on which it was installed. Application of the *Aerotel* steps gave the result:

Stage 2 Identify the contribution: A program which makes a computer operate on other programs faster than prior art operating programs enabled it to do by virtue of the claimed features.

Stage 3 Is that solely excluded matter? No, because it has the knock-on effect of the computer working better as a matter of practical reality.

Stage 4 Is it technical? Yes, on any view as to the meaning of the word 'technical'.

This decision was seen as a further step in the direction of general patentability of computer programs.

Patenting of Software-Related Inventions in the USA

Forty years ago, a presidential commission in the USA concluded that computer programs should not be patentable for the pragmatic reasons that the USPTO had neither the documentation nor the qualified examiners to carry out searches and examination in this field, and that the examination would be too time-consuming and expensive; this remained the practice of the USPTO for at least 15 years. On the same basis, the French Patent Law of 1968 specifically excluded computer programs from patent protection and it is thought that this is the basis for the existing exclusion in the EPC. The USA did not, however, make the mistake of enshrining this practice in its patent law, which has left it capable of adapting to developing technology, whereas Europe has carved the principle in stone in an international treaty that has proved extremely difficult to change.

The initial approach was that a computer program was essentially a 'mathematical algorithm', which was outside the statutory definition of patentable inventions. In 1972, a computer program to convert decimal into binary numbers was held unpatentable by the Supreme Court.[28] But in 1981, the Supreme Court held[29] that although 'laws of nature, natural phenomena, and abstract ideas' were excluded, a process for curing rubber under the control of a claimed algorithm was patentable subject matter. During the 1980s, US practice became that a computer-based process was patentable if a physical process was controlled or if a computer was reconfigured to operate in a new way, or if operations were performed on data representing

[27] *Symbian Ltd v. Comptroller* [2008] EWCA Civ 1066.
[28] *Gottschalk v. Benson* 409 US 63 (Sup Ct 1972).
[29] *Diamond v. Diehr* 450 US 175 (Sup Ct 1981).

physical objects or processes. Thus it was held that a mathematical algorithm enabling a computer to produce a smooth waveform display on a monitor constituted a practical application of an abstract idea and could thus constitute a patentable invention. Whether the computer was claimed in terms of 'means plus function' or whether the process was claimed was immaterial.[30]

Since then, US practice has developed further and now allows claims to 'an article of manufacture comprising a computer usable medium with computer-readable program code means embodied therein',[31] in other words, to a computer program recorded on any conventional medium.

Following this and other court decisions upholding the patentability of software, the USPTO issued new guidelines for examination of computer-related inventions in early 1996. These stated that software that demonstrably controls or configures some computer hardware is patentable, regardless of whether it includes mathematical algorithms. The claims must specifically indicate how the software controls the computer and an enabling disclosure must be given. Patent protection is available for databases in combination with some form of computer-readable memory and, as held in *Beauregard*, software recorded on, for example, a CD-ROM or floppy disc can be claimed as an article of manufacture. This has the major advantage that a person selling the program in this form will be a direct infringer, whereas a conventional method claim is infringed directly only by the end user and the supplier is at most a contributory infringer.

Although a computer program is not regarded as excluded from patentability, it has been held by the Federal Circuit that a signal encoded in a particular manner is not statutory subject matter, because it is transitory in nature.[32]

Most of the recent case law on computer-related inventions in the USA relates to inventions of business methods that are normally, but not exclusively, implemented by computer. This has culminated in the case of *Bilski*, discussed below, which was decided *en banc* by the Federal Circuit in October 2008 and as of writing is pending before the Supreme Court. This may have the effect of limiting the type of processes that are considered patentable, with implications going beyond business methods and software processes, and including, for example, diagnostic processes.

Business Method Patents

In the USA

Up to 1996, the USPTO officially considered business methods to be unpatentable subject matter, although some individual patents to business methods were granted.[33] In 1998, the Court of Appeals for the Federal Circuit (CAFC) held in the

[30] *In re Alappat* 31 USPQ 2d 1545 (Fed Cir 1994).

[31] *In re Beauregard* 35 USPQ 2d 1383 (Fed Cir 1995).

[32] *In re Nuijten* 500 F.3d 1346 (Fed Cir 2007).

[33] For example, USP 4,346,442, granted to Merrill Lynch in 1982.

landmark *State Street Bank* decision[34] that a data processing system for managing mutual fund investments produced a 'useful, concrete, and tangible result' and was therefore patentable. There was no such thing as a separate 'business method exception'. This decision led to a huge increase in the number of applications for computerized business methods, particularly in the area of business conducted over the Internet ('e-commerce'), fuelled by the boom in 'dot.com' companies at that time. Concerns expressed about the misuse of such patents led to the enactment of a form of prior-user defence against patent infringement in the USA, limited only to business method patents.[35]

Indeed, for some time prior to *State Street*, the USPTO had been ill-equipped to deal with the rapidly increasing number of patent applications in the software field and this problem was exacerbated by the new flood of business method patents. Examiners were inexperienced in this area and there was no proper collection of non-patent literature upon which searches could be based. The result, as in the early days of biotech patenting, was that many software patents were issued that had over-broad claims and were of questionable validity.

After the bursting of the dot.com bubble in 2000, patents on Internet technologies were thought to be of little importance, but interest revived two or three years later, and many such e-commerce patents were litigated, sometimes with spectacular results. The most spectacular was probably the jury award of damages of more than US$520 million against Microsoft in August 2003 for infringement of a patent owned by the University of California and a small company called Eolas, relating to a program allowing a web browser to access interactive applications.[36] The litigation was finally settled on undisclosed terms.

Not all business methods were considered patentable after *State Street*, however. In *Comiskey*,[37] the application claimed a method for mandatory arbitration resolution of documents such as wills and contracts. The examiner and the Board had refused the claims on the basis of prior art, but the Federal Circuit held that it was unnecessary to consider prior art issues since the claims failed the preliminary test of patentability under §101.

In the landmark case of *in re Bilski*,[38] the invention related to a method of hedging risk in commodities trading, which could be, but did not have to be, implemented on a computer. The case came to the CAFC after rejection by the examiner and an unsuccessful appeal to the Board of Patent Appeals and Interferences (BPAI). After a panel hearing, but before any panel decision, the Court decided to rehear the case *en banc* and, in October 2008, reached the conclusion, based on Supreme Court

[34] *State Street Bank & Trust Co. v. Signature Financial Group Inc.* (Fed Cir 1998).

[35] 35 USC 273.

[36] *Eolas Technologies Inc & the Regents of the University of California v. Microsoft Corp.* (ND Ill 2003).

[37] *In re Comiskey* 499 F.3d 1365 (Fed Cir 2007).

[38] *In re Bilski* 545 F.3d (Fed Cir 2008).

precedents such as *Gottschalk v. Benson*[39] and *Diamond v. Diehr*,[40] that a process was patentable subject matter only if it:

(a) is tied to a particular machine or apparatus; or
(b) transforms a particular article into a different state or thing.

This became known as the 'machine or transformation' test. The 'useful, concrete, and tangible result' test of *State Street* was disapproved, although *State Street* was not specifically overruled.

The majority decision was strongly criticized in a dissenting opinion by Judge Newman, who considered that the claims should be patentable and rightly pointed out that *Gottschalk v. Benson* did not in fact say what the majority claimed it did. Many commentaries have seen the decision as potentially invalidating many valuable patents such as Google's page-ranking patent and as ruling out patent protection for broad areas of innovation. But Judge Rader thought that the claims were simply to an abstract idea and that no new test was necessary, while Judge Mayer went further and thought that all business methods should be unpatentable. In June 2009, the Supreme Court granted certiorari.

Meanwhile, the BPAI has applied *Bilski* to a number of software applications, finding lack of patentability under §101 in eight out of nine of cases. In some, the same analysis was applied to claims to software on a physical carrier medium, indicating that *Beauregard* claims may no longer be patentable. One case that went as far as the CAFC was *in re Ferguson*,[41] claiming a method of marketing a product with a shared marketing force, a purely abstract marketing strategy not tied to any machine or transforming any product. Other claims were to 'paradigms': for example 'A paradigm for marketing software, comprising [. . .]'. As might have been expected, the Federal Circuit is not yet prepared to patent paradigms. Even Judge Newman concurred in this decision, albeit on the ground of obviousness, rather than §101.

Method claims are also under attack in the diagnostic and biotech fields. It has been pointed out, for example, that the method claims of Myriad's patents, such as involving comparison of a patient's genome with the sequence of the BRCA1 gene, would likely fail the *Bilski* test. In one case of this type, the Federal Circuit considered claims to a method of evaluating immunization schedules, consisting in nothing more than:

(a) immunizing mammals; and
(b) evaluating the effect of the immunization.[42]

39 409 US 63 (Sup Ct 1972).
40 450 US 175 (Sup Ct 1981).
41 *In re Ferguson* Fed Cir 2007-1323 (2009).
42 *Classen Immunotherapies v. Biogen IDEC* (Fed Cir 2008).

This was held not to meet the 'machine or transformation' test of *Bilski*, although it would seem clear that step (a) transforms a non-immunized mammal to an immunized one, and the claimed method is simply blatantly obvious.

In Europe

The practice in the EPO was that, while abstract business methods were rejected under Article 52(2) and (3) as methods of doing business 'as such', and would not be examined further, computer-implemented business methods were regarded as not excluded from patentability, and would be examined for novelty and inventive step. There has since been a significant hardening of attitude and now patent applications to such methods will often not even be searched by the EPO. Even where a search has been performed, such cases will often now be refused on the basis that there is no solution to a technical problem. The Boards of Appeal, however, were generally liberal in their interpretation of the 'methods of doing business' exclusion of Article 52(2). This can be most clearly seen in a case involving computerized inventory and financial control,[43] which from the title alone clearly appeared to be a method of doing business. Nevertheless the Board held that the implementation of the method by the computer involved technical considerations and was not purely conventional, so that patentability was not excluded. Similarly, a computer-controlled method for controlling queuing at service counters, for example in a bank,[44] was patentable when expressed as a claim to apparatus including a number-allocating unit, service-point terminals, and an information display unit. The fact that it could be used in a business such as a bank did not automatically make it a method for doing business.

In the *Pension Benefit Systems* case,[45] the TBA held that the method claims essentially defined steps of processing actuarial information and did not go beyond a method of doing business as such. The product claims defined a concrete apparatus, however, and this was not excluded from patentability even if the use was for a business method; the 'contribution approach' was specifically rejected. Having decided that the programmed computer was not excluded from patentability, the Board went on to consider the question of inventive step, but from a rather surprising basis. Inventive step was assessed from the point of view of a software developer or application programmer, having knowledge of 'the concept and structure of the improved pension benefits system and of the underlying schemes of information processing as set out for example in the present method claims', that is, assuming that the 'new' method was already known. Unsurprisingly, the Board then found the claimed apparatus obvious.

The *Pension Benefit Systems* decision has been criticized on a number of grounds. The UK Patent Office considered that, by holding method claims excluded, but

[43] T 769/92 (unpublished) *SOHEI/General purpose management system.*

[44] T 1002/92 (OJ, 1995 605) *PETERSON/Queuing system.*

[45] T 931/95 (OJ 2001, 441) *PBS PARTNERSHIP/Controlling pension benefits system.*

product claims not, the Board was 'exalting style over substance'.[46] It also pointed out that the Board's dismissal of the 'contribution approach' conflicts with the views of the English Court of Appeal.[47] The reliance of the Board on the term 'technical' in spite of its inability to define the term has been sharply attacked by Greg Aharonian,[48] who considers that 'technical' means 'applied science', and economics is a science, so business methods are technical, QED.

Also of significance is a Board of Appeal decision[49] that held that an invention consisting of a mixture of technical and non-technical features and having technical character as a whole is to be assessed with respect to inventive step by taking account of all those features that contribute to the technical character, but not of features making no such contribution More recently, the Examining Division of the EPO has refused Amazon's famous (or infamous) 'one-click' patent application,[50] although the corresponding US patent[51] was granted and survived re-examination with a minor reduction in scope. Another European patent of Amazon[52] claiming a gift ordering system was revoked in opposition proceedings.

Within Europe, the EU Commission undertook an extensive consultation process on the patentability of computer-implemented inventions, including business methods, and eventually produced a draft Software Patenting Directive. This was intended to harmonize European law on the subject in line with the then current approach of the EPO, but without allowing the claiming of programs per se or on data carriers. In line with the results of the consultation, business methods would remain unpatentable, and mere computerization of known techniques and processes would not be seen as involving an inventive step.

Subsequently, following extensive debate, the draft Directive was amended to permit the claiming of programs per se and on data carriers, in line with the EPO approach. The draft Directive was the subject of much heated but ill-informed press and Internet comment, to the effect that it was a proposal to make software-related inventions patentable for the first time in Europe, rather than merely confirming the long-established practice of the EPO and national patent offices. The parallel with the Biotech Patenting Directive is very noticeable. Unfortunately, the opponents' lobbying of the European Parliament was extremely effective. The European Parliament passed the Directive in September 2003 only after 46 amendments, often mutually contradictory, had been made to the text. The Directive was never very clear to begin with and the amended version was practically incomprehensible.

46 *Hutchin's Application* [2002] RPC 8 (Pat Off).
47 *Merrill Lynch* [1989] RPC 561 (CA).
48 [2003] CIPA 555.
49 T 641/00 (OJ 2003, 352) *COMVIC/Two identities.*
50 EPA 0113935.9.
51 USP 5,960,411.
52 EP0 927 945.

The Commission disagreed with the amended form and, as of writing, no final text has yet been agreed upon.

Summary

There is considerable controversy even within the USA as to whether business method patents are granted too readily and enforced by the courts too uncritically, and, generally whether or not they are good for the economy.[53] On the one hand, many people, ranging from Greg Aharonian to Judge Newman, consider that patenting in this area is necessary to promote innovation in the 'knowledge economy' and that to forbid such patents is to leave the patent system mired in the 20th century, if not the 19th century. Others, including the majority of the CAFC, clearly feel that such patents do more harm than good, should be stopped if possible, and that the use of §101 is as good a way as any of doing so. The latter view is also shared by the Canadian Patent Appeals Board, which, in May 2009, rejected an appeal by Amazon on its 'one-click' patent application, holding that to patent business methods would be such a radical departure from the traditional patent regime that 'clear and unequivocal legislation is required for business methods to be patentable'.

Certainly these are just the sort of patents that are used or misused by patent trolls to make money without contributing anything to useful knowledge, or are misued in the pharmaceutical field by companies such as Myriad and Metabolite. It will be interesting to see what the Supreme Court decides in *Bilski*, but in view of the lack of technical expertise of the Court, perhaps we should not expect too much. *Bilski* may be taking an over-rigorous approach to denial of patentability and it may be that stronger application of the obviousness test would produce more sensible results.

[53] See, e.g., R.M. Hunt, 'You can patent that?' (2001) 1 Business Review 5.

PART IV

PATENTING IN PRACTICE

about patent matters. They may be employed by companies or be in private practice, and they may, in different countries, have different types of qualification. Basically, however, their job is the same and the most important part of the job is to help inventors to obtain valid patents for their inventions. In order to do this, patent practitioners must have both scientific and legal training, so that they can understand both the technical background of the invention, and the patent law and practice of at least their own country.

The British Patent Profession

In the UK, the patent practitioner used to be known as a 'patent agent', a professional title that sometimes led people to believe that, by analogy with an estate agent, he or she dealt only with the buying and selling of patents. Perhaps for this reason, the professional title of 'patent attorney' is now more common. As we have seen in Chapter 2, the profession in the UK is regulated by an autonomous professional body, the Chartered Institute of Patent Attorneys (CIPA, formerly the Chartered Institute of Patent Agents). In the remainder of this chapter, except where the context otherwise requires (for example in discussing the US patent profession), the terms 'patent agent' and 'patent attorney' are used interchangeably. The Institute conducts qualifying examinations, controls matters of professional ethics, and keeps a register of qualified patent agents. The Patents Act 1977 provided that no one who was not on the register could act as a patent agent for gain (a non-qualified person could act as agent for someone else as long as he did not get paid for it),[1] but this prohibition was removed by the Copyright, Designs and Patents Act 1988.[2] Now anyone, qualified or not, may for payment act for another in the filing and prosecution of patent applications; the only restriction is that an unqualified person is not entitled to use the title of 'patent agent' or 'patent attorney'.[3] This deregulation of the profession has in practice changed very little, because most clients are sensible enough not to entrust their inventions to unqualified practitioners. Of more practical importance is that it is now possible to have mixed partnerships of solicitors and patent attorneys, and that patent attorneys can no longer be prohibited from advertising their services. The great majority of registered patent attorneys are also Fellows of CIPA, and may designate themselves FCIPA or CPA (Chartered Patent Agent/Attorney).

The British patent profession is a unitary one, with the same professional titles, the same professional ethics, and the same Chartered Institute for all patent attorneys. In this, it differs from the profession in the USA, where there is a sharp distinction between lawyers and non-lawyers, and that in Germany, where private

[1] Section 115, PA 1977.

[2] Schedule 8, CDPA 1988.

[3] Sections 274–276, CDPA 1988.

practice and industrial practice are kept strictly apart. Most British patent professionals feel that they are fortunate in this respect.

Patent Attorneys as Members of the Legal Profession

As well as being technically qualified, patent attorneys have legal training at least within the specific field of industrial property. Patent attorneys occupy a somewhat anomalous niche in the structure of the legal profession. They may not draw up documents such as deeds, which are the prerogative of solicitors. Nevertheless, although normally a party in High Court proceedings must be represented by a barrister instructed by a solicitor, in proceedings before the Patents County Court (PCC), or proceedings before the Patents Court arising on appeals from the Patent Office, patent attorneys may either instruct a barrister directly or may themselves appear on behalf of their clients. Communications between patent attorney and client on patent matters are privileged documents protected against disclosure in legal proceedings in the same way as solicitor–client communications.

In recent years, FCIPAs have been able to qualify for litigator certificates entitling them to conduct litigation before the Patents Court or PCC, and to conduct appeals from these to the Court of Appeals in respect of any intellectual property litigation. To qualify, a patent attorney must have been on the register for at least three years, have agreed in writing to be bound by special rules of professional conduct, have successfully completed a litigation course approved by CIPA, and have had at least six months' experience under the supervision of a person already entitled to conduct litigation.

Under the provisions of the Legal Services Act 2007, professional bodies such as CIPA are to be restricted to representing the interests of their members, while regulatory and disciplinary functions are to be exercised by newly constituted and independent bodies. This will certainly increase the costs of running a patent attorney practice, while the benefit to the client or 'consumer' is questionable.

The Training of a Patent Attorney

At one time, it was the rule for new entrants to the patent profession in the UK to be accepted as articled pupils to chartered patent agents and to work for little or no salary in exchange for their professional training. Today, this method of entry has died out and most new entrants are science or engineering graduates who work as technical assistants to a chartered patent attorney for a reasonable salary. In industrial practice, it often happens that scientists enter the patent profession after some years in research; their additional scientific experience is helpful in dealing with complex technical fields and in understanding the problems of the inventor. Increasingly, new entrants to the profession may have some formal academic qualification in intellectual property law, such as is offered by Queen Mary Intellectual Property Research Institute in the University of London, or similar courses now offered at Manchester University, Brunel University, Bournemouth

University, and others. This may take the form of a Master of Laws (LLM) degree, a one-year Master of Science (MSc) course in the management of intellectual property, or a three-month full-time course leading to the Certificate in Intellectual Property Law.

Whether in private practice or in industry, the new entrant is an assistant to a particular qualified patent attorney: a partner in the firm, perhaps, or a more senior employee in the company patent department, who has a personal responsibility for the newcomer's professional training. It is not possible to qualify by studying the subject from the outside; on-the-job training is more important than theoretical knowledge. One difficulty with this system is that the amount of on-the-job training that a new entrant receives is up to the firm or company for which he works. Some large industrial patent departments have excellent training programmes; in some small firms, new entrants may be left without real guidance to pick up what they can. In private practice, trainees may have the advantage of more varied experience than is to be found in industry.

The CIPA holds qualifying examinations each year; since 1991, these have been organized jointly with the Institute of Trademark Agents (ITMA). Both those who want to qualify as patent attorneys and those who want to qualify as trademark attorneys take the same set of four 'common foundation papers' in: UK design and copyright law; basic UK trademark law; overseas trademark law and practice; and basic principles of English law. In addition, aspiring patent attorneys need to pass either two foundation papers in UK and overseas patent law and practice, or alternatively a trademark practice foundation paper. Subsequently four advanced-level papers in patent practice must be passed, testing skills in: drafting applications (P3); amending patents (P4); giving opinions on patent infringement and validity (P6); and patent agents' practice (P2). Those who wish to be trademark agents must alternatively or in addition take three advanced-level trademark papers.

The examination is now fully modular, so that basically each paper may be passed separately and in any order (subject to the requirement that the foundation-level patent paper for UK patents or trademarks must be passed before any of the advanced-level patent or trademark papers, respectively, may be attempted); once passed, it need not be taken a second time. Those who have already qualified as European patent attorneys are exempt from papers P3 and P4, and those who have obtained qualifications in intellectual property law from any of the four universities mentioned above are exempted from all of the foundation-level papers. The foundation papers are normally taken after the candidate has been in the profession for about one year and the advanced papers normally after three years, but there are no fixed time limits.

The qualified British patent attorney, therefore, has had at least three years of on-the-job training, and has passed a difficult examination testing professional skills and knowledge of the patent law not only of the UK, but also of other countries. Some patent attorneys may have a degree in law, but all must have an academic qualification in a scientific or technical subject. Nevertheless, the training of British patent attorneys has come under severe criticism, particularly from the Sherr Report of 2002, in which Professor Avrom Sherr concluded: 'The system of education, training and qualification is seriously flawed and undermined by basic weaknesses.'[4] His suggestions were to separate academic patent law training (in his opinion, best dealt with by academic institutions) from practical training (being best assessed by means other than a written examination). At the time of writing, however, no significant changes to the system have been made or even proposed as a result of this report.

The Future of the Profession in the UK

At present, there are approximately 1,750 patent attorneys on the register in the UK. Twenty years ago, the numbers were divided fairly evenly between those in private practice and those employed by industrial companies and government departments, but this is no longer the case: already, in 1997, a survey showed that only a quarter of the total were employed in industry;[5] by 2004, the figure was down to less than 20 per cent; today, it is even smaller. The author of the 1997 survey expressed concern that the membership of the British profession was ageing and that there were not enough young people entering the profession. This is partly attributable to the decline in industrial practice, which has always trained proportionally more new entrants than has private practice. It is, however, part of the larger trend in recent years to have more work done by fewer people; as a result, it is very difficult to hire anyone who not only will not produce much useful work for a year or two, but also, during this period, take up the time of the experienced patent agents who will have to train the newcomer. Thus all employers wish to hire someone who is already at least partly trained, but few are prepared to do the training.

Approximately 18 per cent of the patent attorneys on the register are female and this number is steadily increasing. Many women find that the career is not only interesting and challenging, but also easier to combine with raising a family than, for example, a job in research.

The Patent Profession in the USA

There are two distinct classes of US patent practitioner, known as patent attorneys and patent agents. Both have scientific or technical qualifications, and have passed an examination in US patent law and US Patent and Trademark Office (USPTO) practice, although former USPTO examiners with at least four years' service are

[4] 'Where science meets law' [2002] CIPA 611.
[5] P.G. Mole [1997] CIPA 571.

exempt from this requirement. This examination, known as the 'USPTO registration examination' or the 'Patent Bar examination', is a multiple-choice examination set by the USPTO and based upon the current version of the Manual of Patent Examining Procedure (MPEP). This was previously a paper-and-pencil examination taken at the USPTO and a few other locations once or twice a year, and although this is still possible at the USPTO once a year, most candidates now take a computer-based test, which is administered on behalf of the USPTO by the company Prometric at hundreds of locations in the USA. One hundred questions have to be answered within six hours and 70 per cent must be answered correctly in order to pass. Although candidates have access to an electronic version of the MPEP, there is no time to look up all of the answers, and a thorough knowledge of the MPEP is a prerequisite for success. The pass rate in 2006 was 58 per cent.

The patent agent has no further legal qualification and so is entitled to represent a client in proceedings before the USPTO, but not before any court. A patent attorney, on the other hand, is an attorney-at-law who has a law degree and has been admitted to the Bar of the highest court in any jurisdiction in the USA (for example in any state). Subject to the rules of court and state laws regulating legal practice, the patent attorney can represent clients in the courts, as well as carry out tasks reserved by state law to qualified attorneys, such as giving legal opinions and drawing up contracts. Some patent attorneys are litigators who specialize in pleading cases before the courts, as do barristers in England; the majority are patent solicitors, in the sense that they solicit patents from the USPTO.

In the UK and in most European countries, it is exceptional for a person to have a law degree as well as a degree in a scientific subject, because law may be studied as a first degree course and is therefore usually seen as an alternative to a science degree. In the USA, however, law is always a postgraduate course, so it is quite common for a student to do an initial four-year degree course majoring in a scientific subject and then to go on to study law, and this is the course of study that has been followed by the majority of US patent attorneys.

Although there are professional associations, such as the American Intellectual Property Law Association, which draw up codes of professional ethics for patent agents and patent attorneys, it is the USPTO that exercises disciplinary control over the profession and can disbar agents or attorneys from the Patent Office Bar for deliberate fraud on the USPTO, or for breaches of legal ethics in matters such as advertising and soliciting clients. Patent attorneys are also subject to disciplinary control by the Bars and courts of the states of which Bars they are members.

The wider range of activities open to a patent attorney restricts the patent agent in the USA to a secondary status in the profession, so that a law degree is a prerequisite to professional advancement. The US patent attorney may therefore be expected to have a broader legal knowledge than a British patent agent, but will not have been so rigorously tested in the skills of patent drafting and interpretation, and will not normally know so much about non-domestic patent law.

The Patent Profession in Other Countries

In Germany, the profession is also split into two categories, but here the distinction is rather between the *Patentanwalt* (in private practice) and the *Patentassessor* (employed in industry). Members of both groups have precisely the same qualifications and need not have a law degree, although they must have a university degree in law, science, or engineering. The training involves spending time at the German Patent Office and as a clerk in one of the courts that deal with patent matters, and the qualifying examination, which is controlled by the Patent Office, is essentially a theoretical examination in patent law. Two days of written papers are followed by a grueling oral examination before a nine-member examination board consisting of three Patent Office officials, three patent judges, two *Patentanwälte*, and one *Patentassessor*. The result is essentially a legal qualification that does not test the skills of drafting and interpretation of patents.

Both *Patentanwälte* and *Patentassessoren* may represent clients before the Patent Office, the *Bundespatentgericht* (BPatG, or Federal Patent Court) and on appeals from these to the *Bundesgerichthof* (BGH, or Federal Supreme Court), but must be accompanied by a *Rechtsanwalt* (a general lawyer) in patent proceedings, such as infringement actions, which originate in the *Landgerichte* (district courts). The main distinction is that whereas the *Patentanwalt* may represent any client, the *Patentassessor* may only represent his employer, which may, however, include the employer's associated companies in Germany or abroad. The two sides of the profession have separate professional institutes, so that a *Patentanwalt* who accepts employment in industry must resign from the *Patentanwaltskammer* and call himself a *Patentassessor*.

In Switzerland, there was until recently no regulation of the profession, but at the time of writing, a new law on patent attorneys has been passed by the Swiss Parliament and is expected to enter into force shortly. This law provides for a qualifying examination for patent attorneys, recognition of corresponding foreign qualifications, and a training period of three to four years before a candidate can sit the examination. Switzerland offers an excellent academic diploma course in intellectual property given by the *Eidgenössische Technische Hochschule* (ETH, or Federal Institute of Technology) in Zurich. Switzerland also distinguishes between private and industrial practice, as, indeed, do most of the European professions. Only the Netherlands resembles the UK in having a patent profession that is both unitary and well qualified.

In Japan, there is a strict legal and technical examination for qualification as a patent attorney, but there is no requirement that candidates have any prior experience whatsoever. As a result, large numbers of new university graduates take the examination and the pass rate is extremely low: typically less than 3 per cent. It is recognized, however, that the newly qualified patent attorney will require practical experience before being is in a position to practise independently. Since 2002, a

Japanese patent attorney (*benrishi*) has been able to appear before the court in intellectual property cases without the involvement of an attorney at law (*bengoshi*), if the court approves. Approval requires that the patent attorney has taken training courses in civil procedure and passed an examination, similar to the requirements for a litigator's certificate in the UK.

European Patent Attorneys

The Requirements of the European Patent Convention

An applicant for a European patent who is not a natural or legal person resident or based in a contracting state of the European Patent Convention (EPC) must appoint a professional representative,[6] who must be on the list of professional representatives maintained by the European Patent Officer (EPO).[7] Applicant companies based in an EPC country may be represented by an employee who need not be on the list, but this is interpreted very strictly. Thus if, as often happens, in a group of companies the patent rights are held in the name of a holding company that has no employees, an employee of a different company in the same group cannot act as representative if he is not on the list. It is hard to see what good purpose is served by this legal quibble and, indeed, the EPC provides for the possibility of dealing with this situation in the Implementing Regulations.[8] Unfortunately the obstructive attitude of some parts of the private profession has so far prevented any such regulation being adopted. A professional representative who is on the list is entitled in the UK to use the professional title 'European patent attorney', but, if in Germany, must be careful not to use the German translation *Europäischer Patentanwalt*, since the use of the word *Patentanwalt* is reserved to members of the German *Patentanwaltskammer*. Instead the cumbersome title *zugelassener Europäischer Patentvertreter* (qualified European patent representative) should be used.

When the EPO was set up in 1978, the rule was that anyone qualified to practise before his home country patent office could automatically be entered on the European list. In countries such as the UK, Germany, and the Netherlands, in which a register of qualified persons was kept, then anyone on that register could have his name put on the European list; in countries such as Switzerland, in which the profession was not regulated, then persons who could show that they had acted for clients or employers before the local patent office for at least five years could qualify.[9] The same 'grandfather' rules have applied in turn to each country that has subsequently joined the EPC, during an initial period of one year following accession.[10] Since 1979 there has also been a qualifying examination for entry to the

6 Article 133(2), EPC.
7 Article 134(1), EPC.
8 Article 133(3), EPC.
9 Article 163(3), EPC 1973.
10 Article 163(6), EPC 1973; Art. 134(3), EPC 2000.

European list, and since October 1981 for the original member states or after the initial period for other countries, this is the only method by which new entrants can qualify. Qualification as a patent agent in the UK now no longer gives the right to be entered on the European list; it is clearly most desirable, if not essential, for British patent agents to be qualified both in the UK and in Europe, which means passing both examinations.

There were in April 2009 a total of 9,245 professional representatives on the European register, of whom 33 per cent were from Germany, 19 per cent from the UK, and 9 per cent from France. Over 1,100 were from the states that joined the EPC after 2000 and practically all of these were 'grandfathered' on to the register without passing the European Qualifying Examination (EQE).

The European Qualifying Examination

The EQE tests the candidate in drafting a European patent application, amending an application in response to an office action from the EPO, preparing a case for opposition to a European patent, answering short legal questions, and giving a legal opinion. Two different sets of drafting and amendment papers are set: one in the electrical/mechanical field; the other in the chemical field. The opposition paper normally relates to a simple mechanical invention that even a chemist should be able to understand. The examination may be taken in any one of the three official languages of the EPO, but the candidate must have some knowledge of at least one of the other two. For example, a candidate sitting the examination in English will have to consider at least one document (for example relevant prior art) in French or German. Although the examination papers are not provided in any language other than English, French, or German, a candidate from a country that has an official language other than one of the EPO languages may write his answer in that language, and a translation will be made for the use of the examiners if necessary. But little use is made of this possibility, and indeed if someone is to practise before the EPO, he will, in any case, have to do so in one of the official languages, so he may as well complete the examination in such a language as well. The selection of the language for the examination can also be a problem, for example, for native German speakers who work in a company in which the working language is English and must decide whether to use their mother tongue, or the language in which they are accustomed to writing about patent matters.

The examination is now held each year in March or April simultaneously in Munich and The Hague, and in a number of other centres (for example at the British Patent Office in Newport and the Swiss Patent Office in Bern). There are now over 2,000 candidates taking the examination each year. The examination is of the open-book type, in which candidates may bring any written material that they wish into the examination room. It is best to be selective, however: candidates who bring in dozens of textbooks tend not to be very successful.

The first-time pass rate for the EQE averages around 30–40 per cent, and it is notable that the best results are obtained by candidates having one of the official languages as their mother tongue and by candidates from countries that have a national examination. The overall pass rate is much lower, due to large numbers of candidates who are resitting the examination in whole or part each year, with ever-decreasing chances of success. New Examination Regulations issued in 2009 provide for steeply increasing fees for resitting and for a pre-examination, to be held for the first time in 2012. This would probably correspond to the present paper on short legal questions and passing the pre-examination would be a necessary condition for taking the full examination.

British candidates have consistently had a higher pass rate than those from Germany, probably due to the fact that the European examination is a practical examination much more similar to the British national examination than it is to the German one. The British candidate has the advantage that the same approach works for both the national and European examinations; the German candidate must change from a theoretical approach to a pragmatic one.

Training for European Qualification

Apart from on-the-job training, various organizations offer training programmes designed to help new entrants to the European profession to qualify as European patent attorneys. The *Centre d'Études Internationales de la Propriété Industrielle* (CEIPI, or Centre for Intellectual Property Studies), which is associated with the Université Robert Schuman in Strasbourg provides training for the French national qualification, but also has an international section that acts as a central organizing body for tutorial courses, which are held in cities all over Europe. Three-hour tutorials held every two weeks over a two-year period provide basic training in European patent law and practice, and CEIPI also holds residential week-long seminars in Strasbourg each winter, which provide training directly for the examination. Other tutorials, courses, and seminars are organized by professional bodies in several countries, or by private firms or individuals. The EPO publishes each year a compendium of the examination papers from the previous year, together with examiners' comments and samples of good answers from successful candidates, and this is invaluable as a guide to the type of answer for which the examiners are looking.

The European Institute of Professional Representatives

New EPC 2000 Article 134a specifically gives the EPO Administrative Council power to regulate the profession, including the European Institute of Professional Representatives (epi). Unlike CIPA in the UK, this Institute does not control entrance to the European patent profession. The qualifying examination is under the general control of the Administrative Council, and the direct control of a Supervisory Board and an Examination Board, both comprised of officials of the EPO and members of the European Institute. Membership of the epi is compulsory as a condition of remaining on the European register, and for many years, the epi enforced a stricter 'code of conduct', for example, in matters relating to advertising, than did CIPA. These restrictive practices have now had to be changed at the instigation of the European Commission. It should also be noted that a person who has passed the qualifying examination does not have the right to become a member of the epi or to be entered on the list of professional representatives unless he is a national of an EPC contracting state. Rare exceptions are made for persons of other nationalities who have been working in the patent profession in a contracting state for at least ten years. This is in contrast, for example, to the USA, where any legal resident with the appropriate qualifications can practise as patent agent or patent attorney irrespective of nationality. Fairness would seem to demand reciprocity in this respect.

EPC 2000 mentions for the first time privilege from disclosure in proceedings before the EPO of any communications with a professional representative.[11] This, too, may be regulated by the Administrative Council, although as of writing no such provisions have been adopted. The extent to which any such privilege extends to proceedings before national courts is uncertain, and in many countries, the legal profession strongly opposes sharing any of its rights with patent attorneys.

Patent Attorneys in Private Practice

The type of patent agent with whom the inventor is likely to come into contact in the UK will depend on whether the inventor is independent, or is employed by a large or a small company, a university, or a government department. The independent inventor, or the inventor who works for a small company, will normally deal directly with a firm of patent agents in private practice. Firms of patent agents may legally be limited companies or partnerships, although the latter form is far more common. Firms vary in size from one-man operations to those with 20 or more partners and numerous associates (qualified, but not yet partners), technical assistants (not yet qualified), translators, and clerical staff. Their work will normally be a mixture of work for UK clients and agency work, that is, obtaining UK or European patents for foreign clients on instructions sent by the client's agents in his home country. Conversely, when a UK client wishes to file abroad other than through the EPO, the firm of patent agents will send instructions to agents, with whom they deal and whom they usually know personally, in the countries in question.

Although the private practice section of the patent profession is centred in London, in particular, the area around Chancery Lane and High Holborn, within easy walking distance of where the Patent Office used to be before it was moved to Newport, the larger firms have branch offices in cities outside London, and many smaller firms are based in cities and towns across the country. Generally, wherever there is any significant amount of industry, there will be a firm of patent agents reasonably accessible. Some firms of British patent agents have set up branch offices in Munich from where they can deal in person with the EPO, but except for

[11] Article 134a(1)(d), EPC 2000.

attendance at oral proceedings, physical proximity to the EPO brings no great advantage, and for oral proceedings, it is more important to use the agent familiar with the case than another who happens to be on the spot. In any case, the examiner dealing with the case is now quite likely to be in The Hague and not in Munich. The bulk of the representation of US and Japanese clients before the EPO tends to be shared between the British and German professions, with those in other EPC countries having only a small share of this work.

If the client's invention is in an area such as chemistry or electronics, it will be dealt with by a patent agent or technical assistant who has the appropriate technical qualifications. A patent attorney should never be a narrow specialist in one particular technical field, but some degree of specialization is essential and while any patent agent may feel competent to draft a patent application for a relatively simple mechanical invention, no chemical patent agent would consider tackling an electronic invention, or vice versa.

Larger client companies, which may file several patent applications a year, will normally have built up a working relationship with a particular firm of agents, and often with a particular patent agent within that firm, who will have become familiar with the client's technical field and with its particular problems. Similarly, the company will generally find it convenient to designate a single person as contact person with the patent agents on all patent matters; this may be a part-time activity of the research manager or equivalent, but if the company is big enough, this will be a full-time job for a patent liaison officer who, while not a qualified patent attorney, will know something about patent matters and can, for example, write a description of an invention that the patent agent can use as the basis for drafting a patent application.

Industrial Practice

At some stage, however, as the size of the company increases, it will find it advantageous to employ its own patent agents, who can deal directly with the British Patent Office and the EPO, and with firms of patent agents in other countries. This step can be advantageous, from the company's point of view, not only on cost grounds, but also because it enables the company patent department to choose its own methods of working without being tied to the methods of an independent firm of patent agents. Furthermore, company patent agents are relatively free from the time pressures inherent in private practice, in which work has to be charged by the hour, and they have the advantage of closer direct contact with the inventors, who may even be working in the same building as the patent department. These arguments should outweigh the currently fashionable trend to outsource all activities that are not part of the 'core competencies' of the company. It is all very well to outsource catering, security, and even computer services, but for a research-based company, the protection of its intellectual property rights is crucial to its survival. Nothing is closer to its core and the company's own employees had better be competent in this area.

The direct contact that is of prime importance in establishing a good working relationship between the patent department and the research department can easily be lost in a large company, in which research is being carried on at different sites, if the patent department is allowed to become a remote branch of the head office. It may be desirable to have patent officers at each site in liaison with the central patent department; the solution of having a separate patents department at each research centre is feasible only for the very largest companies and even in this case a central administration system is essential.

A large multinational company may have research centres in several countries and in many of these (for example the UK, the USA, and Germany) it is usually necessary to file the first patent application in the country in which the invention is made. This can be done through local patent agents, through a central patent department that prepares the application and sends it back to the originating subsidiary company or an agent for filing in the country of origin, or by having a separate patent department in the subsidiary company. A European company with a US subsidiary, for example, can have a patent department in its subsidiary company, staffed with US-qualified patent agents or attorneys, which can file applications for inventions made in the USA and also act as agents for filing in the USA in respect of inventions made in Europe. The reverse of course applies for US companies with European subsidiaries. The next logical step is to have a fully globalized patent department, in which a patent specialist will be doing the same sort of worldwide prosecution work whether he is sitting in London, Basel, or Los Angeles, but this does present some practical difficulties.

Patent Strategy in Industrial Practice

It is becoming increasingly important that a company patent or intellectual property department does not only react to requests or instructions from research or marketing, but also proactively becomes involved in developing and implementing strategies for getting the maximum value from patenting activities. Patenting strategy must be aligned with the business needs, and with the research and development strategy of the company, and the patent department must become an integral part of the business. Nevertheless it is usually preferable to have a central patent department organized at the corporate level rather than to have separate small patent departments for each business unit: firstly, important synergies will be lost if the patent function becomes too fragmented; and secondly, if the patent practitioners report directly to line management, there will be the danger that management will be told what it wants to hear rather than what it needs to know. It has become conventional wisdom that service departments, such as patents, should try to satisfy their 'customers' within the company. This is false: patent professionals have clients, not customers, and it is their job to do what is best for the client, not to keep the customer happy.

The Job of the Patent Practitioner

Drafting

Whether the patent practitioner is in private practice, or in a large or small industrial patent department, and whether he is in the UK or elsewhere, his task is essentially the same. First of all, the practitioner must, from the information given, understand what it is that the inventor has done and define what the invention is. To do this, he must above all put a great many questions to the inventor, to explore such matters as which features are essential and which are merely preferred, which features can be generalized and how far, which conditions work and which do not, what are the advantages offered by the invention, and whether all of the advantages are shown by all of the embodiments of the invention.

For example, suppose that the inventor has found that a group of chemical compounds are good flame retardants for a particular material, say, cellulose acetate. At this point, it is clear what the inventor has done, but by no means clear what the invention is, or, indeed, if an invention is there at all. If the compounds are new, the invention may be the compounds per se; if they are known, the invention could be their use as flame retardants. If they are known as flame retardants for other substrates, the invention, if any, can only be their use in cellulose acetate and it will be an invention only if there is some unexpected advantage, because *a priori* a flame retardant for one substrate might be expected to be useful for another.

Suppose that the last of these situations is the true one: the patent agent must find out from the inventor what is clever about using these compounds in cellulose acetate—that is, is cellulose acetate a particularly difficult substrate to flameproof? Are there problems associated with the incorporation of the flame retardant into the substrate? Can the flame retardants be used alone in cellulose acetate, whereas in other substrates they had to be used in conjunction with other flame retardants? Is the quantity of flame retardant used in cellulose acetate less than that used in other substrates, or less than that used for other flame retardants in cellulose acetate? How are the physical properties of the end product affected? Only by asking questions such as these can the inventive idea be determined, and often, it turns out to be other than what the inventor originally thought.

It is very important for the inventor to realize that it is an essential part of the process for the patent agent to play devil's advocate, and to take what appears to be a sceptical and even critical view of the invention. Only by asking the right questions at the beginning can the patent application be properly drafted. If it is not, the questions may be asked by the patent office, at a time when it is too late to put into the application the additional information that may be necessary. Any patent practitioner who simply writes down what the inventor tells him, without asking awkward questions, is doing the inventor a disservice.

Of course, the patent attorney must also decide who should be named as sole inventor or joint inventors, and this is another area in which awkward questions may have to be put and answered. A patent attorney cannot simply accept without question an edict that so-and-so must be named as inventor, because if he does so, he not only compromises his professional integrity, but also runs the risk of obtaining an invalid patent in the USA.

Prosecution

Once the application has been filed, the next part of the patent practitioner's job is to prosecute the application through the patent office of the home country, plus any countries in which it has been foreign-filed; the latter is usually done through intermediary agents in those countries, but may be done directly, for example, at the EPO. Most applications must be reviewed with the inventor from time to time in the light of new prior art that may be cited by patent offices or found internally, and in view of specific objections raised. Normally, the agent will check back with the clients, or with research management, before accepting a significant reduction in the scope of the claims. For the USA, it may be necessary for comparative testing to be carried out and this must be thoroughly discussed with the inventor or other persons who will carry out the testing (see p. 395).

When the patent is granted, the patent attorney will be responsible for defending it against opposition or revocation proceedings in the patent office and for ensuring that renewal fees are paid for as long as the client or employer wishes to keep the patent in force.

Opinion Work

The other major part of the patent practitioner's job consists in giving opinions as to whether actual or proposed activities of his client or employer would infringe anyone else's patent rights and, if so, whether such patent rights are valid. This is another area that can easily give rise to misunderstanding between patent agents and their clients, or between the patent department and the research, production, or marketing departments of a company. One can always tell the difference between a patent agent and a non-patent scientist by the way in which they read a patent: the patent attorney turns first to the claims; the scientist, to the examples. All too often, a patent attorney will have to report that the sale of some new product or the use of some new process would infringe a patent, will meet with the incredulous reply 'but we use acetic acid and they use sulphuric acid!', and will have to point out that what the claims say is merely 'acid'. On a more fundamental level, it is amazing how many inventors and managers firmly believe that, if they have their own granted patent, they cannot possibly infringe that of another party, and the patent practitioner often has an uphill task to convince them of the contrary.

The danger exists that, unless the inventors and managers understand something of patents, the patent attorney may be seen as a person who does nothing but put obstacles in the way of company projects. If this attitude prevails, then the tendency will be either to avoid the obstacle by not keeping the patent attorney informed, or to flatten it by putting pressure on the patent professional to give the 'right' answer.

It is better for the company to accept the fact of infringement and change plans or seek a licence before too much money has been invested, than to go ahead and be faced with an infringement action once the product is on the market. Another problem is that although the issue of whether or not something infringes a patent claim is often relatively straightforward (except where questions of equivalence arise), the question of whether or not that claim is valid (or, more correctly, whether the relevant court would find it valid) is usually much more difficult. Often the best that the patent attorney can do is to weigh the balance of probabilities and indicate whether the risk of infringing a valid claim is low, medium, or high. Managers usually prefer something more definite.

In the USA, one should always obtain a written opinion from a US patent attorney that the third-party patent is likely to be invalid or not infringed before going ahead with a potentially infringing project. Although since *In re Seagate*,[12] this may not be essential to avoid the risk of an award of triple damages for wilful infringement (see p. 202), it is still highly desirable. It is usually better to use an external patent attorney, but the US Court of Appeals for the Federal Circuit (CAFC) has held that a timely, thorough, and competent in-house legal opinion is sufficient.[13] Even outside the USA, a patent agent may obtain an outside opinion in an important case: for example from a patent barrister in the UK. If the main points at issue are technical rather than legal, it is unlikely that the outside opinion will be much more reliable than that produced in-house, but it will put the in-house attorney in a stronger position if the opinion is challenged, or if it is relied upon and things go wrong.

Clearance of Publications

A further task of the patent agent, at least in industrial practice, is to review proposed publications by employees to make sure that they do not give away inventions that are not yet adequately protected and do not contain anything that could adversely affect existing patent rights. For example, it is not uncommon for inventors writing up their work for publication in a scientific journal to analyse it as a sequence of logical steps, and even to use phrases such as 'it was then obvious that . . .'. Meanwhile, the patent attorney may finally have convinced the USPTO that the invention was surprising and non-obvious, and the last thing that he wants to see is the inventor contradicting this in print. This is not to say that the patent attorney's representation to the USPTO was false; simply that inventions may often appear 'obvious' with the benefit of hindsight, to the inventors themselves as well as to others. Scientists may well value scientific papers as much as, or more than, patents and will naturally feel resentment if manuscripts are blocked from publication for patent reasons, unless they understand and accept the occasional need for such measures.

[12] *In re Seagate Technology* 497 F.3d 1360 (Fed Cir 2007, *en banc*).

[13] *SRI International Inc. v. Advanced Technological Laboratories Inc.* 44 USPQ 2d 1422 (Fed Cir 1997).

Licensing

The patent attorney will also be involved if the client or employer wishes to license its own patent rights or to take a licence from another, and whether or not he is directly concerned in the negotiation or the drafting of a licence contract, he will at least review and advise upon any clauses relating to patent matters, as well as report on the extent and validity of the licensed patents. If it should actually come to litigation over patents, the patent agent will confer closely with the solicitors and be involved in the briefing of counsel in all matters relating to patents. It should be mentioned that patent agents may also deal with other industrial property rights, such as trademarks, registered designs, copyright, and domain names, although these are outside the scope of this book.

For more on licensing, see Chapter 22.

Relationships with the Inventor

The interests of the inventor and his employer are best served if there is a good working relationship between the inventor and the patent attorney. Both should be able to contact each other directly on a friendly and informal basis. Ideally the inventor should let the patent agent know when something interesting is coming along, before it is at the stage of a formal invention report, and when he has new developments or learns of new prior art on a case already filed. The patent agent should always be ready to listen to the inventor and should not treat patent matters as some sort of arcane mystery for the initiated, but should always be ready to explain any technicality in straightforward language. The cooperation must be based on a mutual understanding of each other's job and a mutual recognition of each other's professional competence. After all, each needs the other: if there were no inventors there would be no patent agents, and if there were no patent agents, it is hard to see how there could be any valid patents.

18

DRAFTING THE PATENT SPECIFICATION

Ce que l'on conçoit bien s'énonce clairement et les mots pour le dire arrivent aisément. (What one conceives well can be stated with clarity and the words to say it come easily.)

Nicolas Boileau (1636–1711)

Drafting the Scope

When an invention of any kind has been made, the first step towards drafting a patent specification is defining the scope of the invention, or forming a mental picture of what is to be claimed. The second step is putting that mental picture into words that clearly say what they mean. These basic principles are admirably set out in two articles by E.W.E. Mickelthwaite, originally dating from 1946 and republished in 2003.[1] Although the author gives examples exclusively from mechanical inventions, the general rules that he states, and the bad examples of sloppy drafting that he gives, are just as relevant now as they were over sixty years ago.

Also writing about mechanical inventions, P. Cole concluded, 'if a patent is to have any value it has to be written with a full understanding of the technical background known to the inventors and a full technical understanding of the further developments that the inventors have contributed'.[2] This advice is also valid for chemical and biotech inventions.

If the invention is in the field of chemistry, for example if a group of novel and useful compounds has been prepared, the first main factor to be considered is the size of the group of compounds that can reasonably be predicted, on the basis of those already made and tested, to be useful for the desired purpose. The second is the closeness of the prior art.

In those rare cases in which a completely new molecular structure has been invented and the prior art is very remote, the patent attorney can cheerfully draft a very broad scope, including all kinds of derivatives of the basic structure that the inventor thinks may be useful. More usually, however, such flights of the imagination will be cut short by some earlier publication of similar structures, which will force the invention to be redefined in more narrow terms in order to avoid claiming what has already been published.

The scope of protection that it is commercially important to achieve varies from one field to another. At one extreme, in a patent application for a pharmaceutical invention, it is sufficient to have a scope that includes those compounds that have a real chance of being marketed by the applicant and, once the market product has been decided upon, it is sufficient for most purposes to protect only that single compound. The reason for this is that imitators imitate only the actual market product; to take any other product, even an adjacent homologue, would involve carrying out all of the necessary animal and clinical testing to get marketing approval from the regulatory authorities. The imitator does not want to involve himself in this expense any more than he wants to incur research costs, so sticks to the product on which someone else has already done the work. It may, however, be relatively easy for an imitator to get marketing approval for an alternative salt form, which is why it is necessary to cover at least all pharmaceutically acceptable salt forms of the compound and, where appropriate, prodrug forms, such as esters.

If a dyestuff is marketed, however, then it is quite likely that not only its immediate homologues, but also even more remote derivatives, will also be useful dyes. They may differ slightly in shade, in cost, or in fastness properties, but they too could be usefully marketed. Consequently, the patentee of a dyestuffs patent needs a broader scope of protection than a pharmaceutical patentee, but because all other

[1] 'Brushing up our drafting' [2003] CIPA 320; [2003] CIPA 379.

[2] 'Effective specification drafting' [2003] CIPA 482.

dyestuffs manufacturers are in the same position, the desired scope of protection will be even more difficult to obtain.

Where the prior art is close, as it usually is in the dyestuffs field, it is often a pointless exercise to try to carve out a scope that touches the prior art at all points. Such a scope may indeed be novel, but the parts of the scope closest to the prior art will be prima facie obvious and usually will not show any significant advantage over the closest prior art compounds. It is much better to try to define a somewhat narrower scope that has a certain distance from the prior art and which will be much more defensible against obviousness attacks.

The Structure of the Patent Specification

Once the patent attorney has got a clear picture of the correct scope of the invention, he or she can begin the task of drafting the specification and claims of the application. Some prefer to draft the full set of claims first; more usually, the main claim, defining the scope of the invention, will be drafted first, in the form of the statement of invention that is the kernel of the patent specification, then the rest of the specification will be drafted, then the claims.

In whichever order they are drafted, the specification and claims have a certain logical structure, which will be described in the context of the invention of a new chemical compound. A British patent specification begins with a title; this may be kept deliberately vague on filing, but a descriptive title will be required during prosecution if it is not supplied initially. The text of the specification normally begins with a very brief statement of the field of art to which the invention relates, such as 'This invention relates to anthraquinone dyes'.

It is quite possible to go straight from this to the main statement of invention: for example 'The invention provides compounds of formula I, in which R = [. . .]', exactly as in the main claim. In many cases, however, some introductory description of the background of the invention is desirable, particularly if the essence of the invention is the solution of an existing problem or the improvement of an existing process. This is not essential, at least in the UK, but if it is omitted, it will be much harder to understand what the invention is and this makes the prosecution of the application more difficult. If the prior art is discussed, one should not explain the inventive step as a series of steps from the prior art. An opponent may subsequently be able to attack the patent by showing that each step is obvious.

There may be additional statements of invention defining the features of the invention for use as the basis for specific independent claims. These should be located in the appropriate place in the description and should be clearly recognizable as statements of invention. Examples include:

- 'In one aspect, the invention provides [. . .]';
- 'In another aspect, the invention provides [. . .]'; and
- 'The above defined formulations are new and form part of the invention.'

After the statements of invention will generally come an indication of what are the preferred parts of the scope and one or more formulae may be given defining narrower sub-generic scopes.

The specification must then describe how the invention is to be carried out: for example how the new compounds are to be made starting from compounds that are known or which could readily be made by analogy with known compounds. This may be given by a schematic reaction diagram, together with information about suitable reagents and process conditions. It should also be stated whether the compounds may be used as formed, or if they must be purified, and if so, how.

Next comes an indication of what the invention is useful for, a part of the specification usually called the 'utility statement'. In mechanical cases, this may be self-evident from the earlier part of the specification, but for new chemical compounds, the utility is usually not clear from the structure and needs to be spelled out. The utility statement may be quite brief for many uses, but for pharmaceuticals, rather more detail may be required. Any advantageous properties of the compounds should be mentioned at this point.

It is best not to give any theory on how the invention works. If it is felt desirable to give such a theory and if there could be any doubt about its correctness, one should leave room for a change of mind later: for example by wording such as 'Whilst we do not intend to be bound to any particular theory, it is believed that [. . .]'.

In a chemical case, the body of the specification is usually followed by a number of examples giving detailed instructions for the preparation of at least one of the compounds within the scope and for the use of the compounds. Often, there will be rather few fully written out examples, followed by a tabular listing of further compounds within the scope that can be prepared in the same way.

Finally, after the examples come the claims, which in the UK start with 'What we claim is [. . .]' and are written as if each separate claim completes a single sentence starting with these words. There may be any number of claims, but the British Patent Office objects if they are multiplied unduly; 50 or more claims are not at all uncommon, although for relatively simple cases, 20 or so should suffice. For the European Patent Office (EPO), substantial additional claim fees (in 2009, €200) are charged for claims numbering more than 15, and for each claim exceeding 50 in number, the claim fees are not merely substantial, but exorbitant (at the time of writing, €500). The number of claims should therefore be kept as low as reasonably possible, although, for biotechnological inventions, it may often be necessary to have a larger number of claims than in a normal chemical application.

If the specification relates to a category of invention other than novel compounds, the exact structure will differ from this, but the general logical sequence of statement of invention, preferences, instructions, uses, examples, and claims will normally be followed.

Priority and Foreign Filing Texts

No distinction should be made between a specification for a first filing from which priority will later be claimed and one that claims priority from an earlier application. At one time, British provisional specifications used to be very sketchy and did not normally contain any claims, and this did not matter as far as providing a priority date for a later complete specification was concerned. But the ruling in the USA that priority may be validly claimed only from a priority document that meets the full US requirements[3] means that even a first filing in the UK must be drafted as if it were to be filed in the USA. The USA has now made possible the claiming of internal priority from US provisional filings that need not contain claims, but the disclosure requirements for the specification are as strict as ever. It is desirable that all priority filings, even US provisionals, should contain claims. The change to a stricter interpretation of priority claiming in the EPO (see p. 88) also means that the priority application must contain as full a disclosure as possible.

On foreign filing, it is clearly undesirable to draft different texts for different countries in which the language is the same and, indeed, it is impossible to use different versions for different countries if the Patent Cooperation Treaty (PCT) route is used. Since the requirements for the USA are stricter than in most other countries, however, a text that is suitable for the USA will also generally be suitable elsewhere. With the requirements of the USA particularly in mind, we can consider individually the different parts of the specification. Claims may, however, be drafted differently for different countries and will be considered separately in the next chapter.

Background of the Invention and Prior Art

When the background of the invention is to be described, this description may refer to some specific piece of prior art; indeed, a discussion of the prior art in the specification is considered desirable by the US Patent and Trademark Office (USPTO) and by the EPO. The Implementing Regulations of the European Patent Convention (EPC) state that the description shall 'indicate the background art which, as far as known to the applicant, can be regarded as useful for understanding the invention [. . .] and, *preferably*, cite the documents reflecting such art'(emphasis added).[4] Amazingly, the Examining Division of the EPO recently held that not including a reference to the closest prior art known to the applicants made the application incurably invalid. Fortunately, the Board of Appeal reversed this decision, re-emphasizing the standard practice of the EPO that prior art citations need not be present on filing, but may

[3] *Kawai v. Metlesics* 178 USPQ 138 (Fed Cir 1973).
[4] Rule 42(1)(b), IR.

be added if requested.[5] In no country is the citation of specific prior art documents a positive requirement, at least not upon filing, and, except in cases in which it is really necessary for an understanding of the invention, it is best avoided. At the time of filing, the closest prior art is not always known and an elaborate discussion of less relevant prior art serves no useful purpose. Furthermore, the emphasis of the invention may change, or it may even be realized only some years after filing what the invention really is. If there is such a change of viewpoint, a prior art discussion in the text that is based on the original incorrect conception of what the invention was will only create difficulties. Finally, it should not be forgotten that an extensive discussion of the prior art significantly adds to the length of the text, and thus to the translation and other costs on foreign filing.

Of course, whether or not specific pieces of prior art are discussed in the specification, the applicant still has the duty to inform the USPTO of all relevant prior art of which he is aware (see p. 219).

The European Patent Office: Problem and Solution

As has already been mentioned (p. 69), the EPO approaches the determination of inventive step according to the 'problem and solution approach', and the closest prior art will determine what the EPO considers to be the technical problem faced by the inventor, even if the inventor was unaware of that art, saw the problem differently, or did not think in terms of problem and solution at all. The Implementing Regulations require that the invention be disclosed in such a way that the technical problem and its solution can be understood,[6] but specifically say that the technical problem need not be expressly stated as such. The practical consequence is that the specification should be drafted so as to include all of the advantageous effects of the invention that might be needed to demonstrate the 'solution' once one knows what the 'problem' is supposed to be, but, except in very clear-cut situations, should not endeavour to define the problem rigidly at too early a stage.

The USA: Danger of Estoppel

In the USA, admissions that are made in the specification as to what is actually taught by the prior art may be binding upon the applicant. A particular publication may not even be prior art, because it may not exactly meet any of the criteria of 35 USC 102 (novelty), but if it is described as such in the specification, the applicant may be prevented from arguing the contrary later. The safest rule is to admit nothing and say as little about the prior art as possible. Of course, all relevant prior art known to the applicant or his attorney must be brought to the attention of the USPTO, but this does not mean that it has to be mentioned in the specification.

[5] T 2321/08 (unpublished) *SAMSUNG ELECTRONICS/Apparatus and method for receiving/transmitting a signal.*

[6] Rule 42(1)(c), IR.

Selection Inventions

One situation in which some mention of the prior art may be essential is that in which the invention is a selection over an earlier disclosure (see Chapter 11). In this case, some reference to the disclosure and to the advantage of the selection should be given, for example in a form such as:

It is disclosed in British Patent No. [. . .] that compounds of formula [broad group] have activity as beta-blockers. It has now been found that a particular group of compounds, in which [definition of narrow group], which are not specifically disclosed in British Patent No. [. . .], have particularly selective action and are indicated for use as cardioselective beta-blockers.

Addition of Reference to Prior Art

In Germany and in the EPO, the examiner may call for a prior art statement to be inserted if one was not originally present. Care must be taken, however, not to add new matter by such an insertion. In the EPO, the Technical Board of Appeal (TBA) decided at an early stage that an application may be refused unless the applicant inserts a reference to the prior art when requested to do so. The Board stated that this could not reasonably be objected to as adding new subject matter. It was 'not inevitable' that even adding a discussion of the advantages of the invention over the prior art would constitute addition of new matter, but clearly this possibility would have to be guarded against.[7] Normally, a brief reference to the prior art will be sufficient, because the advantages of the invention should be apparent from the original description.

Object of the Invention

For the same sort of reasons, statements of the object of the invention, although very common in the USA, should preferably be avoided. They are not required by law, contribute nothing to the disclosure, and might in some countries give rise to problems if it turns out that some of the objects of the invention are not in fact attained.

US Sufficiency Requirements

A discussion of the prior art is optional, but the descriptive part of the specification must meet the requirements for a sufficient disclosure that are part of the patent law in all countries. Because these are particularly strict in the USA, we shall consider the US requirements first. The US requirements as to what constitutes a sufficient disclosure can be considered separately as the 'written description', 'how to make', 'how to use', and 'best mode' requirements.[8]

[7] T 11/82 (OJ 1983, 479) *LANSING BAGNALL/Control circuit.*
[8] 35 USC 112.

Written Description

The written description requirement is essentially a requirement that each claim should be fairly based on the disclosure. This requirement was very strictly interpreted in the notorious case of *in re Welstead*,[9] in which it was held that in a chemical patent application reduction of the scope of a generic claim to a subscope not disclosed as such in the specification or examples as filed was not allowable because the written description requirement would not be met with regard to the amended claim. Although decided by the Court of Claims and Patent Appeals (CCPA), the precursor to the Court of Appeals for the Federal Circuit (CAFC), *Welstead* is apparently still generally applied and it is therefore important that, in the USA, the original disclosure contains a full disclosure of preferred subscopes to which it may be necessary to limit. It is, however, permissible to create a new subscope by restricting a *Markush* group to a single significance.[10]

Thus, preferred significances should be stated for each variable substituent on the general formula: for example if substituent R_1 is defined as alkyl, alkoxy, cyano, halogen, or hydroxy, one might have a statement such as 'R_1 is preferably alkyl, alkoxy, or halogen, more preferably alkyl'.

In addition, the preferred members of a particular significance may be given: for example 'R_1 as alkyl is preferably $C_{1\text{-}4}$alkyl, more preferably methyl or ethyl, particularly methyl; R_1 as alkoxy is preferably $C_{1\text{-}3}$alkoxy, more preferably methoxy; R_1 as halogen is preferably chloro'.

It will be seen that the word 'preferably' tends to become rather overworked in chemical patent specifications, but the reason is to give as much basis as possible for any restrictions in the scope that may become necessary later. Some US practitioners avoid the word 'preferably' for fear that this may limit the applicability of the doctrine of equivalents (see p. 447). Since there is no reported case in which the scope of a claim to new chemical entities (NCEs) has ever been extended by equivalence, this seems to be an unnecessary concern.

Giving a basis for restrictions should not, however, be carried to the lengths of writing out long lists of individual significances of a generic term: for example by spelling out each individual alkyl group comprised in the term '$C_{1\text{-}6}$ alkyl'. Still less should one make long lists of individual compounds covered by the claims, other than those specifically exemplified. This is sometimes done in fields in which there is strong competition in order to block competitors from later specifically claiming any compound within the generic scope of the invention. The practice verges on absurdity when lists are given of individual numerical values of possible dosages within a range.[11] This 'scorched earth' policy does indeed make it much more difficult for any later selection patents to be granted, but since in normal circumstances

[9] *In re Welstead* 174 USPQ 449 (CCPA 1972).

[10] *In re Driscoll* 195 USPQ 431 (CCPA 1977).

[11] See, e.g., EP 1651270, in *Laboratorios Almirall SA v. Boehringer Ingelheim International GmbH* [2009] EWHC 102 (Pat).

it is far more likely to be the original patentee who makes a selection invention than any third party, the practice is liable to do more harm than good to the patent owner.

As described above (see p. 288), recent case law in the USA[12] has emphasized that the written description requirement is more than only a requirement that the claims correspond to wording in the specification; it is also intended to ensure that the applicant was in possession of the invention as of the filing date and it is quite separate from the requirement for enablement. A claim covering the human insulin gene was held invalid because only the rat gene had been obtained and sequenced, and there was no written description of the human gene. The fact that the disclosure enabled the human gene to be obtained did not overcome this objection. Some judges[13] believe that these cases were wrongly decided and that being in possession of the invention is relevant only in the context of interference proceedings to determine who had first reduced an invention to practice. The CAFC has now agreed to an *en banc* rehearing of *Ariad v. Eli Lilly*, in which this point will be decided. In any event, these decisions are contrary to the law in Europe, where there is no need to be in possession of the invention at the filing date as long as one can give an enabling disclosure of how to perform it.

How to Make

The specification must enable the skilled reader to make the invention that is claimed. For example, for new chemical compounds, a process for the production of the compounds from known or obtainable starting materials must be given for all of the compounds in the scope, in sufficient detail to enable a chemist to reproduce the work without undue experimentation. If a single process can be used for all of the compounds, it is normally necessary to give only that one, even if a number of alternative processes are available.

It will be necessary to give additional processes if no single process can be used to make all of the compounds within the scope. Once the minimum number of processes necessary to meet this requirement has been described, however, it is usually pointless to add more. For the USA, the process itself was previously not patentable unless it was inventive in its own right and, although this practice has now changed, the process is often not claimed at all. In other countries having product protection, process claims, although they may be included, are of little value. For those very few remaining countries having only process protection, it is in most cases a hopeless task to try to cover all possible processes (see p. 78). Adding more processes

12 *University of California v. Eli Lilly* 43 USPQ 2d 1398 (Fed Cir 1997); *Enzo Biochem Inc v. Gen-Probe Inc.* 63 USPQ 2d (Fed Cir 2002); *Ariad Pharmaceuticals v. Eli Lilly* 560 F.3d 1366 (Fed Cir 2009).

13 See Judge Linn's dissents from denial of *en banc* rehearing in *University of Rochester* and in *Enzo*, cited above, in which he considered that the key issue was lack of enablement.

to the text will make divisional applications necessary in some countries and give rise to extra translation costs in others, without any real advantages in return.

How to Use

There must be a utility for the entire scope of the claimed invention and although, in some fields, the utility of a claimed product does not have to be specifically disclosed if it would be obvious to the skilled reader, utility can never be inferred for a pharmaceutical product and must always be stated. A pharmaceutical utility statement should state what type of activity the compound has, how this is demonstrated (that is, a particular test method on a particular animal species), what the dosage should be, and how the compound should be administered. Dosage ranges may be very broad, since the scope will often cover a wide range of compounds of differing activity.

The requirements for patents claiming a new pharmaceutical use are stricter than those for new compounds useful as pharmaceuticals. In the former case, it is highly desirable to insert some data comparing the activity of the compound with that of a standard compound already known for the claimed indication, and this may be useful even for novel compound cases. In fields other than that of pharmaceuticals, however, it is best not to give any comparative data with prior art compounds. At the time of filing, any data available will probably be crude and may be unreliable, and inclusion of such data can cause trouble if they are later found to be incorrect. As we shall see, it is often essential in the USA to prove superiority over a prior art compound, but this comparison need not appear in the specification itself. What should be mentioned, however, are all of the properties for which advantages over the prior art might be able to be demonstrated.

If the claimed compounds are useful only as intermediates, then the specification should describe a process for converting them to end products known to be useful or the use of which is disclosed in the specification.

As discussed in Chapter 15, a utility given to support a claim for a DNA sequence must be 'specific, substantial, and credible', and the specification must support this.

Best Mode

The best mode (see p. 219) of carrying out the invention known to the inventor at the time of filing should be given for each aspect of the invention (best product, process, and use). This requirement can give rise to difficulties if, for example, one process is best for a certain part of the product scope and a different process is best for another, although it may well be that it is necessary to give only the best process for the preferred compound. Also, it is not specified in which respect the process must be 'best'. Thus it need not necessarily be the process giving the highest yield, but could, for example, be the one that is most convenient or most economical to carry out. It is not essential that the best mode should be identified as such; it is enough if it is in the specification somewhere or other.

If a continuation-in-part (cip) application is filed in the USA, the best mode at the time of filing the cip should be given if this is not already present. Since the best mode to be supplied is that *known to* the inventor, not necessarily *invented by* him, it is considered that, if a cip is filed, a best use or best preparative method invented by someone else subsequent to the original filing should be added. Previously, when the inventorship of a cip had to be the same as that of the original application, the newly added best mode could not be claimed in the cip; now, it is possible to claim it and add its inventor as a new co-inventor in the cip application.

A patent specification that meets these four sufficiency requirements is an enabling disclosure that can be used to base a claim of priority for a US application, and which amounts to constructive reduction to practice of the invention, thereby establishing a date of invention in case of interference with another application or patent.

Sufficiency Requirements in the UK and the European Patent Office

Article 83 of the EPC states that the European patent application must disclose the invention in a manner that is sufficiently clear and complete for it to be carried out by a person skilled in the art, and failure to comply with this requirement is a ground for revocation in opposition proceedings.[14] In the UK Patents Act 1977, the same ground of invalidity is given.[15] The reader must be able to carry out the invention based on the specification when read as a whole, including the claims, without any inventive activity of his own. What neither the EPC nor the Patents Act 1977 make clear is what is meant by 'carry out the invention'. Previously, it was generally held that it was enough if the specification contained such instructions as would enable the reader to produce something within each claim. According to this approach, there is no need to describe fully more than one embodiment within the scope of a broad claim, but if further embodiments are specifically claimed, they must be fully described. More recent cases have held that the description must be enabling over the whole scope of the claims, so that a broad claim requires a correspondingly broad disclosure. This topic is discussed in more detail in the next chapter.

It has been held that there is no actual requirement that any examples be present in a UK chemical patent specification as long as the description as a whole gives sufficient instructions for putting the invention into effect.[16] Normally, however, a specification will round off the description by a number of examples, both of the synthesis of new compounds and of their application. These examples will normally illustrate representative members of the entire scope of compounds claimed, but there is no necessity that all of the examples should actually have been carried out.

[14] Article 100(b), EPC.
[15] Section 72(1)(c), PA 1977.
[16] *Mobil Oil Corp.'s Application* [1970] FSR 265 (PAT).

Thus, one frequently finds in new compound applications one or two fully written-out examples including characterizing data such as the melting point of the product, followed by a list of examples in tabular form prefixed by wording such as 'Using similar procedures to those described in Example 1, but with suitable choice of starting materials, the following compounds may be prepared [. . .]'. Note that this does not allege that all of these compounds have been made: those in the list that are given with a melting point or other physical data will have been made; the rest may be 'paper examples' intended to illustrate the scope.

It is good practice always to use the present tense when drafting examples, because use of the past tense may be taken as a representation that the example was actually carried out, and may cause problems if the example was only a paper one. In one extreme case in the USA, the use of the past tense in such a situation was actually held by the CAFC[17] to amount to fraud on the USPTO, which rendered the whole patent unenforceable.

The traditional approach in the UK has always been that there is no need to establish the compound scope by actual synthesis if the inventor feels that he can reasonably predict that the claimed compounds will have the stated utility. If some of the compounds cannot be made or are not useful, the patent may be at least partially invalid. This is less clear-cut now that inutility is no longer a ground of invalidity, but a disclosure in which a significant part of the scope did not work may be regarded as insufficient.

A classic example of insufficiency in a chemical patent in the UK was a case dating from the late 19th century in which BASF claimed a process involving heating the reaction mixture in an autoclave. In fact, although the process worked in the iron autoclaves that were commonly used, it did not work in enamelled ones because, unknown to BASF, ferric ions were a necessary catalyst for the reaction. The patent was invalid for insufficiency because the need for an iron autoclave was not specified.[18] It would not have mattered whether or not BASF understood why iron was necessary, as long as this necessary feature was included. This illustrates the fact that there is no need in a chemical patent, or indeed in a patent in any other technical field, to explain mechanisms of action or to give theories as to why the invention works. Such speculation should be avoided, in case it should be interpreted as limiting the scope of protection in any way and also to avoid making the invention appear obvious.

A general rule of patent drafting is that the specification is its own dictionary. One may use expressions that would otherwise be unclear, such as 'lower alkyl',[19] as long as these are defined somewhere in the specification, and one can also use an expression to mean other than its normal definition, or even coin wholly new terms,

[17] *Hoffmann-La Roche Inc. v. Promega* 66 USPQ 2d 1385 (Fed Cir 2003).

[18] *BASF v. Soc. Chim. du Rhone* [1898] 15 RPC 359 (CA).

[19] The term 'lower alkyl' was held to be unclear in T 337/95 (OJ 1996, 628) *NIHON NOHYAKU/Lower alkyl*.

as long as these are all defined and as long as a 'redefined' term does not depart too greatly from its normal meaning.

Sufficiency Requirements in Other Countries

In many countries, it seems as if a patent is awarded not so much for an inventive idea as for routine experimental work. In such countries, the claimed scope must be based upon real characterized examples. Such requirements are met with, for example in Latin America, Japan, Russia, and China. This was also the rule in most Eastern European countries prior to their accession to the EPC.

When such requirements are rigorously applied, as they tend to be in Argentina or Russia, it is practically impossible to obtain a chemical patent of any reasonable scope at all. The Japanese Patent Office (JPO) demands characterization for all examples needed to support the scope. In the field of dyestuffs, in which melting points often cannot be given, it is not enough to state the colour of the dyestuff; some actual numerical value such as λ_{max} (the wavelength of maximum light absorption) must be given.

Japan also has strict sufficiency requirements for pharmaceutical inventions. For novel pharmaceutical compounds, it is necessary only to show *in vitro* or *in vivo* test data, but where a pharmaceutical use is claimed for a known compound, actual clinical data in humans is required, which may not always be available.

It was originally feared that, in the EPO, the examiners would readily object to compound scopes as being insufficiently exemplified. In some quarters, it was urged that the EPO should allow new examples to be added, as used to be the practice in Germany, but this would of course be adding new matter, and if permitted, would make the whole patent invalid. The practice now is that objections of inadequate support should be made only if the examiner can show prima facie evidence that the invention would not work over all of the claimed scope. If such objections are made, the applicant can submit further examples or data to the examiner, which will go into the file and can be used as evidence in support of the scope, but which will *not* become part of the specification and cannot therefore cause problems due to introduction of new matter.[20]

Special Requirements for Biotech Inventions

Microbiological Inventions

As described in Chapter 14, if a patent application claims a new microorganism, or requires a new microorganism in order to produce the claimed product, then a sample of the microorganism must be deposited, normally according to the provisions of the Budapest Treaty. Not only must the deposit be made, but the specification

[20] Guidelines, C VI 5.3.6.

itself must also refer to the deposition and give the accession number. In the USA, this may be done at any time up to grant, but in the EPO, the deposit must be made no later than the date of filing[21] and the information about the deposit, with accession number, must be given to the EPO before the preparations for publication are complete.[22] Care should be taken over who actually makes the deposit; if the person or company that makes the deposit is not the patent applicant, it is necessary within the same time limit to submit to the EPO a document stating that the depositor has authorized the applicant to refer to the deposit and gives irrevocable consent to the deposit being made available to the public.[23] This requirement has been interpreted less strictly when the depositor was the subsidiary company of the applicant,[24] but it may not be safe to rely upon this decision if the facts are not exactly the same.

Inventions Involving DNA Sequences

If a patent application discloses nucleotide or amino acid sequences, then it is necessary to supply a listing of these sequences in the format prescribed by World Intellectual Property Organization (WIPO) Standard ST 25. These sequence identifiers are listed separately at the end of the application, where they frequently take up tens or even hundreds of pages, but must also be supplied on a data carrier from which they can be read electronically for searching purposes. In the EPO, the procedure is regulated by Rule 30, the standards for presentation of the sequence listings, as defined in a Notice of the EPO in 2007, now being identical with those of the PCT.[25] In practice, the necessary sequence listings can be generated using software named PatentIn, which is available from the EPO and, in almost identical form, from the USPTO. A language-neutral form of sequence identifier, in which text headings are replaced by standard numeric identifiers, is now the standard, enabling the same sequence identifier to be used in all countries where it is required, as well as in PCT applications.

Two points should be kept in mind by the patent agent drafting such a case. One is that standard sequence identifiers must be given even for very short sequences and must be given whether the sequences are actually claimed or are merely mentioned in the description. It is no longer necessary to give sequence identifiers for sequences that are in the prior art, as long as a reference can be given to a publicly available sequence database, including the accession number of the sequence and the version number of the database (see para. I.1.2 of the above-referenced Notice). It can also save a lot of work to give a literature reference to known peptides, nucleotide probes, and similar, rather than mentioning in the description a sequence for which a sequence listing will be required.

21 Rule 31(1)(a), IR.

22 Rule 31(2)(a), IR; G 2/93 (OJ 1995, 275) *UNITED STATES OF AMERICA/Hepatitis A virus.*

23 Rule 31(1)(d), IR.

24 T 118/87 (OJ 1991, 474) *CPC/Amylolytic enzyme.*

25 Notice of 12 July 2007 (OJ, 2007, Special edition No. 3).

Secondly, and more importantly, the patent attorney should obtain the sequence data directly from the inventor in an electronic form that can be read directly into PatentIn and which does not require to be entered into PatentIn by typing on a keyboard. In a sequence of several hundred amino acids or nucleotides, it would be little short of a miracle if no typographical error were made and, as we have seen in Chapter 11, errors of this type cannot be corrected later, at least not in the EPO.

Length of Text

A general rule about the drafting of patent specifications is that they should be kept as short as possible, consistent with a sufficient disclosure of all aspects of the invention, including fall-back positions in case the scope has to be cut back during prosecution. An obvious reason for this is the costs incurred by long texts. Few people realize just how great this factor can be, but, back in 1997, a calculation showed that, for a patent application filed in a list of 45 countries (by no means unusual for an application in the pharmaceutical field), each page of the specification cost a total of well over US$800 in translation costs alone, and if extra printing fees and associated charges in many countries were included, the figure was approximately US$1,000. In the last 12 years, this figure will certainly have risen. Furthermore, a specification of 40 pages not only costs US$60,000 less than one of 100 pages, but will almost certainly be better written. A description that is concise is forced to be clear, and a clearly written specification is more likely to give an enforceable patent.

It is a regrettable fact that the average word count of US patent descriptions in all fields has doubled from 3,500 to 7,000 in the last 20 years.[26] Although the longest British patent on record is in the field of computer technology (it was so long that it had to be printed in four volumes), many biotechnology patents are very much longer than they need to be, particularly because they frequently contain very long discussions of the prior art. Very often, such patent applications come from academic institutions or small companies, particularly in the USA, where there is considerable pressure to submit the results for publication in a scientific journal as soon as possible after the patent application has been filed (or, as sometimes still happens, before). The same document is made to serve both purposes, and often it seems that the inventor has written his paper for publication and given it to a patent attorney who in the time available could do little more than add a set of claims at the end. Sometimes, not even this is done, and the paper is filed as a provisional application in its original state. This is a recipe for disaster: the description of an invention for a patent application serves a completely different purpose from a scientific paper and the two should not be confused. As discussed at p. 88, it is very likely that, when a subsequent US regular application or European application is filed, claims

[26] D. Crouch, *Patently-O* blog, 21 December 2008.

of the scope necessary to protect the invention will not be entitled to the priority of such a first filing and the publication of the corresponding scientific paper during the priority year will be fatal in most countries.

Incorporation by Reference

One way in which texts are sometimes shortened is to refer to a previously published document. Thus, for example, it is permissible to indicate that a starting material may be prepared 'as described in US Patent No. 3,456,789' rather than writing out a description of the synthesis of this material. In the USA, this wording would usually be followed by 'the contents of which are incorporated herein by reference' or a similar phrase, but the EPO requires such phrases to be deleted, unless the contents are essential to give a sufficient description of the claimed invention. If they are essential, and particularly if a feature from the referenced document is to be added to a claim, the relevant part of the document referred to should be incorporated in the description so that the specification is complete in itself.[27] But this may be regarded as adding new matter contrary to EPC Article 123(2) unless the reference clearly identifies the part of the document containing the critical material.[28] Material that is really needed in the description should be in there from the start.

[27] EPO Guidelines, C II 4.18.

[28] T 689/90 (OJ 1993, 616) *RAYCHEM/Event detector*.

19

DRAFTING THE CLAIMS

> U.K. citizenship must be given to the inventor [. . .] without application and his family (wife and son and brother and sister and mother) needed for settlement in U.K. because of claim 24, inclusive claims 1 to 24.
>
> Theodore Po Man Poon, Claim 25 of British Patent Application GB 2 226 494

The Purpose and Nature of Patent Claims

Function of Claims

All countries having patents, as well as the European Patent Convention (EPC) and the Patent Cooperation Treaty (PCT), require at least one claim to define the invention.[1] Claims have the purpose of defining the scope of subject matter for which protection is sought.[2] A competitor does not infringe and cannot be stopped unless he makes, sells, offers for sale, imports, or does something falling within the scope of at least one claim of the granted patent. Although, as we shall see, some countries

[1] For example, Art. 78(1)(c), EPC.

[2] As stated, e.g., in Art. 84, first sentence, EPC.

still interpret claims more broadly than their literal wording, others do not. Therefore, a patent drafter should assume that the claims as drafted will be interpreted literally. If an applicant wants to be able to stop a competitor doing something, he should claim it literally, and not rely upon a court to broaden protection on his behalf.

Conversely, there is little point in claiming something that no one is likely ever to wish to do, or which it will be impossible to prove that a competitor is using, such as an improved process that cannot be deduced from inspection or analysis of the product. One should always try to claim the physical article that the patentee or a competitor (or the competitor's customer) will actually be selling, in the form in which it will be sold. Thus the article should be claimed as if 'on the shelf' and not only as in use. A claim to a pump having ports 'connected to supply and outlet hoses' does not literally cover the pump without the hoses, as it would be sold. It should also be remembered that a multi-step process claim will be infringed only if all of the steps are carried out by the same party. Many claims to business methods in the USA are unenforceable because, for example, some steps are carried out by a seller and others by a buyer.[3]

The claims of a patent do not stand alone, but have to be read in the light of the description[4] and have to be supported by the description, as well as be clear and concise.[5] Thus the description must give basis not only for the claims as originally drafted, but also for claims as they may have to be restricted in view of prior art that may be found and cited during prosecution.

Claim Categories

All claims fall into one of two broad categories: they claim either a product (that is, something tangible, such as a mechanical device, a machine, an electronic circuit, a chemical compound, or a formulation); or a process (that is, a method of making, using, or testing something).

For chemical patents, product claims include claims to chemical substances per se, which may be useful themselves or as intermediates in the production of other compounds, alloys, mixtures (for example pharmaceutical compositions), specific forms of substances (for example optical isomers, crystal forms), or even the product of the use of a substance (for example a textile fabric dyed with a new dyestuff). Process claims include: processes of synthesis, isolation, purification, and extraction of chemical substances; methods of use, including first and subsequent pharmaceutical use claims; methods of medical treatment and diagnosis (where patentable); and testing and assay methods. These different types of claim should all be included in the application, but only to the extent that they give useful protection.

[3] *Muniauction, Inc. v. Thomson Corp. and I-Deal* (Fed Cir 2008).

[4] Article 69, EPC.

[5] Article 84, second sentence, EPC.

Independent and Dependent Claims

Claims may be written in independent form or may refer back to an earlier claim. This may be only a form of shorthand to avoid writing out an entire definition many times over; thus if claim 1 is 'A compound of formula I [. . .] in which R_1 is alkyl, alkoxy, or phenyl, R_2 is [. . .] etc.', then a later process claim may begin 'A process for the production of a compound of formula I, stated in claim 1, comprising [. . .]'.

A true dependent claim, however, is of the same type (for example a compound claim) as the claim to which it refers, includes all of the limitations of that claim, and adds further limitations. Thus if claim 1 is the example given above, further dependent claims could be:

2. A compound as claimed in claim 1 in which R_1 is alkyl.
3. A compound as claimed in claim 2 in which R_1 is methyl.

If a dependent claim is anticipated or obvious, the same must be true of the claim(s) upon which it depends. Conversely, if a main claim is novel and unobvious, then so is any claim depending upon it, even though the extra feature added in the dependent claim may itself be old or obvious.

In most countries, multiple dependencies are allowed; thus a claim may be dependent upon, for example, 'any one of claims 1 to 3', or indeed upon 'any preceding claim'. In the USA, multiple dependencies are usually avoided because they carry high claim fees.

Number of Claims

The main claim of a patent specification will normally correspond to the statement of invention and one may well wonder why anything more than this is needed. For novel compounds, a claim to the compounds is infringed not only by selling the compounds, but also by making or using them. Why, then, claim the preparation or the use separately? In the UK, at least, the practice of multiple claim drafting arose following the Patent Act 1919, which established partial validity for the first time. Thereafter, a British patent could be enforced against an infringer even if partly invalid, as long as at least one claim of the patent was valid and infringed. If there were only one claim, the whole patent would be invalid if that claim was invalid, whereas independent claims to the preparation or use might be valid. For the same reason, it was desirable to have a range of compound claims of decreasing scope, finishing up with claims to individual compounds of particular interest, so that if one or more of the broader generic claims were later found invalid, the narrower sub-generic or species (single compound) claims might remain unaffected. The same considerations still apply in most countries.

The drafting of a set of claims of progressively narrower scope is arguably no longer necessary under the UK Patents Act 1977, since a patent may now be partially valid and enforceable even if the valid part is less than a complete claim. But because enforcement in that situation is still contingent on the Comptroller or the court allowing a validating amendment, the patentee is in a stronger position if he has claims that are completely valid and few patent practitioners would rely on a single claim, even if they may now draft fewer dependent claims than before. In order to be able to enforce a patent as quickly as possible, it is good practice to have a narrow independent claim covering precisely what it is proposed to sell and which is entitled to the earliest possible priority date.

In several patent offices, a large set of claims is discouraged by requiring additional fees for claims in excess of a particular number. In the USA, there are additional fees of US$220 (as of 2009) for each independent claim above three and US$52 for each claim in excess of 20. Each multiply dependent claim costs US$390. At the European Patent Office (EPO), there are quite substantial fees for claims in excess of 15, and fees are extremely high for claims above 50 in number (see p. 359). This applies whether claims are present on filing or added during prosecution, but claim fees, once paid, will not be refunded if the number of claims is later reduced. No claim fees are payable for any claims added during opposition proceedings. In Japan, the number of claims has a large effect on costs, because it increases the fees for requesting examination, for grant, and also the annual renewal fees, which are calculated on the basis of a standard fee plus a supplement for each claim. This effect on patenting costs gives a practical reason for avoiding undue multiplicity of claims.

The EPO has attempted to discourage applications, mostly those originating in the USA, having large numbers of independent claims. Rule 43(2) states that a European application may contain more than one independent claim per category (that is, product, preparation, use) only in one of three exceptional situations—that is, if there are:

(a) a plurality of inter-related products;
(b) different uses of a product; or
(c) alternative solutions to a particular problem.

This rule does not significantly change the previous practice, at least in the chemical and pharmaceutical field, except that the burden of proof has now shifted to the applicant, who is obliged to comply with the 'one claim per category' principle unless he can show that he falls within one of the exceptions.

In the USA, rule changes first proposed in 2006 would have done far more than claim fees to discourage the filing of numerous claims. Draft Rule 75 proposed that if an application contains more than five independent claims, or more than 25 claims in all, the applicant must submit an examination support document (ESD) giving the results of a prior art search, a list of the most relevant references, and an explanation of how each independent claim is patentable over the most relevant references. The cost of having this work done by a patent attorney would be many times the extra claim fees and of course the ESD would be a happy hunting ground for those hoping to invalidate the patent for inequitable conduct. Rule 75 was one of the rules contested in *Tafas v. Dudas*[6] and has now been withdrawn (see p. 162).

[6] *Tafas v. Dudas* 541 F. Supp 2nd 805, 814 (ED Va 2008); *Tafas v. Doll* 553 F.3d 1345 (Fed Cir 2009).

Form of Claims

General

The meaning of claims must be definite, precise, and unambiguous for the skilled reader.[7] Wording such as 'preferably' or 'for example' should not be used in claims, and such extra features should be the subject of separate dependent claims. One should avoid internal codes or names, or trademarks without a generic description. A trademark alone is not a reliable description of a product because there is no guarantee that the same mark will not be used at a later date for a different product. The text of the claims must be clear, and drafters should double check for every possible misunderstanding or ambiguity, and eliminate any that they find. For example, the term 'R is phenyl substituted by halogen or methyl' should not be used if what is meant is really 'methyl, or phenyl substituted by halogen'. Use punctuation, spacing, or reference letters, for example:

R is:
(a) phenyl substituted by halogen, or
(b) methyl.

It is important to use consistent terminology throughout the description and claims, because different words may be construed to have different meanings. All defined substituent groups must be logically consistent and mutually exclusive. Avoid open-ended terms such as 'alkyl', because, for example, compounds containing an alkyl group of over 20 carbon atoms would be insoluble in water and unlikely to be useful as a drug. Also, avoid terms such as 'lower alkyl' unless these are clearly defined in the specification.

Universally used abbreviations may be used: for example 'C_{1-5} alkyl' instead of 'alkyl with one to five carbon atoms'. Labels, for example (a), (b), can be used to identify different processes or features. Terms such as 'compounds of the invention' (when clearly defined) may also be used to cut down lengthy repetition. Additionally other abbreviations usual in the state of art can be used if they are defined when they are mentioned for the first time in the claims and in the description.

Where the invention is a composition of a number of different components, care must be taken both in the description and in the claims to specify which components are essential, and whether additional optional components may be present. The wording 'a composition *consisting of* A, B, and C' means that only A, B, and C, and no other components, may be present.[8] Such a claim may be unduly narrow, since infringement would be avoided by adding a small amount of D. More useful protection is given by claiming 'a composition *comprising* A, B, and C'. This is normally understood to mean that A, B, and C *must* all be present, but further components D, E, etc. *may* also be present. A claim of intermediate (but rather vague) scope is

[7] Guidelines, C III 4.
[8] Agreed by the Board of Appeal in T 711/90 (unpublished) *POST OFFICE/Optical fiber glass.*

'a composition *consisting essentially of* A, B, and C', which covers A + B + C, optionally with additional ingredients that are either present in minor amounts or are inactive, so that the properties of A + B + C are not significantly affected. The Board of Appeal has preferred this term to 'comprising substantially', which was considered to lack clarity.[9]

Care must be taken in the use of the term 'comprising', however, because a 1997 decision in Australia held that 'comprising' means the same as 'consisting of' and does not allow for the presence of any additional components.[10] This was an unpleasant surprise for all who thought that the meaning of the term was clearly understood by everyone, even including judges. Even more surprising was the fact that this was not just a one-off decision: the same position was taken in a later case by the full Federal Court.[11] It has been suggested that for Australia, the term 'including' should be used instead of 'comprising', but because it is desirable to have the same wording in all countries, it may be preferable to include a brief definition of 'comprising' in every patent specification.

Even the use of a simple word such as 'or' may cause difficulties. If a claimed composition is stated to contain 'A or B' as well as other components, is the claim infringed by a composition containing both A and B? In the USA, there have been decisions holding that 'A or B' does not include A+B,[12] and if it is intended to include the combination, one should say 'A or B or both', 'at least one of A and B', or the less elegant 'A and/or B'.

In expressing the proportions of the components of a mixture, use may be made of ratios, parts, or percentages, each of which may be calculated on a weight, volume, or molar basis. Whichever is used, it should be used consistently throughout and it should also be remembered that, in an n-component mixture, there are only (n-1) independent variables to consider.

Many compounds are not used in free (base or acid) form, but in salt form, and care must be taken that the claims cover the salt forms. Basis for this may be given in the description by adding a sentence such as 'In this specification unless otherwise indicated terms such as "compounds of formula I" embrace the compounds in salt form as well as in free (base or acid) form'. No distinction needs be made between acid and base addition salts, since the skilled reader will always be in a position to know which of the two kinds of salts the compounds are likely to form. If desired, a short explanatory statement in the description can specify the kind of salts that may be formed. An example of a specific claim might be 'Cyclohexylamine in free or salt form'. In claims to pharmaceutical compositions, the salts should be limited to 'pharmaceutically acceptable' salts, thereby excluding toxic salts such as cyanides.

9 T 759/91 and T 522/91 (unpublished) *AMOCO/Polyamide*.
10 *General Clutch Corp. v. Sbriggs Pty Ltd* [1997] IPR 359 (Fed Ct Australia).
11 *Abbott Laboratories v. Corbridge Group Pty Ltd* [2002] FCAFC 314 (Full Fed Ct, Australia).
12 For example, *Kustom Signals Inc. v. Applied Concepts Inc.* 60 USPQ 2d 1135 (Fed Cir 2001).

In the USA, a claim to treating osteoporosis with 'bisphosphonic acid' was held to cover the use of the monosodium salt of the acid,[13] but one should not rely on such generous interpretation in all countries.

Where the invention is a new pharmaceutical compound, it is important to include a claim to a pharmaceutical composition containing the compound. Such a claim would still give substantial protection even if the compound were to turn out not to be novel: for example if it were shown to be a metabolite of a known drug (see p. 254).

In the UK

Generally speaking, the same principles of claim drafting that apply in the EPO are appropriate also for the UK Patent Office. There are some differences, however: for example product-by-process claims, such as 'A compound of formula I (structure defined) when obtained by process X', are allowable in the UK and construed to claim the product only when made by the specified process, whereas they are refused by the EPO. In the EPO, the only type of product-by-process claim allowed would be of the type 'The compound (structure unspecified) obtained by process X', which would be allowed only if the product is patentable per se and can be defined only by its method of manufacture.[14] Such a claim is construed as covering the compound however it is produced (see p. 241). In order to avoid confusion, it is better to refer to the first of the above claim types as 'derived product claims' and to reserve the term 'product-by-process claim' for the second type.

There are certain types of claim that are usually encountered only in the UK and in a few countries having similar jurisprudence. One of these is a claim often found in specifications in which the scope of protection has been expanded on foreign filing. This claim, which is narrower than the main claim, claims precisely the same scope as was in the priority document and is intended to ensure that that claim at least will certainly be entitled to the priority date. Such claims are often called *Thornhill* claims, after the case in which the problem first arose.[15]

It has been questioned whether such claims are still necessary in the UK under the Patents Act 1977, but since the EPO Enlarged Board of Appeal (EBA) decision G 2/98 (see p. 88), it is clear that such claims are essential both in the EPO and the UK. Furthermore, such claims are also important in the USA. A claim in a US application claiming two priorities that incorporated features from both priority documents was held to take the priority date of neither of the two earlier filings, but only the filing date of the final application,[16] the same finding as in the old *Thornhill* case. In fact, it is good practice, in the EPO, as well as in the UK and USA, to have one such claim for each priority claimed, not only for the first one.

[13] *Merck & Co. Inc. v. Teva Pharmaceuticals USA Inc.* 68 USPQ 2d 1857 (Fed Cir 2003).
[14] T 150/82 (OJ 1984, 309) *IFF/Claim categories*.
[15] *Thornhill's Application* [1962] RPC 199 (PAT).
[16] *Studiengesellschaft Kohle v. Shell Oil Co.* 42 USPQ 2d 1674 (Fed Cir 1997).

Another typically British claim is the 'omnibus claim', which is a relic of the old days in which there were no specific claims and what was claimed was the contraption 'as herein described'. It is still permissible in the UK to include a claim of this type, generally, as the last of a series of claims. In a mechanical case, it refers to the drawings; in a chemical case, to the examples. Such a claim should preferably be in independent form: for example 'A process for the production of polyethylene terephthalate as described in any one of the examples', or 'A bis-naphthalene azo dyestuff as described in any one of examples 1-20'.

Such claims have two main purposes. Firstly, they represent a last-ditch attempt to save something from the patent if all of the other claims should be invalid; thus the second of the above claims could be regarded as 20 separate claims, one to each example, at least one of which might be valid even if the generic and sub-generic claims fall. This type of narrow construction of an omnibus claim was adopted in an electromechanical case by the House of Lords in 1948: the omnibus claim, which read 'An electric generator for a cycle, constructed and arranged substantially as herein described, with reference to and as illustrated in the accompanying drawings', was the only claim held valid and infringed.[17]

Secondly, they represent a form of insurance against the inclusion of some unnecessary limitation in claim 1. It is possible for an omnibus claim to be broader in some respects than a main claim.[18] Omnibus claims are not allowed in the USA, nor in the EPO, unless the applicant can show that the parameters of the claim cannot be expressed verbally: for example when they can be expressed only by reference to a drawing or graph given in the specification.[19]

In the USA

There was a time when the drafting of US claims, particularly in mechanical cases, was hedged about by a large number of rules of a purely formal nature. For example, integers of a claim had to be recited positively, which meant that one could not use phrases such as 'in the absence of' and that semantic problems arose when claiming an object with a hole in it. Practice has now relaxed considerably and it is now possible to include negative features in US claims.

In the chemical field, a major problem area was the extent to which it is possible to put together a claim listing a group of compounds too closely related to be separately patentable, but too distinct to be conveniently expressed by a single generic term. Because of the formalistic approach to US claim drafting at that time, alternatives were not allowed to be stated as such in a claim: one could not therefore claim 'A, B, or C' per se or as an integer in a process claim. The solution adopted in the case of *Ex p Markush* in 1941 was to claim 'a compound selected from the group

[17] *Raleigh v. Miller* [1948] 65 RPC 141 (HL).
[18] See, e.g., *David Kahn v. Conway Stewart* [1972] FSR 620 (Ch D).
[19] T 150/82 (OJ, 1984, 309) *IFF/Claim categories*.

consisting of A, B, *and* C', the fatal word 'or' thus being avoided.[20] This may seem a mere exercise in semantic triviality, but it produced important effects in US patent practice. Thus it was at one time held that the applicant by putting A, B, and C together in a '*Markush* group' was admitting that the compounds were not patentably distinct, so that prior art disclosing A would preclude the patenting of B or C. There have also been a great many cases on the question of the extent to which the USPTO can object that a *Markush* group is too broad and must be split up for the purposes of examination. *Markush* groups could also be made out of substituent groups in a general formula: for example 'a compound of formula I [. . .] in which R_1 is selected from the group consisting of $C_{1\text{-}4}$ alkyl, nitro, and cyano [. . .]'.

In chemical cases, as in mechanical, the overly formal requirements of 70 years ago have been relaxed and today it is acceptable to say simply 'where R_1 is $C_{1\text{-}4}$ alkyl, nitro, or cyano', although from force of habit this may still be referred to as a *Markush* group.

In August 2007, proposed new Rules on *Markush* claims were published in the Federal Register. These would have made it much more difficult to draft *Markush* claims of broad scope and would have prohibited, for example, nested sets of alternatives within the same claim. As of writing, these Rules have not yet been implemented.

A recent *en banc* decision of the Court of Appeals for the Federal Circuit (CAFC) has held (over the vigorous dissent of Judge Newman) that product-by-process claims in the USA do not cover the product when made by a different process, even if the product is new and inventive[21] (see p. 241).

In Germany

In Germany, as well as Switzerland, Austria, and Scandinavian countries, claims are normally drafted in such a way as to distinguish what is old in a claimed combination from what is new. Thus a chemical process invention would be claimed in a form such as 'a process for the preparation of A by condensation of B and C under acid catalysis characterized in that (*dadurch gekennzeichnet*) orthophosphoric acid is used as catalyst.'

The first part of the claim, the 'pre-characterizing clause', recites what is already known, namely, that one can make A from B and C in the presence of an acid catalyst. Then comes the phrase 'characterized in that', followed by the characterizing clause telling us what is the novel feature of the invention, that is, using orthophosphoric acid as the acid catalyst.

This type of claim is particularly useful in claiming inventions that are clearly an improvement over a well-known product or process, and may be used equally well in the UK or the USA (where they are known as *Jepson* claims).[22] They are not

20 *Ex p Markush* [1925] CD 126 (US Patent Office Board of Appeals).
21 *Abbott Laboratories and ors v. Sandoz Inc. and ors* (Fed Cir 18 March 2009).
22 *Ex p Jepson* [1917] CD 62 (US Patent Office).

particularly suitable for claiming novel compounds, and, indeed, the German Patent Office does not insist on this format for compound claims. Problems can also arise where it is not clear what the prior art is and one should always put as little in the pre-characterizing clause as possible, because by putting a feature in the first part of the claim, the applicant effectively admits that it is old. Thus, in our example, if the applicant later found that the prior art only disclosed basic condensation of B and C to A, he might find himself unable to claim acid catalysis generally. *Jepson* claims are used in the USA much less frequently than formerly. In 1982, over 9,000 patents were granted containing *Jepson* claims; in 2008, only 1,000.

The German style of claim drafting is favoured, but not insisted upon, by the EPO.[23] The Rule uses the wording 'wherever appropriate' and, in many cases, for example chemical product claims and claims in which the closest prior art is not clear, this type of claim will not be appropriate. The EPO has on occasion insisted upon it where the invention is clearly an improvement over one clear piece of prior art.[24] In the EPO, at least, putting a feature in the pre-characterizing clause is not regarded as an admission that the feature is old.[25]

The Scope of the Claims

A responsible patent practitioner has a duty to his clients or employer to obtain the best possible protection for their inventions. Part of this duty is that each invention should be claimed as broadly as possible, taking into account the limitations imposed by the prior art known at the time of drafting and by the technical feasibility of the scope that is claimed. It is always better to start out with claims that are too broad rather than too narrow, as long as basis exists in the specification for the restrictions that may have to be made when new prior art is found, or when the inventor finds that part of the invention does not work.

There will often be cases in which broad claim scopes are fully justified by the pioneering nature of the invention. Even in chemical product cases, it may be possible to have 'pioneer' claims referring to classes of compounds that are not expressed as a general formula in which all substituent groups are listed, but are broadly defined, for example, by the presence of a totally novel grouping, or by a combination of structural and functional elements.

Free Beer Claims

The best interests of the client are not being served, however, if the patent practitioner attempts to obtain claims that are, by any reasonable interpretation, far too broad. Excessively broad claims will make patent prosecution unnecessarily difficult and, if the practitioner succeeds in having such claims granted, the resulting

[23] Rule 43(1), IR.
[24] T 13/84 (OJ 1986, 253) *SPERRY/Reformulation of the problem.*
[25] T 99/85 (OJ 1987, 413) *BOEHRINGER/Diagnostic agent.*

to future inventions based on currently disclosed inventions'. This definition already indicates that such claims are unlikely to be considered patentable.

The simplest example of a reach-through claim is 'A compound testing positive in the assay of Claim 1'. This would almost certainly lack novelty, since many known compounds may have this property. The chance of inherent lack of novelty is reduced by limiting the claim to 'A therapeutic agent identified according to the screening method of Claim 1'. If the disease state associated with the novel receptor can be specified, one could also claim 'Use of a therapeutic agent identified according to the screening method of Claim 1 in the preparation of a medicament for treating disease Y' in the EPO, or a corresponding method of medical treatment claim in the USA. In order to enforce such a claim, it would be necessary to prove that the screening method had in fact been used, but in the USA, this should be possible by means of discovery. Many such claims have indeed, been granted in the EPO and the USPTO: for example 'The use of a compound as identified in the screening assay of claim 15 for the preparation of a pharmaceutical suitable in the treatment of hypertension, said compound being different from aldosterone'.[33] Another example is 'Use of a CRF_2 receptor antagonist for the manufacture of a medicament for treating cerebrovascular disorders, wherein said CRF_2 receptor is encoded by a nucleic acid sequence according to claim 1'.[34]

In future, similar claims probably will not be granted. The EPO will object at the search stage that such claims cannot be searched and therefore lack clarity. The Trilateral Project B3b of the EPO/USPTO/JPO agreed that such claims are invalid. Under the EPC, they are insufficient and/or lack clarity and support; under 35 USC, they do not meet the written description and/or enablement requirements.

A major criticism of such claims, according to the Trilateral Project report, is that they 'cover a genus of compounds defined only by their function wherein the relationship between the structural features of the members of the genus and said function have not been defined'.

But if individual compounds interacting with the receptor have been identified, these specific compounds may be claimed; only if there are enough of these to establish a structure–activity relationship will it be possible to claim a genus.

Pickled onions

A simpler variety of over-broad claim is one in which the enthusiasm of the claim drafter has led him to generalize the invention so far beyond its real scope that the claim reads on to the prior art. For example, for an invention that a particular type of pharmaceutical compound could be stabilized in tablet form by the addition of an organic acid, the claim that was drafted read 'A physiologically acceptable substance stabilized in an acidic medium', which sounds like a good pioneer claim, except that it literally covers a jar of pickled onions.

[33] Claim 16 of EP 287 653 B1.

[34] Claim 37 of EP 724 637 B1.

20

PROSECUTION OF THE PATENT APPLICATION TO GRANT

> [T]he whole unfortunate situation might have been avoided if Albert Einstein had not 'doodled out' his equation $E = mc^2$ in the Swiss Patent Office around 1905 instead of getting on with the work he was being paid to do.
>
> A.P. Pedrick, British Patent Specification 1,426,698

Introduction

The procedure for examination and prosecution of patent applications has already been described in general terms in Chapters 6–8. In this chapter, we shall consider some of the problems that may arise when objections of lack of unity, lack of novelty, or obviousness are raised against a patent application either in the international phase of the Patent Cooperation Treaty (PCT), or before the European Patent Office (EPO) or national patent offices. We shall assume that formal objections to the wording of the specification and claims have been overcome.

Lack of Unity

Objections of lack of unity are more of a nuisance than a real threat. It is always clear that one can avoid the problem by filing one or more divisional applications

what the examiner considers to be a single invention. The international preliminary report on patentability (IPRP), either under Chapter I or Chapter II, will also be issued only for those claims. Almost certainly, when the application goes into the regional or national phase, the applicant will then be faced with the requirement to pay additional search and/or examination fees and probably to file divisional applications, if he wishes to protect all of the initially claimed subject matter. On entering the regional phase at the EPO, the applicant is no longer invited to pay additional search fees for inventions that were not searched in the international phase; instead, the application will have to be limited to the searched invention and any other inventions must be made the subject of divisional applications.[7]

Alternatively, the applicant may pay any or all of the additional search fees under protest,[8] whereupon a three-member panel or other special instance of the ISA must decide whether the protest is justified and, if so, reimburse all or part of the extra fees.

If, however, the ISA considers that there is unity of invention, a national patent office in a designated state should not contest this finding during the national phase.[9] Since the practice of the EPO is more liberal on the subject of unity of invention than is the US Patent and Trademark Office (USPTO), it may be advantageous for US applicants who are designating the USA in a PCT application to use the EPO instead of the USPTO as ISA. The findings of the EPO on unity of invention then have to be accepted by the USPTO.

The European Patent Office as International Searching Authority: Protest Procedure

Under the European Patent Convention (EPC), protests were heard by the EPO Boards of Appeal[10] and their decisions on protests against findings on lack of unity of PCT applications when the EPO acts as ISA were reported separately with the prefix 'W' (*Widerspruch*, or 'protest'). In two concurrent decisions,[11] the EBA held that the EPO, as ISA, could find non-unity *a postiori* in the light of the prior art found in the search, that is, if a unifying broad main claim appeared to lack novelty or inventive step, the remaining claims could be considered to lack unity. It was stressed, however, that the ISA had no power to decide that a claim lacked novelty, and such an *a postiori* finding of non-unity could be based only on a preliminary and non-binding opinion. In cases of real doubt, unity of invention should be found.

An objection of non-unity must be accompanied by a reasoned statement giving the grounds for the finding,[12] a mere listing of the supposed inventions not

7 Rule 164(2), IR
8 Article 40.2(c), PCT.
9 Article 27.1, PCT.
10 Article 154(3), EPC.
11 G 1/89 and G 2/89 (OJ 1991, 155 & 166) *Non-unity a postiori.*
12 Rule 40.1, PCT; see, e.g., W 4/85 (OJ 1987, 63).

being enough. Conversely, any protest will be considered only if it is accompanied by reasoned argument.

Under EPC 2000, a three-member review panel of the EPO will decide on protests in a single-stage process, without the possibility of appeal to the Boards of Appeal. If the review panel finds that the protest was wholly justified, the additional search fees and the protest fee will be refunded. If the panel finds that the protest was justified only in part, for example if it finds that two rather than three inventions were present, it will order a refund of the corresponding additional fees, but not the protest fee.[13]

National Patent Offices

In the UK, examiners who have raised such an objection can sometimes be persuaded to change their minds if the applicant argues strongly that there is a common inventive concept that unifies the parts of the invention that the examiner wished to divide. But recent practice seems to be adopting a more restrictive line, taking the view, for example, that intermediates and final products may be claimed in the same application only if the structures are very closely related, so that they may both be searched under the same classification, or if the intermediates have the same properties as the end products. In the USA, although the USPTO has attempted to liberalize its practice, many examiners are still very liable to issue restriction requirements and very reluctant to be persuaded to withdraw them.

At one time, it did not greatly matter if one had to divide an application in the USA, because the extra costs involved were relatively small (one extra application fee and issue fee, there being then no renewal fees). There were even some advantages: because a divisional application does not need to be filed until the parent case has been allowed, the grant of the divisional may be delayed and, under the old system, the term of patent protection was thereby extended. This is no longer the case, because a divisional application takes the same filing date as the parent and thus expires on the same date. There is still one positive aspect of a non-unity objection: when the examiner requires restriction between two groups of compound within the original scope, he is thereby admitting that he considers the groups 'patentably distinct'. Thus, even if one group is anticipated by prior art, the other groups should not be regarded as obvious over that prior art and prosecution can be made a good deal easier (although if a court were later to disagree with the examiner, the resulting patent could still be held invalid for this reason).

Lack of Novelty

Search Report

Under the PCT system, in the EPO, and in countries such as the UK and Germany, a search report will be available to the applicant before any substantive examination

13 Rule 158(3), IR; Decision of the President of the EPO of 24 June 2007.

takes place. Under the PCT and in the EPO, the search report is accompanied by an initial opinion on the patentability of the invention. In the PCT procedure, the ISA sends the applicant and the International Bureau a non-binding written opinion on novelty, inventive step, and industrial applicability at the same time as the search report,[14] although the written opinion is not published with the search report. In the EPO, the applicant receives an extended European search report, which consists of the search report, together with an opinion on whether the application and the invention appear to meet the requirements of the Convention.[15] As in the PCT, this opinion is not published with the search report,[16] but unlike the PCT procedure, the applicant will be invited to comment on the opinion and to amend to correct any deficiencies. If this is not done within six months, the application will be deemed to be withdrawn.[17]

The applicant will, of course, check the citations that have been made and the opinion on patentability which has been given. He will particularly look out for anything amounting to an anticipation of all or part of the claimed scope. Naturally, if the entire scope is anticipated, there is nothing for it but to abandon the application before any more expense is incurred, but if the anticipation is only partial and the remaining scope is still of commercial interest, the applicant must consider whether to amend the claims voluntarily at this stage in order to cut out the old matter. In the EPO, this is now obligatory, and in most national procedures, it is probably desirable because it will simplify the course of further prosecution.

The Patent Cooperation Treaty International Phase

In the PCT procedure, if the first written opinion indicates that the invention is considered not patentable for lack of novelty or inventive step, the applicant has to decide whether to amend the claims, argue the objection, or simply ignore it. If no action is taken, the written opinion will become the IPRP and since the IPRP is non-binding the applicant may choose to go into the national or regional phase with a negative IPRP, and tackle the situation there. Although the IPRP is non-binding on the designated offices, and some designated offices (such as the USPTO) will simply ignore it, some designated offices, particularly those the patent offices of which are not capable of carrying out their own search and examination, will tend to rely on the IPRP. A negative IPRP is a bad start to prosecution in those countries and should be avoided if possible.

Within Chapter I, there is no formal procedure for applicants to respond to the written opinion, but they may submit informal comments on it directly to the International Bureau, which will communicate such comments to the designated offices together with the IPRP. The PCT also gives applicants the opportunity to

14 Rule 43*bis*, PCT.
15 Rule 62(1), IR.
16 Rule 62(2), IR.
17 Rule 70a, IR.

amend the claims within two months of the transmittal of the search report and written opinion.[18] The amendments will be made and the comments will be on the record, but the IPRP will remain unchanged.

By filing a demand under Chapter II, the applicant has the opportunity to amend the entire specification[19] and to try to obtain an IPRP—known in Chapter II as the international preliminary examination report (IPER)—that is more favourable than the written opinion, and those are the only reasons why a demand should ever be filed. Amendments and comments on the written opinion are to be sent to the international preliminary examining authority (IPEA), preferably at the same time that the demand is filed. If arguments and possibly amendments are made, the examiner will normally issue a second written opinion, on which the applicant may comment once. The IPER is then issued. The IPER must be issued within 28 months from the priority date, so the time is short and the examination is superficial. There will not be time to enter into long discussions about inventive step and there will certainly not be time in which to carry out comparative testing with the prior art.

A possible compromise is to argue any finding of lack of novelty that is not clearly correct, but to leave matters of inventive step to the national phase. This strategy may be summed up as follows.

(a) If the initial written opinion is positive, do nothing.
(b) If the written opinion indicates lack of novelty, then:
    (i) if this is correct, amend claims, make comments, but do not file a demand; or
    (ii) if this is incorrect, file demand with arguments.
(c) If the initial written opinion indicates that the invention is novel, but lacks inventive step, it is probably best to file informal comments arguing for inventive step, but not to file a demand.

For more on the PCT international phase, see Chapter 6.

The European Patent Office and National Procedure

When the examiner makes an allegation of lack of novelty, this must first be closely examined to check if it is correct. It is possible that the examiner may have misread the prior art, or failed to notice a limitation in the claims that distinguish them from what has been cited. In such cases, the response will be to argue that this objection should be withdrawn. It may be appropriate to argue that the objection is really one of lack of inventive step rather than lack of novelty and whereas in most cases it will hardly matter whether a claim is anticipated or blatantly obvious, the matter becomes critical where the prior art citation is a prior unpublished application, which cannot be used to attack inventive step. It must also be investigated whether

[18] Article 19, PCT.
[19] Article 34, PCT.

the document cited by the examiner is indeed prior art. Thus, for example, in the UK and the EPO, a scientific paper published before one's filing date would not be prior art if one could rely upon a priority date earlier than the publication. Similarly, another application claiming an earlier priority date would not be prior art under the whole contents approach if one could show that, for the relevant subject matter, one's own application was entitled to its priority date and the cited application was not.

When the publication date of a citation is critical, it should always be checked carefully. Thus many scientific journals carry a publication date that may be earlier than the true date of publication for patent purposes, which is the date upon which it was made available to the public, for example by being received by individual subscribers or by a library. But many journals are now published electronically on the Internet even before their printed publication date and the electronic publication date may be the true date of publication for patent purposes.

In the USA, a publication less than a year before the actual US filing date or the international filing date of a PCT application designating the USA may be overcome by relying on one's priority date, showing that the publication originated from the inventors, or 'swearing back' by means of an affidavit or declaration to establish an invention date in any World Trade Organization (WTO) country before the date of the reference.

The question frequently arises as to whether it is enough to destroy the novelty of a compound that its name or structure has been published, even if there is no indication of how it may be made or whether it is useful for anything. In the USA, it seems to be the law that such a publication is enough, provided that the compound could be made by an average competent chemist.[20] It is immaterial in the USA whether or not a compound has actually been made, so that 'paper examples' are just as effective anticipations as real examples (see p. 367).

If the reference states, however, that the named compound could not be made, or if working the suggested process does not give the named compound and no other obvious method of preparation is known, then the disclosure may not be an anticipation. In the latter case, however, it is very difficult to convince the USPTO that a method disclosed in the prior art does not work, particularly if the prior art is a US patent, which is legally presumed to be workable. If the later applicant gives evidence to the effect that he was unable to repeat the process of the earlier patent, he may be suspected of not having tried hard enough, particularly if only one or two attempts were made and no attempts were made to make changes that would have been within the ability of one of ordinary skill in the art. Evidence from an independent expert may be more credible.

In the EPO, naming a compound is anticipation only if the document itself, taken together with common general knowledge where appropriate, enables the

[20] *In re Donohue* 207 USPQ 196 (CCPA 1980).

substance to be prepared and isolated.[21] Note that the common general knowledge must be that as of the date of publication of the document, not the filing date of the application. If, for example, the compound could be made at the filing date thanks only to general knowledge developed after the publication date of the prior art document, then that is properly a matter of inventive step, not novelty.

In the UK, there have been decisions to the effect that the mere naming of a compound, without a method for its preparation, is anticipation. This would lead to the logical conclusion that a computer programmed to print out the structural formulae of all possible chemical compounds could prevent all future patenting of chemicals. But the Court of Appeal decision in *Du Pont v. Akzo* (see p. 234), which was approved by the House of Lords,[22] went to the other extreme, holding that 'a compound which has never hitherto been made cannot be a known substance' and therefore cannot anticipate a later claim to the substance. On this basis, even a full written disclosure of how to make a named compound would not be novelty-destroying in the absence of proof that the compound had, in fact, been produced. Since, in the patent literature, it is usually impossible to tell which examples are real and which are paper examples, the general adoption of this approach would lead to a chaotic situation.

Restriction of Scope

If the cited document does genuinely amount to an anticipation, action must be taken to limit the scope of the claims so that they no longer claim what is old. In this case, a distinction must be made between prior art that discloses the same or a similar utility for the compounds as does the application, and prior art that discloses either a different utility or no utility at all. In the latter case, the prior art may be regarded as an 'accidental anticipation', since it did not relate to the same inventive idea as that of the application; it is possible in the UK and the EPO to avoid the prior art by means of a disclaimer, which does not have to have any basis in the specification as filed. This practice has been challenged, but has now been upheld by the EBA (see p. 239).

But whereas, in the case of an accidental anticipation, one can go right up to the boundary of the prior art, where the anticipation is in the same field of inventive activity, one must leave a gap between what one claims and the prior art, and there must be basis for the new claim scope in the specification. In the USA, by contrast, although claims on filing may incorporate provisos to disclaim individual compounds, difficulties will be encountered if one tries to do this during prosecution, because basis in the specification will normally be required and will usually not be present. It does, at least, now seem to be possible to restrict to a preferred significance of one substituent while leaving other substituents unchanged. Previously, following

[21] Guidelines, C IV 7.3a.

[22] *Du Pont v. Akzo* [1982] FSR 303 (HL).

Welstead[23] (see p. 363), the reduced scope was considered to be new matter because it was not described as such in the specification as filed. In some recent cases, disclaimer of specific compounds has been allowed and, in general, US practice seems to be much less formalistic in this respect than it was some years ago.

If one must restrict to a scope for which there really is no basis in the specification, one can always do this in the USA by filing a continuation-in-part (cip) application containing the new scope. The problem is that this new matter may not be entitled to the filing date of the parent case, with possibly fatal effect upon validity.

It must also be remembered that, in the USA, since the *Festo* decision[24] (see p. 449), any restriction of scope made for patentability reasons will normally mean that it will not be possible to allege later that something falling within the abandoned part of the scope infringes the limited claim under the doctrine of equivalence.

Lack of Inventive Step

Apart from questions of whether or not a citation really is prior art and whether or not it has been interpreted correctly, objections of lack of novelty are a relatively straightforward matter: either the claim is anticipated or it is not, and if it is, it must be limited accordingly. Obviousness or lack of inventive step, however, is a subjective matter and for this reason by far the greatest part of the work involved in patent prosecution is in arguing that the invention is not obvious over the citations that have been made.

As described above, although objections of lack of inventive step may be raised during international preliminary examination (IPE) of a PCT application, the short time limits of this procedure do not usually allow full argumentation to be given and the matter is best left until the regional or national phase.

Although the UK Patent Office is empowered to raise objections of obviousness under the Patents Act 1977, applicants are not usually required to prove non-obviousness by submitting comparative test results, as is often necessary in the USA. An objection of obviousness can always be argued and, if the argument is prima facie reasonable, it will be difficult for the examiner not to accept it. The arguments given by the applicant will of course be on the record and if they can be shown to be incorrect or misleading, the patent will be very vulnerable to any subsequent attack. If an obviousness objection appears to be at all well founded, the sensible course is to limit the scope of the claims by adding further features not found in the cited prior art, assuming of course that there is basis for such limitation in the specification or claims as filed.

23 *In re Welstead* 174 USPQ 449 (CCPA 1978).

24 *Festo Corp. v. Shoketsu Kinzoku KK et al.* 62 USPQ 2d 1705 (Sup Ct 2002).

The practice in the USA, however, is well established and follows rules that are simple enough in concept, although they may give rise to great difficulty in individual cases. The examiner may cite under 35 USC 103 any combination of prior art references in order to allege the obviousness of a claim. In a simple case, for example in which the claim covers compounds of a certain formula with a C_2–C_6 alkyl substituent and the prior art discloses the methyl compound, for the same utility, then that single reference will suffice. If, however, the prior art disclosed the ethoxy compound, then the examiner might cite that as the primary reference, together with a second reference showing that in similar but less closely related compounds that were also useful for the same purpose, both alkoxy and alkyl substituents were disclosed. He would then argue that, in view of the secondary reference, it would be obvious to substitute ethyl for ethoxy in the compound of the primary reference, so obtaining one of the claimed compounds.

The examiner may combine more than two references, but when he does so, it is a clear sign that his argument is not a strong one. The invention cannot really be very obvious if it can only be reconstructed with the benefit of hindsight, by selecting bits and pieces from three or more different publications.

The first line of defence in response to an obviousness rejection is to argue the matter, if necessary, combining this with a suitable limitation of the claims to features not found in the prior art. The argument should first of all point out the differences (the more the better) between the prior art and the claimed invention. If the claimed utility is not disclosed in the prior art, it can be argued that the person skilled in the art would have no reason to suppose that compounds similar to the prior art compounds would have that utility. If two or more references have been combined, it can often be argued that it would not have occurred to anyone to read both references together (for example because they relate to different technical fields); one can also allege that, even if they were combined, they would not amount to the invention.

It is a strong argument against obviousness if it can be said that the prior art 'teaches away from' the invention: for example by indicating that poorer results were obtained from the compounds closest to the scope now claimed.

Showings in the USA

If none of these arguments are effective, it may be necessary to submit a 'showing', that is, a declaration or affidavit (merely different legal forms for essentially the same thing), which gives the results of comparative testing to show that the compounds of the invention have unexpectedly different, preferably superior, properties to the closest prior art compounds.

The basis for such showings is that although a claimed compound may be prima facie obvious in view of the prior art because of close similarity in chemical structure, it may nevertheless be legally unobvious and patentable if not merely its structure, but also its properties as a whole, are taken into consideration and unexpected differences, particularly advantageous ones, are found among these properties.

In carrying out a showing, the number of compounds to be compared depends upon the size of the scope that is claimed and on the nature of the prior art. If the scope is narrow, it may be enough to compare one compound from the scope with a prior art compound; a broader scope will require two or more pairs to be compared. Even a narrow scope may require comparison with more than one prior art compound if there are prior art compounds that may be considered equally close (for example a close homologue and a positional isomer).

The compounds chosen for comparison should be selected using the following procedure. Firstly, find the specifically disclosed or exemplified compound from the prior art that is closest in structure to the claimed scope. Having done this, find the compound in the claimed scope closest to the chosen prior art compound, whether or not this compound is specifically disclosed or exemplified. If two prior art compounds are equally close, either may be chosen, but a comparison involving both might carry more weight. If it is not clear which compounds constitute the closest prior art, it is desirable to discuss the proposed showings with the examiner before the work is started to ensure that the examiner agrees that the closest prior art will be tested and to get an indication of the number of comparisons considered by the examiner to be acceptable for the claimed scope.

Only in rather exceptional circumstances should this procedure vary. For example, where the prior art scope includes a compound not specifically disclosed, but closer to the claimed scope than any specifically disclosed compound, it may be necessary to choose that compound if the prior art points to it, for example, by including it in a narrow sub-scope. It may sometimes be allowable to compare against a compound that is not structurally the closest: for example where the prior art discloses that such a compound has better properties than the closest compound.

Having selected the compounds for comparative testing, the next question is what tests to apply. In some cases, this is fairly clear: if the compounds are supposed to have a pharmaceutical utility, then a test for that activity will normally be necessary. If the compounds are dyestuffs, however, a wide range of possible properties could be compared: for example fastness to light or to various wet treatments; dyeing properties; migration; compatibility with other dyes; stability to pH changes; and many more. Care should be taken in the selection of properties to test, since the duty of candour to the USPTO means that all test results obtained must be disclosed. One cannot carry out a whole battery of tests and select only the one or two in which good results have been obtained.

There are no absolute rules as to how many properties must be shown or how great the superiority must be. The only criterion is that the superiority should be unexpected and relatively small improvements may be dismissed by the examiner as being within the normal expected range of variation. The properties tested must be disclosed or inherent in the specification, but if a property relied upon in a showing is not mentioned or inherent, it may be added by means of a cip without loss of priority date. An example of a property that is inherent is, for a compound having a pharmaceutical utility, reduced toxicity. A showing may be based on this property even though toxicity has not been mentioned in the specification, because it is inherent that a drug should have as low toxicity as possible. If, for a dyestuff, it is intended to base a showing on improved wet fastness, this property must be mentioned in the specification, although it is not necessary to state that the claimed compounds are actually superior in this respect. It is for this reason that dyestuff patent applications often contain 'laundry lists' of properties relevant to the class of dyestuffs to which the claimed compounds belong.

The person who signs the declaration or affidavit should be the person who either personally carries out the tests reported or supervises directly the person who does. This will often be the inventor and this is acceptable as long as the declaration sticks to reporting facts. In so far as it gives opinions, the opinions of the inventor are often given little weight by the USPTO and thus if, for any reason, an opinion declaration is needed, this should be done by someone other than the inventor and preferably by an outside expert.

Declarations may sometimes be used to establish what are sometimes called 'secondary criteria of non-obviousness': for example that there had been a long-felt want in the trade that the product met and that this gave rise to commercial success.

In many cases, the prior art will be so close that it is quite clear that the application will never be allowed without a showing; in others, the prior art is remote and a showing is clearly not needed. In between is a large grey area, in which the examiner alleges obviousness and in which a showing may not be strictly necessary, but would make allowance of the application much easier.

In such cases, it is advisable to avoid a showing if at all possible and to do everything in one's power to convince the examiner by argument. There are three reasons for this

(a) a showing requires the investment of time and effort by research workers, which costs money and which diverts them from their main job of making new inventions;
(b) submitting a showing could be taken as an admission that the invention was prima facie obvious and required a showing to establish non-obviousness; and
(c) perhaps most importantly, a showing presents a point of attack to a competitor who may want to challenge the patent later.

If a showing has been submitted, it can be argued that the patent has been granted only because of the contents of that showing. If these contents can be discredited, the patent may well be held invalid. Worse, if it can be shown that there was any deliberate concealment of relevant facts, or any misrepresentation of the results, the patentee will have been guilty of inequitable conduct, which, as we have seen, can have consequences beyond the invalidity of the patent. The same patent granted as a result of argument rather than a showing has a much better presumption of validity, since the examiner has been convinced that the invention is not prima facie obvious, and that conclusion is difficult to rebut.

For these reasons, submissions of a showing should, where there is any doubt, be regarded as a last resort. The same goes for declarations about commercial success and such matters. It is far better if the patent attorney can work out why there was commercial success and relate this to the nature of the invention in such a way as to be evidence of prima facie unobviousness. It will in many cases be preferable to go on appeal rather than carry out a showing, although this must be balanced against any possible negative effects of delayed issuance of the patent.

Problem and Solution in the European Patent Office

As discussed in Chapter 4, the EPO adopts the 'problem and solution approach' to obviousness: the invention is seen as the solution to the problem of getting from the closest prior art to the advantageous new result. It is for this reason that the EPO wishes the applicant to include a discussion of the prior art in the specification, so that the problem and the solution will be apparent to the reader. This is all very well in theory, but in practice the inventor often does not know what is the closest prior art and the 'problem' is a fictitious one arrived at by analysis with the benefit of hindsight.

The EPO Guidelines[25] instruct the examiner to take a three-step approach in assessing inventive step:

(a) identify the closest prior art;
(b) identify the technical problem, based on the differences between the closest prior art and the invention, whether or not that was what the inventor considered to be the problem; and
(c) consider whether there is any teaching in the prior art as a whole that would prompt the skilled person, faced with the technical problem, to modify the closest prior art, taking account of such teaching, so as to arrive at what is claimed.

From the point of view of the applicant or the representative, an argument against an objection of lack of inventive step should be directed to one or more of these three points. Thus one can argue that the prior art document considered by the examiner to represent the closest prior art is not indeed the closest, that the objective technical problem is not that stated by the examiner (it need not be that stated in the application, either), or that the necessary conditions for the third step are not met. This might involve arguing that the skilled person would have had no reason to combine the references relied upon, or that, even if they were combined, the teachings would not make the invention obvious to the skilled person.

It may also be argued that the invention has surprising advantages, or that it overcomes a technical prejudice in the art. Other secondary considerations of inventive step, such as long-felt want and commercial success, may be argued, but whereas it

[25] Guidelines C IV 9.8.

is considered evidence of inventive step if the invention solves a problem that workers in the field have been trying unsuccessfully to solve for some time, commercial success is given little weight, because this may be due to unrelated factors, such as good marketing.

Note that if claims are cancelled or limited in scope so as to overcome an objection such as lack of inventive step, the EPO takes the position that the excluded subject matter has been abandoned, unless the applicant specifically states that he reserves the right to pursue the excluded matter in a divisional application.[26]

Interviews with the Examiner

During prosecution, both in the EPO and the USPTO, it can often be very useful to have an interview with the examiner rather than to carry out the entire prosecution by written communications. Often, minor matters can be quickly and easily clarified by a telephone conversation, and this may be instigated by the examiner, particularly in a situation in which the application requires only a few formal amendments in order to be allowable, and the examiner wants to get the file off the desk and to be able to record another application disposed of. In more difficult situations in which there is a serious difference of opinion about the relevance of a piece of prior art, or in which the patent attorney feels that the examiner has not understood the invention, a personal face-to-face interview may be very useful; in the latter situation, it may be very helpful to bring the inventor along to explain the invention. In this way, clarification can often be obtained in a few minutes of a point that would take months to sort out by exchange of letters.

It should also be borne in mind that many examiners welcome an interview with an attorney, particularly if an inventor is present, as a break from the normal routine of their work and may be more favourably inclined to listen to an argument presented at an interview than the same argument in a letter. But an examiner will justifiably resent an interview that has no real purpose and which may be seen as a waste of time. In the EPO, although the applicant has a right to oral proceedings, the grant of an informal interview is at the discretion of the examiner. The applicant's attorney must be careful that, in requesting an interview, the right to formal oral proceedings is not relinquished. Wording such as 'In the event that the examiner is considering the refusal of the application, the applicant requests an interview with the examiner as a preliminary to oral proceedings' may be used.

In both the EPO[27] and the USPTO, there should be a written record of an interview and any substantial outcome must be recorded in writing. In practice, however, such records are often cryptic in the extreme, which makes life difficult for a third party who may later wish to challenge the validity of the patent. Whereas a file

[26] J 15/85 (OJ 1986, 395) *Abandonment of claim.*

[27] Guidelines, C VI 6.2.

inspection will show in detail all written arguments made by the applicant and why the examiner rejected or accepted them, when one comes to a half-page record of an interview that says something such as 'The representative of the applicant argued that the invention showed inventive step over document A, and after discussion it was agreed that inventive step was present', the reader has no idea what it was that persuaded the examiner to allow the application. This situation may be changing somewhat in the USA, where the increasing importance of the written prosecution history may lead examiners to write longer interview reports, and even to revisit and renegotiate the result of the interview.

PART V

COMMERCIAL EXPLOITATION OF PATENTS

invention to the UK could be considered to be the inventor. It may at first appear a simple matter to decide who the inventor is, but this is really so only in those rare cases in which only one person is involved in the matter from start to finish, and can come to the patent attorney with a complete working invention and say: 'This is all my own work.' Real-life situations are usually more complicated than this. In a commercial research organization, inventions are seldom stumbled upon by accident (although this can happen).[2] More usually, they arise out of planned research in which research scientists are set goals by their management, and are helped to achieve them by laboratory assistants and other technical staff who may run tests, carry out analyses, and perform more or less routine operations. Just where in the sequence of operations the invention occurs is something that must be determined specifically for each case, although there are guidelines that may be used to help.

The first basic rule is that invention is the mental act of conception of the inventive idea. The person who makes this mental step is the inventor and will be the sole inventor if he alone conceived the idea in its full operative form, even though it may require the work of others to put it into effect. If the invention can be put into effect, or reduced to practice, by routine work once the idea is there, then the person who does that routine work may be regarded as the 'extended technical arm' of the inventor and is not himself an inventor.

Suppose a chemist thinks that compound X may be useful as an insecticide, and instructs his lab technician to prepare it by some standard synthetic method and then sends it for routine screening. If it does turn out to be an insecticide, the invention (which may be claimed in the form of the compound per se, the process for making it, an insecticidal composition containing it, a process for killing insects by applying it, or any combination of these) is the invention of the chemist alone and is not the joint invention of him and one or both of the people who made the compound and tested it for the hoped-for use.

Of course, it may happen that unexpected difficulties arise in the synthesis. The reaction may not go in the expected way, and it may be, for example, that the technician finds a solution to the problem that goes beyond mere routine adjustment of reaction conditions. In such a case, the technician will have made an inventive contribution and will be a co-inventor.

Similarly, if the person who carries out the tests finds that compound X is not an insecticide, but has the idea of testing it as a fungicide, then, if X is a fungicide, the person who discovers this use is also a co-inventor. Note that he must have the idea himself to do the additional testing; if a new compound is put through a battery of standard tests for a whole range of possible utilities, then it is the person who sent the compound for testing, and not the tester, who is the inventor (if there is any invention at all; of course if the compound is not useful for anything, no invention

[2] The invention of Teflon® at Du Pont is a good example.

has been made). To be a co-inventor, one must contribute to the idea, rather than to the work.

It may be argued that the writing down of a list of compounds to be made and tested may be just as routine a job for a chemist as the running of a standard reaction or the carrying out of a standard test method. Indeed, one could program a computer to print out formulae of random compounds, and one could then synthesize and test the novel ones. If one of them were useful, novel, and non-obvious, there would be an invention—would the computer then be the inventor?

The answer is that there is no necessary correlation between the inventive step needed for a patentable invention and the amount of mental effort required to produce it. The writing down of the compound by the chemist may be the product of a great deal of thought or may be done more or less at random: this does not affect the fact that this is the actual point of invention. Indeed, it is quite possible for the actual inventive step to involve the least mental effort of any of the steps necessary to attain a working invention and it is a defect of most schemes of compensation for employee inventors that they reward only the persons who according to patent law are the inventors, even though others who are not inventors may have contributed more.

In the illustration given above of a computer-generated compound being a patentable invention, the inventor would presumably be either the computer programmer or, if there is such a person, the one who set up the whole project, instructed the programmer, and chose which tests should be carried out. This brings us to the question of whether a scientist's supervisor or manager will be a co-inventor. The answer depends on the extent to which the supervisor suggests solutions as well as sets problems. It is not usually an inventive act merely to pose a problem for someone else to solve, but if a particular line of approach is proposed that proves fruitful, that suggestion would probably be an inventive contribution.

It is an oversimplified picture to portray inventions being made only in a hierarchical system of supervisor–scientist–technician. More commonly, two or more researchers will work on a single problem, and they will be constantly exchanging ideas with each other, with their assistants, and with their supervisors. It is often difficult in such circumstances to determine who contributed to a particular invention, and the task becomes almost impossible when inventions arise out of team meetings or brainstorming sessions after which the participants often cannot remember who said what. Danish law previously got around this problem by allowing the company itself to be named as inventor in such cases, but this was unique to Denmark and the provision has since been abolished; all countries now require an inventor to be a real person. Correct designation of inventorship is important for several reasons. In the USA, the validity of the resulting patent may be attacked if the patent has not been granted to the correct inventor or inventors, and an earlier invention made by a different 'inventive entity' may be prior art in a situation in which one made by the same inventors would not be. In all countries, the ownership of the patent rights will be affected by the inventorship unless all of the possible inventors are such that their inventions would be owned by the same person (usually their employer), and in some countries, including Germany and the UK, employee-inventors may qualify for extra compensation from their employers.

In the UK, the inventor has the right to be named as such on the patent, whether or not he owns any rights in it.[3] Before the patent application is published, Form 7 must be filed naming all of the inventors,[4] and unless all of the inventors are also applicants or the application is one claiming Convention priority from an earlier foreign application, enough copies of this form must be provided for the Patent Office to send a copy to each of the inventors. Presumably this is intended to prevent patenting by an employer without the knowledge of the inventors, since it is no longer necessary for the inventors to sign the application form. There are provisions for any person who considers that he should be named as an inventor to apply to the Patent Office for his name to be added.

Conversely, a person who is an inventor may choose not to be mentioned as such on the patent and in any event the home address of the inventor need not be given. Regrettably, scientists in the UK who work on projects in which animal testing is involved are liable to be physically attacked by criminals who think that their concern for animals gives them the right to injure humans; such scientists may understandably be reluctant to have their names and addresses on a patent in which animal test methods are mentioned.

Inventorship in the European Patent Office

The European Patent Convention (EPC) provides that the inventor must be designated in a European application,[5] that the inventor has a right vis-à-vis the applicant to be so designated,[6] and that the name of the designated inventor will be mentioned in the published European application and the European patent specification, unless he waives the right to be mentioned.[7] The European Patent Office (EPO) will not, however, verify the accuracy of the designated inventorship,[8] and any dispute as to who is or is not the real inventor must be settled by national law.[9] Incorrectly designated inventorship may be corrected later under the provisions of Rule 21, according to which the consent of the wrongly designated person is required. Thus to add an inventor requires only a request (by the applicant or with his consent), but to remove someone as an inventor requires the consent of that person. There was one

[3] Section 13(1), PA 1977.
[4] Section 13(2), PA 1977.
[5] Article 81, EPC; r. 19(1), IR.
[6] Article 62, EPC.
[7] Rule 20(1), IR.
[8] Rule 19(2), IR.
[9] This seems to follow from Art. 74, EPC—see below.

case in which the responsible division of the EPO refused to add one new inventor to a list of co-inventors without the consent of all, on the basis that, because the initial inventorship as incorrect, everyone was wrongly designated. This over-formalistic decision was later overturned by the Legal Board of Appeal (LBA), which held that the consent of an already-named inventor is not required in order to be able to add a further inventor.[10]

Inventorship in the USA

In most countries in the world, although the inventorship of a patent may be relevant to the question of who owns the rights, it is irrelevant to the validity of the patent, at least as long as the applicant had the right to apply. In many countries, indeed, the inventor need never be mentioned and his or her name does not even appear on the patent. For the USA, however, it is important to have the inventorship correctly determined according to US law. Each US applicant must sign an oath or declaration that that person believes himself to be the (or a) true inventor, and a patent for which the inventorship is incorrect may be invalidated. Although it is not essential for the same inventors to be named in the USA as in other countries, any discrepancy between the stated inventorship for the US patent and for equivalent patents in other countries presents a point of attack that could be used by others to allege invalidity; it is therefore desirable to name the same inventors in all countries. This in effect means that US practice on inventorship is applied to inventions made in the UK and other countries, at least if there is any possibility that the case will be foreign-filed in the USA. One advantage of the UK Patents Act 1977 is that it is no longer necessary to indicate the inventors at the time of first filing; the matter can now be left to be finally determined at the foreign-filing stage one year later.

For the USA at least, inventorship is a purely legal question, and not one of emotion or company politics. Most patent attorneys have come across the research manager who insists that his name must go on every patent arising from his group. In some cases, perhaps in most, he may actually be a co-inventor, but the facts must be determined for each case separately. Political choice of inventorship is also evident in many patents originating from research institutes in Russia or Japan, which have been known to carry 15 or more names. The all-time record may well be a US published patent application[11] with no fewer than 75 inventors, corresponding to 1.67 inventors for each word in the main claim. Those applicants clearly have the policy to include the whole research group as inventors, which may or may not be a fairer system, but which certainly does not correspond to the legal concept of inventorship in the USA, any more than the Danish 'company invention' did.

10 J 8/82 (unpublished) *FUJITSU/Designation of inventors.*
11 US 20030028685.

Prior to 1953, it was not possible to change an incorrect inventorship on a US patent; nowadays, this can be done, as long as the original designation was made without deceptive intent. It was originally the case that at least one inventor had to be the same after the change, that is, one could change A + B into A + B + C, A + C, or A alone, but not into C alone. It is now even possible to substitute one single inventor by another, but this does not mean, of course, that the patent owner or his attorney is relieved of the responsibility of trying to get it right first time.

An issued US patent is presumed to name the correct inventors, so any inventorship challenge must bring 'clear and convincing evidence' that a newly named inventor contributed to the conception of the claimed invention. This must be corroborated by contemporary documents or by independent witnesses.[12]

The inventorship of a patent or patent application in the USA may have other consequences. For example, a patent application filed by B before the invention date of A, and subsequently granted, is prior art against a later application of A. In the USA, the inventive entity (A + B) is legally different from A or from B; this means that an earlier patent to (A + B) may be prior art against a later application by A alone or by B alone in circumstances (for example where there is not common ownership of the applications) in which the earlier patent to (A + B) could not be prior art to a later joint application by A and B together. Another consequence of joint inventorship for inventions made in the USA is that a co-inventor cannot be used to corroborate an invention date for the purposes of an interference. If the whole research group are co-inventors, there will be no one left who can give corroboration.

It used to be the case that if it was desired to incorporate a new development into a pending US application by means of a continuation-in-part (cip) application, this could be done only if the inventorship for the new development was the same as that for the original application. Since 1984, it has been possible for a cip to have different inventorship from the parent case, as long as the cases have at least one inventor in common.

Joint Inventorship in the USA

The same legislation relaxed considerably the previously strict requirements for joint inventorship. Previously, there had to be real collaboration between the co-inventors and some courts had held that, in a joint application, all inventors had to be co-inventors for each and every claim, considered separately. This absurd idea was laid to rest by the amended law, which states clearly that inventors may apply jointly even though they did not physically work together or at the same time, each did not make the same type or amount of contribution, or each did not make a contribution to the subject matter of every claim of a patent.[13] Nevertheless, there

[12] *Oren Tavory v. NTP* (Fed Cir 2008) (non-precedential).

[13] 35 USC 116.

must be some collaboration between the individuals for them to be co-inventors. A patent attorney cannot file a single patent application covering the inventions of two or more inventors, whether employees of the same or different companies, who had no contact with each other at the time that they made their respective inventions.

Since as a rule US courts are more critical of omitting a real inventor than of including a non-inventor, it is generally desirable in case of doubt to name joint inventors rather than a single inventor. One should not go too far, however: it is *a priori* unlikely that more than three or four people could really have contributed to a single inventive step and, incidentally, one frequently notes that the more inventors named, the more trivial the invention. Really good inventions tend to be the work of one or two people, not a committee.

A recent case in the USA[14] has set out in detail the criteria for determination of inventorship and is worth reading if a more complete picture is required. All disputes about US inventorship must, however, be settled by a qualified US patent attorney.

Ownership

The ownership of the rights in an invention, whether or not it is patentable, may be regulated by common law, by contract, or by statute law. In the USA, only the first two of these are relevant; for inventions in the UK, the statutory provisions of the Patents Act 1977 now apply. The EPC provides that national law applies to European applications treated as property.[15] It does, however, contain provisions as to the right to be granted a European patent (see below).

Common Law Provisions

Under the common law, the first premise is that the rights in the invention, including the right to apply for and be granted a patent, belong to the inventor. This presumption no longer holds good, however, if the inventor was being paid or commissioned to make the invention on behalf of someone else. Then the principle followed was that the rights belong to the person who paid for the work to be done. When the inventor was an employee, the rights were held to belong to the employer if it was part of the duty of the employee to make inventions or if the employee was in so senior a position (for example a general manager or director) that he had a general duty to further the interests of his company.

If the employee was not in either of these positions, then he retained the rights to the inventions, even if these were made using the facilities of the employer and related to the employer's business. In the USA, but not in the UK, it was held that if

[14] *Board of Education et al. v. American Bioscience Inc. et al.* 67 USPQ 2d 1252 (Fed Cir 2003).
[15] Article 74, EPC.

the employer's facilities were used, then although the employee owned the invention, the employer had a 'shop right' to use it. A shop right was equivalent to a royalty-free non-exclusive licence, which could be transferred with the business, but not transferred separately.

Contracts of Employment

Because the terms of a contract could be used to overrule the provisions of common law, more and more companies began to insert into their contracts of employment clauses in which the employee promised in advance to assign all inventions to the employer. Such clauses became the rule for all employees having individual contracts of employment, whether or not they would normally be expected to invent in the course of their duties. Sometimes such clauses were drawn up so broadly that they required the employee to assign all inventions, even those having no connection with the employer's business, and even those made after termination of the employment.

Both in the USA and the UK, the courts frequently had to review such contracts as a result of disputes between employers and (usually) ex-employees. Some of the most restrictive terms were sometimes held to be unenforceable as being in unreasonable restraint of trade. For example, a clause requiring assignment of future inventions after termination of employment would, if enforceable, make the ex-employees unemployable anywhere else. Such 'trailer clauses' are, in the USA, considered enforceable only for a limited time (say, up to one year) and only if the invention arises out of the former employment.

In the UK, there was an important case on employee inventions in 1977 that was decided under the old law.[16] Hudson was employed as a storekeeper by Electrolux, and invented an adaptor that allowed a cheap paper dust bag to be used on an Electrolux vacuum cleaner, instead of the more expensive bag that Electrolux supplied. His standard staff contract of employment stated that the company owned all inventions made by staff relating to 'any articles manufactured and/or marketed by the company or its associated companies in the United Kingdom or elsewhere'. The judge held that this clause was unenforceable: firstly, Hudson was a storekeeper, not a research worker, and was not paid to invent; secondly, even a research worker should not be required to assign inventions that might have no relation to the field in which he was employed. No US court has gone so far as this in overturning the terms of a contract of employment, and agreements to assign all inventions in the USA are probably enforceable.

Statute Law in the UK

The Patents Act 1949 made it possible to apportion rights between the employer and the employee.[17] This approach failed, however, because, the House of Lords

[16] *Electrolux v. Hudson* [1977] FSR 312 (Ch D).

[17] Section 56(2), PA 1949.

held[18] that this provision could apply only when the invention did not, as a matter of law, belong wholly to one or to the other. Because, in the great majority of cases, the common law or contract law could be used to give one party or another the whole rights, the apportionment statute was a dead letter.

The Patents Act 1977 made some fundamental changes, theoretically to the benefit of the employee-inventor. As regards ownership of patent rights, the old common law provision was restated: the invention belongs to the employee unless it was made in the course of his normal duties or in the course of duties specifically assigned to him, and the circumstances in either case were such that an invention might reasonably be expected to result from the carrying out of those duties, or the employee had a special obligation to further the interests of the employer's undertaking. What is more, any contract that purports to limit the employee's rights in any future inventions is unenforceable.

In other words, no contract of employment for persons employed in the UK can now legally require the employee to assign to the employer all inventions that he may make. The only inventions that belong to the employer are those made in the course of the employee's duties, if such duties are expected to result in inventions. It is clear that a research scientist will still have to assign to his employer all inventions that he makes that are relevant to the research area assigned to him, but if he makes an invention not relating to his own work, even though it be in an area of interest to the employer, the invention is his, no matter what it may say in his employment contract. A person employed in a position, such as storekeeper, in which he would not be expected to make inventions, will own any invention that he may make, while the position of people in customer service, marketing, design, and production, will generally depend upon how their normal duties are defined.

Clauses requiring assignment of all inventions made by the employee must be dropped from employment contracts in the UK, but in order that there should be a reasonable degree of certainty about which inventions do belong to the company, thought must be given to a definition of the normal duties of the employee, and this should be reviewed from time to time as the employee moves to different positions or higher levels of responsibility within the company. It is quite reasonable that sales managers or production engineers, if they make an invention relating to a product that they are responsible for selling or producing, should have to assign such an invention to the employer, but their contracts of employment should make it clear that their normal duties include making such job-related inventions. No matter what is written in the contract, if an invention would not be expected to arise in the course of his duties, any invention that the employee makes is his own.

Some light is shed on this point by one of the first cases on employee inventions decided under the Patents Act 1977.[19] Harris was manager of the Wey valve

[18] *Patchett v. Sterling Engineering Co.* (1955) 72 RPC 50 (HL).

[19] *Harris' Application* [1985] RPC 19 (Pat Ct).

department of Reiss Engineering Co., which was UK distributor for the Swiss manufacturer of Wey valves. Harris had no written contract of employment. In the period between being given notice by Reiss and leaving the company, he made an invention relating to Wey valves and filed a patent application in his own name. Reiss sued for a declaration that it owned the patent. It was held that Harris was not expected to make inventions as part of his normal duties (which consisted mostly of sales and customer service), because the facts showed that no inventive activity at all went on within Reiss Engineering; all design problems relating to the valves were referred back to the Swiss company. Accordingly, Harris retained ownership of the invention. In another case, a doctor employed as a clinician in a Glasgow hospital who spent one day a week carrying out research work was allowed to retain ownership of an invention arising out of the research because only his clinical work was considered to be the normal duties of his employment.[20]

A further consideration is that the right of an employee-inventor to his own inventions does not remove his duty of confidentiality, and he would not be free to file a patent application for the invention if this could not be done without revealing confidential information owned by the employer.[21] The employee does not, however, infringe copyright or design rights of the employer merely by filing a patent application for his invention.[22]

Statute Law in Germany

The German law on employees' inventions dates from 1957 and, in 1959, directives were issued to provide guidelines to the amount of compensation to be paid. An Arbitration Board adjudicates disputes between employer and employee, both as to ownership of inventions and as to the amount of compensation. The law recognizes two types of invention: service inventions, which belong to the employer; and free inventions, which belong to the employee-inventor. An invention is a service invention if it arises in the course of the employee's duties, or if it is based upon experience or activities within the firm. Thus even if the employee is not paid to do research, his invention may belong to the employer if it is closely related to the firm's business: for example an improvement or development of a product of the company.

An employee who makes a service invention is obliged to report it in writing to the employer. The employer then has a period of four months in which it may lay claim to the invention. If it does so, the rights automatically pass to the employer; if it does not, the invention becomes a free invention and the inventor may dispose of it as he pleases, including, for example, selling it to a competitor. If the employer does lay claim to it, however, it is obliged to file a domestic patent application (unless it is necessary to keep the invention as secret know-how, or the inventor

[20] *Greater Glasgow Health Board's Application* [1996] RPC 207 (Pat Ct).

[21] Section 42(3), PA 1977.

[22] Section 39(3), PA 1977.

agrees that no application need be filed), and it must pay compensation to the inventor. Further, if the employer decides not to file patent applications abroad, or to abandon any applications or patents once filed, it must give the inventor the opportunity to file in his own name, or to take over the patent rights.

The employee-inventor is also obliged to report to his employer any invention that he considers to be a free invention, unless it is obviously incapable of being used in the employer's business. The employer can agree that the invention is a free one, or may allege that it is a service invention; in case of dispute, the Arbitration Board can be used. The employee cannot exploit his free invention while he remains employed by the company without offering the employer at least a non-exclusive licence on reasonable terms, but the employer must take up this offer within three months. Thus an invention that is originally a free invention cannot be exploited so freely by the inventor as a service invention that becomes free by the employer failing to claim it. Like the new UK law, the provisions of the German law on employee inventions are mandatory and cannot be set aside by a contract.

Academic Inventors

In Germany, university professors, lecturers, and assistants were outside the provisions of the law on employee inventions, so that any invention that they made in the course of their employment by the university was a free invention. Since February 2002 this been changed, and inventions made by academics after that date will normally be owned by the university. Within the UK, the same provisions that apply to the employee of a company or an individual apply also to employees of government departments, so that scientific civil servants will also be able to keep for themselves inventions made outside of their normal duties. The position of academic scientists is less clear. They will generally be employees in as much as they will have a contract of employment with a university, but the position is often complicated by the receipt of research funds from the government or from private industry. The obligations of the academic inventor to all of these parties should be clearly set out in written agreements; otherwise, considerable confusion can result. It would seem, however, that a contract entered into between a company providing money for research and an academic scientist who is not an employee of the company is not subject to the limitations placed by the Patents Act 1977 on contracts of employment. A consultant may certainly be required to assign in advance any inventions that he may make in the field of his consultancy.

In the UK, in cases in which government research funding is involved, the government used to require assignment of the patent rights to the National Research Development Corporation (NRDC), which tried to commercialize them, usually by licensing to industry. This is no longer obligatory, however, and the funding body, for example the Medical Research Council, may use the British Technology Group (successor to the NRDC), may patent the results itself, or may leave the rights to the inventors. As far as the universities themselves are concerned, the majority now assert rights to inventions made by their academic employees. In the USA, universities seem to be more patent-conscious, and the majority of them now have their own technology transfer offices to commercialize their inventions and may also employ their own patent lawyers. The situation in which US government funding is involved used to be that the government retained rights and granted only non-exclusive licences to industry. Because non-exclusive rights were not very attractive, there were few takers. In 1978, the US government owned 28,000 patents and licensed only 5 per cent of them.[23] In the same year, Senators Bayh and Dole introduced a Bill to enable universities to retain patent rights to inventions funded with federal money. The Bill became law in 1980 and, since then, licensing of inventions made by academic scientists has expanded enormously. The US government retains certain rights, including 'march-in' rights in the event that licensees do not adequately exploit the licensed technology by making the invention available to the public on reasonable terms. To date, these rights have never been used. For example, in 1997 a request by CellPro Inc. that the National Institutes of Health (NIH) exercise its 'march-in' rights to force Johns Hopkins University to license CellPro under stem cell selection patents invented with the help of NIH funding was denied. Between 2004 and 2007, the well-known activist James Love, acting through an organization called Essential Inventions Inc., petitioned the US government to use march-in rights to force compulsory licensing of certain pharmaceuticals alleged to have been developed with the help of federally funded research and alleged to be overpriced. None of these petitions have been granted.

In continental Europe outside Germany, academic scientists seem to have greater freedom to own their own inventions and universities seldom assert title. In Switzerland, inventions funded by the Swiss National Science Foundation (SNSF) may be owned by the inventors; in the event that the invention is a commercial success, the SNSF seeks repayment of the funding and some share of the financial benefits.

Compensation for Employee-Inventors

The UK

The other main change brought about by the Patents Act 1977 is the provision for compensation for employee-inventors. The British law differs from that of Germany in some important aspects.

The provisions apply both in cases in which the invention is owned by the employer and those in which it is owned by the employee but the employer has been granted an assignment or an exclusive licence for a lump sum or royalty payment, or other consideration. If the invention is owned by the employer, compensation is to be paid by the employer to the employee-inventor when a patent has been granted and the invention or the patent is, having regard to the size and nature of the

23 Former Senator Birch Bayh [1996] Special issue *The MIT Report* 3.

employer's undertaking, of outstanding benefit to the employer.[24] When the invention is owned by the employee and assigned or licensed exclusively to the employer, additional compensation is payable if the benefit to the employee from the contract of assignment or licence is inadequate in relation to the benefit derived by the employer from the invention or the patent.[25] In this case, it seems that the patent does not have to be of outstanding benefit. It is not possible for the employee to contract out of his right to compensation.

This is all rather complex, but the main point is that, in the UK, compensation is not due as a matter of course on all commercialized employer-owned inventions, as it is in Germany, but only on ones of 'outstanding benefit to the employer', whatever that means. As of writing, only four court cases on this subject have been reported, and in only one of them did the court consider that there was any 'outstanding benefit' or that any compensation should be paid. In that case,[26] the outstanding benefit to the employer was valued at £50 million, and the two inventors were awarded 1 per cent and 2 per cent, respectively, of this sum.

The system may still operate unfairly in a number of ways. Given that the object is to provide additional compensation for employees who give the company something above and beyond what is expected of them, it is worth considering whether this end will be achieved in view of the following problems.

(a) Only *inventors* can benefit. As we have seen, in the work involved in commercializing an invention, the actual inventive step may involve the least effort.
(b) A patent must be granted. If the invention is kept as secret know-how, or if a patent is refused because of prior art, no compensation will be paid, no matter how important the invention.
(c) Because the size of the company is taken into account in considering the question of 'outstanding benefit', it seems that inventors in a large company are less likely to get compensation than those in a smaller one.
(d) The compensation is due only from the employer at the time that the invention was made; if the employer sells the business, the employee has no claim against the new owner.
(e) The commercial success of an invention bears no necessary relationship to its inventive merits.

The 1986 edition of this book took the view that:

> The only sensible way to regard the British law on compensation for employee-inventors is that it gives the inventor a ticket in a lottery. He may be lucky or he may not, but at least he is better off than he was before, when he did not even have the ticket.

Finally, 23 years later, the first winning number has been drawn.

24 Section 40(1), PA 1977, as amended by s. 10, PA 2004.
25 Section 40(2), PA 1977, as amended.
26 *Kelly and Chiu v. GE Healthcare* [2009] EWHC 181 (Pat).

Germany

If a service invention is used by the employer, then compensation is payable according to the system set out in the 1959 directives. The first step is to determine the invention value (IV), which is usually assessed on the basis of the royalty that the employer would have to pay for an exclusive licence if the invention were a free invention and it had to negotiate for it at arm's length, less overheads such as the cost of patenting. Less frequently, it may be based on the savings made to the company (for example by a process improvement that cuts costs), reduced by a factor to allow for the company's overheads and investment. An IV may also be assigned to a defensive patent that is not actually worked. Once the IV has been determined, the next step is to calculate the participation factor (PF). This is done on a points system based on consideration of how the problem was posed, how it was solved, and what was the employee's position in the company.

(a) In the first category, the scale goes from 1 (the employer posed the problem and indicated the approach to be taken) to 6 (the employee posed himself a problem falling outside of his normal range of duties).
(b) In the second category, the scale is also 1–6 and depends on how much contribution was made by the employer, including facilities and assistance, towards solving the problem: the less help from the employer, the higher the point rating.
(c) Finally, the position of the employee is rated on a 1–8 scale: a research director making an invention gets one point, a research chemist, four points, and an unskilled worker, eight points.

These three categories are totalled to give a point count of from 3 to 20, and the PF is read off as a percentage from a table: for example three points gives a PF of 2 per cent; eight points gives 15 per cent; 15 points, 55 per cent; and 20 points, 100 per cent. Multiplying the IV by the PF gives the compensation due to the inventor. In the majority of cases, the PF lies between 15 and 21 per cent, so that the compensation is 15–20 per cent of a net notional royalty.

The German system applies to all inventions that are of commercial value and not merely to those of exceptional benefit. The philosophy applied here is essentially different. The UK and US case law talks of employees being 'hired to invent'; by German thinking, this is not possible, because no one can guarantee to produce inventions. Inventions may be made in the course of an employee's duty and they will then (if claimed) belong to the employer, but they are always regarded as something extra over and above what the employee is being paid to do. From the point of view of the employer, the German system has the advantages that service inventions are more broadly defined than the equivalent employer-owned inventions in the UK Patents Act 1977 and that the employer must be offered a licence for free inventions.

Other Countries

Germany and the UK are by no means the only countries that provide for compensation for employee-inventors. Austria has provisions similar to those of Germany, but without the detailed implementing regulations, the amount of any award being at the discretion of the courts. The laws of the Scandinavian countries, the Netherlands, Japan, and France, while differing in the categories of invention that belong to the employer, all provide for additional compensation for an inventor employed to do research, at least if the invention is of unusual commercial importance. In Japan, a Supreme Court decision[27] in 2003 held that an employer can legitimately own an employee's invention only if the employee has been given 'fair compensation' over and above his salary for contributing to the invention. Most Japanese companies provide fixed and rather small sums by way of compensation, but the decision states that this may not be enough and the employee has the right to sue for additional payments. The case in question resulted in an award of only 1.4 million yen (¥)—that is, US$19,000—less than 1 per cent of what was claimed, but still more than any employee inventor had previously received in Japan.

More recently, much higher awards have been made: the Tokyo High Court ordered Hitachi to pay ¥163 million to an ex-employee for his invention on optical disc technology.[28] Even this was dwarfed by an award of ¥20 billion (US$190 million) by the Tokyo District Court against the firm Nichia to the inventor of the blue-emitting diode, which was responsible for 60 per cent of the sales made by Nichia in 2001.[29] On appeal, the Tokyo High Court reduced this figure to ¥600 million, and the rule now seems to be that employee compensation cannot exceed 5 per cent of the profits due to the invention.

Conclusion

In Germany, a very great deal of time and effort has to be expended within company patent departments in order to resolve questions of inventorship and to deal with the complex administration of the law on inventor remuneration. Determination of inventorship can be difficult enough when all that is involved is the honour of having one's name on a patent: when money is at stake, it can only make things worse.

As some German commentators have remarked, even after 40 years of the German Law on Employee Inventors, there was no real evidence that it had effectively encouraged inventive activity or significantly benefited inventors.[30] The best that can be said for it is that it gives a clear and reliable set of rules to regulate the transfer of ownership from the inventor to the employer. If this is all that is needed, a minor amendment to the German Patent Law, without the massive edifice of the regulations on compensation, would do as well.

Suggestions are made from time to time that the German system on employee-inventor compensation should be extended throughout the European Union (EU).

[27] *Olympus Corp. v. Shumpai Tanaka Heisei* 1256 (Jap Sup Ct, 22 April 2003) (JYU).

[28] *Hitachi Ltd v. Seiji Yonezawa* 14 (NE) 6451 (Tokyo High Ct, 29 January 2004).

[29] *Nichia Corp. v. Shuji Nakamura* 13 (WA) 17772 (Tokyo Dist Ct, 30 January 2004).

[30] B. Hansen and P. Klusmann, '40 years of employed inventors law in Germany: A time to celebrate?' (1996) February *Patents and Licensing* 37.

If, as seems likely, it would bring extra work and costs without significant increase in inventive activity or significant additional benefits to the employee, the German liking for rules and regulations is not enough to justify such a proposal.

The Right to Apply for a Patent and to be Granted a Patent

In the UK

In the USA, the inventor, whether or not he owns the patent rights, must apply for the patent. In the UK, it seems that anyone may apply for a patent, although on the application form the applicant is required to state either that he is an inventor or that a statement of inventorship on Form 7 is, or will be, supplied. Presumably, someone not entitled to the grant of the patent could apply and sort out the ownership later, but this is not clear. The Patents Act 1977 is more specific about who has the right to the grant of the patent: this is either the inventor himself, or a person who, at the time the invention was made, owned the rights in it (for example an employer), or anyone to whom either of the first two have assigned the rights.[31] The Act provides for the Patent Office to settle disputes about who has the right to the grant (see below) and it is a ground of invalidity if the patent has been granted to the wrong person.[32]

In the European Patent Office

The EPC, like the British Patents Act 1977, is vague about who may file (any person, according to Article 58), but is more precise about the right to be granted the European patent, which belongs to the inventor or his successor in title.[33] It is made clear, however, that unlike the USA it is not the first inventor who has the right, but rather the inventor who has the earliest date of filing, provided that the application is published.[34] A person who files a later application for the same invention is entitled to a patent if the earlier application is abandoned or withdrawn before publication, whereas if the earlier application is withdrawn only after publication, it remains novelty-destroying prior art under EPC Article 54(3). For employee-inventors, ownership is determined according to the law of the country in which the employee is mainly employed. Although there is no prima facie limitation on who may file, it is required that, if the applicant is not the sole inventor, the designation of inventor must indicate the origin of the applicant's right to the European patent.[35] It is enough to say 'by contract', even if no contract exists at the time, because an incorrect statement may be corrected later; it is merely necessary to say *something* on the form. There may of course be joint applicants and one party may be applicant

[31] Section 7(2), PA 1977.
[32] Section 72(1)(b), PA 1977.
[33] Article 60(1), EPC.
[34] Article 60(2), EPC.
[35] Article 81, second sentence, EPC.

in respect of some designated states, while the other is applicant for the remaining states.[36]

Co-Ownership

In all countries the possibility exists that a patent may be jointly owned by two or more parties, and the question arises what are the rights of the co-owners, and in particular to what extent the co-owners can operate independently of each other. In most countries, the joint owners each have an equal undivided share in the patent, and it is not possible, for example, for one party to own 25 per cent of a patent and the other 75 per cent, although they could, of course, agree between themselves to split any profits from the patent in whatever proportion they wish. In the UK and in most other European countries, the position, in the absence of a contract with different provisions, is that each joint owner may work the patented invention independently without accounting to the others, but cannot assign his rights nor grant a licence without the consent of the others.[37]

In the USA, however, the situation is different: each joint owner may grant licences independently of the others.[38] This means that licensees under a US patent must be very careful that all persons who may possibly have rights in the patent are parties to the licence agreement. If this is not the case, nasty surprises can be in store, as Ethicon Inc. found to its cost[39] when it took an exclusive licence from an independent inventor who was sole inventor and owner of a US patent, and then sued its competitor US Surgical Corp. for infringement. Surgical found that an unpaid assistant of the inventor had made an inventive contribution to four of the fifty-five claims. It made a deal with this assistant, obtained a licence under his rights, and succeeded in having the assistant added as inventor. When this was done, the infringement suit had to be dismissed, because Surgical now had a valid licence under the whole scope of the patent, not only to the small part invented by the assistant.

The third revision to the Chinese patent law, which entered into force on 1 October 2009, makes Chinese law on co-ownership of patents the same as the US law in this respect. Foreign companies entering into joint ventures in China must consequently be careful to avoid co-ownership if possible.

Disputes as to Ownership

In the UK

Disputes may arise at any time as to the correct ownership of an invention, either before or after the patent is granted. The Patents Act 1977 gives the Comptroller

[36] Article 59, EPC.

[37] See, e.g., s. 36(2) and (3), PA 1977.

[38] 35 USC 261, 262.

[39] *Ethicon Inc. v. US Surgical Corp.* 45 USPQ 2d 1545 (Fed Cir 1998).

powers to decide these matters and to make the appropriate orders.[40] In the former case, the dispute may come before the Comptroller even if no patent application has yet been filed; in the latter case, a question as to ownership may be referred to the Comptroller within two years of grant by any person claiming a proprietary interest in the patent. The Comptroller may also determine entitlement to foreign applications and, for example, applications under the Patent Cooperation Treaty (PCT),[41] although this section can be effective only if both parties agree to accept the Comptroller's jurisdiction. In all cases, the Comptroller may refer the dispute to the court if this seems more appropriate: for example if witnesses have to be cross-examined upon oath. In any event, the Comptroller's findings may be appealed to the courts.

If, as result of a dispute as to ownership, an additional co-owner is added, licences granted by the previous sole owner or joint owners remain in force, but if the patent is transferred to a completely new owner or owners, existing licences are cancelled unless the licensee has already begun to work the invention.[42]

Between 2005 and 2007, there were a number of high-profile UK cases on entitlement. In *Markem v. Zipher*,[43] the Court of Appeal took the view that the entitlement to a granted patent was a different matter from the right to apply for a patent and that it must depend on 'some other rule of law', such as contract law or breach of confidence.The Court took the same line in the case of *Yeda v. Rhone-Poulenc*, but on further appeal to the House of Lords, it was held that, although the actual decision on ownership in *Markham* was correct, the principle set out by the Court of Appeal in these two cases was wrong.[44] As stated by Lord Hoffmann, section 7(2) and (3) of the Patents Act 1977 provides an exhaustive code for determining who is entitled to the grant of a patent. Entitlement depends upon inventorship, and 'there is no justification, in a dispute over who was the inventor, to import questions of whether one claimant has some personal cause of action against the other'.

Accordingly, the correct approach in a dispute of this kind is firstly to identify the inventive concept, and secondly, to determine who 'came up with', or 'contributed to the formulation of' this concept. Only then need one consider who else (such as an employer) may be entitled under the other provisions of section 7.

At first sight, this seems to leave open the possibility that a completely independent inventor might claim entitlement to a patent granted to someone else. In practice, this would not be allowed to happen, but it is hard to see exactly why not. In the EPC, it is stated explicitly that if there are two or more independent inventors, it is the first to file who has the right to the European patent, but the UK law merely relies upon a second filing being invalid for lack of novelty.

40 Sections 8 and 37, PA 1977.

41 Section 12, PA 1977.

42 Sections 11 and 38, PA 1977.

43 *Markem Corp. v. Zipher Ltd* [2005] RPC 31 (CA).

44 *Yeda Research and Development Compand Ltd v. Rhone-Poulenc Rorer Inc.* [2007] UKHL 43.

In the European Patent Office

Questions of ownership of a European patent application must be decided according to the appropriate national law, but the question of which national court has jurisdiction is a separate issue that is dealt with in the Protocol on Recognition, attached to the EPC. This provides that, in proceedings between an applicant for a European patent and another party claiming to have the right to the grant of the patent:

(a) if the applicant has his residence or principal place of business in an EPC contracting state, proceedings must be brought in that state;
(b) if (a) does not apply, then the courts of the contracting state in which the person claiming the right has his residence or principal place of business have jurisdiction; and
(c) if neither (a) nor (b) apply, the German courts have jurisdiction.

The term 'courts' includes other authorities having the power to decide such issues: for example the Comptroller in the UK.[45] There are some possible exceptions: for example if the dispute is between employer and employee, or if the parties have agreed upon the courts of a particular country.

This can give rise to some odd situations. In one case, there was a dispute as to inventorship and ownership between scientists engaged in collaborative research in two London research institutes. It was alleged that one party had wrongly filed a European patent application without involving the other party, who had learned of the situation only when the European patent application was published. By then, however, the application had been assigned to a Swedish company, so that, according to the Protocol on Recognition, the Swedish courts were competent. The claimant tried to get the English High Court to agree that it was entitled to hear and decide the issue of beneficial ownership of the rights, as distinct from the formal matter of the name that would appear on the granted European patent, and was successful at first instance. The Court of Appeal, however, held that this would simply evade the clear provisions of the Protocol, and reversed the decision.[46] Thus, a Swedish court had to apply English law to decide the rights of two sets of inventors resident in England to an invention made in England—or would have had to, were it not for the fact that the case was settled out of court.

Suspension of Proceedings

If a person starts entitlement proceedings before the appropriate court and gives proof of this to the EPO, the EPO will suspend proceedings on the application (or opposition proceedings on a granted patent) until a final decision has been given.[47] The proceedings must, however, continue as far as publication of the application.

[45] Article 1(2), Protocol on Recognition.
[46] *Kakkar v. Szelke* [1989] FSR 225 (CA).
[47] Rule 14, IR.

After the notification to the EPO, the applicant is not allowed to withdraw the application or any designations.[48] If the final decision is in favour of the claimant, the claimant has the option of taking over the continued prosecution of the application, filing a new application taking the same filing date as the original, or requesting that the application be refused.[49]

If a new application is filed, it is treated essentially as if it were a divisional application and it was always supposed that, as for a divisional, the original application had to be still pending at the time that the new application was filed. A case in which the original application had been abandoned before the rightful owner became aware of it came before the Enlarged Board of Appeal (EBA), which decided that co-pendency was not essential.[50] This was one of the very few decisions of the EBA in which a dissenting opinion was published. The dissenting Board members agreed that it was unjust that the claimant should have no remedy only because the wrongful applicant had allowed his application to be abandoned (the abandoned application was novelty-destroying prior art against the real inventor's own later application), but thought that this was outweighed by the interest of the public in being certain that an application that was dead could not be revived years later.

Recordal and Transfer of Ownership

In the UK, the ownership of a patent application or a patent is among the information recorded in the Register of Patents,[51] but it must be remembered that the entry in the Register of a person as patent proprietor is evidence, but not proof, that that person really does own the rights. A subsequent assignment to another party may have been made but not registered, for example. Registration of a transfer of rights is not compulsory, but certainly desirable, in view of the fact that only registration can really protect the assignee from the claims of another party who has a purported assignment from the original proprietor of later date.[52]

Similarly, in the EPO, the transfer of a European patent application may be recorded on request of one of the parties on production of documents attesting to the transfer: for example a copy of the written and signed assignment.[53] Registration is not compulsory, but the transfer is not effective vis-à-vis the EPO until the registration has been made.

[48] Rule 15, IR.
[49] Article 61(1), EPC.
[50] G 3/92 (OJ 1994, 607) *LATCHWAYS/Unlawful applicant.*
[51] Section 32, PA 1977.
[52] Section 33(1)(a), PA 1977.
[53] Rule 22(1), IR.

22

COMMERCIAL EXPLOITATION OF PATENTS

> [This trend], which everyone knows about, and [that trend], which is so incredibly arcane that you probably didn't know about it until just now, and [this other trend over here] . . . when all taken together lead us to the (proprietary, secret, heavily patented, trademarked and NDAed) insight that we could increase shareholder value by [doing stuff].
>
> Neal Stephenson, 'Business plan of Epiphyte Corporation', *Cryptonomicon* (2000)

Patents to Exclude the Competition: The Pharmaceutical Industry

There can be no doubt that patents are of greater commercial importance in the fields of chemistry, and particularly pharmaceuticals and biotechnology, than in other fields such as engineering or electronics. A new chemical compound may be imitated more easily and with less investment than a complex new machine or semiconductor device, and the patent protection available for the compound is more readily enforced, because it is normally easier to establish infringement. For pharmaceuticals, the value of patent protection is even more important than for chemicals in general. To bring a new pharmaceutical on to the market requires a vast amount of investment, the major part of which is spent in testing the compound for safety and efficacy, and only a very minor part on developing the synthesis of the product. It may be possible to produce the compound quite cheaply, but it will still be necessary to charge a high price in order to recover the money spent on testing not only on the compound that is finally marketed, but also on the others that do not reach this stage. In the absence of patent protection, an imitator who has to carry none of the costs of research and development could offer the compound at a much lower price and still make a handsome profit. Accordingly, the pharmaceutical industry, more than any other, is interested in using patents for their classical purpose, that is, that of excluding the competition for a limited time.

The pharmaceutical industry develops products that take a very long time to reach the market, but which may then continue to be sold for decades. Thus the industry usually has no particular interest in obtaining early grant of patents, but is nonetheless very concerned to obtain the maximum possible effective patent term. It is perhaps worth looking in some detail at the drug development process, in order to understand why it takes such a long time.[1]

Development of a New Pharmaceutical

The following is a summary of the drug development process.

1. **Lead finding** (One to two years)
   - 1.1 *Research planning* This may include classical structure–activity correlations, rational drug design, and high-throughput screening of libraries obtained by combinatorial chemistry or from natural sources, etc.
   - 1.2 *Obtaining test compounds or samples* This involves laboratory scale preparation, the preparation of the compound or sample libraries, a determination of *in vitro* or animal models to test activity, and the setting up of high-throughput screens.
   - 1.3 *Screening* This stage includes basic pharmacological and biochemical screening, and the selection of 'hits' and identification of active compounds. It is at this stage that the patent application for the compound will normally be filed.
2. **Preclinical trials** (Four to six years)
   - 2.1 *Preclinical trials stage I* This will involve animal testing for acute toxicity, detailed pharmacological studies (main effect, side effect, duration of effect), analytical methods for active substance, and stability studies.
   - 2.2 *Preclinical trials stage II* This will involve pharmacokinetics (absorption, distribution, metabolism, excretion), subchronic toxicity, teratogenicity, mutagenicity, the scale-up of synthesis, the development of the final dosage form, and the production of clinical samples.

[1] See, e.g., H.P. Rang (ed.), *Drug Discovery and Development*, Elsevier, 2006.

3. **Clinical trials** (Four to six years)
   3.1 *Phase I* This will test tolerance in healthy volunteers, bioavailability and pharmacokinetics in man, and supplementary animal pharmacology.
   3.2 *Phase IIa* Exploratory trials in small groups of patients will be designed to give preliminary evidence of efficacy and safety, sometimes referred to as 'proof of concept' (POC).
   Phase IIb Trials in larger groups of patients will confirm efficacy with statistical significance, and determine optimal dose and dosing regimen. There will also be chronic toxicity and carcogenicity studies in animals.
   3.3 *Phase III* A large-scale trial at several centres will finally establish the therapeutic profile (indications, dosages and types of administration, contraindications, side effects), proof of efficacy and safety in long-term administration, demonstrate therapeutic advantages in comparison with known drugs, and clarify interactions with other medication.
4. **Registration, launch, and sales** (Two to three years)
   4.1 *Registration with health authorities* This will demand the documentation of all relevant data, expert opinions on clinical trials and toxicology, preparation for launch, preparation of information for doctors, wholesalers, and pharmacists, the training of sales staff, the preparation of packaging and package inserts, and the dispatch of samples.
   4.2 *Launch and sales* This will include the production and packaging of the final form, along with quality control.

The period from the first patent filing to the marketing of the new drug is therefore typically anywhere from seven to 14 years. It is of course clear that if a strongly negative result is obtained at any stage of this process, the entire project must be abandoned. It is estimated that approximately 5,000 compounds are found to have activity in early testing for every one that finally is marketed; even for compounds that go into Phase I clinical testing, only one in five reach the market. In view of the high numbers of drug candidates that fail in full clinical trials (about 50 per cent), considerable efforts are being made to streamline the drug development process by running more activities in parallel and by introducing early POC studies to eliminate poor candidates at an earlier stage. Nevertheless, and despite all of the advances in high-throughput screening, genomics, and other technologies, the productivity of pharmaceutical research is declining steadily. In 2008, the US Food and Drug Administration (FDA) approved a total of only 21 new chemical entities (NCEs).

Estimates of the costs of drug development are contentious, because there is no basic agreement on how these costs should be calculated. The simple out-of-pocket expenditure on the development of one successful drug (typically US$100 million–US$200 million) is only the starting point. One must also take into account the amount spent on compounds that failed to reach the market. A survey reported in

2003[2] found that, up to 1999, on the basis of the total research and development expenditure on NCEs (excluding research and development on new formulations and uses of existing drugs), the average cost for a new drug product was US$403 million, and that taking into account the cost of capital in view of the long period between research expenditure and a positive cash flow for the product, this figure doubled to US$802 million. Another study, based on the period 2000–02, came up with the figure of US$1,700 million, but this figure is not comparable, because it includes elements such as marketing costs on launch.[3] Many people are under the impression that the majority of new drugs are discovered by university and government researchers, or that drug development costs are largely government funded. In fact, of the 196 NCEs approved in the USA between 1981 and 1990, 181 (92 per cent) were discovered by private industry[4] and only a few drugs for rare diseases received significant government funding for their development. Some people even appear to believe that a new drug is a public good that it is the duty of the pharmaceutical industry to supply to everyone at cost price and that any attempt to recover the huge development costs is a form of exploitation.

This enormous investment is finally recouped on the sales of a compound that may be simple enough that any good chemist could produce kilogram quantities of it in a backyard laboratory. In the absence of patent protection for the compound, sales by imitators with no research overheads would destroy any possibility of the innovator recovering its investment and consequently the investment would not be made. The value of patent protection to the research-based pharmaceutical industry may be seen by the effect of patent expiry upon the market, when the introduction of generic drugs in the USA often drops prices by 80 per cent. It is estimated that, in the year following the patent expiry of Zantac®, Glaxo Wellcome lost US$1,650 million in sales worldwide;[5] when Eli Lilly's patent for Prozac® expired in the USA, 60 per cent of sales were very quickly lost to generics, amounting to US$1,500 million per year, or over US$4 million per day in the USA alone.

Effective Term of Pharmaceutical Patents

The point in the development programme at which a patent application is filed will vary somewhat from company to company, but will normally be at an early stage in the process, when the substance has been made and been shown to be active in early screening. As discussed in Chapter 9, this means that, for a patent with a nominal term of 20 years from filing, the effective term during which the patentee has exclusive rights to a marketed product may easily be as short as seven years or even less. This explains the importance attached by the pharmaceutical industry to provisions

[2] J.A. Di Masi et al., 'The price of innovation: New estimates of drug development costs' (2003) 22 Journal of Health Economics 151–185.

[3] Bain & Co study, reported in the *Wall Street Journal*, 8 December 2003.

[4] I. Kaitin, N.R. Bryant, and L. Lasagna [1993] 33 Journal of Clinical Pharmacology 412.

[5] *Financial Times*, 27 July 1998.

to extend the patent term, whether directly, as in the USA and Japan, or indirectly, by way of the supplementary protection certificate (SPC) in Europe, in order to compensate for this loss of effective patent term. It also explains the importance to the industry of the minimum 20-year term guaranteed by the TRIPs Agreement. When India gave a patent term of only seven years for pharmaceutical inventions, this was, in effect, giving no patent rights at all. The provisions of TRIPs that equate importation with local working are also extremely important. Countries that require that inventions must be locally worked within a short period, such as three years from grant, are denying the possibility of any exclusive rights, because a pharmaceutical product can seldom, if ever, be brought on the market in that time.

The Structure of the Pharmaceutical Industry

In the above discussion, references to the pharmaceutical industry should be understood to mean the research-based pharmaceutical industry, that is, companies that invest in research and which develop innovative new products that they protect by patents. Typically, such companies spend about 15–20 per cent of their total sales revenue on research and development, as compared with less than 4 per cent for industry overall; they spend between 1 and 3 per cent of their research and development expenditure on patent protection. These companies, however, form only part of the whole industry. Apart from companies providing services, such as contract research organizations, and apart from biotechnology companies, which we shall discuss in the next section, there are two other types of company that manufacture and sell pharmaceutical products, which may be roughly classified as the generic manufacturers and the imitators. Finally, there are firms that are primarily parallel importers, which do not manufacture products, but rather buy them cheaply in one country and export them to a country in which they can be sold at a higher price.

Generic companies specialize in manufacturing and selling compounds for which any patent protection has expired. There is of course nothing wrong with that—it is, after all, a basic principle that, after expiry of the patent, the invention is free for anyone to use—but there is usually intense competition between generic companies to be the first on the market with a generic copy after the patent on an innovative drug expires and, in order to do this, activities are often undertaken by the generic company prior to patent expiry that the patentee will consider to be infringement. There are consequently frequent legal conflicts between research-based and generic companies. It is unusual for the basic patent on a marketed product to be challenged by generic competitors, but, once this has expired, the originator normally relies on so-called 'life cycle management' (LCM) patents covering salt or crystal forms, galenic formulations, uses, and other subsidiary aspects. Because such patents are often somewhat weak and because the exact scope of protection that they give may not be clear, these patents are frequently the subject of litigation, particularly in the USA.

The pharmaceutical industry is frequently criticized for filing and enforcing LCM patents, a practice that is referred to disparagingly as 'evergreening'. These criticisms are sometimes justified. There is no reason why a genuine invention, such as a new crystal form that increases stability, or a new galenic formulation that improves bioavailability of a drug that is poorly absorbed in conventional formulations, should not be protected by patents. But when companies, for example, attempt to stop generic sales of a drug by means of later patents on its metabolites, this is something that cannot be justified on any reasonable interpretation of patent law and only brings the industry into disrepute. Some countries, such as India, have tried to prevent this perceived abuse by excluding novel salt or crystal forms, or new uses, of known compounds from the category of patentable inventions. A much better approach is to apply the inventive step requirement strictly during examination. If a novel LCM invention has unexpected advantages, it should be patentable; if not, then patentability should be denied for lack of inventive step.

Imitator companies, often referred to as 'pirates' by the research-based companies, produce innovative drugs during their normal patent life, manufacturing in countries (such as India) in which there is still no effective patent protection for pharmaceuticals, and selling there or in other countries with weak or non-existent protection for existing drugs. Of course, this is also perfectly legal, if less ethical.

Parallel importers are not manufacturers at all, but traders that exploit price differentials between one country and another by buying the innovator's genuine goods in a cheap country and reselling where a better price can be obtained.

Of course, these categories overlap to a considerable extent. Many research-based pharmaceutical companies are active in the generics business, directly or through affiliate companies; some of the less ethical generic companies imitate patented products when they think that they can get away with it, and some generic and imitator companies also act as parallel importers.

At the very bottom of the scale are the counterfeiters, who pass off their products as the genuine goods of the originator, deliberately infringing trademarks as well as patents. This amounts to criminal activity, even if the quality of the product is good. Much more often, however, the product is substandard, and in many cases, the dosage forms contain less active ingredient than they should, or even contain no active ingredient at all. This is not only severely damaging to the reputation of the originator company, but, more importantly, it puts the lives of patients at risk.

The area of the world in which imitators can freely operate is becoming severely limited as the TRIPs Agreement is implemented and classical patent piracy should become less of a problem in the future. The research-based industry still faces competition from generic manufacturers and parallel importers; because of these competing activities, two of the most important issues facing the industry are so-called '*Bolar* exemptions' (see Chapter 9) and the 'international exhaustion' of patent rights.[6]

[6] See, e.g., C. Heath, 'Parallel imports and international trade' [1997] 28 IIC 623.

There is little that can be done in respect of *Bolar* provisions, for the reasons already discussed, but international exhaustion is still a major issue.

International Exhaustion

Whereas, in the UK, Japan, and the USA,[7] a contract for the sale of patented goods may legitimately place limitations on the purchaser's freedom to export or resell the goods, in the absence of such a restriction, the law is generally taken to be that the rights acquired by the purchaser include the right to resell, even in another country. In civil law countries, the theory of exhaustion of patent rights is often taken to mean that no such restriction is permissible. Nevertheless, the question of whether the first sale of a product under a patent in country A can exhaust the rights of the patentee under his patent in country B (so-called 'international exhaustion') is by no means clear, and may vary from country to country. In Germany[8] and in Switzerland,[9] for example, there are cases holding that there is no such international exhaustion of rights, whereas, in Japan, the opposite is true (see p. 187). As we shall see in Chapter 25, within the European Economic Area (EEA), the principle of free movement of goods takes priority over the territorial nature of patent rights. This is often referred to as exhaustion of rights, but this term is misleading, since it applies even where there is no patent right in the country of first sale. In any event, it is not possible for a patentee in an EEA country to use the patent in that country to prevent importation and sale of genuine goods that were put on the market in another EEA country by the patentee or with his consent and, in view of the artificial price differentials between EEA countries, this causes considerable financial harm to research-based pharmaceutical companies.

In the USA, a long series of cases held that international exhaustion applied and that goods sold by or with the consent of the US patentee without contractual restriction in a foreign country could be imported into the USA without infringement, whether or not there was a patent in the country of sale. But in 2001, the Court of Appeals for the Federal Circuit (CAFC) held[10] that 'to invoke the protection of the first sale doctrine, the authorized first sale must have occurred under the United States patent'. The decision is open to criticism,[11] because, in the Supreme Court decision cited in support,[12] the goods were not in fact sold with the consent of the US patentee and because the issue of domestic versus foreign sales was not argued by the parties, but raised by the Court on its own initiative. Nevertheless, the Supreme Court refused certiorari and the CAFC precedent remains in force.

[7] For example, *Adams v. Burke* 84 US 453 (Sup Ct 1873).

[8] For example, *Re Tylosin* [1977] 1 CMLR 460 (BGH).

[9] *Kodak II* (BG, Switzerland, 1999).

[10] *Jazz Photo Corp. v. International Trade Commission* 264 F.3d (Fed Cir 2001).

[11] See, e.g., M. Barrett, 'A fond farewell to parallel imports of patented goods: The United States and the rule of international exhaustion' [2002] EIPR 571–578.

[12] *Boesch v. Graff* 133 US 697 (Sup Ct 1890).

the company running until it can show some positive results from its research. At this stage, the company may be able to enter into a research collaboration agreement with a major pharmaceutical company, under which the pharmaceutical company receives a licence or an option for future products in return for upfront payments and research funding, with milestone payments and royalties if a product is ever commercialized.

This may then be a good time (depending upon the current stock market conditions) to make an initial public offering (IPO) of stock. If this is successful, the stock held by the original venture capitalists will be worth many times the amount of their initial investment, while that held by company officers and employees may suddenly become worth millions of dollars. It is a very risky business, however: often, IPOs have to be withdrawn, and once the company stock is publicly traded, its value is at the mercy of every unsuccessful clinical trial or unsubstantiated rumour. The management's real aim, in many cases, is to ensure that the start-up company is acquired at a good price by a larger company, perhaps that with which the first licence agreement was made.

The bursting of the bubble in Internet shares and the general bear market of 2001–02 also had a severe effect on the biotechnology industry: venture capital sources dried up, IPOs were impossible, and a number of small companies went bankrupt or were acquired cheaply by large ones. By 2003, however, good news about the efficacy of some new biotechnology products and about some profitable deals with large pharmaceutical companies had caused renewed optimism in the sector, and a total of over US$16,000 million in biotechnology financing was raised in that year. In the recession of 2008–09, at the time of writing, we are seeing the same situation as in 2001–02, but the position is worse, because the entire economy is affected and not only the high-technology sector.

The emerging biotechnology industry in Europe has been disadvantaged by comparison with that of the USA because of the difficulty in raising venture capital (the main source of initial funding in Europe being historically banks, which, as is well known, will lend money only to someone who can prove that he does not need it) and because of excessive governmental regulation. The Green movement in Europe, with its strong bias against anything to do with gene technology, has also been a negative factor.

The major UK biotechnology companies are Celltech, which was set up with the help of government investment, and British Biotechnology. The latter, after a brief period in which, in terms of market capitalization, it was one of the biggest companies in the country, crashed spectacularly when its lead products failed in clinical testing and there were accusations that early bad results had been concealed. In 2003, it merged with Vernalis, which became the name of the merged company. More recently, Celltech itself has been taken over by the Belgian company UCB. Neither Celltech nor Vernalis have yet brought a therapeutic product to market. The company PPL Therapeutics, best known for its production of Dolly, the first cloned sheep, was acquired by a US company for a near-nominal price in 2003. There are still supposedly about 500 biotechnology companies in the UK,[16] most of which are very small. Investors seem to have been scared away by the British Biotechnology fiasco and the prospects for the UK biotechnology industry do not look good.

Elsewhere in Europe, there are still relatively few pure biotechnology companies, although there are some medium-sized companies in Germany and Switzerland. The major European pharmaceutical companies are either active in the field themselves, or else are involved in licensing or research cooperation agreements with US biotechnology companies or academic research centres.

In Japan, there are few, if any, small biotechnology companies, but in contrast to the USA and Europe, there was already in the 1980s heavy investment in biotechnology both by government agencies and by large established firms. These were not only pharmaceutical companies, but also companies previously experienced in the traditional fermentation technology of brewing and baking, such as Kirin Breweries, Morinaga Milk Co., and Suntory (whisky), and even companies with major interests in textiles (for example Teijin) and engineering (for example Mitsubishi). These Japanese companies have generally not made a notable success of biotechnology, although the jointly owned company Kirin-Amgen has been very successful with the product erythropoietin (epo).

Patent Conflicts in the Biotechnology Industry

As we have seen in Chapter 14, there have been a large number of patent conflicts in the field of biotechnology, particularly for recombinant DNA products and considerably more so than for 'classical' pharmaceutical products. A number of different reasons may be given for this situation.

In the field of classical pharmaceutical chemistry, companies usually pursue quite different research leads. If a patent conflict should arise, it is generally recognized at an early stage in the development of the compound; if it cannot be resolved by licensing, the party who is in the less favourable position can simply drop the compound without much fuss. After all, there are plenty more compounds to investigate and the chances are that it would have died in any case, as most development compounds do, for reasons of toxicology or lack of efficacy. For the products of recombinant DNA technology, however, the situation was very different, for three main reasons.

(a) A large number of companies were chasing essentially the same shopping list of one or two dozen interesting polypeptide products, so that a great deal of duplication of research effort was inevitable.
(b) Because these compounds occur naturally in the body, they could, once isolated, be used directly in clinical trials without long years of toxicity testing in animals. The compounds were therefore often already in the clinical testing stage before the patent conflicts were fully appreciated.

[16] *The Guardian*, 27 November 2003.

(c) Perhaps most importantly, many of the companies concerned had no products and indeed no assets other than their patent rights. It therefore was of vital importance to them to be able to claim that they had valid patent protection in certain areas, because this was the main way in which they could attract money from licensees and investors. In such circumstances, a company that abandoned a project because another firm had a patent application of earlier date would very soon have nothing at all.

In addition, the extreme complexity of the subject matter made it very difficult to resolve those conflicts that did arise. Faced with numerous overlapping patent applications of differing degrees of sufficiency, rapidly changing state of the art, fluctuating standards of examination in the patent offices, and a lack of court decisions to serve as precedents, the patent attorney called upon to give clear and reliable advice to his client or employer was in a difficult position.

This combination of circumstances necessarily led to litigation. In the event of litigation between major pharmaceutical companies, out-of-court settlement is common because defeat is annoying and expensive, but not fatal. For a small biotechnology company with only one or two products in development, it could be a matter of life or death, so that litigation was pursued to the bitter end and a great deal of the investors' money went to pay for lawyers instead of research scientists. By the time that a court gives a final ruling upon validity and infringement of a patent, 15 years or more may well have passed since the invention was made. Case law made in recent years has been about the classical protein products of early biotechnology, that is, tissue plasminogen activator (tPA), epo, human growth hormone, alpha-interferon, insulin, etc., for which the work was done 20 years ago. There is no reason to suppose that more recent technology will give rise to less litigation than the old.

Patents as a Source of Royalty Income: Universities

As discussed in Chapter 21, most universities in the UK and the USA now assert ownership of inventions made by their academic scientists. Universities are not themselves in a position to exploit these inventions commercially: direct commercial activity is generally regarded as incompatible with the purpose of a university and universities usually have special tax benefits that they would lose if they were to become profit-making businesses. The strategy is therefore to license out the inventions to industrial companies in exchange for royalties and (possibly) up-front payments. Sometimes, the companies in question are spin-off companies set up by the academic scientists themselves, often with the assistance of the university. Similar to universities are non-profit research institutions.

Many universities in the USA now have their own patent attorneys and deal directly with their own patents, which may be assigned either to the university or to a foundation such as the Wisconsin Alumni Research Foundation. The University of California heads the list with 390 US patents granted in the USA in 2005 (the last year for which statistics are available), nearly three times as many as granted to the second listed MIT, and twice as many as all British universities put together. It has been estimated[17] that the total royalty income of all US universities and research institutions in 2001 was over US$1,000 million. Stanford University has been particularly successful in licensing out its inventions, including the basic Cohen/Boyer recombinant DNA patents, jointly owned by Stanford and the University of California.

Stanford's licensing of the Cohen/Boyer patents was successful, because licences were offered on a non-exclusive basis to all-comers for annual minimum royalties of US$10,000 and royalties on commercial products of 1 per cent of net sales. This meant that, although the validity of the patents was by no means certain, it would have cost each licensee more to challenge the patent than to pay the royalties. As a result, the patents were never challenged and expired in December 1997 after generating over US$200 million in royalty income. When the patents expired, Stanford's annual royalty income dropped overnight from US$50 million to US$15 million.

This may be contrasted with the unsuccessful licensing strategy adopted for the Harvard OncoMouse®. Firstly, instead of giving non-exclusive licences to everybody, Harvard granted exclusive rights to Du Pont. Du Pont then attempted to sublicense on the basis that it would receive reach-through royalties of 2 per cent of net sales on all drugs found using the OncoMouse as an animal model, but found no takers. This is hardly surprising: the Cohen/Boyer patents, if valid, covered all recombinant DNA technology, and for hundreds of biotechnology companies, the only alternatives were to pay the reasonable royalties or challenge the patent. In contrast, there are dozens of available animal models for testing cancer drugs, and if the price for using the OncoMouse is set too high, companies will simply use one of the alternatives, even if (which is not necessarily the case) it is less advantageous. Du Pont (and Harvard), in effect, lost a potential source of income by being too greedy.

Columbia University also raises large amounts of money in royalty income (over US$109 million in 2004), in part by using questionable methods, such as submarine patenting and lobbying for special privileges.[18]

Increasingly, the trend is for UK universities to follow the example of those in the USA and set up their own technology transfer departments, or enter into agreements with specific companies set up to exploit inventions made in the university (for example Isis Innovation for Oxford University). But the levels of royalty income are minimal in comparison with the US figures. It was estimated[19] that the total annual royalty income in 2006–07 of all British universities together amounted

[17] A. Stevens, 'Twenty years of academic licensing: Royalty income and economic impact' (2003) Sept. *LES Nouvelles* 133.

[18] 'Ownership at too high a price?' (2003) 21 Nature Biotechnology 953.

[19] *Higher Education Business and Community Interaction Survey*, 2007.

to about £40 million (then US$80 million), much less than that of Columbia University alone. It is claimed, however, that British universities are more efficient than their US counterparts in starting spin-off companies. In 2006–07, 226 such companies were formed in the UK, the research expenditure per spin-off being £20 million, as compared with a figure of £44 million in the USA. The relative success of the US and UK companies is not known.

Research Tools

Many of the inventions made by academic inventors, as well as by inventors working for small companies, cover techniques, methods, assay systems, and target molecules that may be useful in research, in general, and in drug discovery, in particular. Previously, the majority of such methods were published and made freely available to all researchers, but, increasingly, patent protection is being applied for and obtained for inventions of this type.

There is nothing wrong with patenting a research tool, and if such a tool is patented, it is only right that those who wish to use it should pay for doing so. Payment may be made, for example, by purchasing from the patentee or his licensee the reagents necessary to carry out the procedure, even if they may be obtained more cheaply elsewhere. This is the situation for the thermostable enzymes used in the polymerase chain reaction (PCR) process. Alternatively, a fee could be paid for using a patented target or assay, either on an annual basis, or based upon the number of screens in which it was used. A number of owners of such patents make the technology freely available for non-commercial research and charge only for use in a commercial context.

What is not acceptable is that the patentee should expect to receive 'reach-through' royalties as a percentage of net sales of drug products discovered or developed with the help of the patented research tool, except perhaps in the very rare situation in which there is no real alternative and the discovery could not have been made in the absence of the original invention. Much more usually, there are always alternatives available and a drug initially discovered using one particular assay might just as well have been found using a different one. After all, one can patent an electric drill, but one does not expect to get royalties on everything in which it has bored a hole, because other machines for boring holes are known. Reach-through claims that attempt directly to claim drugs found using a patented assay are generally invalid (see p. 384) and claims to a process of testing cannot cover products identified by the test, because these are not the direct product of a patented process.

This has been confirmed for the USA by the case of *Bayer v. Housey*. Housey had a broad US patent covering cell-based screening assays and Bayer used such assays outside the USA. Housey alleged that Bayer infringed the patent[20] by importing into the USA the results of the assay and compounds identified by the assay. In summary

20 Under 35 USC 271(g).

judgment proceedings, the CAFC upheld the District Court, holding, firstly, that derived product protection applied only to physical products and not to information, and secondly, that the products were not 'made by' the patented process: 'The process must be used directly in the manufacture of the product, and not merely as a predicate process to identify the product to be manufactured.'[21] Subsequently, the Housey patent was held unenforceable for fraud on the US Patent and Trademark Office (USPTO), the inventor having concealed the contributions of others and having relied on the 'results' of experiments that were never made.[22]

One problem is that, faced with the alternatives of paying a significant fee for using a patented research tool or being able to use it without cost for the promise of future royalties, many managers will be tempted to accept the royalty option. This saves on immediate research costs; it can be argued that, in the unlikely event that a commercial product arises out of the research, the money to pay the royalties will be available and in any case the manager will probably have retired or moved on by the time that any royalties are actually due. The danger, however, is twofold. Firstly, it will often be the case that a number of different tools or technologies have contributed to the drug development, and whereas a single royalty of 1 or 2 per cent may be an acceptable burden, an accumulation of such royalties soon adds up to an unacceptable amount. Secondly, if a research tool patentee later sues a pharmaceutical company for royalties on an end product, his position will be enormously strengthened if he can show that a number of companies have contractually agreed to pay such royalties.

Most pharmaceutical companies are not interested in commercial exploitation of research tools invented within the company, but only in maintaining their own freedom to use them. Keeping them secret hinders research generally, and leaves open the danger that another party may independently invent the tool and file patents that would block the first inventor. Publishing or filing patents purely for defensive purposes are alternative strategies to deal with this threat. Because defensive patenting has certain advantages, particularly in the USA, some companies adopt this strategy, filing patent applications that will be published, but which are seldom, if ever, prosecuted to grant. This has the disadvantage that it contributes to the general perception that there are 'thickets' of patents that will block all research. It is wrongly assumed that all published Patent Cooperation Treaty (PCT) applications are 'patents', and that they can and will be enforced. Publication, on the other hand, is clearly donating the invention to the public.

Patents as Lottery Tickets: Individual Inventors

A large number of patents, particularly in the mechanical field, are applied for by individual inventors. As discussed in Chapter 24, such inventors must find a

[21] *Bayer AG v. Housey Pharmaceuticals Inc.* 68 USPQ 2d 1001 (Fed Cir 2003).

[22] *Bayer AG v. Housey Pharmaceuticals Inc.* (D Del, 4 December 2003).

suitable licensee before they run out of money and without making any non-confidential disclosure of the invention, at least until a patent application is filed. This is not an easy task and the odds are so heavily against an individual inventor making any significant amount of money out of his inventions that the patents may well be compared to lottery tickets.

Nevertheless, it does happen that individuals become wealthy on the basis of their inventions: classic examples in the UK are the inventors of the Workmate® adjustable workbench and that of the Dyson centrifugal vacuum cleaner. In the USA, the situation of the individual inventor is made somewhat easier (at least in so far as protection in the USA is concerned) by the existence of the one-year period within which he may file a valid patent application after disclosing the invention, and the fact that pending US-only applications are not published and can be kept pending almost indefinitely.

Patents as Bargaining Chips: The Electronics Industry

In the electronics and telecommunications industry, products that are marketed are typically interdependent and have to meet common industry standards. Thus, for example, many companies manufacture CD players, but all must be capable of playing the same standard CD. The situation in the early days of VCR technology, in which there were three mutually incompatible systems on the market, was not one that was ultimately beneficial for the consumer, even if the consumer could be said to have had more choice.

Similarly, in telecommunications, many companies manufacture mobile phones (cellphones), but each must be able to place a call to a mobile of a different manufacturer in a different country. As long as the industry was regulated and each national telephone company had a statutory monopoly, there was less pressure to adopt international standards. Now, deregulation has led to real competition, but such competition must take place within the framework of systems that can be used by all. In such a situation, there is no room for individual monopolies to different technologies, or for the use of patents to block the competition. There are so many patents that cover various aspects of the standard technologies that, if they were used in this way, it would be impossible to develop and market any products at all.

In this industry, therefore, the practice has long been to use patents basically as bargaining chips in cross-licensing and patent pooling agreements. In the days of regulation and limited competition, it seems that it was common practice simply to count the numbers of patents held by each company to determine the financial conditions of the cross-licensing; nowadays, the actual value of individual patents receives much more attention. Nevertheless, electronic companies still have very much larger patent portfolios than pharmaceutical companies of comparable size. In 2008, the top ten companies in terms of US patents granted were all in the electronics field, headed by IBM with a staggering 4,169 granted patents, compared with fewer than 200 for most large pharmaceutical companies.

Competition authorities, such as the EC Commission and the US Justice Department, have generally been highly suspicious of patent pools (see Chapter 25), but the Justice Department at least now recognizes that they can be, on balance, pro-competitive and bring benefits to the consumer in this industry. Thus, in June 1997, its Antitrust Division approved a patent pooling arrangement (the MPEG Licensing Administrator LLC Agreement) entered into by eight companies and a university, which, between them, owned 27 patents necessary in order to meet the international standard for video compression, a technique used in digital transmissions of all kinds.[23] This provides a package licence that is freely available to all; the potential licensee does not have the trouble and expense of negotiating separately with each of the nine patent holders, and the patent holders share the royalties among themselves in proportion to the value of the patents that they contribute to the pool.

It has been suggested that patent pools of this type could be useful in the biotechnology industry to avoid having to carry out multiple negotiations with owners of potentially dominating rights, including those to expressed sequence tags (ESTs) and research tools.[24] This seems unlikely, however, in the absence of the pressure for industry standards and in the absence of any consensus as to whether the holders of such patents are entitled to downstream royalties at all.

Patents as Tools of Extortion: Patent Trolls

The term 'patent troll' has become much used in the last few years, and is usually applied to an entity that has no business activities except asserting and litigating patent rights. Of course, being a 'non-practising entity' is itself not a problem; after all, most universities do not work their own patents. But what earns a company the pejorative label of 'patent troll' is that it asserts its patents primarily against large companies that are not imitators and not only does it not work its own patents, but it also does not license them, except to those non-imitating companies that are forced to take licences as a result of litigation or the threat of litigation. In Nordic mythology, a 'troll' is an evil monster that hides under a bridge and devours innocent travellers who try to cross over it; similarly, the patent troll lurks in the shadows until a victim comes along, who is then sued for infringement of a patent often of broad scope and dubious validity.

Neither the patent troll nor his patent makes any useful contribution to society, because the people who actually make useful products would have done so in the absence of the patent. Jerome Lemelson was one of the earliest and most successful patent trolls, although the term was not invented during his lifetime. Lemelson was a prolific inventor who, among other things, filed patent applications in the 1950s that

[23] A.C. Brunetti, 'Wading into patent pooling' (1997) *Intellectual Property*, November.
[24] For example, L.M. Sung (1998) *The National Law Journal*, 22 June, C2.

issued 30 or 40 years later as patents that were alleged to cover barcoding and other technologies. Although his patents were highly dubious and now are all likely to be held invalid or unenforceable (see p. 161), many companies found it cheaper to settle than to litigate, with the result that Lemelson (who died in 1997) and his successor, the Lemelson Foundation, extracted hundreds of millions of dollars over the years. Lemelson developed 'submarine patenting' into a fine art: when the 'submarine' patent surfaces, the industry, believing itself to be operating in a patent-free environment, or at least one in which the relevant patent rights are known, suddenly finds itself confronted with demands to pay royalties under the new patent.

Today's patent trolls generally no longer have the advantage of submarine patenting, but the Lemelson strategy of asserting obscure patents against established industries is still in use.

23
HOW TO CATCH THE INFRINGER—AND HOW NOT TO BE CAUGHT

> [T]he *Festo* litigation suggests, with all respect to the courts of the United States, that American patent litigants pay dearly for results which are no more just or predictable than could be achieved by simply reading the claims.
>
> *Kirin-Amgen v. Hoechst* [2004] UKHL 46, *per* Lord Hoffmann

From the Viewpoint of the Patentee

When the owner of a patent becomes aware of commercial activity by a competitor that is in the area of his invention, he will naturally want to know if the patent can be used to stop these activities. The first and most important question is whether what the competitor is doing amounts to an infringement.

This may sometimes be a very easy question to answer, particularly when the patentee has product per se protection for a group of chemical compounds and a compound clearly within the claims is being sold without the patentee's permission. It may be that determining whether or not there is infringement is essentially a problem in analytical chemistry: for example where the competitor is selling a complex mixture that may or may not contain a patented compound. If the presence of the compound can be detected, the rest is simple. More complex analysis may be required when the patent claims a specific crystal form of the compound that may or may not be present in a pharmaceutical product, such as a tablet.

The Interpretation of Claims

In many cases, however, the question of whether or not there is infringement is one that requires careful study. The problem is one of analysing the scope of the claims of the patent in the country in question, and although practically all countries having patents require claims, there are considerable differences from country to country in the way in which claims are interpreted.

There are indeed some general principles of claim interpretation that hold good in all, or nearly all, countries. The claim may be analysed into distinct features or integers (for example parameters of a process, parts of a device, components of a mixture, or substituent groups on a molecule), the specification being used as a guide to the meaning of terms used in the claim. If, then, a process, device, mixture, or compound has all of these essential features, then its unauthorized manufacture, sale or use is an infringement of the claim. To take a simple example, a claim to 'a mixture comprising A, B, and C' is not infringed by a mixture of A and B, but is by a mixture of A, B, C, and D: infringement is not avoided by adding extra features (except perhaps in Australia, see p. 377). If a patent contains a number of claims, some may be infringed and others not. As we have seen, claims are often written in dependent form, and a dependent claim incorporates all of the features of an earlier claim and either further defines one or more of these features, or adds some extra features. For example, we may have:

1. A process for the manufacture of X, comprising the step of reacting Y and Z in an inert solvent.
2. A process as claimed in claim 1 in which the solvent is water.

Because claim 2 adds an extra feature (sometimes called an 'extra integer') to claim 1, it is narrower in scope and it follows that a process that does not infringe claim 1 cannot infringe claim 2. The use of alcohol as a solvent would, however, infringe claim 1, but not claim 2.

Sometimes, the presence of a dependent claim may assist in the interpretation of the claim on which it depends. Thus, given a claim having 'a fluid' as an integer, it may not be clear whether this really does include a gas as well as a liquid, particularly if only liquids are exemplified. But if there is a dependent claim adding only the feature 'in which the fluid is a liquid', then because there is a legal presumption that different claims have different scopes, the earlier claim must include fluids that are other than liquids, that is, gases.

Difficulties in interpretation tend to arise not so much when integers of the claim are added or omitted, but when they are substituted. Does a mixture of A, B', and C infringe our claim to a mixture comprising, A, B, and C, where B' is very similar to B? It is this question of 'equivalence' that is approached differently by different countries. For European countries that are contracting states of the European Patent Convention (EPC), claim construction of European patents (and by analogy national patents) is governed by EPC Article 69 and the Protocol on Interpretation of Article 69, but in order to understand this part of the EPC, we need to look at the situation in these countries before the Convention went into effect.

The UK In the UK, the practice has always been to interpret claims somewhat literally. The attitude has been that it is up to the patentee to define his own claims and it is his own misfortune if he fails to do so broadly enough. In the classic statement by Romer J, for there to be infringement, the alleged infringer must take the invention claimed by the patent, 'not that which the Patentee might have claimed if he had been well advised or bolder'.[1] In a case 109 years later,[2] very similar views were expressed by Laddie J: 'The Courts are not a branch of social services whose job it is to help the infirm or unwise.' But the courts have also recognized that it is unjust that someone should be able to take the benefit of a patentee's invention just because of a minor technicality in claim drafting. Under the old system, claims were construed literally, the words of the claim being given their normal dictionary meaning, and reference was made to the specification only if it was necessary to do so in order to resolve an ambiguity. But there was still the possibility of infringement being found where the courts were convinced that the alleged infringer had taken the essence of the invention (the 'pith and marrow', to use a phrase common in the old case law),[3] even if no claim was literally infringed.

The leading English case on claim interpretation involved a mechanical invention relating to steel door lintels.[4] The claim called for the back wall of the lintel to be vertical, whereas, in the alleged infringements, it was at an angle of 6° or 8° from the vertical. In the House of Lords, Lord Diplock did not distinguish between literal and 'pith and marrow' infringement, but simply construed the claim broadly enough to cover immaterial variants, in this case, deviations from the vertical small enough not to affect the function of the article significantly. The person skilled in the art, to whom the specification was addressed, would understand that, by using the word 'vertical' in this context, the patentee would not mean that strict verticality was essential. This is sometimes referred to as 'purposive construction' of a claim, meaning that a person may be taken to mean something different when he uses words for one purpose from that which he would be taken to mean if he were using them for another. There was a difference between what a person would reasonably

1 *Nobel v. Anderson* (1894) 11 RPC 115, 128 (Ch D).

2 *Merck v. Generics (UK)* [2003] EWHC 2842 (Pat).

3 First used by Lord Cairns in *Clark v. Adie* [1877] 2 App Cas 315.

4 *Catnic Components Ltd v. Hill & Smith Ltd* [1982] RPC 183 (HL).

be taken to mean by using the word 'vertical' in a mathematical theorem and by using it in a claimed definition of a lintel for use in the building trade.

The *Catnic* approach may only be used when the term to be construed is a descriptive word. Where numerical limits are given in the claims, these are to be taken as the patentee's statement of an essential feature of the invention and there is no room for considering anything falling outside the range as a 'variant', material or otherwise.[5]

Germany and the Netherlands In Germany, claims tended to be very broadly interpreted by the courts The practice distinguished two types of equivalents to claimed integers. The first, *glatte Aequivalente* (plain equivalents), were those that are immediately obvious as equivalents (a nail and a screw, for example); *nicht glatte Aequivalente* (not plain equivalents) were equivalents the recognition of which requires careful thought, although no inventive activity. These concepts were then applied to define three scopes of increasing breadth:

(a) the direct subject matter of the invention, which is the literal wording of the claims;
(b) the subject matter of the invention, which is the literal wording of the claims plus plain equivalents; and
(c) the general inventive concept, which also includes the *nicht glatte Aequivalente*.

Anything coming within the broadest of these scopes might be held to be an infringement, although the extent of the range of equivalents depended upon the degree of inventiveness. Thus the claims were treated only as a point of departure (*Ausgangspunkt*) in determining the extent of protection, for which the criterion was the inventive achievement (*erfinderische Leistung*) disclosed by the specification as a whole.

In the Netherlands, claims were generally construed broadly by the courts, much in the same way as in Germany. The role of the claims before 1977 was extremely modest and what mattered was the 'essence of the invention', or general inventive concept.

The European Patent Convention Under the old practice in the UK and Germany, a simplified view of the situation was that, in the UK, the Patent Office would grant relatively broad claims that the courts would interpret literally, while in Germany, narrow claims were granted that the courts would interpret broadly. The end result was that the patentee got much the same scope of protection in both countries. The problem with the European patent is that there is a single set of claims to be interpreted by the national courts of the member states, the practice of which in the interpretation of their own patents ranged, as we have

[5] *Auchincloss v. Agricultural & Veterinary Supplies Ltd* [1997] RPC 649.

seen, from essentially literal interpretation in the UK to wide-ranging equivalence in Germany. Although infringement of European patents remains a matter for the national courts, it was clearly undesirable to accept a situation in which a European patent would have a broader scope of protection in some countries than in others. For this reason, the EPC includes provisions on the extent of protection given by a European patent. Article 69(1) states that the extent of protection shall be determined by the claims, and that the description and drawings shall be used to interpret the claims. A Protocol on the Interpretation of Article 69 was also added to the Convention.

The Protocol attempts to find a compromise position between those of the UK and Germany by presenting an extreme view of the UK system, on the one hand, and an extreme view of the German system, on the other, and stating that both of these extremes are to be avoided. The result is supposed to combine 'a fair protection for the patentee with a reasonable degree of certainty for third parties'. Because, in both countries, there had already been some movement away from the earlier extreme positions, both national courts tended to take the view that their current practice was in line with the Protocol. That the Protocol did not immediately even out differences between national claim interpretation in Europe was shown by the *Epilady* cases. These were infringement actions by Improver Co. against Remington in which the patented article was a depilatory device that plucked out hairs by trapping them in the coils of a rotating spiral spring; the claims specified a spring as an integer. The alleged infringement had a flexible slotted plastic rod in place of a spring, the result being the same. Improver sued for infringement in the UK and Germany, among other countries.

We should not be too surprised if courts in different countries were to reach different conclusions in a case such as this; after all, it is common for an appellate court to disagree with the court of first instance within the same country. What was rather depressing in this case was that, in the UK, where interlocutory injunctions are normally fairly easy to obtain, the case was struck out in interlocutory proceedings in the Patents Court because the judge considered it to be inconceivable that the plaintiff could succeed. In Germany, however, where interlocutory injunctions (*einstweilige Verfügungen*) are normally less readily granted, the Düsseldorf *Landgericht* (District Court) granted an injunction because the accused device was so clearly an infringement. The situation was then reversed in both countries on appeal, the English Court of Appeal granting an injunction,[6] whereas the *Oberlandesgericht* discharged the injunction granted by the lower court. In the UK, on the full trial of the action, the patent was held not to be infringed.[7]

A new Article 2 was added to the Protocol by EPC 2000. This states that 'due account shall be taken of any element which is equivalent to an element specified in the claims'.

[6] *Improver Corp. v. Remington Consumer Products Ltd* [1989] RPC 69 (CA).

[7] *Improver Corp. v. Remington Consumer Products Ltd* [1990] FSR 181(CA).

National Courts and the Protocol The Protocol applies not only to European patents, but also to British national patents granted under the Patents Act 1977.[8]

In the *Improver* case mentioned above, Lord Diplock's approach in *Catnic* was reformulated by Hoffmann J, who held that, where a feature of an alleged infringement fell outside the literal meaning of the corresponding part of a claim, the court should ask itself the following three questions in order to determine whether or not this 'variant' fell within the true scope of the claim:

1) Does the variant have a material effect upon the way the invention works? If yes, the variant is outside the claim. If no:
2) Would this (i.e. that the variant has no material effect) have been obvious at the date of publication of the patent to a reader skilled in the art? If no, the variant is outside the claim. If yes:
3) Would the reader skilled in the art nevertheless have understood from the language of the claim that the patentee intended that strict compliance with the primary meaning was an essential requirement of the invention? If yes, the variant is outside the claim.

These questions were applied in subsequent cases[9] and usually referred to as the '*Improver* questions' or the 'Protocol questions'.[10] But giving the judgment of the House of Lords in *Kirin-Amgen v. TKT*,[11] Lord Hoffmann (as he had then become) considered that too much importance had been given to the Protocol questions, which were not laws of construction, but guidelines, more useful in some cases than in others. He said:

> The determination of the extent of protection conferred by a European patent is an examination in which there is only one compulsory question, namely that set by [A]rticle 69 and its Protocol: what would a person skilled in the art have understood the patentee to have used the language of the claim to mean? Everything else, including the Protocol questions, is only guidance to a judge trying to answer that question. But there is no point in going through the motions of answering the Protocol questions when you cannot sensibly do so until you have construed the claim. [...] The Protocol questions are useful in many cases, but they are not a substitute for trying to understand what the person skilled in the art would have understood the patentee to mean by the language of the claims.

Since *Kirin-Amgen*, the courts have generally refrained from using the Protocol questions and have relied instead on the general principles enunciated by Lord Hoffmann. The *Catnic* approach is held to be in conformity with the Protocol, and since Article 69 and the Protocol make it clear that protection cannot extend beyond the claims (however they are construed), there is no doctrine of equivalents in English law.

[8] Section 125(3), PA 1977.
[9] For example, in *Pharmacia Corp. v. Merck & Co Inc.* [2002] RPC 777 (CA).
[10] First by Aldous LJ in *Wheatley v. Drillsafe* [2001] RPC 133 (CA).
[11] *Kirin-Amgen Inc. v. Transkaryotic Therapies Inc.* [2005] RPC 9 (HL)—see p. 291.

The highest German court has said that the effect of Article 69 is to give the claims a 'central role', that is, that the claims are no longer merely a point of departure, but the decisive basis for determining the extent of protection.[12] In issuing five decisions on the same day (sometimes referred to as the 'decision quintet'), the *Bundesgerichtshof* (BGH, or Federal Supreme Court) defined a three-step scheme to determine the scope of protection of a patent claim beyond its literal meaning. According to the BGH, the first question is whether the 'modified means' objectively solve the same technical problem as the claimed means. The second question asks whether a skilled person could have understood at the priority date that the modified means have the same effect as the claimed means. Finally, it has to be examined in the third question whether the considerations taken into account by the skilled person to identify the modified means as having the same effect are such that he would have taken into account the modified means when reading the patent claim.[13] These questions are very similar to the Protocol questions and indeed the BGH reviewed the English decisions at some length.

The courts in the Netherlands have taken a similar position, with more emphasis on the claims and less on the inventive concept. Thus, for example, the Court of Appeals of The Hague held, in 2000, that a claim to a vehicle with a fixed upper loading platform and a moveable lower loading platform was not infringed by a vehicle in which the lower platform was fixed and the upper was moveable.[14]

The USA As we have seen in Chapter 8, since the *Markman* decision of the Supreme Court, claim construction is an issue of law to be decided by a judge, not a jury. This has been restated by the US Court of Appeals for the Federal Circuit (CAFC) in *Phillips v. AWH*,[15] which also held that the meaning of terms in a claim should be interpreted with reference to the specification itself rather than to external sources such as dictionaries and encyclopedias. Nevertheless, the question of whether an accused infringement falls within the claim remains a jury question, and it seems that either a judge or a jury may extend the scope of the claims beyond their literal meaning by application of the 'doctrine of equivalents'. Originally developed by the courts in considering mechanical inventions,[16] the doctrine of equivalents also applies in chemical cases. The leading case is *Graver Tank v. Linde Air Products*, in which the Supreme Court held that infringement could be found when an integer of the claim is replaced by a different integer that 'performs substantially the same function in substantially the same way to obtain

[12] *Batteriekastenschnur* [1989] GRUR 903, 904 (BGH).

[13] *Schneidmesser I, SchneidmesserII, Custodiol I, Custodiol II, Kunstoffrohrteil* [2002] GRUR 511 (BGH).

[14] *Van Bentum v. Kool* (CA, The Hague, 30 March 2000).

[15] *Phillips v. AWH Corp.* 376 F.3d 1382 (Fed Cir 2004, *en banc*).

[16] For example, *Royal Typewriter Co. v. Remington Rand Inc.* 77 USPQ 517 (2nd Cir 1948).

the same result'.[17] In this case, manganese silicate replaced magnesium silicate in a welding flux composition. The same principle is applied to numerical limits: for example a claim to a dentifrice composition containing 1–10 per cent of urea (plus other ingredients) was held infringed by a composition containing 13 per cent urea, which gave the same result in the same way.[18] This result would not be possible in the English courts, where the patentee would have been bound by the numerical limits that he had chosen to set to the claim.

One restriction upon the extent to which the doctrine of equivalents can be applied in the USA is that the claim cannot be extended to cover anything that is old; consequently, the claims will have a greater range of equivalents in the case of a pioneer invention than for one that represents a small advance in an already well-worked field. Another limitation is that a patentee cannot allege that matter disclosed, but not claimed, in the patent can infringe under the doctrine of equivalents.[19] A further type of limitation arises from the fact that, in the USA, unlike the UK, the claims are interpreted not only in the light of the specification, but also in the light of what happened during the prosecution of the patent application. This is a matter of public record in the 'file wrapper' of the patent and is sometimes referred to as the 'file history', or 'prosecution history'. If this shows that, during prosecution, the patentee had to limit his claims in a relevant respect and argued that the limitation made the claims patentable over the prior art, then he will not be able to extend his claims by the doctrine of equivalents to recover the ground that he gave up during prosecution. This is usually referred to as 'file wrapper estoppel' or 'prosecution history estoppel'.

In 1997, the Supreme Court reviewed the whole issue of the doctrine of equivalents in a case that was very similar in some respects to the dentifrice example discussed above.[20] The patent claimed a process that was to take place at a pH of 'from approximately 6.0 to 9.0'. The alleged infringer used a pH of 5.0, a jury in the District Court found infringement by equivalence, and the CAFC confirmed. The Supreme Court agreed to hear the case and it was thought that the Court might severely restrict or even abolish the doctrine of equivalents. In the event, however, the doctrine of equivalents remained alive, but with the limitation that it must be applied to each and every integer of the claim, and not to the scope of the claim as a whole. The issue of file wrapper estoppel was relevant in this case, because the claim as originally filed did not contain any pH limitation, and this was added during prosecution to avoid prior art that specified a pH over 9.0. The patentee could not therefore have claimed that a process at a pH of 10.0, for example, would infringe, but the Supreme Court could not determine why the lower limit of 6.0 was set and remanded the case back to the lower courts to settle this point. The Court

17 *Graver Tank & Manufacturing Co. v. Linde Air Products Co.* 339 US 605 (1950).
18 *University of Illinois Foundation v. Block Drug Co.* 112 USPQ 204 (7th Cir 1957).
19 *Johnson & Johnson Associates Inc. v. R.E. Service Co. et al.* 62 USPQ 2d 1225 (Fed Cir 2002).
20 *Warner-Jenkinson Co. Inc. v. Hilton Davis Chem. Co.* 41 USPQ 2d 1865 (Sup Ct 1997).

stressed that, if the claims were limited during prosecution, there should be a rebuttable presumption that this was done to ensure patentability over the prior art.

The matter was carried further in the (in)famous *Festo* case. An *en banc* decision of the CAFC[21] held, following remand from the Supreme Court, that prosecution history estoppel applied not only to situations in which the claim was limited in response to a rejection over the prior art, but also to any reasons relating to the statutory requirements for a patent, and included voluntary claim amendments, as well as those made in response to a rejection, and any amendments for which no reason was given in the file history. Thus, wherever there had been a restriction in respect of an integer of the claim, no range of equivalents was available for that integer.

The immediate response of many US patent attorneys was to advise their clients that as many claims as possible should be drafted, so that instead of amending a claim, it could simply be cancelled, relying on the next claim of greatest scope. Because this is exactly equivalent to amending the claim, it should have been clear that this would, sooner or later, be treated by the courts in exactly the same way (as indeed it was in the *Honeywell* case).[22] A better response would be to draft claims more expertly and not to rely on the doctrine of equivalents to do what a properly drafted set of claims should do in the first place.

The CAFC decision then went again to the Supreme Court,[23] which agreed with the CAFC that 'Estoppel arises when an amendment is made to secure the patent and the amendment narrows the patent's scope' (the patent, please note, not the claim). But it considered that the CAFC had gone too far in completely ruling out the doctrine of equivalents in this situation. There was no reason why a narrowing amendment should be deemed to relinquish equivalents unforeseeable at the time of the application, or aspects of the invention that had only a peripheral relation to the reason that the amendment was submitted. The onus should be upon the patentee to show why the whole area surrendered by amendment should not be excluded from the application of the doctrine of equivalents.

The case was remitted to the CAFC, which heard it *en banc* for the second time in 2003.[24] The Court considered that, to determine whether or not an equivalent was unforeseeable, evidence outside the prosecution history would be needed. For the other possibilities identified by the Supreme Court ('tangential relationship' and 'other reasons'), however, the issues must be decided on the prosecution history alone. Judge Newman pointed out, in her dissenting opinion, that this, in effect, limited the possibilities for avoiding estoppel to the single issue of unforeseeability. To European eyes, it seems a strange result that, if there has been a narrowing amendment, an inventive improvement to a patented invention is more likely to infringe by equivalence (because unforeseeable) than a close copy that just avoids

21 *Festo Corp. v. Shoketsu Kinzoku KK et al.* 56 USPQ 2d 1865 (Fed Cir 2000) (*Festo VI*).
22 *Honeywell International v. Hamilton Sundstrand Co.* 370 F.3d 1131 (Fed Cir 2004, *en banc*).
23 *Festo Corp. v. Shoketsu Kinzoku KK et al.* 62 USPQ 2d 1705 (Sup Ct 2002) (*Festo VIII*).
24 *Festo Corp. v. Shoketsu Kinzoku KK et al.* 68 USPQ 2d 1321 (Fed Cir 2003) (*Festo X*).

literal infringement. Because, by this time, everyone had forgotten what the original facts of the case were, the CAFC remitted it to the District Court for a determination as to whether or not it met the new criteria for unforeseeability of the equivalent. The District Court considered that it did not and accordingly held that there was no infringement.[25] The CAFC agreed,[26] bringing this long saga to an end.[27]

Nevertheless, the doctrine of equivalents is not wholly dead in the USA, as shown by a recent case relating to catheters,[28] in which the claim included the requirement that the catheter be engaged 'along a line' of the aorta. The District Court, however, found that the claim was not limited to linear portions of the aorta. On appeal, the CAFC confirmed the construction, stating that 'the context in which a term is used in the asserted claim can be highly instructive'. This sounds very much like application of the English *Catnic* principle.

Japan In Japan, there have been few cases in which the doctrine of equivalents has been applied, but in 1998, the Japanese Supreme Court, in a mechanical case,[29] set out five conditions under which a product or process not literally covered by an element of the claim could nevertheless infringe. These basically are:

(a) the element is not an essential feature;
(b) the variant achieves the same effect in the same way;
(c) the variant was an obvious one at the time of the infringement;
(d) there is no *Gillette* defence (that is, the accused product was not identical to or obvious over the prior art as of the filing date); and
(e) there is no file wrapper estoppel.

These conditions are certainly more restrictive than those of the USA, and are closer to the UK approach, as set out by Hoffmann J.

Conclusion In any event, if the competitor's activity does not fall within the literal wording of the claims, so that the only way in which he can be caught is by relying upon equivalence, then it is essential to obtain the opinion of an expert in patent law in the relevant country before proceeding any further. Only someone thoroughly familiar with local law and practice can give any useful opinion as to whether a court would hold the patent to be infringed.

Process Patents and Derived Product Protection

If the patent is for a process rather than a product, it is, of course, more difficult to establish infringement. If the process is a process of manufacture of a chemical

[25] *Festo Corp. v. Shoketsu Kinzoku KK et al.* (D Mass 2005) (*Festo XI*).
[26] *Festo Corp. v. Shoketsu Kinzoku KK et al.* (Fed Cir 2007) (*Festo XIII*).
[27] 'US patent attorneys raise cheers/Each time the name *Festo* appears./Do equivalents topple/By file wrapper estoppel?/It'll keep them all busy for years.' Winning entry, *CIPA* limerick competition, 2003.
[28] *Dr Voda v. Cordis Corp.* 536 F.3d 1311 (Fed Cir 2008).
[29] *Tsubakimoto Seiki v. THK Co.* (Japanese Sup Ct, 24 February 1998).

product, then the manufacturer will certainly do its best to keep secret the process that it is using, and one may have to rely upon circumstantial evidence, such as analysis for unchanged reactants or by-products that may characterize the process used. It may be possible, during an infringement action, to obtain information under one form or other of legal compulsion, but the trouble is that one cannot fully evaluate the chances of success before filing suit.

If the product is being produced abroad and imported, the product will infringe if it is the direct product of the patented process. The meaning of the term 'direct' was considered by the Court of Appeal in the *Pioneer* case,[30] which dealt with the production of CDs. This is a multistep process in which a 'master' disc is used to produce a 'father' disc, which, in turn, is used to make a number of 'mother' discs, from which, after one further step, the final CDs are produced. The patent claimed a process for making the 'master' or 'father' discs, and did not mention the further, conventional, steps between them and the final CD product. Nevertheless, it was alleged to be infringed by the sale of the final CDs.

The Court of Appeal, upholding the decision of the Patents Court to strike out the action, confirmed that the old *Saccharin* doctrine (see Chapter 10) was no longer good law. Under *Saccharin*, if any of the process steps would have infringed the patent, sale of the final product would have been an infringement. Now, the law applies only to the 'direct' product of the process,[31] and in order to determine what is 'direct', the courts must look at the corresponding provisions of the EPC.[32] There,[33] the corresponding word in the German text is *unmittelbar*, which has the meaning of having no intermediate step between two points. The Court adopted the 'loss of identity' test from German law to formulate a three-step enquiry, which can be summarized as follows.

(a) Construe the process claim according to the Protocol on Article 69 of the EPC.
(b) Identify the product resulting from the final process step of the claim.
(c) Consider whether the alleged infringement shares the same identity (that is, the same essential characteristics) as the product in step (b).

In the case in question, the 'master' discs did not share the same identity as the CDs (in particular, they could not be played on CD players) and there was accordingly no infringement. The lesson from this is that applicants should always draft process claims so as to include the last step that gives the commercial product, even if this step is conventional.

[30] *Pioneer Electronics Capital Inc. v. Warner Music Manufacturing Europe GmbH* [1997] RPC 757 (CA).

[31] Section 60(1)(c), PA 1977.

[32] Since s. 60 is specifically mentioned in s. 130(7), PA 1977.

[33] Article 64(2), EPC.

This case was followed in a controversial biotech case in which Monsanto, having a patent relating to RoundUp®-resistant soy plants and in particular a method for producing resistance by inserting DNA for a certain enzyme, attempted to stop importation of soy flour produced from resistant plants.[34] It was held that this was not the direct product of the process, which was limited to the first-generation transgenic plants (see p. 310)

Contributory Infringement

If the claimed process is a process of use of a non-patented product, then it will be directly infringed only by users of the process who buy the product from an unauthorized competitor. In Chapter 10, we discussed the nature of contributory infringement, which gives the patentee the possibility of taking action against the competitor himself, at least if the product in question has no substantial non-infringing use. *Rohm & Haas v. Dawson*[35] in the USA, referred to briefly in that chapter, gives a good illustration of a typical contributory infringement situation and is worth considering here in more detail.

The case related to the compound propanil, which can kill weeds in rice fields without harming the rice. The compound was known and could not be claimed per se, and Rohm & Haas had a patent for its use as a selective herbicide, that is, a process for the control of weeds in rice by applying propanil. Propanil had no other substantial use and thus was a 'non-staple', so that although the patent could be directly infringed only by farmers, unauthorized sale of propanil to farmers was contributory infringement. Rohm & Haas refused to grant licences, but farmers purchasing propanil from the company received an implied licence to use the patented process. Dawson also sold propanil, and when sued by Rohm & Haas, admitted contributory infringement, but alleged that the conduct of Rohm & Haas constituted patent misuse. The Supreme Court decided in favour of Rohm & Haas, although by only a five-to-four majority.

An interesting point is that Dawson was selling propanil with full instructions for its use as a selective weedkiller, that is, with instructions to infringe Rohm & Haas's process patent. This should constitute direct infringement according to US patent law, because the law says 'whoever actively induces infringement of a patent shall be liable as an infringer'.[36] This point was not argued or decided in the *Rohm & Haas* case, presumably because Dawson admitted that it was liable as contributory infringer if its defence of patent misuse failed. But it is of vital importance in the more common cases in which the product does have other non-infringing uses.

Take, for example, a patent for an improved dyeing process. The process can be applied to old, unpatented dyestuffs sold by a number of manufacturers, all of which can, of course, be used in many old, unpatented dyeing processes. If a rival

[34] *Monsanto Technology LLC v. Cargill International SA* [2007] EWHC 2257 (Pat).

[35] *Dawson Chemical Co. v. Rohm & Haas Co.* 206 USPQ 385 (Sup Ct 1980).

[36] 35 USC 271(b).

manufacturer sold its dyestuffs with trade literature describing how to carry out the patented process, then it is not a contributory infringer in US law, because the product is a staple article of commerce, but should be liable as a direct infringer.

An ironic twist to the story of *Rohm & Haas* is that, in subsequent litigation,[37] Rohm & Haas's patent was held to be invalid. Nevertheless, this does not affect the value of the Supreme Court's decision on the issue of contributory infringement.

In the UK, selling a product with instructions to infringe a patented process will amount to supplying for the purpose of inducing infringement and so fall within the definition of infringement under the Patents Act 1977.[38] Infringement would not be avoided simply by calling the customers' attention to the existence of the patent.

Contributory, or indirect, infringement was recognized by German case law before 1981, but it was a rule that there could be no indirect infringement without direct infringement. Thus, because a claim to a cleaning process using a particular composition would not be directly infringed by a housewife (since private non-commercial use was not infringement), it followed that sale of the composition as a domestic cleaner was not indirect infringement. Since 1981, the law has defined infringement so as to include sale of a product the sole or main use of which falls within the patent claim, or its sale with instructions to infringe. It is no longer necessary to prove that direct infringement has occurred, and the law does not now make any distinction between direct and indirect infringement.

Exhaustion of Rights

The patentee must also remember that if the 'infringer' is selling or using goods that originate from the patentee himself, then such sale or use can generally not be prevented. In the absence of a binding agreement to the contrary, the original purchaser of the goods from the patentee will have an implied licence to use or resell them, which passes with the goods to subsequent purchasers. Furthermore, within the European Economic Area (EEA), sales of imported goods, even though they are infringements under national patent law, cannot be prevented if the goods were first put on the market in another EEA country by the patentee or with his consent.

Process patents may also be subject to exhaustion of rights. In the USA, LGE licensed Intel to produce chip components, which it sold to Quanta, which used the products in a process patented by LGE. The licence to Intel specifically disclaimed any licence to third parties to practise the patents by combining licensed products with other components, but the Supreme Court held that Intel's licensed sales to Quanta exhausted the patent rights, irrespective of the contractual situation.[39]

[37] *Rohm & Haas v. Crystal Chemical Co.* 220 USPQ 289 (Fed Cir 1983).

[38] Section 60(3), PA 1977.

[39] *Quanta v. LG Electronics*, 128 Sup Ct 2109 (2008).

Obtaining Evidence

If it is not clear whether or not activities of a competitor amount to infringement, it may be necessary to carry out investigations to determine this. This may be possible without recourse to any form of compulsion: for example by buying a competing product on the open market and carrying out an analysis to determine its composition, or discover how it was manufactured. In some countries, it may also be possible to obtain evidence by means of a court order. In the UK, search orders (formerly *Anton Piller* orders) may be obtainable on an *ex parte* basis, allowing an alleged infringer's premises to be searched without warning and evidence secured. These are, however, normally used in copyright and trademark cases, in which there is a strong prima facie case of infringement and counterfeit goods need to be secured before the trader disposes of the goods or absconds. They are seldom granted in patent infringement cases.

In France, however, orders for *saisie contrefaçon* are commonly used to obtain evidence in patent cases. These are available to patentees or applicants for patents as of right, without any prima facie case having to be made. A court-appointed bailiff, accompanied by one or more independent experts, makes an unannounced visit to the premises of the alleged infringer, and may make notes, take photographs, etc., but goods may not be removed without payment. Although the patent owner's attorney may be present, no employees of the patentee are allowed. Material claimed to be confidential is made available to the judge, but not to the patent owner. An important point is that, if an infringement suit is not lodged within a relatively short time, the evidence collected in the *saisie* cannot be used in any future litigation. Similar orders are available in Belgium and Italy.

In Germany, it was formerly impossible to obtain evidence by court order, but this was changed by the new Civil Code of Procedure in 2002, as applied in the *Faxkarte* case in 2003,[40] and has been further developed since the entry into force of the European Union (EU) Enforcement Directive[41] (see p. 205). It is now possible to use the so-called 'Düsseldorf solution' to implement the relevant provisions of the Directive.[42] Application, which may be *ex parte*, should be made to the court providing as much evidence for infringement as possible, explaining why the patent should be considered valid, making a detailed request for inspection of the infringer's premises, and appointing an independent expert to give an opinion on the question of infringement. The defendant is obliged to tolerate the inspection passively (he is not obliged to be cooperative) and has a two-hour grace period in which to involve his own lawyers before the inspection begins. The inspection is carried out by the appointed expert, accompanied by an officer of the court, and by the plaintiff's patent attorney, who is, however, bound to secrecy vis-à-vis his client

40 *Faxkarte* [2002] GRUR 1046 (BGH).

41 Directive 2004/48/EC of 29 April 2004 on the enforcement of intellectual property rights.

42 Articles 6 and 7, Directive 2004/48/EC.

unless and until the expert opinion is released by the court. Unlike the French *saisie*, there is no requirement that the results be used in litigation within a specific time.

To Sue or not to Sue

Given that the competitor's activity is infringement and that there are no legal barriers to the enforcement of the patentee's rights, the patentee must still consider carefully what options are available. Normally, the first step would be to write a warning letter bringing the patent rights to the competitor's attention. In some countries, this may be a necessary preliminary to any subsequent claims for damages, but, at this stage, the main object is to find out the infringer's reaction. Care must be taken to avoid liability for a threats action in the UK (see p. 198). The infringer may agree to stop infringing activities, or may ask for a licence. If the patentee is willing to grant licences, negotiations about terms will naturally follow; if not, or if the infringer denies infringement or alleges that the patent is invalid, then the patentee has to decide whether or not to sue the infringer.

In making his decision, the patentee must weigh up all of the many factors that are involved, one of the most important being the degree of confidence that he has in the validity of his patent. A patent of questionable validity may be respected by the majority of the patentee's competitors, but if, as a result of attempting to enforce it against one competitor, the patent is found invalid, then all of the other competitors will be free. The indirect effect upon corresponding patents in other countries must also be considered. Furthermore, infringement actions are expensive, and if the infringement is small, what would be gained by a successful action might not justify the costs.

Nonetheless, it is important in order to deter future infringement for it to be known that the patentee is willing to enforce his rights. Certainly, a reputation for timidity in such matters can become an open invitation to infringers and it is probably worth even the occasional unsuccessful action in order to avoid such a reputation. Finally, of course, the patentee may have granted one or more licences by which he is contractually obliged to enforce the patent against infringers, in which case, he may have no choice in the matter.

Where to Sue

If sales are taking place in more than one country in which patent rights are in force, the patentee may decide to sue the infringer in all of them simultaneously, or to select one country in which to take the initial action and to await the results before deciding what to do in the other countries. The choice may be determined by business considerations, such as the size of the market or the damage done by the infringement in the various countries, or by patent considerations, such as the scope or strength of the patent protection, or the speed or predictability of the legal procedure. If a single European patent is being infringed in more than one EPC country, the scope and strength will usually be the same everywhere (unless, for example, there is Article 52(4) prior art or prior national rights in some countries, but not others), so the most important issues will often be how likely the national court is to grant an interlocutory injunction to stop further infringement, and how rapidly and cheaply it will do so. If the infringer is stopped in one country, he is much more likely to settle out of court and agree to cease infringing activities in other countries as well.

Given a free choice of jurisdictions in which to sue for infringement in Europe, the three that mainly come in question are the UK, Germany, and the Netherlands. As compared with the others, on the one hand, the English courts perhaps do a more thorough job, in which testing the reliability of witness evidence by cross-examination is taken more seriously. On the other hand, costs in England are higher (one has to estimate at least £750,000 for a full-scale pharmaceutical infringement action at first instance) and there is a certain perception that the English courts are more likely to hold a patent invalid than their German or Dutch counterparts (validity being dealt with by a separate court in Germany). If you have a strong patent, the English court is a good place to enforce it, but the problem is that the judge may take a different view of the strength of the patent than you do.

When a company, or a group of related companies, is infringing the same European patent in a number of different EU countries, the provisions of the Brussels Regulation[43] on jurisdiction and enforcement become important. The Regulation, which replaces the earlier Brussels and Lugano Conventions, is intended to simplify the litigation of multinational disputes by setting rules as to which court has jurisdiction and ensuring that the judgment will be implemented in all EEA states. It was also enacted into English law in 2001.[44] The basic rule is that a defendant is to be sued in his country of domicile,[45] but there are a number of exceptions to this.

- Article 5(3) states that, in matters relating to tort (for example patent infringement), the defendant may also be sued in the state in which the tort occurred.
- Article 6(1) provides that, where there are a number of defendants in what is essentially the same action, then, if it is expedient to hear the cases together to avoid the risk of irreconcilable judgments resulting from separate proceedings, they may all be sued in a country in which any one of them is domiciled.
- Article 22(4) states that, where the case is concerned with the validity of a registered intellectual property right, the suit should be brought in the country in which the right is registered.
- Article 27 provides that, if proceedings involving the same cause of action and the same parties are brought in different states, the court in which the litigation was first started must consider whether it has jurisdiction, and the other courts

[43] Council Regulation 44/2001/EC of 22 December 2001 on jurisdiction and the recognition and enforcement of judgments in civil and commercial matters.

[44] Under the Civil Jurisdiction and Judgments Order, SI 2001/3929.

[45] Article 2, Brussels Regulation.

must stay proceedings until this is determined and decline jurisdiction if the first court asserts it.

- Finally, under Article 31, the courts of one member state may grant provisional measures, such as an interim injunction, even if the court of another state has jurisdiction as to the substance of the action.

Cross-Border Injunctions

The Regulation, like the earlier Convention, provided considerable opportunities for 'forum shopping' by the patentee, and the Dutch courts proved to be a desirable forum. For one thing, the Dutch courts were the first to apply the Brussels Convention to grant cross-border injunctions,[46] that is, as long as one defendant was domiciled in the Netherlands, the court would be prepared to grant an injunction not only against infringement of a Dutch patent in the Netherlands, but also against infringement of a British patent in the UK or a German patent in Germany, at least if the patents were all based on the same European patent. Furthermore, the Dutch *kort geding* (short proceedings) are a form of preliminary hearing that is both rapid and cheap, and which gives a preliminary judgment that is immediately enforceable, on penalty of heavy fines. It is therefore a moot point whether or not the judgment is enforceable in other countries: as long as one of the defendants has assets in the Netherlands, these can be attached to compel compliance with the cross-border injunction.

The English courts were not so ready to grant cross-border injunctions as the Dutch. In the UK, it was held[47] that, as soon as validity became an issue (which it almost always would be), the provisions of Article 16(4) applied. The judge therefore refused to consider matters dealing with infringement of German and Spanish patents.

Even in the Netherlands, it became more difficult to obtain a cross-border injunction. The Court of Appeal of The Hague held[48] that the Dutch courts have jurisdiction under Article 6(1) only if the Dutch defendant is the headquarters of the relevant group, in that the 'common design' for the Europe-wide infringement emanates from the Dutch entity. In effect, this means that cross-border injunctions would be available only against Dutch-based companies and not against all multinational companies that happen to have a Dutch affiliate. Nevertheless, in a suitable case, this remained a very effective procedure. For example, in 2003, the District Court of The Hague issued a preliminary injunction against a total of 30 defendants in no fewer than 14 countries.[49]

[46] The first being in *Lincoln v. Interlas* (1989) 404 NJ 1597 (Hoge Raad) (a trademark case).

[47] *Coin Controls v. Suzo* [1997] FSR 660.

[48] *Expandable Grafts v. Boston Scientific* (CA, The Hague, 23 April 1998).

[49] C-2/3065 *Regents of University of California v. ev3 et al.* (DC, The Hague, 22 October 2003).

Torpedoes

A parallel development has been for the potential defendant to begin proceedings for a declaration of non-infringement and invalidity of the European patent in a member state with a notoriously slow court system, typically Belgium or Italy, while, at the same time, seeking a declaration of non-infringement of all corresponding European patents. If the patentee then sues for infringement in a faster or more pro-patentee jurisdiction, such as the Netherlands, the defendant would then claim that the Dutch court lacked jurisdiction in view of Article 27 of the Brussels Regulation and must stay the action.

The Brussels Court of Appeal has, however, taken the view that such a 'torpedo action' fails if the Belgian defendant was sued for the sole purpose of being able to summon the other defendants before the same court.[50] In Italy, torpedo actions were admitted by the courts and the question was whether or not courts in other states would decline jurisdiction as required by the Regulation. In one case in France, the court refused to give effect to an Italian torpedo, considering the whole thing an abuse of process.

In 2003, the Italian Supreme Court effectively put an end to the Italian torpedo. The whole principle of torpedo actions is based on the premise that an action for a declaration of non-infringement is a 'matter relating to tort' according to Article 5.3 of the Regulation. In the case of *BL Macchine Automatiche v. Windmoeller*,[51] the Supreme Court held that, by seeking a declaration of non-infringement, the party is denying that any tort has been committed, so that Article 5(3) cannot apply. Consequently, the Italian courts did not have jurisdiction.

The final blow to cross-border jurisdiction in Europe has been given by the European Court of Justice (ECJ) in its 2006 judgments on two cases, both delivered on the same day. In *Gat v. Luk*,[52] a dispute between two German companies involved a challenge to the validity of a French patent. The ECJ held that the German courts had no jurisdiction, even if the effect of the decision were to be limited to the parties. In *Roche v. Primus*,[53] it was held that Article 6(1) could not be applied in the situation of multiple infringements of corresponding patents in different countries, since, because national laws on infringement were different, there could be no risk of 'irreconcilable judgments'. The only solution is a new agreement about patent litigation in Europe.

From the Viewpoint of the Potential Infringer

Few companies set out deliberately to infringe the patents of others, but every company that puts a product on the market or uses a process is potentially an infringer

[50] *Roche v. Wellcome Foundation* (CA, Brussels, 2001).

[51] *BL Macchine Automatiche v. Windmoeller* [2004] *Rivista di diritto industriale* 2 (*Corte di Cassazione*, 19 December 2003).

[52] C-4/03 *Gat v. Luk* (ECJ, 2006, referral from *Oberlandesgericht* Düsseldorf).

[53] C-539/03 *Roche v. Primus* (ECJ, 2006, referral from *Hoge Raad*, The Hague).

of one or more of the tens, or even hundreds of thousands, of patents presently in force. It is irresponsible, to say the least, to market a new product without carrying out a thorough search for any patents that might be infringed.

Infringement Searches

A full infringement search requires the joint efforts of a professional searcher familiar with the technical field, who can find the patents that may be relevant, and a patent practitioner, who can give an opinion on whether any of the relevant patents would actually be infringed by what it is proposed to do.

Such a search will normally be based on a variety of sources, and may turn up a mixture of patents and published patent applications from various countries. The first step will be to correlate them with the countries in which it is actually proposed to market. For example, if it is proposed to sell in the UK and USA, and a search finds a relevant Japanese published application, then it is necessary to check whether granted equivalents of this exist in the two countries in question. Various databases give lists of equivalent patents, that is, patents that all claim priority from the same original application. It will then be necessary to look at these to check what claims have been granted in each country.

The claims of any relevant patents should then be checked for possible infringement as described above, bearing in mind the possibilities of infringement by equivalence or contributory infringement in some countries. Whereas the patentee is concerned with possible infringement of a particular patent, the potential infringer will often have a number of relevant patents to consider. Nevertheless, he at least does not have the same problems of chemical analysis, because he presumably knows what it is that he proposes to sell.

There may be situations in which it is not clear whether the proposed activity would amount to infringement or not. Apart from the whole question of equivalence, discussed above, there may be problems, for example, when the patent claims a multi-step process and it is proposed to carry out the steps in a different order, or to carry out the early steps in one country and the later steps in another. The answers will, of course, depend upon the precise wording of the claims and on the law of the country in question, but, in general, unless the claim stresses that the steps must be carried out in a particular order, then altering the order in which the steps are stated in the claim would not avoid infringement. The situation in the second case would probably be that there would be no infringement unless the entire claimed process was carried out within one country.

If the patent is near to expiry and no commercial sales will take place until after the patent has expired, one may still have to consider whether necessary experimental work, such as clinical or field trials that would have to be conducted while the patent is still in force, would amount to infringement and could be stopped by interlocutory injunction as in *Monsanto v. Stauffer*.[54]

[54] *Monsanto v. Stauffer* [1985] FSR 55 (Pat Ct)—see p. 192.

Is the Patent not yet Granted?

Very often, what will be found by the search will not be a granted patent, but a published European or Patent Cooperation Treaty (PCT) application. Here, it will be necessary to follow the progress of prosecution in the countries of interest, where this is possible, and to check whether it is likely that patents will be granted with claims that would cover the proposed activity. If feasible, oppositions may be lodged. In any event, it may well be that the proposed activity would be of short duration and would be concluded by the time that any patent would be granted. For example, a patent application may claim a test method that might be used only for a short time and then superseded by another before the patentee had any enforceable rights. In such a case, the party that had used the method without permission would, in Europe, be liable for no more than reasonable compensation for the use after publication of the application[55] (assuming that translations had been filed as required), and in the USA, not even that.

Is the Patent in Force?

As a result of a rapid check, many apparently relevant patents can be dismissed from consideration and the potential infringer may be left with a small number for which a serious question of infringement arises. The next step is to check whether these patents are still in force, which can be determined relatively quickly in most countries from the national patent office and usually can be determined instantly from online file inspection. This check should always be done at an early stage: many patent practitioners have, at some time or another, wasted time in detailed study of a patent, only to learn later that the renewal fees had not been paid for the last ten years.

Is the Patent Valid?

If a relevant patent is in force and would be infringed by the proposed activity, the potential infringer should then consider whether or not the patent is valid. The original infringement search will have already found patents that are prior art to the patent in question and the scientists responsible for the proposed activity may know of more. A further search may still be necessary directed specifically towards finding prior art that could anticipate or render obvious the claims of the patent.

Internal grounds of invalidity, such as insufficiency, should also be considered and, in countries in which this is available, the prosecution history of the patent should be studied carefully. The potential infringer should also consider whether or not his proposed activity would have been patentable at the priority date of the patent. In the UK, at least, it is an effective argument against a charge of infringement to say:

> What I am doing was, at the priority date of the patent, either old or obvious in view of the prior art. Therefore either what I am doing does not fall within the claims of the patent, and

[55] Article 67(2), EPC.

there is no infringement; or the claims cover what is old or obvious and therefore are invalid. What is more, it is not even necessary to determine which of these alternatives applies.

This argument is known as a '*Gillette* defence' from the name of the case in which it was first mentioned.[56]

This defence is not available in this simple form in the USA. Although it is old law in the USA that 'that which infringes if later, anticipates if earlier',[57] nevertheless, the CAFC has recently held that it is not a defence to infringement to plead that one was practising the prior art; anticipation must be proved by showing clear and convincing evidence that each element of the claim is found in a single prior art reference.[58]

Are there Rights of Prior Use?

As well as considering questions of validity, the potential infringer should check whether he has any rights of prior use. Even if the product has not yet been put upon the market, in many countries, making effective and serious preparations to do so before the priority date of the patent can give the right to continue such activity after the patent is granted.[59]

Is there a Research Exemption?

It should also be considered whether or not the proposed activity falls under a statutory or common law research exemption. In the USA, this seems to be excluded, except if the activity relates to obtaining Food and Drug Administration (FDA) approval, but in Europe, if what one is doing is research *on* the invention, for example trying to improve upon it or find new uses for it, as opposed to research *with* the invention, such as the use of a research tool for its intended purpose, the research exception may well apply. Of course, commercialization of the results of the research is a different situation, which must be independently evaluated.

Can one Design Around the Patent?

If, after all of this, the potential infringer is faced with a patent that would be infringed, which is in force, and against which there are neither grounds of attack nor rights of prior use, the next step must be to consider whether or not it is possible to 'design around' the claims of the patent.

This is hardly possible in the field of chemistry if the potential infringer wishes to sell a single compound that falls within the scope of the claims, but with patents covering mixtures of compounds, the question arises whether one of the components can be replaced by another compound not covered by the claims, or whether

[56] *Gillette Safety Razor Co. v. Anglo-American Trading Co.* (1913) 30 RPC 465 (HL).

[57] *Peters v. Active Mfg* 21 F 319 (WD Ohio 1884).

[58] *Zenith Electronics v. PDI Communications* 522 F.3d 1348 (Fed Cir 2008).

[59] For example, in the UK, under s. 64(1), PA 1977.

the proportions of the components can be varied outside the claimed range without too great a loss of advantageous properties. He must, of course, be careful not to be caught by the doctrine of equivalents in countries in which this applies. He must also be careful not to infringe other industrial property rights, such as copyright; in the UK, there may be, in the mechanical field, copyright in engineering drawings that could possibly be infringed without infringing the claims of a patent, or even after the patent expires.

Although deliberate copying of another's efforts may be regarded as wrong, there is nothing unethical about designing around a patent that stands in the way of one's own developments. No patentee is entitled to any more rights than the law gives him, and has no cause for complaint if another manages to avoid infringement and still to have a commercially successful product.

Is a Licence Available?

If all else fails, the potential infringer must decide whether to abandon his plans, seek a licence from the patentee, or to go ahead without a licence. It is usually worthwhile to ask for a licence, unless it is absolutely clear that there is no chance of one being granted. It may be that a mutually acceptable cross-licence may be negotiable: for example if the potential infringer has a patent for a selection invention within the scope of an earlier dominating patent. In this situation, the possibility of an application for a compulsory licence in the UK and certain other countries should not be ignored; the same is true if the patent that would be infringed is not being worked.

In cases in which the parties are hostile and no licence will be forthcoming, then, if there would clearly be infringement, the potential infringer has little choice but to abandon the project. But in a situation in which there is some real doubt as to the validity of the patent or whether there would be infringement, it may be an acceptable business risk to go ahead. Factors such as the remaining life of the patent, the difficulty of analysis of the product, the costs involved if the product were to have to be taken off the market once launched, and the possible expense of having to fight an infringement action would then have to be considered.

24

PATENT ASPECTS OF LICENSING

> In France, British duds falling behind German lines bore the tiny stamp KPz96/04, 1896 being the year Vickers first licensed Krupp's fuse patent and 1904 the year the agreement was renewed.
>
> William Manchester, *The Arms of Krupp* (1968)

Patent Conflict Licences

A manufacturer who finds that a product that it wishes to sell or a project that it wishes to develop appears to be covered by someone else's patent rights may have to obtain a licence from the other person in order to be free to go ahead. As discussed in the previous chapter, the first necessary step is to carry out a full evaluation of the patent situation. If the patents are in force and may still be kept in force beyond the likely date of introduction of the project—bearing in mind any possible extensions of term such as supplementary protection certificates (SPCs)—then an assessment of their validity is of prime importance.

If the patent rights appear to be valid, an ethical company will respect them and approach the patentee to seek a licence; the project must be abandoned unless a licence can be obtained on commercially acceptable terms, which will often be possible between research-based companies. The licence required is a non-exclusive licence under the specified patent rights, or (as may be preferable) a covenant not to sue in respect of the project in question under any patent rights held by the other party. The legal effect is the same in both cases, at least in the USA, where sale of a patented product by a party with whom a covenant not to sue had been entered into by the patentee was held to be an authorized sale giving rise to exhaustion of rights.[1] The terms may, for example, involve payment by means of a lump sum or a running royalty on sales, or it may be necessary to grant a cross-licence in exchange. A special case of this type of licence occurs frequently in the USA when two parties find that patent applications assigned to them are involved in interference proceedings. The parties may decide to enter into an interference settlement agreement under which they agree to decide between themselves which party has the earlier invention date and agree in advance that the winning party will grant a licence on reasonable terms to the other. The interference proceedings in the US Patent and Trademark Office (USPTO) can then be terminated by consent, a patent for the claims involved in the interference being granted to the party agreed to have the earlier date.

Technology Transfer Licences

A completely different type of licence is one that is entered into not to gain freedom to pursue one's own project, but to take over a project or a product that originated elsewhere. Whereas the first type usually involves a non-exclusive patent licence or immunity from suit, without any know-how being involved, the second type may be broadly described as transfer of technology, which normally involves rights to confidential know-how, as well as under patents, such rights often being exclusive. Trademarks and other forms of intellectual property may also be involved. The most common categories are:

- in-licensing (that is, passive licences), in which the company acquires rights from another party;
- out-licensing (that is, active licences), in which rights of the company are given to another party; and

[1] *TransCore v. Electronic Transaction Consultants* (Fed Cir, 8 April 2009).

- research cooperation, in which the company and another party jointly develop technology.

The first two are clearly opposite sides of the same coin, and differ only in the viewpoint of the parties, that is, whether the party is licensee or licensor. The third needs to be considered separately.

Potential Licensees

What types of company wish to license in technology and under what circumstances? One type is the small company that lacks the facilities to do basic research of its own and wishes to buy the products of others' research. It is not easy, however, for a small company to find suitable products. If the licensor is another small company or a university, the licensor will normally seek a licensee with the market strength and technological facilities to develop the project rapidly. If the licensor is a large company, the small licensee will be in danger of being swallowed up or becoming a mere distributor.

A second type of potential licensee is a larger, research-based company that wishes to expand its product line or investigate areas new to it. No matter how good a company's research department may be, it cannot investigate everything and will naturally concentrate its activities in certain areas of particular interest. If the company wishes to branch out into other fields, there will inevitably be a long lead time before its own research activities, starting from scratch, can produce anything of commercial interest. This gap can best be bridged by licensing in a project that is already on the market, or nearly so. Even if a completely new field is not involved, many large pharmaceutical companies are under great pressure to have a full pipeline of products at various stages of development and there is intense competition to find good compounds to fill the pipeline.

A third type of situation arises when a company, large or small, is established in a particular business and an invention is made by another company of such basic importance to that business that all companies involved practically must take a licence (assuming that the patentee is willing to grant licences). Naturally, this does not occur very often, but a good example is the patent on oil-extended synthetic rubber for vehicle tyres, held by General Tire Co., which was licensed to a number of other tyre companies and infringed by most of the others (an infringement action against Firestone was successful in the UK).[2] A more recent example is the Cohen/Boyer patent that was basic to recombinant DNA technology (see p. 283).

Potential Licensors

Taking again the example of the large research-based company that wants to license in projects to supplement its own research: where should the company look for such projects? One possibility is from university research, either from individual

[2] *General Tire v. Firestone* [1972] RPC 457 (CA).

academic scientists, in so far as they own rights in their own inventions, or from the universities or other organizations that do own the rights. Such projects will often be in a relatively early stage of development, and will require considerable effort and investment by the licensee.

A further possibility is licensing in from other large companies. In this case, however, a certain amount of caution is needed, and the potential licensee should always try to find the answer to the obvious question: 'If this project is so great, why isn't the other company doing it itself?' Often, there is a satisfactory answer, particularly if the licensor company has decided that it has no further interest in a particular area in which it was previously involved, and all projects within that area, no matter how promising, are being offered for licence. Another situation that now arises more frequently than before is when two pharmaceutical companies wish to merge and competition authorities, such as the European Commission or the US Federal Trade Commission (FTC), consider that the merged company would have excessive market power in some area. As a condition for approval of the merger, it may be necessary for one or both of the companies to divest certain products, which may then be licensed or sold to a third party. It may be, however, that the project offered for out-licensing is one that has been dropped because it has some defect; in this case, unless the potential licensee knows that he can cure the defect, he is better to decline the offer with thanks.

A good source of late-stage licence projects for a multinational company is the smaller, national or regional, research-based company that can cope with its home market adequately, but which does not have the resources to launch a product worldwide. Such a company, having a project that looks as if it may be a commercial success, often wishes to enter into a licensing agreement whereby it retains all rights in its home territory and grants an exclusive licence to a multinational company for all other countries. Since no Japanese pharmaceutical companies are yet in a position to sell directly in Europe or the USA (except through a small number of subsidiary or joint venture companies with local partners), Japanese companies have been an important source of licence projects. The number of research-based national pharmaceutical companies in Europe is small, and is in danger of declining further as the cost of developing new drugs continues to rise and the number of acquisitions and mergers within the industry increase. More and more licensing deals tend to be made between multinational companies and small biotech-based companies, mainly situated in the USA. It is, however, unusual for such companies to have projects in an advanced stage of development, so that research cooperation agreements are more common.

The In-licensing Process

Confidential Disclosure Agreements

Although at the initial stage of discussions between potential licensor and licensee some information may be exchanged on a non-confidential basis, as soon as things begin to get serious, the first step will be the signing of a confidential disclosure agreement (CDA). In a CDA, the disclosing party agrees to give confidential proprietary information to the receiving party, who undertakes not to disclose it to other parties and to use it only for the purpose of evaluation of the proposed licence. Normally, the recipient agrees to return or destroy copies of the information if no licence agreement is entered into at the end of an agreed evaluation period.

Most large companies have their standard forms of CDA, and if the other party agrees to them, everything is fine. More usually, however, either the other party will propose changes, or it will produce its own standard agreement for the potential licensee to sign. In general, from the point of view of the recipient of confidential information, any CDA should include:

- standard exceptions to the confidentiality obligations (excluding, for example, material that is or will later become part of the public domain, or can be shown to be independently discovered by the recipient);
- a fixed term of no longer than ten years;
- the right to keep one copy of the information for record purposes; and
- the right to give the information to affiliates and consultants.

It should *not* include, for example:

- requirements that recipients of the information should be named or should have to sign individual confidentiality agreements;
- prohibition on making copies; or
- clauses admitting that, in case of breach, damages are not a sufficient remedy.

Due Diligence

Once real interest has been established and more detailed information is available about the technology, a more complete patent situation must be prepared. This will normally involve obtaining and studying copies of the file history of relevant granted US and European patents, and carrying out and evaluating an infringement and validity search, that is, a search designed to find prior art documents that could be damaging to the validity of the patent rights, as well as third-party patents that could be infringed by making and selling products within the technology offered for licence.

It must be remembered that the scope of valid protection may vary from country to country, and particularly that because of the first-to-invent principle of US law, equivalent patents may be valid in the USA but invalid elsewhere. This still occasionally happens when an academic inventor in the USA publishes his own work before filing a US patent application.

Once a CDA has been signed, the potential licensor should be required to give a full list of all relevant patent rights that it owns or under which it is licensed, and specifically to disclose any pending or threatened litigation, any third-party attacks on the validity of the patent rights, and any possibly dominating rights of third parties. This information should be checked against the information obtained by independent searches and any discrepancies resolved.

Responsibility for the Licensor's Patents

From the viewpoint of the licensee, the ideal arrangement is that the licensor retains full responsibility for the licensed patent rights, and is obliged to prosecute them to grant and keep them in force for their full term in each country of the licence territory, including applying for and obtaining all possible extensions, SPCs, etc., all at the licensor's expense. And where the licensor is a large company, this will often be the agreed situation, particularly when the licensor retains marketing or co-marketing rights in some countries. At most, the licensor may be allowed to abandon a licensed patent in a country if the licensee is given the opportunity to maintain it at his expense, and the royalty for sales in that country is reduced or abolished. The licensee will, however, be required to cooperate with the licensor, for example, in giving information about registration approvals so that SPCs and other extensions may be applied for on time.

For many small companies and academic institutions, however, the financial burden of the prosecution and maintenance of patent rights in what may be a large number of countries is not one that they can easily bear, particularly before royalty revenues start to flow in. Such licensors will therefore often demand that the licensee pay the patenting costs, at least outside of the licensor's home country, as part of the consideration for the licence. In principle, this may be acceptable, but the licensee must think carefully before entering into such an obligation. An unlimited obligation to pay patenting costs is simply handing over a blank cheque to the licensor, or rather to his patent attorneys: no invoice sent to the licensor will ever be queried, because the licensee is paying.

Many companies as licensees therefore take the position that they will pay for the licensor's patenting activities only if they directly control these activities. For one thing, costs incurred by a large company are normally substantially less than those incurred by a small company for the same work, since not only can they do much of the work in-house, but they can also secure discounts from outside attorneys because of the large volume of work that they give them. For another thing, if a patent that is within the licensed patent rights turns out not to cover a licensed product, the licensee can normally decline to maintain it, whereupon the licensor must then decide either to let it lapse or to maintain it himself. If he does maintain it, the patent will normally be excluded from the licensed patent rights, but *ex hypothesi* this does not matter, because no licence for that patent is required.

The difficulty is that there is a potential conflict of interest when a licensee carries out patent work for a licensor. Normally, royalties are payable only on products the sale, manufacture, or use of which would infringe one of the licensed patent rights in the relevant country, and if the scope of a licensed patent has been reduced during prosecution by the licensee so that one of the products is no longer covered, the licensor may suspect that this has been done deliberately so as to avoid payment of royalties. In fact, both licensor and licensee normally have the same interest in obtaining the best possible patent protection, and it would be a foolish licensee who, by deliberately weakening patent protection to reduce royalty payments, allowed a competitor free access to the market. Nevertheless, situations may arise in which a licensed patent is broad enough to cover a licensed product, but too weak to give effective protection against infringers. In such a case, it is clearly in the interest of the licensee, but not of the licensor, to reduce the scope.

Accordingly, arrangements of this kind can work well only when the licensee's in-house patent attorneys act in a professional manner as the agents of the licensor and take instructions from the licensor, particularly where any limitation of the claims may be necessary. If there is real disagreement, the case in question must be handed over to an independent patent attorney, at the licensee's expense. Even when the licence is exclusive for the whole scope of the patent, great care is needed to avoid possible conflict of interest; if the licence is non-exclusive, or exclusive for less than all of the scope or fields of use, then prosecution by the licensee's in-house attorneys is not appropriate, because rights of third parties may also be affected. Alternatively, the potential for conflict of interest will be greatly reduced if the contract provides for a royalty upon sales that is independent of the patent situation (to the extent that this is permitted by competition law, see the next chapter).

Enforcement of the Licensed Patent Rights

Again, the best situation for the licensee is that in which the licensor has a binding obligation to enforce the patents against infringers throughout the licensed territory. If this cannot be negotiated, the licensor should at least have a strong incentive to enforce the patents: for example in an exclusive licence under a provision whereby the royalty on sales in a country is substantially reduced in the event that unlicensed competition takes more than a certain percentage of sales in that country. In a situation in which there are reasonably strong patents in most major markets, but in some countries in which the product may be marketed the patents are weak or non-existent, then rather than setting different royalty rates at the outset for patent and non-patent countries, the licensee may agree to pay the same royalty on all sales throughout the territory, subject to such an unlicensed competition clause. After all, what the licensee is paying for is exclusivity and as long as he has it in a particular country, the absence of a patent there is irrelevant. Similarly, if he loses exclusivity, it is immaterial to him whether this is because a licensed patent is not being enforced, or because there is no patent that can be enforced, and the consequences should be the same.

Another approach, which may be combined with the above, is to allow the licensee to take action against infringement if the licensor has not taken action within a certain time. Alternatively, the responsibility to take action may be split on a territorial basis: for example where the licensor is a small US company and the licensee is a European multinational, it may be agreed that the licensor has the prime responsibility to sue infringers in the USA and the licensee in the rest of the world. It should be remembered that, in many countries, an exclusive licensee can take action in its own name only if the licence is registered and that there may be limitations in damages, or other disadvantages, if the licence is registered late, that is, only shortly before proceedings are started.[3]

The licence agreement clauses relating to enforcement need particularly careful drafting, so that not only the responsibility for lodging suit is clear, but also questions of apportionment of the costs of the suit and the sharing of any benefits such as damages. If the licensee has responsibility, does this extend to defending the patent against a counter-claim of invalidity, and if so, can he reach a settlement without the consent of the patentee? All of these and other points must be considered, and the agreement reached must be accurately reflected in the contract. It is not good practice to regard an enforcement clause (or any other substantive clause, for that matter) as standard 'boilerplate' to be inserted as a mere formality.

Out-licensing: Patent Strategy for the Licensor

Large Companies

Most large research-based companies are not normally in the business of licensing out, but this strategy is often used to try to recover the investment made on a project that has been abandoned and will not be further developed within the company. Naturally, if the project has been dropped because it does not work (for example a pharmaceutical product that kills the mice on which it is tested), it cannot be licensed out, but often promising projects are dropped because other projects are even more attractive and resources for development are limited. Sometimes a strategic decision may be taken to get out of some particular therapeutic area completely, in which case a number of advanced development projects may be made available for licensing.

The patents covering the project that has been abandoned should be kept in force in order to be able to offer exclusive rights to a potential licensee. There should, however, be clear time limits set within which licensees should be found, otherwise patents will never be abandoned and costs will not be kept under control. For example, if no licensee is found within one year, the patents could be dropped in all except major countries, and if none is found within a further two years, the patents could be abandoned in all countries. If a licensee is found, then ideally the licensor should transfer all responsibility for the patents to the licensee. It is perfectly possible to assign patents in return for a running royalty on sales during the full lifetime of the patents and this is the easiest solution.

Small Companies

For many small start-up companies, particularly in the biotech field, their patent portfolio is practically their only asset, and their commercial success depends upon their finding one or more large companies to license and develop the technology

[3] For example, s. 68, PA 1977.

covered by their patents. Patents are thus even more important to them than they are to a large company. But cash is scarce and every dollar spent on patenting is one dollar less for research to make new inventions. Cost-effectiveness is therefore essential. Small companies should therefore:

- *Pick the right time to file* In highly competitive fields, undue delay can be fatal; if filing is done too early, the specification may not be sufficient to base useful claims. Courts in the UK and USA are growing increasingly intolerant of broad claims based on speculative disclosure.
- *File follow-up applications before the first case is published* A disclosure that is insufficient to base useful claims may nevertheless make such claims obvious if they have a later priority date.
- *Stop scientists from publishing* Commercial organizations are in business to make money, not to do academic research or to bolster the egos of their research staff.
- *Use Patent Cooperation Treaty (PCT) procedure* The PCT procedure gives the maximum flexibility. If it is necessary to enter the national phase before a licensee is found, the country list should be restricted at that stage.
- *Keep texts short* When filing in a number of countries in which translations are needed, every page of the specification adds massively to the costs. Most patents would be better, as well as cheaper, if they were half the length. (This is also good advice for large companies.)

Universities and Academic Institutions

The importance of licensing for these organizations is discussed in Chapter 22.

The Individual Inventor

This brings us to the role of the private individual inventor as licensor. In the chemical field, the chances of a 'backyard' inventor being able to make a significant invention are somewhat remote, but this is not necessarily the case in the mechanical field, in which the resources required for experimentation are much smaller and many ingenious devices, some of which are actually useful and potentially commercial, are invented by individuals in their own time. Even in the chemical field, individual inventors may own the rights to their own inventions, whether they be totally independent, academic scientists whose university does not lay claim to their inventions, or employees whose inventions belong to them under the provisions of the relevant national law. Any of these individuals will usually lack the means to develop the invention themselves and will be interested in finding a suitable licensee.

The first step in the process, *before* publishing the invention, is for the inventor to consult a patent attorney and get a patent application on file, thereby obtaining a priority date for the invention and protecting himself against the consequences of non-confidential disclosure. This is also a protection for any company that may look into the possibility of taking a licence. Suppose that the company itself had a similar project under development and that it was in the process of filing a patent application. The information that it received from the outside inventor might be of no use to the company, but if it were to decline the offer and then shortly afterwards file a patent application for something closely similar, then the outside inventor would certainly feel that the company had stolen the invention. To avoid the danger of being unjustly accused of theft or breach of confidence, many companies (particularly those making mechanical devices, which frequently receive unsolicited offers from inventors) refuse to look at any invention submitted to them unless it is already the subject of a patent application. In the USA, some large companies try to protect their interests by accepting information only on a non-confidential basis. This is unfair to the inventor, because not only does it leave him without redress if the company should steal his ideas, but also it will prevent him from obtaining valid patent protection outside the USA unless he has already filed a US patent application.

The problem is that, from the date of filing the priority application, only nine months or so remains before a decision on foreign filing must be taken and substantial sums of money must be spent on patenting. This is scarcely enough time for the inventor to contact companies that may be interested, or for the companies to evaluate the idea. In case it should be necessary to abandon and refile the application, all communication of the invention should preferably be done under conditions of confidentiality; this of course would preclude the inventor from demonstrating the invention at exhibitions, or advertising it generally to the public in any way. If no licensee has been found by the end of the priority year and there has been any non-confidential publication, the inventor will either have to abandon hopes of foreign patent protection or else find the money for foreign filing from his own resources. Again, the use of the PCT filing procedure will be useful in this situation.

Option Agreements

If, as is usual, the potential licensee will require some time to evaluate the project to the point at which it can decide if it really wants a licence, it is common practice to enter into an option agreement, whereby the company has a limited period of time (perhaps one to two years) in which to make up its mind and by the end of which it must either enter into a licence agreement or leave the inventor free to offer his invention elsewhere. In return for the option rights, the company may pay the inventor a lump sum or may, for example, offer to pay the costs of foreign filing the patent application, or, if the company has its own patents department, offer to carry out the foreign filing itself, in the inventor's name. Academic inventors may be interested in obtaining funds for further research in exchange for option rights and essentially any agreement that meets the needs of both parties may be entered into.

Research Collaboration Agreements

In research collaboration agreements, technology is not simply transferred, but is generated during the agreement by the joint efforts of both parties. Typically the partners will be a small start-up company A and a large company B, which agree to cooperate in a defined field of technology for a specific time. During this time, A will provide existing expertise in the technology, B will provide research funding, but both will carry out research work, often under the supervision of a joint steering committee or similar body on which both parties are represented. Afterwards, B will develop and market products arising out of the research, generally in return for milestone payments and royalties to A.

Often, A will already have relevant patent rights in the field and for these the same considerations apply as discussed above in relation to in-licensing. What has to be negotiated is what happens to inventions made during the joint research—that is, who owns them, who is responsible for patenting, and what rights each party has, both if the agreement continues and it if is terminated. Here, there are two basic options: cross-licensing and co-ownership.

Cross-Licensing

In the cross-licensing scenario, ownership of inventions and of the corresponding patent rights is predetermined, usually on the basis of inventorship. If only employees of A are inventors, the invention belongs to A; if only of B, then to B; and if employees of both A and B are co-inventors, then the invention belongs jointly to A and B. In the event of commercialization of a product, B will be exclusively licensed under patents owned by A and under A's share of jointly owned patents; under certain circumstances, A may be licensed under rights of B.

Each company will typically be responsible for its own patents, but it will be necessary to agree on how to handle jointly owned patents. A typical solution might be that the priority filing would be made by the party more closely connected with the invention, that both would consult on foreign filing, and that patenting would be done by A in its home country and by B elsewhere. The same considerations about possible conflict of interest also apply in this situation.

A case in the USA[4] has highlighted a possible pitfall in agreements of this type. This is that, in the USA, an earlier patent arising out of a cooperation between A and B, and owned by A, will be prior art against a later patent owned jointly by A and B. Additionally, problems can arise even when there is no earlier patent. Suppose that, in the context of the research collaboration agreement, Smith and Jones, employees of A, develop certain subject matter. Under the agreement, rights to this belong to A. Then, Smith discusses the matter with Robinson, an employee of B, and Smith and Robinson jointly make an invention. Rights to this are owned jointly by A and B. The subject matter developed by Smith and Jones is prior art against this invention, and may be combined with other prior art to render the invention of Smith and Robinson prima facie obvious. At the time of writing, the only way to avoid this possibility is to ensure that there is common ownership of both inventions at the time that the later invention is made. A Bill is, however, currently before the US

[4] *OddzOn Products Inc. v. Just Toys Inc.* 43 USPQ 2d 1641 (Fed Cir 1997).

Congress—that is, the Cooperative Research and Technology Enhancement Bill (CREATE)[5]—which would legislatively overrule *OddzOn*, and this welcome proposal seems likely to be implemented.

Co-Ownership

An alternative approach is to agree that all inventions arising out of the joint research shall be jointly owned, irrespective of inventorship. This 'one basket' approach encourages the parties to be completely open with one another and to collaborate fully, because there are no advantages to be gained by keeping a good idea to oneself in the hope that it will give rise to a fully owned, instead of a jointly owned, invention. It presupposes that the research efforts of the two parties will be approximately equivalent and that there will be a good working relationship between the scientists on both sides.

It should be pointed out, however, that joint ownership also has disadvantages. Not only is it administratively much more complicated (for example signatures of both parties are required on many documents and it may be more difficult for the case to be dealt with by in-house attorneys), but also very careful contract drafting is necessary to specify the rights of the parties both during the term of the agreement and after termination. In the absence of specific provisions to the contrary, each co-owner has full rights to commercialize the invention without accounting to the other and, in some countries, including the USA and China, may grant licences without the consent of the other party (see Chapter 21).

In this situation, patenting for all inventions must be regulated between the parties in the same way as for joint inventions under the previous scenario. It should be noted, however, that co-ownership of patent rights does not necessarily mean that the patents must be filed and granted in the joint names; exactly the same legal consequences can be achieved by contractual provisions as by co-ownership on the face of the patent. Nevertheless, most companies are reluctant to allow their names to be omitted, because public co-ownership of patents is more likely to impress shareholders and potential investors.

Single-Party Ownership

From the patent point of view, the cleanest situation is to have a single party owning all of the patents and licensing the other as appropriate, but when each party would like to be the sole owner, this may be difficult or impossible to negotiate. Particularly if one party is a small company, it may be difficult or impossible for it to agree that patent rights for its inventions should not be in its own name. Ownership by the other party may, however, be acceptable for a large company in certain situations: for example if the other party already owns relevant patent rights, or if it is likely

[5] There seems to be quite an industry in finding uplifting names for legislation in the USA: e.g., the 'American Inventors Protection Act', the 'PATRIOT Act', etc.

that the other party will make most of the inventions, or if the large company is not expected to pay for patenting the other party's inventions, or in the case of collaborations with companies developing galenical forms or drug delivery systems for a specific product of the large pharmaceutical company, in which it is agreed that the other company may subsequently offer the system to non-competing products of other pharmaceutical companies.

Contract Research Agreements

A simpler situation may exist when only one party will be carrying out research and the other will be paying for it. Some companies make it their business to carry out research for payment and are willing to assign the results totally to the company that pays. Such contract research is usually not highly innovative, but more usually routine screening, animal or clinical testing, or formulation work. Contracts for this type of situation are usually very straightforward, the main issue being how to secure confidentiality of the results.

For more innovative research, it may be difficult to find a partner willing to carry out research on a contract basis. Universities may consider that this would be detrimental to their academic freedom and it may not be possible in the USA if any of the facilities or equipment used have been paid for with public funds. Small companies may be unattractive to investors unless they can ensure participation in the success of a research project by means of royalties on sales of the final product. It is therefore not unusual for 'research collaboration agreements' to be entered into even when there is no real collaboration and all of the research is done by one of the parties.

Funding Agreements

Large companies sometimes enter into long-term agreements with a university or an academic research institute by which, in exchange for funding general research work rather than specific projects, the company has an option to a licence to all or some of the inventions made at the institute, upon pre-agreed terms. The extent of the option rights, as well as the level of royalties on the licence, will depend upon the level of funding.

Such agreements can be very complex, and must go in detail into the question of how inventions are disclosed, how and when the option must be exercised, who controls the patenting, and what happens if an option is exercised, but the invention is not developed further.

Compound Purchase Agreements

One might imagine that making a contract to purchase one or more compounds for research purposes would be as simple as placing a purchase order for a laboratory reagent, but it can be more complex than this. If the compounds are patented, no licence is required, because purchase gives the purchaser the right to use the materials, but the purchaser must clearly understand that he obtains rights only to the actual samples of the compounds that are physically transferred, and not to the compounds in general. Also, the purchaser does not automatically have the right to make, use, and sell derivatives if these should fall under the seller's patent rights, and the purchaser may wish to ensure that he receives a non-exclusive licence or freedom from suit under any such patents. The seller, however, will want a clause denying any warranty that the compounds are useful for any particular purpose, or that they do not infringe third-party patent rights, and requiring indemnification if third parties suffer damage as the result of the purchaser's use of the materials. He may in addition wish to impose conditions on the sale: for example that the compounds should not be resold to third parties.

The situation is more complex if what is being purchased is a compound library, which can be used in the purchaser's screening assays. Unless agreed otherwise, the seller is free to sell the same library to other companies, and if lead compounds are found, there may well be conflicting rights.

Licence Contracts

When it comes to entering into a licence agreement, whether between two multinational corporations, or between a small company and an individual, a suitable contract must be drafted that accurately expresses what has been agreed in the negotiations between the parties. The production of a first draft may be no more than a stage in the negotiations and gives a certain tactical advantage to the party that writes the draft. The other party may require numerous amendments, or produce a draft of its own; at any rate, when all disputes have been resolved, the final version must be clear and unambiguous.

Drafting a Licence Agreement

There are different styles of contract drafting in different countries, and a simple contract covering exactly the same terms might be five pages long in Germany, 15 in the UK, and 50 in the USA. A British view is that continental practice expresses the intent of the agreement in terms that are often somewhat general and imprecise, while US practice runs to excessive legal jargon and unnecessary verbiage. A contract should be written in language that is clear, grammatical, and as simple as possible, without losing the main object of precision. Because the facts of each licensing situation are different, 'boilerplate', or standard clauses, should be avoided as much as possible.

The worst fault that a contract can have is ambiguity. An agreement is clearly on a very shaky footing if it could be reached only because a certain clause in the contract was vague enough to be interpreted one way by one party and differently by the other. This is standard practice in diplomatic joint communiqués, but is a recipe for disaster in a business relationship, which will fail unless there is trust on both sides. It may also happen that a clause is recognized to be ambiguously worded, but it is decided to leave it as it is because the negotiators of both sides agree fully what is meant by it. That is all very well at the time, but ten years later, the negotiators may have changed jobs, retired, or died, and the problem will rear its head when no one remembers what the original intention was.

Although standard boilerplate clauses are generally to be avoided, they can be useful for situations that are not specific to the facts of the agreement, but which can be applied to any contract. For example, definitions of terms such as 'affiliate', 'net sales', and the like can be standardized, as can confidentiality clauses, severability clauses, etc. These are normally not controversial, but it cannot be excluded that there may be disagreement, for example, about the duration of a confidentiality obligation and, of course, the other party may have its own standard versions that it wants to use.

A checklist of what sorts of clause to include in a licence agreement may be very useful. The contract should provide not only for what is intended to happen, but also for what may happen if things go wrong—that is, for what happens if the parties change ownership, the patents are infringed, the licensee makes no sales, and any other hazards that can be foreseen. It is good practice to provide for clear exit points enabling either party to terminate if clear criteria are not met. A typical straightforward patent licence agreement, without trademarks being involved and with no provision for equity investment, supply of materials or technical cooperation between the parties, might have a structure something like the following.

A. Preamble

1. *Parties* Names; addresses; legal status.
2. *Reasons* Brief explanation, often in the form of one or more 'whereas clauses', reciting that A wishes a licence from B and B is willing to grant it.
3. *Definitions* For example of terms such as 'affiliate', 'associate', 'territory' (whole world, UK only, etc.), as well as the technical field involved. If the licence is for one particular compound, 'compound' can be defined by its chemical name and formula.
4. *Schedule of patents* Normally referred to here and added as an annex to the agreement. Should contain pending applications as well as granted patents.

B. Grant

1. *Extent* Is the licence to make, to use, or to sell, or more than one of these? Is the whole scope of the patent licensed or just a part of it (perhaps only one compound)?
2. *Exclusivity* Is the licence exclusive, sole, or non-exclusive?
3. *Sublicensing rights* Can the licensee grant sublicences? If so, to anyone, or just to his affiliates or customers?

4. *Know-how* May take the form of operating manuals, test results, registration documents, loan of technical staff, etc. What provisions are there for documenting this know-how and keeping it confidential?

C. Consideration

1. *Down-payment* A licensor may want a down-payment sufficient to cover his patenting costs and perhaps a percentage of the development costs.
2. *Success fees (milestone payments)* Lump sums may be payable when certain goals are reached. How much? What event triggers payment? Can a given milestone be paid more than once?
3. *Royalty*
   (a) A lump sum or running royalty?
   (b) How calculated: price per unit of production, percentage of profit, percentage of net selling price? Should percentage royalty decrease or increase with increasing sales?
   (c) Should there be a minimum annual royalty requirement? If so, how much and for how long?
   (d) Are royalty payments terminated or reduced when licensed patents expire, or when an imitator appears on the market and cannot be stopped?
   (e) Are reach-through royalties payable on products not directly covered by the licensed patents? (Not a good idea for the licensee.)
4. *Mechanism of payment* When payable (quarterly, half-yearly), where, and in what currency? Provisions for exchange rate fluctuations. Licensee should keep accounts, open to inspection by accountant appointed by licensor.

D. Patent provisions

1. *Maintenance in force* Does the licensor undertake to maintain the patents in force, or does he give the licensee the option of taking over any patents that he may wish to drop?
2. *Policing of infringement* See comments above.
3. *Third-party rights* A licensor will not normally be willing to indemnify his licensee against possible infringement of third-party patents. A possible compromise is for the royalty to reduce by the amount that may have to be paid to third parties, but only up to an agreed limit.

E. General provisions

1. *Exploitation* The licensee should use his best endeavours to exploit. If he fails, there may be provision for the licence to terminate or to go from exclusive to non-exclusive. Minimum royalty requirements may have the same effect, but the licensor should not be satisfied with a minimum royalty indefinitely and may provide for termination or non-exclusivity if only minimum royalties are paid for two or three years running.
2. *Quality control* Will the licensor have rights of inspection to control quality of goods manufactured under the licence? Should the licensee mark the goods with the licensor's name, patent number, or trademark?

3. *Developments* Are there cross-licensing rights to developments made by either party during the agreement? Must the parties communicate all developments to the other, or give clearance for any publications?
4. *Most favoured licensee* If the licence is non-exclusive, such a clause promises that, if any subsequent licences are offered on more favourable terms, the first licensee will be offered the same terms.
5. *Duration and termination* Is the duration for the life of a single patent, the last to expire of a series, or for some defined longer or shorter term? Under what conditions (for example default, change of ownership of a party) can the licence be terminated earlier? What obligations (for example confidentiality of know-how) continue after termination?

F. Legal framework

1. *Language* If there are texts in more than one language, which one is authentic?
2. *Applicable law* Under the legal system of which country (state, canton) will the contract be interpreted? What courts have jurisdiction to hear disputes?
3. *Arbitration* Should there be provisions for arbitration, for example by the International Chamber of Commerce?
4. *Severability* A clause to spell out the intention that, if any clauses be found inoperative or illegal, the remainder of the agreement will not be affected.
5. *Approval of authorities* If approval from or registration with a national authority is required, which party should apply for it?

Such a list should be no more than a guideline and it should always be remembered that, in many countries and within the EU, parties are not wholly free to reach whatever agreement seems best to them, but are subject to anti-trust laws, European Commission Regulations, and rules of national authorities. These problems are dealt with in more detail in Chapter 25.

25

PATENTS AND COMPETITION LAW

Even such a tax was less intolerable than the privileges of monopolies, which checked the fair competition of industry, and, for the sake of a small and dishonest gain, imposed an arbitrary burden on the wants and luxury of the subject.

Edward Gibbon, *Decline and Fall of the Roman Empire* (1776–88), ch. XL

Charges Against the Patent System

From the days of the Emperor Justinian onwards, monopolies have been seen as an economic evil; from the time of James I, patent monopolies have been recognized, more or less grudgingly, as a limited exception to this rule. The distinction that because a patent gives a monopoly for a new invention, it cannot take anything away from the public that it already has is not always appreciated. Accordingly, patents have been blamed for a whole series of economic ills, real or imaginary, including high prices, the erection of barriers to free trade, the foreign domination of national economies, the exploitation of developing countries, and the suppression of worthwhile inventions.

Some of these charges are obviously false: for example the patent system, because it involves publication of the invention, clearly cannot be used to suppress inventions. A patent monopoly, even if it is not in itself objectionable (as a monopoly on existing products would be), is nevertheless capable of being used to restrict competition in ways that may be regarded as undesirable or even illegal by various national governments and supranational organizations, such as the European Union (EU). Most countries, as well as the EU, have developed a body of competition law to provide countermeasures against what are regarded as abuses of monopoly by patentees: these are sometimes referred to collectively as 'anti-trust laws', although this term properly applies only to the US Sherman and Clayton Acts (see below).

Competition Law: The UK

It has already been mentioned that, in the majority of countries, it is regarded as abuse of monopoly to use a patent simply to exclude others, while refusing to work the invention oneself. This is dealt with by provisions, codified in the Paris Convention, enabling compulsory licences to be granted if the patent is not worked within a certain time, normally three years after grant, or even by the lapse of a patent if it is not worked. Under the TRIPs Agreement, of course, the market may be supplied by importation rather than local manufacture, but refusal to supply the market at all would still be considered an abuse of monopoly rights. In the UK, compulsory licences may also be granted if the patentee is preventing the working of a dependent patent by refusing to grant a licence. In this case, the owner of the dependent patent has to be prepared to grant a cross-licence. Compulsory licensing in the UK may also result from an investigation of the UK Competition Authority into a monopoly situation or merger that is considered to be against the public interest, and in which a market in patented products is involved.

The Patents Act 1977 prohibited tie-in clauses in contracts: for example a licence contract entitling the licensee to manufacture a patented compound could not compel him to purchase unpatented starting material from the licensor, or forbid him to obtain it from others.[1] It was also a defence in an infringement action to show that a

[1] Section 44(1), PA 1977.

contract containing such a clause had been made by the patentee, even if the defendant was not a party to the contract.[2] Clauses in a patent licence agreement requiring continued payment of royalties after patent expiry were stated to be unenforceable.[3] These provisions were abolished by section 70 of the Competition Act 1998, under which such restrictive clauses might render the contract itself void, but would have no effect upon the enforceability of the patent.

At one time, apart from these specific prohibitions, UK law left patentees and licensees free to work out their own agreements and put any clauses that they liked into them. Now, any licence agreements must conform to EU competition law, if trade between EU member states is affected, or the provisions of the UK Competition Act 1998, if the effects of the agreement are only within the UK.

Competition Law: The European Union

The Organization of the European Union

The European Community (EC), although a supranational organization, has many of the features of a national government, and the closest analogy to the EU is probably the government of the USA. The US government is based on a written Constitution and has separate legislative, executive, and judicial branches, each of which can act as a check on the others. In particular, the most important role of the judiciary is that of interpreter of the Constitution, and it may declare a law passed by the legislature or an act of the executive as unconstitutional and thus void.

In very much the same way, the EU has an executive branch, the Commission, and a judicial branch, the European Court of Justice (ECJ), assisted by the European Court of First Instance (CFI), which the Lisbon Treaty has now renamed as the General Court (GC).[4] The role of legislature is shared by two bodies: the Council of Ministers and the European Parliament. Originally, the European Parliament played a very minor role and all really important decisions have been taken by the Council, composed of ministers of member state governments, but the importance of the Parliament is increasing. The Commission is presided over by Commissioners appointed by the member states and staffed by EU civil servants; it is divided up into various Directorates-General, including one dealing with competition within the EU. The Commission implements the policies of the Council, but also has considerable powers to initiate policy on its own and has the sole power to initiate Community legislation.

The combined 'constitution' of the EU is the Treaty of Rome, signed in 1962 by the six original member states, together with the Treaty of Maastricht of 1993 and the Libson Treaty of 2007, ratified by all member states in 2009. In 1973, the UK,

2 *Chiron v. Organon* [1996] RPC 535 (Pat Ct, CA).

3 Section 45(1), PA 1977.

4 Established by Council Directive 88/591.

Ireland, and Denmark acceded to the Treaty, followed by Greece in 1981, Spain and Portugal in 1986, Sweden, Finland, and Austria in 1996, Cyprus, the Czech Republic, Estonia, Hungary, Latvia, Lithuania, Malta, Poland, Slovakia, and Slovenia in 2004, and Romania and Bulgaria in 2007, with others possibly to follow. The provisions of the Treaty of Rome, as subsequently amended as well as Regulations made under it by the Council and the Commission, are directly binding upon member states and, in cases of conflict, take precedence over national law. Directives, on the other hand, must be implemented by national legislation before they become effective. It is the main task of the ECJ to interpret the Treaty of Rome and other Community legislation.

Relevant Provisions of the Treaty of Rome

The most important provisions of the Treaty, from the point of view of competition law, are Articles 30–34, 36, 81, and 82. Articles 30–34 prohibit quantitative restrictions on imports and exports within the EU, and although Article 36 specifically exempts from such prohibition restrictions that are justified on certain grounds including the protection of industrial property, the last sentence of Article 36 states that such restrictions, which are in principle allowed, 'shall not constitute a means of arbitrary discrimination or a disguised restriction on trade between Member States'. Article 81 (until 2001, numbered as Article 85) prohibits agreements and concerted practices between undertakings that have the effect of limiting competition, and Article 82 (previously Article 86) prohibits the abuse of a dominant position to limit competition. Thus Article 81 applies only when two or more independent companies are involved, while Article 82 can apply to a single company, but only if it has a dominant position in the relevant market.

The Role of the General Court and the European Court of Justice

The ECJ has jurisdiction only in certain types of proceeding. The two that primarily concern intellectual property are: complaints that an act of the Council or Commission is illegal, or that the Council or Commission has failed to act where it should; and requests referred by a national court or tribunal for preliminary rulings concerning points of interpretation of the Treaty of Rome, or the interpretation and validity of other Community legislation, such as Regulations and Directives.

The first type of proceeding may be brought by a member state, the Commission, or the Council, but also by any natural or legal person affected. One major case relating to intellectual property and involving a member state was the challenge to the Biotechnology Patenting Directive lodged by The Netherlands (see p. 310). More common are complaints that decisions of the Commission, for example imposing fines on undertakings for violations of Articles 81 or 82, are illegal. Such cases go initially to the GC, from which a further appeal lies, on points of law only, to the ECJ. Also, any legal action against a Community institution or an appeal against decisions of Community authorities, for example a rejection of a Community design or trademark by the Office of Harmonization for the Internal Market (OHIM) in Alicante, or a violation of the regulatory data exclusivity provisions by the European Medicines Evaluation Agency (EMEA), are initially heard by the GC.[5] In the second type, a national court might ask for a ruling, for example, on whether a national patent could be used to prevent importation of goods under certain circumstances. In this situation, the GC is not involved.

At present, these are essentially the only two types of case involving industrial property—including copyright, design protection, trademarks, and supplementary protection certificates (SPCs), as well as patents—that can come before the Court. For the Treaty of Rome to be relevant, there must be an effect on trade between member states, so that questions of validity of a national patent, or its infringement where no importation is involved, can normally not be heard by the Court. The Court is not a supranational court of appeal on general patent matters, although this could change if the Community patent were ever to come into effect. Because the European Patent Office (EPO) is not an EU organization, its decisions cannot be challenged in the ECJ.[6] But SPCs are granted in EU states on the basis of an EU Regulation and there have been many cases in which points of interpretation of this Regulation have been decided by the Court. Also, the ECJ may have a stronger influence on patent law in the future via questions on the interpretation of the Enforcement Directive[7] or the Biotech Patenting Directive (BPD).[8]

Parallel Importation

Many of the cases that have come before the ECJ have been concerned with the question of whether the holder of an industrial property right in one member state can use that right to prevent importation from a second member state of goods covered by that industrial property right that were legally on the market in the second member state, particularly when the goods were put on the market there by the holder of the intellectual property right or his licensee. Such 'parallel importation' is particularly a problem in the pharmaceutical industry: firstly, because the high value of pharmaceuticals per unit weight makes it worthwhile to transport this type of goods from one country to another to take advantage of even relatively small price differentials; and secondly, because government control of pharmaceutical pricing in EU countries results in very large price differentials between certain countries.

[5] But the GC has no jurisdiction on alleged patent infringement by Community Institutions: see T 295/05 *Document Security Systems, Inc. v. European Central Bank (ECB)* [2007] ECR II-2835, in which the CFI declined jurisdiction on a claim of patent infringement against the ECB.

[6] See Enlarged Board of Appeal (EBA) Decision 2/06, in which the applicant's request to refer questions to the ECJ to obtain an interpretation of r. 28, IR, which corresponds to provisions of the EU BPD, was refused.

[7] Directive 2004/48/EC of 29 April 2004 on the enforcement of intellectual property rights.

[8] Directive 98/44/EC of 6 July 1998 on the legal protection of biotechnological inventions. The first such referral from the Dutch Courts on the interpretation of the BPD was in C-428/08 *Monsanto Technologies v. Cefetra*—see Chapter 15.

Unless the Community patent ever becomes reality, patent rights are essentially national in character and stop at national boundaries. The doctrine of exhaustion of patent rights, which has long been recognized in Germany and other EU countries (although not in the UK, see Chapter 10), should not extend to the proposition that the sale of goods under a patent in one country exhausts the patentee's rights under an equivalent patent in another country. Still less can the sale of the goods in a country in which there is no patent be said to exhaust rights in a patent that does exist elsewhere.

This principle of territoriality of industrial property rights conflicts with the objective of the Treaty of Rome to promote the free movement of goods within the EU. The ECJ has concluded that although the *existence* of industrial property rights, such as patents, is not affected by the Treaty of Rome, the *exercise* of these rights may in certain circumstances be incompatible with the Treaty and therefore prohibited.[9]

In 1974, Sterling Drug sued Centrafarm for infringement of a Dutch patent by importing into the Netherlands a patented pharmaceutical produced and marketed in the UK by the UK subsidiary of Sterling under an equivalent British patent. The Dutch court found infringement, but referred to the ECJ the question of whether national patent laws could be used to stop parallel importation in these circumstances. The Court, applying Articles 30–34 of the Treaty, held that they could not: once patented goods had been put on the market in any EU country by the patentee or with his consent, then those same goods could be resold freely anywhere in the EU and national patent rights could not prevent this.[10] In other words, the effect is the same as regional exhaustion of patent rights within the EU.

In the case of *Merck v. Stephar*,[11] the source of the parallel imports was Italy, where there was no patent because Italy did not grant pharmaceutical patents at the relevant time. It was argued that there could be no exhaustion of patent rights where there was no patent to be exhausted, but the ECJ held that if the patentee chooses to market his product, directly or through a licensee, in an EU country where he has no patent, he must accept that the principle of free movement of goods enables the product to be parallel imported freely into any other EU country.

It is at least clear that where a product has been put on the market by the holder of a compulsory licence, this is not to be regarded as with the consent of the patent holder, and that parallel importation into another EU country can be prevented.[12] It has been argued at various times that a patent holder who chooses not to apply for a patent in a particular EU country is giving implied consent to third parties to market there and therefore that importation (of infringing copies, not genuine goods) into

[9] First applied in *Costen & Grundig v. Commission* [1966] CMLR 418 (ECJ) (a copyright case).

[10] *Centrafarm v. Sterling Drug* [1974] CMLR 1 (ECJ).

[11] *Merck v. Stephar* [1981] CMLR 463 (ECJ).

[12] *Pharmon v. Hoechst* [1985] 3 CMLR 775 (ECJ).

other EU countries could not be prevented.[13] This would mean that patent protection within the EU would be ineffective unless patents were obtained in every EU country and no 'hole' were left. The question came up before the German courts and a reference to the ECJ was refused, because it was considered *acte claire* that importation could be stopped in these circumstances.[14] It might have been better to have let the ECJ itself put an end to this 'hole in the EU' theory. Now, with the expansion of the EU, this discredited idea is raising its head again, and patentees are being led to believe that they may have a serious loss of rights throughout the EU unless they file in Malta and Estonia, which is not the case.

If free market conditions apply within a common market area, then the principle of free movement of goods is pro-competitive and brings benefit to the consumer by levelling out price differentials. The pharmaceutical market in the EU is not a free market, but is totally distorted by price control of varying degrees in different countries. In these circumstances, parallel imported goods are sold in the country of import at the controlled price and the consumer (the patient or the health insurance system) obtains no benefit, as confirmed by a 2004 study by the London School of Economics (funded by a pharmaceutical company, but plausible all the same).[15] The sole result is the transfer of money from the innovator to the parallel importer, which means that less money is available for research into new pharmaceuticals.

In this situation, it is hard to see why both the ECJ and the Commission are so keen to encourage parallel importation in the pharmaceutical area. That this is the case, however, is shown by the refusal of the ECJ to overrule *Merck v. Stephar*[16] and by the zeal with which the Commission pursues any pharmaceutical company that attempts to make parallel importation of its products more difficult.

The transitional provisions on the accession of new member states in 2004 and 2007 provide that the parallel importation of a product from a new member state into an old one may be prevented by a patent or SPC in the old state if, at the time that patent was filed, it was not possible to obtain patent protection for that product in the new state.[17] The wording is less than clear, but it seems generally accepted that the relevant date is the introduction of product per se protection for pharmaceuticals in the new accession state. It is unclear what is the situation for a product protected only by a formulation patent. Another anomaly is that these transition provisions do not apply to Cyprus and Malta, possibly because it has been possible, in theory at least, to obtain product patents in these countries at any time in the last 20 years. In Cyprus it was possible to register British patents cheaply and most pharmaceutical companies did so; in Malta no one ever filed patents because of the

[13] For example, B. Redies, 'Liberties and risks in the present system of patent protection in the EC' [1989] 6 EIPR 192.

[14] *Patented Feedingstuffs* [1989] CMLR 902 (BVerfG, Germany).

[15] P. Kanavos, J. Costa-I-Font, S. Merkur, and M. Gemmill, *The Economic Impact of Pharmaceutical Parallel Trade in the EU Member States*, London School of Economics & Political Science, 2004.

[16] In *Merck v. Primecrown* [1997] 1 CMLR 83 (ECJ).

[17] Annex IV(2) of the Act of Accession signed on 16 April 2003.

tiny size of the market. It is to be hoped that this omission will not give rise to unforeseen disadvantages now that Malta has become a member state of the EU.

Meanwhile, the Commission had been busy imposing fines on companies that tried to prevent parallel importation of their products. Although unilateral action by a company should not be touched by Article 81 and should fall under Article 82 only if the company has a dominant position, the Commission has in a number of cases held that even unilateral imposition of anti-competitive terms of sale violate Article 81 if the customer nevertheless continues to buy the goods, the buyer–seller relationship then amounting to a concerted practice. Thus an Italian subsidiary of a Swiss firm was fined simply for using invoices bearing the phrase 'not for export', even though there was no evidence that any actual restriction on exports had occurred. On appeal, the ECJ reduced the fine somewhat, but upheld the principle.[18]

The Commission pushed this principle to the extreme in its action against Bayer, relating to the parallel importation of its antihypertensive product Adalat® from France and Spain. Bayer adopted a policy of refusing to supply wholesalers in these countries with more product than they needed to supply their national markets. Far from agreeing to a 'continuing business relationship' on these terms, these customers complained bitterly to the Commission. Although Adalat was a major product, Bayer did not have a dominant position in the antihypertensive market, so Article 82 could not be applied. The Commission therefore fined Bayer €3 million for violation of Article 81, for what could by no stretch of the imagination be described as a concerted practice. Bayer appealed to the CFI, which held in favour of Bayer, essentially saying that a company is not obliged to make life easy for the parallel importer or to supply the whole EU from the cheapest country. The case was appealed to the ECJ, which upheld the decision of the CFI.[19] The Commission has retired to lick its wounds and to think up new strategies—perhaps by defining markets so narrowly that every seller will be in a dominant position. Currently, parallel trade constitutes about 5 per cent of the pharmaceutical market in the EU and its initial rapid rate of increase has slowed down considerably.

A further ECJ case relating to parallel imports originating in Greece presented more of a problem for the patent holder (Glaxo SmithKline, or GSK), because, in this situation, GSK did have a dominant position in the relevant market, so that Article 81 was applicable. The fact situation was similar to that in the *Bayer* case: GSK refused to supply wholesalers with large quantities of its products, and instead set up a company to supply hospitals and pharmacies directly. After an initial reference to the Court was rejected because the referring body was administrative and

[18] *Sandoz Prodotti Farmaceuti v. Commission* [1989] 4 CMLR 628 (ECJ).

[19] *Bayer v. Commission* (ECJ 2004).

not a court or tribunal,[20] an admissible reference was finally made by the Athens Appeal Court.[21]

The Court ruled that, although it is an abuse of a dominant position for an undertaking to refuse to supply wholesalers in order to put a stop to parallel importation, an undertaking can stop supplying, in order to protect its own economic interests, if orders are placed that are out of all proportion to those previously made by the same wholesalers to meet the needs of the market in that member state. It is for the courts of the member states to determine whether orders are out of proportion to the needs of that member state's market. While not going as far as the decision in *Bayer*, this judgment represented a further setback for the Commission.

Meanwhile, in three consolidated trademark cases,[22] the ECJ had held that the importation into the European Economic Area (EEA) of genuine goods first placed on sale in a non-EEA country could be prevented if the trademark owner had not consented to the importation. The importers argued that placing the goods on sale outside the EEA without contractual restriction amounted to implied consent, but the Court held that consent could not be implied from mere silence, lack of marking of the goods, or sale outside the EEA without imposing contractual restrictions prohibiting importation into the EEA. Most importantly, the burden of proof is on the importer to show consent, not on the trademark holder to prove its absence. Although the decision was based on a Community Trademark Directive and there is as yet no comparable legal instrument for patents, the draft Community Patent Regulation envisages Community-wide exhaustion, not international exhaustion. There seems no reason why this case should not equally apply to parallel importation of patented goods.

Patent Licence Agreements in the European Union

The other main impact of EU law upon matters relating to patents concerns restrictions upon the terms of patent licence agreements. A licence agreement is an agreement between undertakings that falls within the general scope of Article 81 of the Treaty of Rome. Paragraph 1 of this Article declares that agreements that prevent, restrict, or distort competition within the common market are prohibited. Paragraph 2 states that agreements so prohibited are automatically void, and paragraph 3 provides for exceptions in certain cases.

Although Article 81(1) gives some examples of objectionable agreements, for example those that fix prices, control or share markets, or have tie-in provisions, no detailed guidance is given as to which clauses in a contract for the licensing of

20 C-53/03 *Syfait and Ors* [2005] ECR I-4609.

21 Joined cases C-468/06–C-478/06 *Sot. Lélos Kai Sia EE (and Ors) v. GlaxoSmithKline AEVE Farmakeftikon Proionton.*

22 C-414/99 & C-416/99 *Zino Davidoff v. A&G Imports; Levi Strauss v. Tesco; Levi Strauss v. Costco* (ECJ 2001).

industrial property might contravene the Treaty. Similarly, Article 81(3) is framed in very general language and says, in effect, that agreements that *a priori* violate Article 81(1) may be exempted if they contribute to the production or distribution of goods or to technical progress, give the consumer a fair share of the benefits, and do not include unnecessary restrictions or eliminate competition.

Notification and Negative Clearance

Initially, there were no detailed rules as to what conditions were or were not considered permissible in patent licence agreements, and companies had to notify licence agreements individually to the Commission.[23] At the same time as notification was made, negative clearance could also be requested. Negative clearance was a declaration by the Commission that the agreement did not conflict with Article 81(1), whereas notification was a request for exemption under Article 81(3) of an agreement that might fall under 81(1).

Block Exemption for Patent Licence Agreements

A Regulation granting a block exemption for a certain category of agreements, such as patent licence agreements, has the effect that agreements in that category that meet certain clear conditions and do not contain specified objectionable clauses are automatically exempted under Article 81(3), without the need for individual notification. In 1965, the Council empowered the Commission to grant by Regulation a block exemption to patent licence agreements between two parties. But it was not until 1 January 1985 that the first such Regulation[24] came into force, for a period of ten years.

The most important parts of the Regulation were the first three articles, which listed what are usually referred to as 'grey', 'white', and 'black' clauses, respectively. The grey clauses of Article 1 restricted competition and therefore fell under Article 85(1) of the Treaty of Rome, but were, by the Regulation, exempted under 85(3). The white clauses of the Regulation Article 2 were considered not to restrict competition and not to violate Article 85(1) of the Treaty of Rome, while the black clauses listed in Article 3 of the Regulation, for example price fixing, no-challenge clauses, and exclusive grant-back of improvements, violated Article 85(1) of the Treaty and could not be exempted. Following the reasoning of the ECJ in the *Maize Seeds* case,[25] Article 1 of the Regulation allowed complete exclusivity for the lifetime of the patents as between licensor and licensee, and allowed the licensor to keep part of the EU as his own exclusive territory.

The Regulation gave retrospective exemption to existing agreements meeting its conditions and applied also to assignments of patents for a royalty rather than a lump sum. It did not apply to agreements between more than two parties, cross-licensing

[23] Under the provisions of Council Regulation 17/62.

[24] Regulation 2349/84.

[25] *Nungesser & Eisele v. Commission*, sub nom *Maize Seeds Case* [1983] 1 CMLR 278 (ECJ).

agreements, patent pools, or joint ventures. If an agreement of the kind to which the Regulation applied contained restrictions on competition beyond those allowed in Articles 1 and 2, but did not contain any of the black clauses of Article 3, it was possible to obtain individual exemption by an accelerated 'opposition' procedure. An exemption given under the Regulation could be withdrawn if the agreement turned out, in practice, to have anti-competitive effects.

Technology Transfer Block Exemption

A similar block exemption for know-how agreements came into effect in 1989,[26] but there were often difficulties in knowing which of the two Regulations applied in the common situation in which both patents and know-how were involved; the choice was important because the allowable periods of exclusivity differed. It was therefore desirable to combine the two into a single block exemption for technology transfer agreements generally, and the Commission attempted to do this before Regulation 2349/84 expired on 1 January 1995.

In drafting the new Regulation, the Commission took note of a number of decisions in which the ECJ had held that certain restrictive conditions found in the black list of Regulation 2349/84 were not necessarily contrary to Treaty Article 85(1)[27] and accordingly reduced the number of black clauses. It was proposed, however, that the exemption would not apply if either of the parties had a market share of 20 per cent or more, or if the market was 'oligopolistic', whatever that meant. This 'market share threshold' criterion was based upon a case in which the CFI agreed with the Commission that an agreement that was exempted by Regulation 2349/84 might nevertheless lead to abuse of a dominant position.[28] In later drafts, the proposed threshold was raised from 20 per cent to 40 per cent, and finally removed altogether as a criterion preventing the grant of an exemption and retained only as a possible ground whereby the benefit of the Regulation could be withdrawn upon later review. The new Regulation 240/96 was adopted on 31 January 1996 and entered into force on 1 April 1996, again for a period of ten years, the validity of Regulation 2349/84 being extended to 31 March 1996 to fill the legal gap. The Know-how Regulation 556/89 was revoked as of the same date. The main difference between this Regulation and those that it replaced is that a number of restrictive clauses were no longer blacklisted and, for example, the right of the licensor to terminate if the licensee challenged the validity of the patent was now specifically in the white list. It will come as no surprise to the reader that restrictions on parallel imports remained blacklisted.

26 Regulation 556/89.

27 For example, *Ottung v. Klee & Weilbach* [1990] 4 CMLR 915 (ECJ) (payment of royalties after patent expiry); *Bayer v. Süllhofer* [1990] 4 CMLR 182 (ECJ) (no-contest clause).

28 *Tetra Pak Rausing v. Commission* [1991] 5 CMLR 334 (CFI).

Individual Notification
Individual notification of some agreements was still possible, but the work involved for the company notifying the agreement was considerable. The actual form (Form A/B) on which the request for negative clearance or exemption was made was simple enough, but answers to very detailed questions about the companies involved, their market shares, and the projected markets for the licensed products, and much more had to be supplied with it.

Notification was never compulsory, but if the agreement was of doubtful legality, it was very useful as a form of insurance. A company entering into an agreement that violates Article 81(1) is liable to heavy fines, but no fines could be imposed in respect of any period between notification and the issuance of a Decision by the Commission. Usually the result of a notification was neither a Decision giving exemption or negative clearance nor a Decision that the agreement violated Article 81(1) and could not be exempted, but rather a letter indicating that the Commission intended to take no further action. The only value of such a 'comfort letter' was that even though the comfort letter might be subsequently withdrawn, the parties could not be fined for anything done before the date of withdrawal. It was generally supposed that, even after a comfort letter had been issued, the Commission could reopen the case whenever it liked. But the CFI held in 2000[29] that the Commission could not do this unless it had fresh evidence of a change in the situation—and it could not look for such evidence by asking the parties to provide it.

New Approach to Implementation of Articles 81 and 82
Council Regulation 17/62 had given the Commission the sole power to grant exemptions under Article 85(3) (now Article 81(3)), either by individual notification or by block exemption regulations. In 1962, the Community had only six members, and a centralized administration of competition law was (just about) workable. With 25 members, a continuation of the centralized system would mean that the Commission would no longer have the resources to deal with more important issues such as cartels and mega-mergers.

Under Council Regulation 1/2003, effective as of 1 May 2004, it is no longer possible to notify individual agreements, and there will be no more comfort letters and no more automatic protection from fines. It will in future be up to undertakings to make their own assessment of whether or not their proposed agreements violate European competition law and rulings will be made only if complaints or disputes arise. There will, however, continue to be block exemptions and these should give guidelines as to what is or is not permissible.

The other major change is that the national courts and national competition authorities now also have jurisdiction to apply EU competition rules, including the power to decide that an agreement should be exempted under Article 81(3). This of

[29] *Stork Amsterdam BV v. Commission* [2000] CMLR 31 (CFI).

course carries the danger that there will be inconsistent interpretation of the law, and although the Regulation gives guidelines that aim to prevent this, it is not clear how well these will work in practice. The immediate result is likely to be increased confusion, particularly in the new member states, and a large increase in the number of referrals to the ECJ from national courts on competition law issues. It is possible that the GC may have to take some of these referrals (at present, it hears only appeals from decisions of the Commission or other EU authorities).

New Technology Transfer Block Exemption Regulation

In line with Regulation 1/2003, the Commission produced a new Technology Transfer Block Exemption Regulation (TTBER),[30] which also came into effect on 1 May 2004. Under the new Regulation, there are no longer any 'white clauses', but only black and grey, now renamed 'hardcore restrictions' and 'excluded restrictions', respectively. If a hardcore restriction is present, the entire agreement falls outside the TTBER; if there are excluded restrictions, then the individual clauses containing these restrictions fall outside the TTBER, but the agreement as a whole is not affected.

A major problem in applying the TTBER is that the market share of the parties is relevant, and that the rules are different for competitors and non-competitors, as well as for reciprocal and non-reciprocal agreements. The block exemption applies only where, for competing parties, the combined market share is less than 20 per cent and, for non-competing parties, neither party has a market share above 30 per cent. Having market share below these limits is sometimes called a 'safe harbour', but this is far from true, because, if the parties are below these limits, they still must avoid black and grey clauses to come within the TTBER, while if they are above these limits, they are outside the TTBER however clean the agreement is. Furthermore, if the market share of the parties increases during the term of the agreement (as might be expected if the licensed product is really innovative), an agreement originally within the TTBER may fall outside of it later.

As well as market share criteria being different for competitors and for non-competitors, the lists of black and grey clauses are different, so that it is generally possible to have more restrictions if the parties are non-competitors. For example, a restriction on the licensee's ability to exploit his own technology is black if the parties are competitors, but only grey if they are not. Where the parties are competitors, it also becomes important whether the licence is reciprocal or non-reciprocal, more contractual restrictions being permitted for non-reciprocal agreements.

Further complications are caused by the fact that the many of the key definitions are unclear. The definition of 'market share' includes the relevant technology market (whatever that is), as well as the relevant product market. The definition of 'competing undertakings' is particularly incomprehensible, including potential

30 Regulation 772/2004.

competitors, as well as actual ones. A reciprocal licence may be granted 'in the same or separate contracts', but no time limit is given, leaving open the possibility that a non-reciprocal agreement may become reciprocal at a later date.

The Regulation is supposed to give legal certainty to undertakings, but instead gives such a tangled mess of complex requirements that companies may well wonder whether the best course might not be simply to avoid hardcore restrictions (as most licence agreements do in any case) and otherwise to ignore the Regulation completely. The Commission frequently states that licensing agreements are generally pro-competitive and the Guidelines to the TTBER clearly say that there is no presumption of illegality even if the block exemption does not apply. The Commission will not check old or new licence agreements on its own initiative, cartel and merger issues being considered much more important, and national competition authorities are also likely to be overworked.

The risk is that an agreement may be attacked as anticompetitive by a third party excluded from the market, or indeed by the other party to the agreement, if there is a dispute. Such complaints would be made to the national competition authorities, which may levy fines or decide that the agreement is void. There do not appear to be any cases in which fines have been applied in respect of a patent licence agreement falling under Article 81(1), but having a licence agreement for a major product held to be void is a bigger threat.

Euro-Defences

Although, under the TRIPs Agreement, there is normally no obligation to grant licences as long as the market is being supplied, there have been cases in which refusal of a license under intellectual property rights upon reasonable terms has been considered a violation of Article 82, the result being a form of compulsory licensing. A leading case decided by the ECJ[31] involved copyright, not patents, and concerned the refusal of television companies owning the copyright for their weekly schedule of television programmes to allow any other company to publish programmes on other than a day-to-day basis. In the English courts, attempts to defend against accusations of patent infringement by asserting as a 'Euro-defence' that the patentee was abusing a dominant position or was willing to license only on terms that would violate Article 81 generally were struck out at a preliminary stage in the proceedings, but in 2002 the Court of Appeal[32] overturned a summary judgment of the Patents Court, holding that it could not be said that there was no prospect of success at trial. The Decision of the Commission in 2004 that Microsoft had abused a dominant position, confirmed by the CFI in 2007 and not appealed further to the ECJ, will lead to compulsory licensing of some Microsoft intellectual property relating to interoperability standards.

[31] C-241/91 & C-242/91 *Telefis Eirann v. Commission (Magill)* (ECJ 1995).

[32] *Intel v. Via* [2003] FSR 574 (CA).

Moreover, the German Federal Supreme Court (*Bundesgerichtshof*, BGH) confirmed, in a case involving a patent on an improved method to read out data from rewritable CDs in the 'Orange Book standard', that the patentee (Phillips) could lose the right to a permanent injunction under certain circumstances: for example if the patentee has a dominant position in the relevant market and the infringer can show that the patentee refused in a discriminatory way an unconditional offer of the infringer to take a licence. The infringer further needs to show that he has already fulfilled all of the obligations that a licensee could expect to be imposed by the patentee, including paying royalties and providing the patentee with information to calculate them.[33]

The Powers of the Commission

Council Regulation 1/2003 has given the Commission very extensive powers to investigate violations of Articles 81 and 82 of the Treaty of Rome. Article 81 does not deal only with written contracts, but also with concerted practices, such as unwritten agreements to fix prices or share out markets. It is, of course, clear that, if such illegal practices are to be stopped, there must be some powers of investigation, since the companies concerned are certainly not about to notify this type of agreement to the Commission of their own free will.

There are basically two different ways in which the Commission can obtain information. Under Article 18 of the Regulation, the Commission may, by simple request or by decision, require any undertaking (not necessarily the one being investigated) to provide information. If the information is not provided or is incomplete, the Commission may, by Article 23, impose fines. Furthermore, by Article 20, the Commission may send investigators to the premises of a company to examine business records, take copies of documents, and ask for oral explanation. Under Article 21, inspections may even be made at the homes of directors and senior managers.

The ECJ ruled, in 1980,[34] under Regulation 17/62, that the company does not have to be given prior warning and the opportunity to comply voluntarily, but can be faced with a 'dawn raid' in which the investigators bring with them a Commission decision that must be complied with on penalty of heavy fines. The same applies under Regulation 1/2003 and, subject to national law, national competition authorities may also make such dawn raids, either on their own initiative or at the request of the Commission. In contrast to discovery proceedings in US civil litigation, there is no attorney–client privilege for in-house lawyers with regard to EU Commission investigations, because only communications with external counsel exclusively prepared for defending a specific anti-trust investigation itself are privileged.[35]

[33] *Orange Book Standard—Compulsory License*, NJW-RR 24, 1047 (2009) K ZR 39/06 (BGH).

[34] *National Panasonic v. Commission* [1980] 3 CMLR 325 (ECJ).

[35] T 125/03 & T 253/03 *Akzo Nobel Chemicals and Akcros Chemicals v. Commission* (17 September 2007, presently under appeal as C-550/07).

There has been considerable concern that, in proceedings before the Commission, the fundamental legal rights of the party whose conduct is being investigated may not be fully safeguarded, and there have been a number of decisions of the CFI and the ECJ that have instituted some safeguards in this respect.[36] In particular, the Commission cannot use as evidence any document that it is not free to disclose to the accused party: the accused party must be able to see what exactly is being used as evidence against him.

Pharma Sector Inquiry

In January 2008, the EC Directorate-General of Competition started a sector inquiry on the pharmaceutical industry in the EU. Sector inquiries are carried out by the Commission when there is a suspicion that competition is not working properly. In contrast to proceedings relating to concrete violations of competition provisions by individual companies or persons, sector inquiries aim to learn how a particular industry sector functions and to identify possible obstacles to competition that need to be addressed in future legislation. This particular inquiry was started because of the observation that fewer and fewer innovative pharmaceutical products with new active ingredients had been reaching the market in recent years, and because there was an impression that generic competition was being delayed even after expiry of the relevant patents. Because patents play a key role for many pharmaceutical products, the functioning of the patent system and how it is used was at the centre of the inquiry. It seems likely that the decision to begin the inquiry was influenced by cases in which pharmaceutical companies were considered to have abused the system: for example the cases involving Hässle (p. 174) and Servier (p. 244).

The focus of the sector inquiry was to study not only competition between originator and generic pharmaceutical companies, but also possible competition between originator companies in developing and bringing alternative or competing products for the same indication onto the market.

The sector inquiry started with on-site inspections (dawn raids) at the premises of several generic and originator companies all over the EU, the first time that this power had been used in an industry sector inquiry. Besides collecting data and information by on-site inspection, the Commission sent out two main questionnaires and numerous supplemental questionnaires to a total of around 200 organizations in the pharmaceutical sector, including parallel traders and health insurance funds. The majority of the questions addressed what influence the patent system has on the strategy of originator and generic pharmaceutical companies, and on the competition between them. Hundreds of thousands of data were collected in relation to patent status, litigation, and development projects of around 200 specified active ingredients. Due to very short time limits and often not very clearly formulated

[36] For example, *AEG v. Commission* [1984] 3 CMLR 325 (ECJ).

questions, the efforts by the companies to respond to those questionaires were very significant, certainly thousands of man-hours per company.

In November 2008, the commission issued a preliminary report[37] alleging that originator companies have developed a 'tool box', in the Commission's view, to prevent timely generic competition and hamper competition. According to the Commission, the tool box includes elements such as:

- building up 'patent thickets' by filing numerous patent applications around a marketed product and thus preventing others from developing a competing product;
- intitiating vexatious patent suits having no merits, the only reason being to threaten competitors;
- interfering with health and pricing authorities based on patents or often unjustified safety concerns, in order to prevent them from granting a marketing authorization or price for the generic product;
- persuading generic companies into settlements that include a value transfer from the originator to the generic company, which, in exchange, agrees not to come to the market until a later time (that is, 'reverse payments', see p. 499); and
- switching a successful originator product shortly before the key patents expire to a newer version of the product that is subject to a longer patent protection.

Additionally, the Commission said that earlier generic competition could have amounted to savings of about €3,000 million for the health systems during 2000–07 for the chosen sample of medicines facing patent expiry in seventeen member states.

The process and findings of the preliminary report were heavily criticized from different sides, for example by Lord Justice Jacob[38] and by the EPO, although the final version of the EPO's comments[39] was less sharply critical than the original, and a large number of observations and comments were received by the Commission. In July 2009, the final report was published.[40] This was an improvement over the preliminary report, because it acknowledged the importance of a well-functioning patent system to foster innovation and recognized that there are often other reasons not related to patents or exclusivity periods why generic products reach the market later than expected. It further identified some areas in which it is possible to improve the present system: for example by more harmonization through a Community patent, and a unified European patent litigation system and in accelerating proceedings at the EPO, in particular, opposition proceedings. Despite its shortcomings, the

[37] See http://ec.europa.eu/competition/sectors/pharmaceuticals/inquiry/index.html [accessed 28 August 2009].

[38] [2009] CIPA 380.

[39] See EPO comments of 10.03.2009 (final), available online at http://ec.europa.eu/competition/consultations/2009_pharma/european_patent_office.pdf [accessed 28 August 2009].

[40] See http://ec.europa.eu/competition/sectors/pharmaceuticals/inquiry/communication_en.pdf [accessed 28 August 2009].

report is a unique source of statistical information on patents and patent litigation in the EU.[41]

It remains to be seen what influence the findings of the report will have on future legislation or competition law practice of the Commission or of national competition authorities. But what the sector inquiry has shown once more is the inherent tension between patent law and competition law: the latter trying to ensure that there is sufficient competition, while the patent system, by its function, inherently excludes competition for a limited period of time.

Competition Law: The USA

Although the USA does not oblige the patentee to work his patented invention and there are no compulsory licence provisions, US law has generally taken a stricter view than does UK law of attempts to extend the patent monopoly beyond its legal scope and has placed more restrictions on the freedom of a licensor. There are two main sources of US law in such matters: one is the common law doctrine of patent misuse, which as we have seen stems from the equitable 'clean hands' concept; the other is the statute law, particularly the Sherman and Clayton Anti-trust Acts (see below). The fact that US judges tend to be more familiar with anti-trust law than with industrial property law has, at times, led to very narrow interpretations of the rights given by a patent.

The Anti-Trust Acts

The period after the Civil War in the USA was one of unregulated capitalism in which trusts and cartels were set up in all major industries to fix prices and exclude unwanted competition. Public reaction against these abuses finally reached the point at which the anti-trust forces were able to pass legislation to curb them. This took the form of the Sherman Anti-trust Act,[42] passed in 1890, §1 of which declares any contracts or combinations in restraint of trade to be illegal, and §2 of which makes any attempt to monopolize any part of interstate or foreign trade a criminal offence.

It will be seen that whereas §1 of the Sherman Act requires mutual action of at least two parties to make an illegal act, a §2 offence may be committed by a single party. There are close parallels between these two sections and Articles 81 and 82 of the Treaty of Rome. This is no mere coincidence: these parts of the Treaty of Rome were based upon the post-war German competition law, which, in turn, was modelled on the Sherman Act. A further piece of US anti-trust law is the Clayton

[41] For example, that the granting rates of preliminary injunction requests in pharmaceutical patent litigation in Belgium are the highest, at 77 per cent, and the lowest are in Portugal, at 0 per cent—see the report's Technical Annex, p. 230, Fig. 84.

[42] 15 USC 1-7.

Anti-trust Act,[43] §3 of which prohibits tie-ins by a party having market power, while §4 gives individuals injured by any act forbidden by the anti-trust laws the right to sue the person responsible and §7 prohibits the acquisition of any assets of another company if this would create a monopoly or substantially lessen competition.

All of these provisions are expressed in very vague and general language, and it is not spelled out exactly what acts are to be forbidden. The Sherman Act does not mention patents at all, but quite clearly a patent licence agreement may be a contract contravening §1 of the Sherman Act, and patents may be among the assets the acquisition of which may violate §7 of the Clayton Act.

The US patent law is equally vague about what types of restriction may legally be made when a patent is licensed, although it does clearly state that a patent may be exclusively licensed for the whole or part of the USA.[44] Exclusive licences and territorial limitations, as such, are therefore allowed by statute in the USA and the Sherman Act, which is of equal status to the patent law, cannot overturn this.

There is, of course, a conflict between the patent law, which grants limited monopolies to inventors, and the anti-trust laws, which broadly condemn monopolies and restrictions on competition. Because the anti-trust laws are so broadly drafted, it is perhaps inevitable that they should be seen as constituting a general principle, to which the patent law forms a strictly limited exception. Thus, considering the statutory right to grant a territorially limited licence, it should be noted that whereas a simple licence granting the licensee exclusive rights in, say, the USA west of the Mississippi is legal, a cross-licensing agreement whereby A licenses B in the West and B licenses A in the East could constitute a division of markets that would be illegal under §1 of the Sherman Act.

The Hart-Scott-Rodino Anti-trust Improvements Act of 1976 required any acquisition by one company of another or assets of another to the value of US$15 million or more to be notified to the US Federal Trade Commission (FTC). Failure to report, although not a criminal offence, carries heavy administrative fines (US$10,000 per day of delay). Until the mid-1990s, no one thought that this had anything to do with patent licences, but at that time the Justice Department stated its opinion that Hart-Scott-Rodino applies to exclusive patent licence agreements under which the total consideration for the licence, including expected milestone payments and royalties, exceeds US$15 million. This figure is small by today's standards; if corrected for inflation since 1976, it should be more like US$55 million.

Patent Misuse and Anti-Trust

Apart from the anti-trust laws, under which certain acts may be illegal or even criminal, the other main restraint upon the patentee is that provided by the concept

[43] 15 USC 12-27.
[44] 35 USC 261.

of patent misuse. A court may consider patent misuse to exist when a patentee attempts to extend the scope of his monopoly to areas outside of the bounds of the patent: for example by tying a licence to the purchase of unpatented goods.

Until recently, it was considered that any tying of unpatented goods by a patentee was automatically patent misuse, even in a situation in which the patentee did not, in fact, have the market power necessary to compel his customers to accept the tying provisions. Thus, in a situation in which there were several alternatives to the patented product on the market, there could be no violation of the anti-trust laws, but there would be considered to be patent misuse because it was presumed that a patent automatically gave market power. This anomaly was corrected by the Patent Misuse Reform Act of 1988, which abolished this presumption and applied the market power test to patent misuse in the same way that it was applied to Sherman Act violations. It has been held that this applies equally to tie-out clauses in licence contracts as to tie-in clauses.[45]

Patent misuse, when found, does not result in patent invalidity, but disentitles the patentee to equitable relief for infringement and may be used as a defence by an alleged infringer. If the misuse is purged, the patent may, once again, be enforced against infringers. Anti-trust violations, in contrast, give rise to a cause of action by the US government or by any person who has suffered damage. If proven, the penalties may include an injunction against the action found to be illegal, the unenforceability of any relevant contract, the unenforceability of the patent, damages to the extent of three times the actual damage sustained, and in particularly flagrant cases, the possibility of a jail sentence. Clearly these are not the sort of consequences to be taken lightly.

Inequitable conduct before the US Patent and Trademark Office (USPTO) may give rise to breach of the anti-trust laws, if all of the elements of such a violation are established. When a patent has been obtained by 'knowing and wilful fraud', then any attempt to enforce the patent may amount to an illegal attempt to monopolize, in violation of §2 of the Sherman Act,[46] the patent will be unenforceable, and the patentee may be liable for triple damages to the person against whom the patent was enforced. It could even be alleged that although the patent was not enforced, its mere existence kept competitors off the market and was in violation of the Sherman Act. Challenges to patent validity arising out of anti-trust suits by the US Justice Department were at one time used frequently to bring about what was in effect compulsory licensing in a number of areas.

Reverse Payments

There have been a number of cases within the last few years in which a pharmaceutical company, faced with a Paragraph IV certification against one of its patents by

[45] *In re Recombinant DNA Technology* 30 USPQ 2d 1881 (SD Ind 1994).

[46] *Walker Process Equipment, Inc. v. Food Machinery & Chemical Corp.* 382 US 172 (Sup Ct 1965).

a generic competitor (see p. 189), has reached an agreement in which the generic company agrees to withdraw its certification or change it to one under Paragraph III (no launch until after patent expiry) in exchange for a monetary payment, sometimes together with supply of the product. Both the Justice Department and the FTC have taken the view that such an agreement amounts to a per se anti-trust violation and a number of anti-trust cases have been brought before the courts to challenge these agreements.

One recent such case concerned the antibiotic ciprofloxacin and involved reverse payments by Bayer to Barr Labs Inc. and others, in which the District Court's grant of summary judgment against the challengers was upheld by the Federal Circuit in October 2008.[47] The decision was reached on the very simple ground that the settlement agreements were within the exclusionary zone of the patent and therefore were outside the scope of federal anti-trust law. In the words of the Court, 'a patent by its very nature is anticompetitive'. A petition for certiorari was filed in March 2009 and, at the time of writing, remains pending. The Supreme Court has, however, denied certiorari in four earlier cases of this type.

Patent Licence Agreements

The legality of specific types of restriction that may be placed upon a licensee in a patent licence agreement is determined by rules that have developed through case law. A restriction may be regarded as per se illegal, or as subject to the 'rule of reason'. In the first category are restrictions that constitute anti-trust violations that, in the words of the Supreme Court,[48] because of their 'pernicious effect on competition and lack of any redeeming virtue', are conclusively presumed to be 'illegal without elaborate inquiry as to the precise harm they have caused'. Application of the rule of reason requires an enquiry into the overall effect of the agreement. The restriction is not illegal if it is ancillary to the lawful primary purpose of the agreement, if its scope and duration are no greater than necessary, and if it is otherwise reasonable in all of the circumstances.

Clauses that, at one time or another, have been considered per se illegal include tie-ins, 'tie-outs' (clauses forbidding the licensee to deal in competitors' products), clauses requiring continued payment of royalties after the expiry of the patent, and package licences in which the licensee is compelled to accept a licence under a whole package of patents when in fact he only needs a licence under some of them, or in which the licensee is compelled to pay royalties upon total sales of one type of product only some of which fall under the licensed patents. Although resale price maintenance agreements are per se illegal, there is an old Supreme Court decision[49] allowing the licensor to fix the first sale price by the licensee. This case is still good law, although it has been given a very restricted interpretation by the courts

47 *In re Ciprofloxacin Hydrochloride Antitrust Litigation* (Fed Cir 2008).
48 *Northern Pacific Railway Co v. United States* 356 US 1 (Sup Ct 1958).
49 *United States v. General Electric Co.* 272 US 476 (Sup Ct 1926).

in later cases. In 1997, the Supreme Court confirmed that agreements fixing a *maximum* resale price were not per se illegal.[50]

Patent pool agreements were subjected to rule of reason analysis by the Supreme Court in 1931,[51] but the Justice Department has always regarded them with deep suspicion. Recently, however, patent pool agreements in the telecommunications industry have been approved (see p. 439).

Validity Challenges by Licensees

One important change in licensing law in the USA has related to the question of whether a licensee can challenge the validity of the patent under which he is licensed. The law always used to be the same as in England: that is, that a licensee was estopped from so doing, because it was not just that he should be able to enjoy the protection of the patent for as long as it suited him, and then be able to turn around and attack it. In 1969, however, this was overturned by the Supreme Court,[52] which felt that the public interest in invalid patents being challenged outweighed the possible injustice to the patentee, and ruled that, notwithstanding any clause in the agreement to the contrary, a licensee could, at any time, challenge the patent's validity.

Now, a licensee may at any time cease to pay royalties and seek a declaratory judgment of patent invalidity. If he wins, he has no further obligation to pay royalties, but cannot recover any royalties already paid. But if the patent is invalidated by a third party, the licensee's freedom from royalty payments begins only when the patent is finally held invalid, unless the licensee himself stopped paying royalties earlier and made some contribution towards the resolving of the issue of validity. The licensee may even be able to seek declaratory judgment of invalidity while still continuing to pay royalties, as was held by the Supreme Court in *Medimmune v. Genentech*.[53]

As in the EU, it is acceptable to include a clause allowing the licensor to terminate the licence if the licensee challenges validity. If the licensee's challenge is unsuccessful, he will then be liable as an infringer for anything done after he stopped paying royalties. The rule of *Lear v. Adkins* does not apply to patent assignments, even those made in return for a running royalty, nor does it apply to know-how agreements. In one case,[54] a licensee who had signed a contract obliging him to pay royalties indefinitely on a know-how agreement (a patent was applied for but not granted) could not terminate the payment of royalties even though the know-how had entered the public domain. Thus, in this case, the licensor was in a better position without a patent than she would have been with one.

[50] *State Oil Co v. Khan* 1997 WL 679424 (Sup Ct 1997).
[51] *Standard Oil (Indiana) v. United States* 283 US 163 (Sup Ct 1931).
[52] *Lear v. Adkins* 162 USPQ 1 (Sup Ct 1969).
[53] *Medimmune v. Genentech* 549 US 118 (Sup Ct 2007)—see p. 216.
[54] *Aronson v. Quick Point Pencil Co.* 201 USPQ 1 (Sup Ct 1979).

Justice Department Guidelines

In the 1970s, the Anti-trust Division of the Justice Department took an attitude similar to that of the EU Commission in trying to impose more and more severe controls upon clauses of patent licence agreements, and issued guidelines spelling out nine types of clause, some of which were previously subject to the rule of reason, which it regarded as per se illegal, although it was not clear that the courts would take the same view. These restrictions became known as the 'nine no-nos'. ('That's a no-no' was, at the time, a catchphrase in a popular TV comedy show.)[55]

Under the Reagan administration, the 'nine no-nos' were repudiated, and the official line was that the rule of reason should be applied in all cases. With the Clinton administration, the pendulum swung back a little and the Justice Department recovered a little of its old aggressiveness. In 1995, the Justice Department issued new anti-trust guidelines that were very similar to the then current draft of the EU Technology Transfer Block Exemption Regulation. These guidelines indicated that the rule-of-reason approach would be followed in the great majority of cases, per se illegality being reserved for 'naked price fixing', restrictions on output, and market division. In all other cases, the Justice Department will now raise objections only if a rule-of-reason analysis indicates that market power is present and that the anti-competitive effects of the restrictions are not outweighed by increased efficiency benefitting the consumer. It will not be assumed that patents automatically give market power, and it will be presumed that no market power exists when the combined market share of licensor and licensee is less than 20 per cent. These guidelines, which have not been superseded and are still in effect in 2009, were not enforced with any enthusiasm during the Bush administration, and it remains to be seen what attitude the Obama administration will take.

There remains some concern about the introduction into the guidelines of vague concepts such as 'technology markets' and 'innovation markets'.[56] This seems to be in line with FTC practice, where, for example, in one pharmaceutical merger case,[57] there was an extended analysis of the gene technology market, although there were no gene technology products on the market at the time (or, indeed, since).

Not only the Justice Department, but also the FTC, has got in on the act. In 2003, it published a study entitled *To Promote Innovation: The Proper Balance of Competition and Patent Law and Policy*. Its main recommendations were essentially to make the patent-granting process more strict and to facilitate post-grant opposition by third parties in order to promote competition. It is an interesting phenomenon that not only in the USA, but also in the UK, Switzerland, and other countries, other government departments (as well as quasi-governmental organizations, and self-appointed committees of the great and the good) take it upon themselves to tell the national patent offices how to do their job. It would be interesting

55 *Rowan & Martin's Laugh-in.*

56 Anti-trust Guidelines [1995] 3.2.2 & 3.2.3.

57 *Sandoz/Ciba Geigy* Federal Register 62 240 65706 (FTC, 15 December 1997).

to see a USPTO report on the functioning of the Justice Department, or a UK Patent Office report on the workings of the Nuffield Commission.

Collaboration Between the US Justice Department and the European Commission

The EU Commission and the US Justice Department, as the respective competition authorities in Europe and the USA (together with the FTC), have concluded an agreement on the application of positive comity principles in competition matters, providing for exchange of information in merger and anti-trust matters. It must be remembered, however, that the starting positions are not the same, because the US Sherman Act gives the Justice Department power to protect the foreign trade of US companies, while the Commission has no such power to assist European firms. As can be seen from the Justice Department's International Anti-trust Guidelines, the US has no qualms about asserting jurisdiction over the activities of foreign companies in foreign countries, if these activities can adversely affect US exports. There can hardly be comity if there is not reciprocity in such a basic matter as this.

GLOSSARY

absolute novelty a system whereby any prior publication anywhere destroys the novelty of a patent.
abstract a summary of the disclosure of a patent **specification**, written by the applicant.
acceptance (UK) the formal decision by the Patent Office that a patent should be granted.
account of profits alternative to damages, based on profits made by infringer.
Administrative Council the governing body of the European Patent Office, consisting of a committee of its competitors.
advisory action (USA) an **official action** issued by the US examiner when a **final rejection** is being maintained.
allowance (USA) *see* **acceptance**.
amendment alteration made to a patent **specification** during prosecution or after grant.
analogy process a non-inventive chemical process.
Andean Pact an association of South American countries with common economic policies.
anticipation **prior art** that destroys the novelty of a **claim** by giving an enabling description of something falling within it.
anti-trust laws or regulations to prevent abuse of monopoly.
appeal brief (USA) a summary of arguments why the invention should be patentable, submitted to the Board of Patent Appeals and Interferences.
assignment transfer of ownership of a patent.
attorney fees (USA) award of costs in US litigation (unusual).

belated opposition (UK) formerly, an application to the Patent Office for revocation of a newly granted patent.
best mode requirement (USA) the obligation to describe the best way of carrying out the invention.
biopiracy pejorative term applied by Green organizations to **bioprospecting** in developing countries.
bioprospecting looking for useful compounds from biological sources, such as plants and microorganisms.
Biotechnology Patenting Directive (BPD) (EU) a **Directive** to harmonize patenting of biotech inventions in the Community; argued over for nearly twenty years and not yet properly implemented.
black box application an application claiming a pharmaceutical per se, filed under TRIPs Agreement, Art. 70.6, in a country that did not, at that time, provide such protection.
block exemption (EU) a **Regulation** exempting a category of agreements from Art. 81(1) of the Treaty of Rome if certain conditions are met.
blocking patent *see* **defensive patent**.
Board of Appeal (EPO) the body that hears appeals from the Examining and Opposition Divisions.
boilerplate standard clauses in contracts.
***Bolar* exemption** a provision corresponding to the legislative overruling of the US case *In re Bolar*, allowing preparations during the lifetime of a patent for commercial use on its expiry.

Budapest Treaty the international treaty regulating the deposition of microorganisms for patent purposes.

Cartagena Agreement the treaty setting up the **Andean Pact**.

caveat a request to a patent office to be informed when some future event (for example, the grant of a patent on another's application) occurs.

certiorari (USA) a petition requesting the Supreme Court to review the decision of a lower court.

Chapter I the first, compulsory, stage of the **international phase** of an application under the Patents Cooperation Treaty (PCT), up to issuance of the international search report (ISR) and publication of the application.

Chapter II the second, optional, stage of the **international phase** of a PCT application, including issuance of the **international preliminary report on patentability (IPRP)**.

characterizing clause the part of a German or European-style **claim** that indicates the novel features of the invention.

chartered patent agent (UK) a British-qualified **patent agent** who is a fellow of the Chartered Institute of Patent Agents.

chartered patent attorney (UK) a **chartered patent agent** who prefers the more impressive title.

citation a document identified by a search report as being relevant **prior art**.

claim the part of a patent **specification** that defines the **scope** of protection.

claimant (UK) since 1999, the person seeking relief in a court action (formerly the plaintiff).

claim form (UK) since 1999, the formal document initiating court proceedings, for example, for **infringement**.

Clayton Act (USA) one of the two major US **anti-trust** Acts.

collateral estoppel (USA) the principle that a judgment of patent invalidity is binding in subsequent **infringement** actions.

comfort letter (EU) formerly, a letter from the Commission stating that no action would be taken in respect of a notified agreement.

Community patent a single unitary patent covering all EU member states; does not yet exist.

Community Patent Convention (CPC) an international treaty to establish a **Community patent**, never ratified.

compulsory licence a **licence** that government authorities or courts force the **patentee** to grant to another party.

complete specification (UK) formerly, a full description of an invention filed after an initial brief description.

conception (USA) the mental part of the inventive process.

constructive reduction to practice (USA) the filing of an adequate patent application as evidence of having fully worked out the invention.

continuation application (USA) a new filing of a US application with unaltered **specification** to allow presentation of new **claims**.

continuation-in-part (cip) application (USA) a new filing of a US application, with alterations or additions to the **specification**.

contributory infringement **infringement** by supplying another with the means to infringe a patent.

Convention country a state that is a member of the Paris Convention for the Protection of Industrial Property.

Convention on Biological Diversity (CBD) an international treaty about protection of, access to, and benefit-sharing from biological resources.

Convention year a period of 12 months from a first application in a **Convention country** within which applications having the effective date of the original filing may be filed in other Convention countries.

copycat an infantile and pejorative term often applied by US patent owners to accused infringers.

cross-border injunction an **injunction** granted by a court in one country (for example, the Netherlands) forbidding **infringement** of a patent in another country.

declaratory judgment (USA) in the context of patent law, a judgment on patent validity by a US court, initiated by a party in dispute with the **patentee**.

defensive patent a patent that does not cover what the **patentee** is doing, but which he hopes will keep competitors away from his area of interest.

deferred examination a system in which the examination of a patent application may be postponed for several years until requested by the applicant.

delivery up (UK) a court order compelling an infringer to deliver infringing goods to the **patentee** for alteration or destruction.

dependent claim a **claim** incorporating all of the features of an earlier claim to which it refers.

depositions (USA) in a US **infringement** suit, the pre-trial stage of taking down evidence on oath.

designation (EPO; PCT) the naming of the countries for which a European or PCT application may give rights; now automatic for all contracting states.

Directive (EU) a legislative instrument adopted by the Council with the agreement of the Parliament requiring individual implementation by all member states.

disclaimer the exclusion of specific subject matter from the **scope** of protection claimed.

discovery a court order to parties in an **infringement** action to produce relevant documents.

divisional application an application claiming part of the subject matter originally in an earlier pending application.

doctrine of equivalents (USA) *see* **equivalence**.

Doha licence a **compulsory licence** granted to allow export of patented pharmaceuticals to a developing country.

double patenting when the same invention is claimed in two different patents of the same owner; *see* **terminal disclaimer**.

due diligence the process of checking the basis of a proposed contract (for example, the patent situation, for a **licence** agreement) before signing.

early publication publication of a patent application before it is examined.

elected state (PCT) a contracting state for which the procedure of **Chapter II** has been chosen to apply.

election of species (USA) restriction to a single invention in response to an objection of **non-unity**.

enabling disclosure (USA) a description sufficient to enable the reader to carry out the invention.

equivalence the extension of the **scope** of a patent to cover something outside the literal wording of the **claims**.

Enlarged Board of Appeal (EBA) (EPO) the body that resolves difficult points of **European Patent Convention (EPC)** law at the request of a **Board of Appeal** or the president of the European Patent Office (EPO).

estoppel the barring of a person from claiming a legal right because of an earlier act incompatible with his claim.

European patent a patent in any member state of the EPC, granted by the EPO.

European patent attorney the English title for a professional representative before the EPO.

European Patent Convention (EPC) the international treaty establishing the **European patent** system.

European Patent Office (EPO) the body responsible for granting **European patents.**

European Patent Organization a body consisting of the **Administrative Council** and the EPO.

European route obtaining patents in EPC countries by a single application at the EPO.

evergreening derogatory term for **life cycle management (LCM) patenting**.

examiner's answer (USA) written in response to an appeal brief.

exclusive licence a **licence** that excludes all others, the **patentee** himself among them.

exclusive marketing rights the right of a person filing a **black box application** to exclude competitors until his application is granted or refused, if certain conditions are met.

exhaustion of rights the principle that, when patented goods have been sold by the **patentee**, he has no further control over them.

ex parte administrative or judicial proceedings in which only one party is involved.

extended technical arm (USA) someone who helps to reduce an invention to practice, although not an inventor.

extension prolongation of a normally fixed **term**, such as the life of a patent.

fairly based [of a **claim**] supported by the descriptive part of the **specification** or by a **priority document**.

file history/file wrapper (USA) the dossier containing all papers relevant to the prosecution of an application.

file wrapper estoppel *see* **prosecution history estoppel**.

final rejection (USA) an **official action** severely limiting further prosecution.

fingerprint claim a **claim** characterizing a new product of unknown structure in terms of its properties.

first to file the system for determining priority of invention in all countries except the USA.

first to invent the system used to determine priority of invention in the USA.

foreign filing filing in countries other than the country of first filing.

forum shopping legal manoeuvring to try to have a suit heard by a court thought to be favourable to one's own side.

fraud on the Patent Office (USA) obtaining a patent by concealment or misrepresentation of relevant facts; now usually called 'inequitable conduct'.

free beer claim a **claim** that covers all possible ways of reaching an obviously desirable goal.

free invention (Germany) an invention made by an employee that does not belong to the employer.

further processing (EPO) recovering the situation after losing rights by failure to meet an office time limit; no excuse needed.

generic claim a **claim** that covers a number of compounds defined by common structural features.

generic drug a drug having the same active ingredient as an earlier marketed product, and requiring for marketing approval only bioequivalency studies and a reference to the earlier data.

genus the group of compounds claimed in a **generic claim**.

grace period a period of time before the filing date of an application during which certain types of publication do not invalidate the application.

grant-back a clause in a **licence** agreement granting the licensor rights in developments made by the licensee.

***Gillette* defence** (UK) the argument that one cannot infringe a valid patent because what one is doing is essentially the same as the **prior art**.

Hatch-Waxman Act (US) law providing for **extension** of patent **term** to compensate for regulatory delay, and allowing **generic drugs** to be marketed immediately upon patent expiry.

Implementing Regulations (EPO) rules for the implementation of the EPC, made by the **Administrative Council**.

importation invention (UK) formerly, an invention known abroad, but brought into the UK for the first time.

infringement doing something forbidden by the grant of a patent to another.

injunction a court order to cease doing something (for example, infringing).

innocent infringer (UK) an infringer who could not be expected to know of the existence of the patent and who does not have to pay damages.

insufficiency a ground of invalidity of a patent, if the description does not enable the skilled reader to work the invention.

integer a distinct feature of a **claim**.

interdict (Scotland) *see* **injunction**.

interference (USA) proceedings in the US Patent and Trademark Office (USPTO) to determine priority of invention.

interlocutory injunction a court order to stop alleged **infringement** pending the trial of the action.

internal priority the possibility of filing, within 12 months, an application claiming priority from an earlier application in the same country.

international application an application made under the provisions of the PCT.

International Bureau (PCT) the part of the World Intellectual Property Organization (WIPO), situated in Geneva, that deals with PCT applications.

international exhaustion the theory that sale of a patented product in one country exhausts the patent rights in all countries.

international phase (PCT) the period between filing an **international application** and entering the national or regional patent offices.

International Preliminary Examining Authority (IPEA) (PCT) the patent office that carries out the examination prescribed in PCT, **Chapter II**.

international preliminary report on patentability (IPRP) a non-binding opinion on the patentability of an invention, issued during PCT, **Chapter II**, proceedings.

International Searching Authority (ISA) (PCT) the patent office that prepares the search report on an **international application**.

inter partes administrative or judicial proceedings involving two or more contending parties.

inutility a ground of invalidity of a patent, if the claimed product does not give the promised results.

inventive step (EPO) what an invention has if it is not **obvious**.

inventor's certificate formerly granted by the USSR, primarily to Soviet inventors, the rights in the invention being owned by the state.

***Jepson* claim** (USA) a **claim** in the German style that distinguishes the new from the old features.

junior party (USA) the party in an **interference** who has the later US filing date.

know-how unpatented and unpublished technical or commercial information.

kort geding (Netherlands) expedited preliminary **infringement** proceedings in which **cross-border injunctions** could be granted.

label licence (USA) a statement that the purchase of the labelled goods gives a **licence** to use them in a patented process.

laches undue delay in bringing suit, which may prevent grant of relief.

language of the proceedings (EPO) the language (English, French, or German) in which a **European patent** application is filed or into which it is translated.

letters patent an 'open letter' from the sovereign to notify a grant, such as of a monopoly for a new invention.

licence permission to use technology, usually patented, normally upon payment of a fee.

licences of right an endorsement on a patent to the effect that anyone may have a **licence** upon reasonable terms.

life cycle management (LCM) patenting obtaining patents on new uses, formulations, etc., that remain in force after the basic patent for a drug expires.

local novelty the principle that a patent can be invalidated by prior publication only if the publication was in the country granting the patent.

maintenance fees annual fees payable to keep a patent application pending or a granted patent in force.

manner of manufacture (UK) the term used in the **Statute of Monopolies** for the proper subject of a patent.

***Markman* hearing** (USA) a hearing before the judge in the absence of a jury to determine the true literal construction of the **claims**.

***Markush* group** (USA) a group of compounds or substituents defined for the purpose of a patent **claim** and lacking a common generic description.

mixed novelty the principle that a patent can be invalidated by a prior printed publication anywhere in the world, but by **prior use** only in the country granting the patent.
mosaicing the combination of several different pieces of **prior art** to put together all of the features of claimed invention.

national phase (EPC) registration of a granted **European patent** in the designated countries, including usually the filing of a complete translation.
national (regional) phase (PCT) submission of a PCT application to national or regional patent offices for the grant procedure.
national route (EPC) obtaining patents in EPC countries by separate applications at national patent offices.
negative clearance (EU) formerly, a statement by the European Commission that an agreement does not appear to contravene Art. 85 (1) of the Treaty of Rome.
new matter material entered into a **specification** by **amendment** describing something not previously disclosed in it.
no-contest clause a clause in a patent **licence** agreement in which the licensee agrees not to attack the validity of the licensed patent.
nominal working an attempt to avoid **working requirements** by offering **licences** under the patent.
non-Convention filing an application that does not claim priority from an earlier application, although one exists.
non-exclusive licence a **licence** that does not exclude the possibility of further licences being granted.
non-staple an article of commerce the only significant use of which is in a patented process.
non-unity the condition of an application claiming more than one invention.
novelty the essential condition for patentability—that is, that what is claimed is new.

obiter (dictum) a judicial expression of opinion that does not constitute a precedent.
object clause part of patent **specification** stating the object to be achieved by the invention.
obligation of cando(u)r (USA) the obligation of the applicant and his attorney to inform the USPTO of all relevant facts.
obvious capable of being performed by the average skilled person in possession of the **prior art**.
official action (USA) communications from the USPTO examiner raising objections to a patent application.
official letter (UK) communications from the Patent Office examiner raising objections to a patent application.
Official Gazette (USA) publication of the USPTO giving information about newly granted patents and other matters.
Official Journal (EPO) publication giving information on changes to the **Implementing Regulations** and important new case law of the Boards of Appeal.
omnibus claim (UK) a **claim** claiming the invention as described with reference to the drawings or examples.
opposition proceedings before a patent office in which a third party raises objections to the grant or validity of a patent.

oral proceedings (EPO) a hearing before a Division of the EPO, or a **Board of Appeal**.

paper example an example in a chemical patent **specification** that has not, in fact, been carried out.

parallel importation the unauthorized importation of genuine patented goods from a country in which the goods are legally on the market.

partial validity the principle that a patent may be valid and enforceable in part, even if part of its **scope** is invalid.

patent agent (USA) a US patent practitioner who is not a lawyer; (UK) *see* **chartered patent agent**.

Patent and Design Journal (UK) publication of the Patent Office giving information about newly granted patents and other matters.

Patentanwalt (Germany) a German patent practitioner in private practice.

Patentassessor (Germany) a German patent practitioner employed in industry.

patent attorney (USA) a US patent practitioner who is admitted to the bar; (UK) *see* **chartered patent attorney**.

patentee the owner of a granted patent.

patent misuse (USA) an attempt to extend the **scope** or effect of a patent beyond that granted by law.

patent of importation/revalidation patents granted by certain countries on the basis of patents already granted elsewhere.

Patents Bulletin (EPO) publication of the EPO giving information about newly granted patents and other matters.

person skilled in the art the hypothetical person to whom the patent **specification** is addressed; in the USA, sometimes called the 'person having ordinary skill in the art' (PHOSITA).

pipeline protection granted by some countries when product protection is first introduced to allow protection for products already patented elsewhere.

pith and marrow (UK) formerly, the essential features of an invention.

pre-characterizing clause the part of a German or European-style **claim** that recites the features of the invention that are already known.

presumption of validity the rebuttable legal presumption that a granted patent is valid.

prior art all public knowledge before the **priority date** (invention date in the USA) that could be relevant to the **novelty** or **inventive step** of an invention.

prior claiming the claiming of essentially the same invention in an earlier application that was not published at the **priority date**.

prior use the use of an invention before the **priority date** of an application claiming it.

priority date the date on which an invention was first disclosed to a patent office in a patent application or in an earlier application from which it validly claims priority.

priority document a patent application from which subsequent applications claim priority.

priority year *see* **Convention year**.

product-by-process claim (UK and USA) a **claim** to a product when made by a specified process; (EPO) a per se claim to a product of unknown structure defined by a process used to make it.

product per se claim a **claim** to a product irrespective of how it is made.

prosecution history estoppel (USA) a principle forbidding the **patentee** from asserting a **scope** broader than the literal wording of the **claims** if the claims were narrowed during prosecution.

prosecution laches (USA) the principle that a patent can be invalid or unenforceable because of undue delay in prosecution; a depth charge against **submarine patents**.

Protocol on Recognition (EPO) an annex to the EPC setting out which national courts have jurisdiction in disputes on ownership of **European patent** applications.

provisional application (USA) a US priority application that must be converted into or replaced by a regular application before it can lead to a granted patent.

provisional specification (UK) formerly, a brief description of the invention filed with an application, to be followed by a **complete specification**.

purposive construction (UK) the principle that a **claim** is to be interpreted on the basis of what the **person skilled in the art** would understand the **patentee** to intend by the language used.

reach-through claim in a patent for an assay or screening method, a **claim** to any compound found using the method; invalid.

reach-through royalties royalties paid on the sales of a drug found using a screening method covered by a patent, with or without **reach-through claims**; agreeing to pay such royalties is not a good idea.

Receiving Office (PCT) the competent patent office at which to file an **international application**.

Receiving Section (EPO) the branch of the EPO that deals with formal examination of applications.

reduction to practice (USA) completion of the inventive act by carrying out the invention and finding a use for it.

re-examination (*ex parte*) (USA) a procedure whereby the **patentee** or a third party can ask the USPTO to review a granted patent in the light of new **prior art**; a third party requester is not a full party to the proceedings.

re-examination (*inter partes*) (USA) a procedure whereby a third party can ask the USPTO to review a granted patent in the light of new **prior art**; the requester is a full party to the proceedings.

refiling abandonment of a first-filed application and filing a new application with essentially the same **specification**.

regional phase (PCT) the entry of a PCT application, after the **international phase**, into a regional patent office, such as the EPO.

registration obtaining patent protection in certain territories on the basis of a granted British or other patent.

Regulation (EU) a legislative instrument issued by the European Council or Commission, directly applicable in all member states.

reissue (USA) a procedure whereby defects in a granted US patent may be corrected on application by the **patentee**.

renewal fees *see* **maintenance fees**.

request for continued examination (RCE) (USA) a simplified form of **refiling** to enable prosecution to continue.

request for reconsideration (USA) a request to the Board of Appeals to reverse its own adverse decision.

res judicata an issue finally decided by a court, which cannot be reopened by the parties.

restoration proceedings to revive a patent that has lapsed by non-payment of **renewal fees**.

restoration of rights (*restitutio in integrum*) (EPO) recovering the situation after losing rights through failure to take some action; a good excuse is needed.

restriction requirement (USA) a demand by the examiner that the applicant limit his or her application because of **non-unity**.

reversal of onus the principle that, when a patent claims a process for making a new compound, the compound will be presumed to be made by the patented process unless proved otherwise.

revocation proceedings proceedings before a patent office or a court to have a patent declared invalid.

right of prior use the principle that even a valid patent cannot be used to stop someone from continuing what he or she was doing before the **priority date** of the patent.

***Saccharin* doctrine** (UK) the former principle that importation of a product made abroad from an intermediate infringes a UK patent for that intermediate.

saisie contrefaçon (FR) a court-ordered inspection to determine whether a patent is being infringed.

scope the total field encompassed by a **claim**.

Search Division (EPO) the branch of the EPO that carries out searches on **European patent** applications.

search order (UK) a court order allowing the premises of an alleged infringer to be searched for evidence of **infringement**; usually granted ***ex parte***, without prior warning (formerly, an *Anton Piller* order).

second (pharmaceutical) use claim a **claim** to protect the invention that a compound already known as a pharmaceutical has a new, unrelated pharmaceutical use.

secret use (UK) the former ground of invalidity that the invention had been used (although not in public) before the **priority date**.

selection invention an invention that selects a group of individually novel members from a previously known class.

semi-exclusive licence a **licence** exclusive except that the **patentee** retains the right to use the invention.

senior party (USA) the party in an **interference** who has the earlier US filing date.

sequence identifier a separate part of a biotech **specification**, listing amino acid or DNA sequences in standard format.

serial number (USA) formerly, a six-figure number given to an application on filing; now known as the application number.

service invention (Germany) an invention made by an employee that belongs to the employer (if claimed by it).

Sherman Act (USA) the most important of the US **anti-trust** laws.

shop right (USA) the right of an employer to a free **licence** under a patent belonging to an employee, if the invention was made using the employer's facilities.

showing (USA) evidence of unexpected superiority filed to overcome an objection of obviousness over **prior art**.

sole licence *see* **semi-exclusive licence**.

sovereign immunity (USA) the principle that a state (and thus a state university) cannot be sued for **infringement** without its consent.

species claim (USA) a **claim** to a single chemical compound, a member of a claimed **genus**.

specification the description of the invention filed with a patent application.

state of the art the total information in the relevant field known to the hypothetical **person skilled in the art**.

statement of invention the part of the **specification** corresponding to the main **claim**, summarizing the broadest aspect of the invention.

Statute of Monopolies (UK) the early English law banning monopolies except for those for new inventions.

sub-claim *see* **dependent claim**.

submarine patent (US) a patent granted without **early publication** after a long prosecution, claiming as an invention something commonly used in industry (for example, bar codes).

substantive examination examination for patentability, as distinct from formal matters.

supplementary protection certificate (SPC) (EU) an intellectual property right effectively extending the protection for a pharmaceutical or agrochemical product by up to five years to compensate for regulatory delays.

suspension of proceedings (EPO) putting prosecution of a European application on hold until a court decides who is the rightful owner.

swearing back (USA) presenting evidence of an invention date earlier than the publication date of a cited reference.

Swiss-type claim a second pharmaceutical use **claim** in the form 'The use of compound X in the preparation of a medicament for the treatment of disease Y' as approved by the EPO.

synergism what happens when two plus two makes five; in chemistry, the interaction of two or more compounds to give a superadditive effect; in mechanics, meaningless.

technical effect (EPO) a requirement for patentability applied by the EPO that is neither found in the EPC nor capable of meaningful definition.

technical progress formerly, a requirement for patentability in some countries according to which an invention had to show advantages over the **state of the art**.

term the lifetime of a patent (20 years from filing in most countries).

terminal disclaimer (US) in a **double patenting** situation, preventing a second patent from extending the **term** of protection by limiting its term to that of the first patent.

***Thornhill* claim** (UK) a **claim** with exactly the same **scope** as that of a specific **priority document**.

threats action a suit brought by a person threatened with an **infringement** action.

tie-in clause a clause in a **licence** agreement requiring the licensee to buy unpatented materials from the licensor.

tie-out clause a clause in a **licence** agreement requiring the licensee not to buy materials from competitors.

torpedo a suit for a declaration of non-infringement in one EU country intended to pre-empt an **infringement** action in another.

tort a wrong for which remedies may be obtained by suit in the civil courts.

traditional knowledge useful information—for example, about medicinal plants—held by indigenous peoples; not patentable, but potentially protectable by special forms of intellectual property.

trailer clause a clause in a contract of employment giving the employer rights to inventions made by the employee for a limited time after he leaves his job.

troll a patent owner who does not practise or out-license the invention, but uses the patent to sue established businesses.

unclean hands doctrine (USA) the principle that equitable relief such as enforcement of patent rights cannot be granted to one who has acted in bad faith.

utility statement the part of the **specification** that states what the invention is useful for.

venue (USA) the part of the USA in which the Federal District Court has jurisdiction to hear a patent case.

whole contents approach the principle that the whole contents (not only the **claims**) of an unpublished application may destroy the **novelty** of a later application.

working requirements provisions that a patent will be subject to compulsory licensing or lapse unless the invention is operated commercially in the country in question.

writ (of summons) (UK) formerly, the formal document initiating a suit.

written description requirement (US) the separate requirements that there be written basis for the **claims** in the disclosure and that the description shows that the inventor was in possession of the invention at the filing date.

INDEX